Historical Perspectives in the Conservation of Works of Art on Paper

Readings in Conservation

Historical Perspectives in the Conservation of Works of Art on Paper

Edited by

MARGARET HOLBEN ELLIS

THE GETTY CONSERVATION INSTITUTE
LOS ANGELES

The Getty Conservation Institute

Timothy P. Whalen, *Director*
Jeanne Marie Teutonico, *Associate Director, Programs*

The Getty Conservation Institute works to advance conservation practice in the visual arts, broadly interpreted to include objects, collections, architecture, and sites. It serves the conservation community through scientific research, education and training, model field projects, and the broad dissemination of the results of both its own work and the work of others in the field. And in all its endeavors, it focuses on the creation and dissemination of knowledge that will benefit professionals and organizations responsible for the conservation of the world's cultural heritage.

Published by the Getty Conservation Institute
Getty Publications
1200 Getty Center Drive, Suite 500
Los Angeles, California 90049-1682
www.getty.edu/publications

Tevvy Ball, *Project Editor*
Sheila Berg, *Copy Editor*
Jeffrey Cohen, *Designer*
Amita Molloy, *Production Coordinator*

Front cover: Strickland Lowry, *Trompe l'oeil with an engraving of the* Self Portrait of Sir Joshua Reynolds. Courtesy of Sotheby's.
Back cover: Charles Vernet, *François Delpeche's Print Shop*, ca. 1820. See pp. 86–87.

Printed in China

Library of Congress Cataloging-in Publication Data
Historical perspectives in the conservation of works of art on paper /
edited by Margaret Holben Ellis.
pages cm—(Readings in conservation)
Summary: "An anthology of important texts tracing the historical development of the conservation of works of art on paper"—Provided by publisher.
Includes index.
ISBN 978-1-60606-432-0 (pbk.)—ISBN 1-60606-432-0 (pbk.)
1. Prints—Conservation and restoration. 2. Drawing—Conservation and restoration. 3. Paper—Preservation. 4. Cultural property—Protection. I. Ellis, Margaret Holben, editor. II. Getty Conservation Institute, issuing body. III. Series: Readings in conservation.
NE380.H57 2014
676.028'8—dc23

2014030490

Unless otherwise noted, readings originally in German were translated by Russell Stockman; those in French were translated by Sharon Grevet.

Contents

Part VII Treatment: Limits and Limitations 365

To Eugene Victor Thaw
with deepest respect and affection

MHE

Foreword

The Getty Conservation Institute's Readings in Conservation series, which was first envisioned in the 1990s as a way for the GCI to help meet the education and training needs of the conservation field, gathers seminal texts on conservation history, theory, and practice into carefully edited anthologies of essential writings. The first book in the series, *Historical and Philosophical Issues in the Conservation of Cultural Heritage*, was published in 1996. Since then the GCI has published Readings volumes on paintings, photographs, textiles, archaeological sites, and preventive conservation. Intended initially for students and teachers, these books have found a larger audience in the conservation community. We are now pleased to introduce the seventh volume in the series, *Historical Perspectives in the Conservation of Works of Art on Paper*.

Since its invention nearly two thousand years ago, paper has served as a medium for sharing ideas and relaying information, and for creative expression. Because so much of our heritage—including prints and drawings—is paper-based, the importance of these artifacts is undeniable, and the need to conserve them is critical. This volume addresses issues, including paper conservation history, techniques of prints and drawings, the causes and effects of deterioration, and conservation treatments and ethics, that will help inform today's paper conservators as they carry out their work.

Considerable effort is invested in the creation of the books in this series, and *Historical Perspectives in the Conservation of Works of Art on Paper* is no exception. Our distinguished colleague Margaret ("Peggy") Holben Ellis was the perfect person to lead this project. The author of the influential book *The Care of Prints and Drawings*, Peggy is the Eugene Thaw Professor of Paper Conservation at the Conservation Center of the Institute of Fine Arts, New York University (NYU), and director of the Thaw Conservation Center at the Morgan Library & Museum in New York. Her graduate seminar "Readings in Paper Conservation" at NYU's Institute of Fine Arts served as a laboratory for this book, with classroom discussions leading to new discoveries that helped fine-tune the volume.

We are grateful to Peggy for her intelligence, leadership, and characteristic good humor throughout the preparation of this volume. We also extend our gratitude to the members of her editorial advisory group, whose broad professional experience developed and shaped the book, including the provision of many difficult-to-locate texts.

This volume will be an important tool for conservators, curators, and all others who care for and love this ubiquitous medium.

Timothy P. Whalen
Director
The Getty Conservation Institute

Preface

This book is addressed to students of conservation and art history, curators, artists, and collectors—that is to say, anyone with a serious interest in prints and drawings. It contains passages excerpted from a variety of historical texts, some never before translated into English and many originally published only in obscure journals, as well as contemporary ones. The readings range from popular writings to poetry and academic articles and are meant to be provocative, informative, and entertaining. The selection process for this volume was collaborative: after submissions were gathered from an international group of paper conservation educators and an advisory committee composed of John Krill, Irene Brückle, and Judith Walsh, several iterations of the present compilation were "test driven" at six seminars offered to graduate students at the Conservation Center of the Institute of Fine Arts, New York University. After lively and lengthy debate, an unwieldy mass of literature was distilled into slightly fewer than one hundred selections that describe the media, techniques, and paper of prints and drawings, and prompt discussion of the effects of deterioration and change, approaches to treatments, trends in conservation ethics, and the role of professional paper conservators today. Each reading reflects a specific time and place in the history of paper conservation, for example, how prints and drawings were collected, categorized, used, and valued and the changing reputation of "paper cleaners." Issues discussed include the meaning and significance of materiality and facture, the shifting definition of an "original" work and its ramifications for prints and their preservation, taxonomies for describing paper, and the limits and limitations of conservation treatments. *Readings* is intended to affirm the rewards of enhanced looking and to trace the development of the highly refined hand skills and sensitive aesthetic judgment of paper conservators. Above all, it is intended to acknowledge their many interdisciplinary contributions.

Some readers will no doubt miss their favorite author or article or be frustrated by the abbreviation of others. In this respect, this volume cannot be considered a paper conservation textbook; long gone is the day when budding paper conservators considered Anne Clapp's *Reading List* the fastest path to the acquisition of all that is known in paper conservation.[1] At best, it will place certain recurrent topics in historical context and lead to a better understanding and appreciation of how paper conservation, and its now-popular corollary, technical art history, has evolved.

Also, readers will no doubt notice the preponderance of English-language citations. While the literature on Western paper conservation is international, especially that concerning print and drawing connoisseurship, even the readings suggested by European conservators remained stubbornly English based; however, several French, German, and Italian submissions were most gratefully received. Fortunately, by the time of this publication, many important publications will have been translated into English. Even now, two important German-language periodicals, *Restaurator* and *Restauro* (IADA *Journal of Paper Conservation*), are published in English. No doubt adding to the bias in favor of English, the two largest organizations of professional paper conservators are in Great Britain and the United States. We hope that this situation will not discourage international collaboration, which today would seem to be more financially than language driven.

Finally, a word concerning terminology. The items under discussion here are variously described as works of art on paper, paper-based material culture, documents, artifacts, and so on, but, physically, all are simply marks on paper. Thus the line between library and archive conservation and paper conservation in this volume, at least, is a fine one. Regardless of whether an item is found in an art museum, the special collection of a library, an archive, or a historical society, emphasis is placed on its unique role in creative expression and the history of ideas. Nonetheless, most of the prints and drawings under discussion in this volume have been accorded the imprimatur of "artwork." It is my hope that these readings will inspire the questioning of societal attitudes toward all works on paper, whether "original" artworks or historical artifacts, and will promote their close study and preservation.

Notes

1 Anne F. Clapp, *Suggested Reading for a Four Year Apprenticeship in Conservation of Works of Art on Paper* (Washington, D.C.: AIC, 1975); Anne F. Clapp, Roy L. Perkinson, Antoinette King, Marjorie B. Cohn, Paul N. Banks, and Jose Orraca, *Reading List for Students in Conservation of Historic and Artistic Works on Paper and on Photographs* (Washington, D.C.: AIC, 1980).

Acknowledgments

A well-run and successful papermaking mill, artist's studio, or paper conservation lab depends upon the skill, creativity, intelligence, cooperation, commitment, and hard labor of its workers—the very same characteristics of the team that produced *Historical Perspectives in the Conservation of Works of Art on Paper*. As with any ambitious project, and especially for a volume in the Readings in Conservation series, credit is due to many people.

First, I would like to recognize the foresight and trust of the Getty Conservation Institute editorial board: Catherine Antomarchi, Cevat Erder, Debra Hess Norris, Frank Matero, Carol Mancusi-Ungaro, Colin Pearson, Clifford Price, Gionata Rizzi, Luiz Souza, Jeanne Marie Teutonico, and Tim Whalen. My advisory committee, comprising John Krill, Judith Walsh, and Irene Brückle, shouldered an enormous responsibility, sifting through hundreds of possible selections and making the case for their own favorites; to them I give my utmost thanks. Members of International Paper Conservation Educators, an ever-changing group tirelessly corraled and nurtured by Jean Brown, as well as my own book and paper cohorts in the American Institute for Conservation of Historic and Artistic Works, the International Institute for Conservation of Historic and Artistic Works, and the Institute of Conservation, generously shared their favorite literary sources; I am indebted to Thea Burns, Fred Bearman, Nancy Bell, Deborah La Camera, Bas van Velson, Shannon Zachery, Theresa Zammit Lupi, Holly Houston Krueger, Konstanze Bachmann, Roy Perkinson, Ingeliese Nielsen, Marie-Christine Enshaian, Timothy Barrett, and many others.

I would also like to recognize the Institute of Fine Arts of New York University, in particular its former director, Mariet Westermann, and its current director, Patricia Rubin, for granting me invaluable research leave and sabbaticals. Michele Marincola likewise provided continued support, encouraging me to offer "Readings in Paper Conservation" as pilot seminars. Other Conservation Center faculty members, including Norbert Baer and Hannelore Roemich, and staff—particularly the extraordinary librarians Dan Biddle and Robert Stacy (now retired)—shared their expertise and ferreted out obscure information; Kevin Martin and Catherine Lucaszewski cheerfully provided much-needed technological support.

I am very grateful to my curatorial colleagues at the Morgan Library & Museum and the conservation staff of the Thaw Conservation Center—Patricia Reyes (retired), Reba Snyder, Maria Fredericks, Frank Trujillo, Lindsey Tyne, and James Donchez—for their sustaining friendship.

The staff at the Getty Conservation Institute deserve recognition and enormous gratitude, especially Cynthia Godlewski, who patiently shepherded this project from its earliest days; former department heads Jemima Rellie and Kristin Kelly, who offered encouragement and support; and Anna Duer, who handled permissions and photo research. At Getty Publications, I would like to thank the project editor, Tevvy Ball, who oversaw the editorial preparation of the manuscript despite its somewhat unconventional format, for which I am very appreciative; and the consultant, Sheila Berg, who copyedited the book. Thanks to their efforts the usual Getty style has been combined with the expected Getty substance. I would also like to acknowledge the fine work of Jeffrey Cohen, who oversaw the design of the book, and Amita Molloy, who coordinated its production.

Most of all, I am indebted to my always inquisitive and challenging students: Gawain Weaver, Melissa Cacciola, Rebecca Capua, Camille Moore, Mary Oey, Laura Brill, Melissa Buschey, Jennifer Badger, Erin Jue, Lisa Conte, Lisa Nelson, Katie Sanderson, Lindsey Tyne, Eliza Spaulding, Justine Ellis, Morgan Adams, Amy Hughes, Quinn Ferris, Maggie Wessling, Emily Lynch, and Saira Haqqi. They were the ones who "test drove" the material in this volume in six seminars at which these selections were dissected, accepted, and abridged, or discarded outright. This publication would not have been possible without their insight and honesty.

Finally I thank my family for standing by me all this time.

Note to the Reader

The readings in this volume come from a wide range of historical and contemporary sources and comprise a variety of genres. In preparing these texts for publication here, a number of conventions were observed.

Every attempt has been made to respect the textual integrity of the original material. Capitalization has been standardized in heads and subheads, and footnotes in the original publication appear here as endnotes. With texts originally written in English, in all other stylistic matters—spelling, punctuation, capitalization—the original style has been retained. Footnotes and references have not been edited or completed with additional information but appear as they did in the original publication.

Editorial additions, restorations, and corrections to the original text appear within brackets. Editorial annotations to the original text appear as numbered endnotes, placed within brackets and followed by "—Ed"; if the original text has its own notes, the editorial annotations, in brackets, are folded into the existing notes and note sequence. When the original texts have been editorially abridged for publication here, deleted passages within the body of the excerpted text are indicated by ellipses in brackets; cross-references to elided material have been deleted. Elisions of more than one paragraph that occur immediately following a head or subhead are indicated by the symbol —∿—, placed on a separate line between the head and the first line of text. Ellipses appearing in the original text are indicated by ellipses without brackets. If there are numbered notes, and some notes have been elided, the remaining notes have been renumbered consecutively. Similarly, if there are numbered illustrations, and some have been deleted, the remaining illustrations have been renumbered consecutively. Type garbled in the original has been corrected in brackets. To avoid redundancy, where the original text contains information that has been editorially extracted and used in the chapter title (author name, book or chapter title, etc.), it has been silently deleted from the text itself.

The readings have been dated and, for the most part, arranged thematically in each of the book's eight parts. Dates appear as part of the title of each reading. In cases where a text's date of composition differs significantly from its date of publication, and the date of composition is known, the date of composition is given in the title. In cases of previously published translations, the publication or composition date of the text in its original language, if known, is given. For all readings, the date of publication and the original title of the work, as well as the edition and page numbers from which our selection has been taken, appear in the source note at the bottom of the first page.

Note to the Reader

For texts originally written in a language other than English and translated for this volume, every effort has been made to produce a translation that is both accurate and, as far possible, stylistically faithful to the original. Particular attention has been paid to the specific meanings of conservation terminology. Care has been taken to provide translations that are contemporaneous with the text and reflect the meanings of the historical period in which it was written.

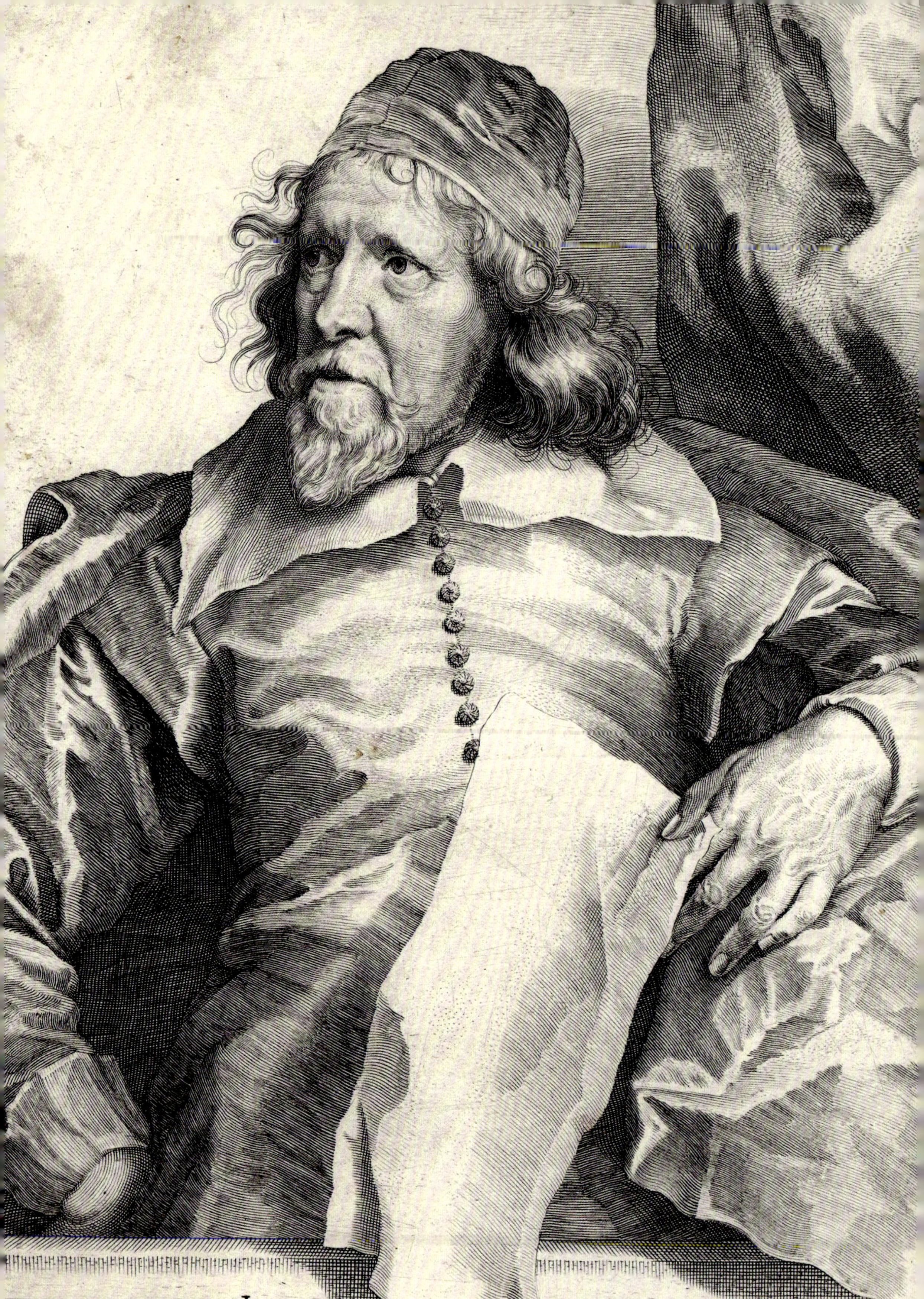

CELEBERRIMVS VIR INIGO IONES PRÆFECTVS ARCHITECTVRÆ
MAGNÆ BRITTANIÆ REGIS ETC.

Ant. van Dÿck pinxit
R.V. Vorst sculp.

Cum priuilegi

PART I

The Powers of Paper

The primary power of paper is its ability to physically preserve the marks written, drawn, or printed on its surface for transmission from generation to generation.[1] In this context, it seems entirely appropriate that daydreaming scholars in the Library of Congress lift their gaze to read *"Litera scripta manet"* in its lofty rotunda: words and images recorded on paper will endure.[2] Conservators strive to accurately identify, describe, and stabilize through treatment the strokes of pens, the scribbles of pencils, and the imprints of letterpress made on paper, as well as the paper itself—all efforts taken up in subsequent parts of this book.

Paper conservators now also perceive a broader use, value, and meaning of paper.[3] This is its second and, many will argue, more significant power. Transcending its cellulosic substance, paper also serves as an incorporeal transmitter of ideas and information. This less tangible power has resulted in recognition of the important role of paper-based artifacts in society, the urgent need to preserve them, and the expanded jurisdiction of paper conservation as a result.

The powers of paper are epitomized by transformation. The process of making paper is one of transformation: worthless rags are transformed into a valued commodity. Transformation is seen in paper's uses as both a physical preserver and a metaphysical messenger. And it is the theme of transformation that runs through Part I—from the manufacture of paper through its final use. The ideas transmitted by paper effect social, political, and moral transformations and go full circle, as in the case of the vatman Kellock (reading 6), whose sullied conscience prevents him from making "pure" paper. The terms *papier vierge*, *feuille blanche*, and *blank page* allude to redemption or an opportunity to start afresh. Indeed, a portion of a poem by Paul Valéry (1871–1945), "La feuille blanche," expresses this potential:

> In truth, a blank sheet
> Declares by the void

Inigo Jones, by Robert van Voerst; after Sir Anthony Van Dyck, line engraving, 1635 (circa 1632–1635).

That there is nothing as beautiful
As that which does not exist.[4]

On handy desk blotters and paper napkins, high-priced artist's papers with fetishized watermarks and imitation deckles, transformation begins with the creation of the first mark. Or perhaps the paper support has a power in and of itself, as in the words of the social commentator Harold Acton (1904–1994): "So often is the virgin sheet of paper more real than what one has to say, and so often one regrets having marred it."[5]

In this age of electronic information, one must wonder if the power of transformation and redemption, symbolized so perfectly by paper and papermaking, will be mirrored in digital media and their technologies.

Notes

1 The title of Part I is borrowed from *Pouvoirs du papier, Cahiers de médiologie* 4 (1997), essays on the cultural significance of paper collected by Pierre-Marc de Biasi (see reading 9) and Marc Guillaume.

2 "*Litera scripta manet*" is usually preceded by "*Vox audita perit*" (The audible voice perishes, the written word endures). However, today's technologies ensure that spoken words and moving and still images are captured for posterity. The painted inscription is located in the Librarian's Room of the Thomas Jefferson building of the Library of Congress.

3 See Irene Brückle, "Aqueous Treatment in Context," in *Paper and Water: A Guide for Conservators*, ed. Gerhard Banik and Irene Brückle (Oxford: Elsevier, 2011), 432–34.

4 Paul Valéry, "La feuille blanche" (1922), trans. Peter Dale, in *Charms and Other Pieces* (London: Anvil Press Poetry, 2007).

5 Harold Acton, *Memoirs of an Aesthete* (London: Methuen, 1948), 253.

Reading 1

Verner W. Clapp

The Story of Permanent/Durable Book-Paper, 1115–1970 (1972)

Having spent most of his career at the Library of Congress, Verner W. Clapp (1901–1972) went on to become the first president of the Council on Library Resources.[1] *Clapp was one of the foremost proponents of microfilming and other duplication processes. His faith in the ability of technology to solve space and preservation problems was manifested in the work of William Barrow but was later savagely attacked by Nicholson Baker (see Part V).*

In this selection from the first installment of a two-part article, Clapp describes paper's early competition with parchment in the race to become the more trustworthy substrate in the fifteenth century. The ascendancy of paper was ultimately won, not because of its superior permanence or durability, but because of its economy, portability, and compactness—all properties conducive to the widest dissemination of ideas. For Clapp, who once proposed an automated page-turning device and a closed-circuit television viewing system for those in far-flung locations, transformation through access was the ultimate power of paper. Despite losing out to paper's practicality, parchment retained its prestige. Similarly, in the face of the Kindle, some predict that the physical book will acquire cult status.

Litera scripta manet. That the written word endures may be assumed to be an article of faith with all whose interests meet in the pages of this journal. The fact is, however, that by itself the written word has no claims to permanence. For this it is wholly dependent upon the substratum upon which it is inscribed. Through the centuries the substratum has taken many forms—the face of a cliff, a marble shaft, or a tablet of clay; a papyrus scroll or an animal skin; a bronze plaque or a sheet of paper; and many more.

From Verner W. Clapp, "The Story of Permanent/Durable Book-Paper, 1115–1970," *Restaurator*, Supplement 3 (1972): 1–5. Used by permission of DeGruyter.

This recitation suggests that the choice of paper as a universally preferred writing material was not made on grounds of lastingness. The truth is that we ask many qualities in a writing material, and to some we give a higher priority than to its permanence.[1] Specifically, in its ability to accept a message readily and to render it truthfully, in its compactness, manipulability and portability, and perhaps above all its cheapness, paper has proved itself so far superior to all its competitors that we are lucky if we obtained any permanence at all in the bargain. Actually, some permanence was thrown in, but not enough. If it had been an all-or-none proposition—if all paper had been permanent, or none—the matter would undoubtedly have been disposed of one way or another long ago. Instead, indeterminacy has exacerbated the problem for more than 850 years.

1115–1500 Paper vs Parchment

Paper entered the European scene with a black mark against its name, written—it may be suspected—on parchment. In 1115 parchment was achieving dominance in European chancelleries, having almost entirely superseded papyrus. Paper, introduced from Africa into Spain and Sicily by the Arabs, had been gaining ground for a century. We can only guess today the evidence on which parchment was preferred to paper in the twelfth century, but in 1115 Roger II of Sicily (then Count) renewed on parchment a charter which his father, Roger I, had granted the abbey of Santa Maria di Terreti in Calabria on paper only 25 years before.[2] Again, in an assembly in Palermo in 1145, Roger renewed on parchment four charters granted on paper to the Basilian monastery of St Philip of Fragala from 1102 to 1124.[3] The reason for the preference is not explained in either case.

The Emperor Frederick II (Roger II's nephew) was more explicit. In the constitutions which he ordained for the Two Sicilies in 1231 he required that all public documents should be written on parchment 'so that they may bear testimony to future times and not risk destruction through age.'[4]

Whether the parchment industry has ever erected a monument to this monarch, I do not know. It should most certainly recognize him as its patron, for Frederick's edict, which has worked greatly to its advantage, has lasted longer and more effectively than the acts of most lawgivers. (Perhaps because it was written on parchment!)

Thus, until very recent years, parchment has been specified for documents whenever permanence was in view. Not only were such national muniments as Magna Carta and the Declaration of Independence engrossed on parchment, but the walls of the western world have within human memory been papered with the certificates of educational achievement known as 'sheepskins.' By contrast the Charter of the United Nations is on paper, as are all but a very elite few of the current product of the diploma-mills.

Ironically, when parchment began to get expensive, it was discovered that a 'vegetable parchment' could be made from—of all things—paper, by dipping it into—of all things—sulphuric acid! This improbable material has taken the place of parchment for many documental uses, while 'parchmentized' paper is widely used for wrapping butter. All that would be needed to complete the irony would be to watermark this product with the figure of Frederick II!

In spite of the imperial discouragement, the manufacture of paper progressed amain over the centuries following 1231; so much so that it was able to support a very lively publishing industry when printing came into existence in the second half of the fifteenth century. It is true that fastidious book-purchasers continued to insist on parchment, and the Gutenberg Bible, c. 1455 was printed in parchment as well as paper copies. The next book, the Mainz Psalter 1457, was printed exclusively in parchment copies. But as the century wore on, the use of parchment became less and less frequent.

Of course, the printing press rang the knell of technological obsolescence on the scriptorium. The scriptorium did not, for that reason, immediately go out of business; it lingered for years. When it didn't have manuscripts to copy, it sometimes copied printed works, and Dr Curt Buehler[5] of the Pierpont Morgan Library has amused himself by collecting instances of manuscript copies—sometimes slavish reproductions, colophon and all—of printed books.[6] Such copying required justification. One rationalization argued that copies printed on paper were perishable, and needed to be replaced by transcripts on more durable parchment. We owe to the formidable Johann Tritheim—theologian, historian, bibliographer, cryptographer, and director of a scriptorium in the Benedictine monastery at Sponheim where he was abbot from 1483 to 1506—perhaps the best wording of this rationalization. 'If writing,' said he in *In Praise of Scribes*, 'is inscribed on parchment, it will last a millennium. But if on paper, how long will it last? Two hundred years would be a lot.' He consequently urged scribes 'to perpetuate in writing the useful products of the press.'[7]

In spite of this distrust of paper, Tritheim was not averse to seeing his own works become useful products of the press. He seems, however, to have taken the precaution that at least some copies of his books were printed on parchment. Of no fewer than 117 copies of just those of his works that were published before the end of the fifteenth century which have found their way into the collections of North American libraries, no fewer than nine are on parchment![8] I have not seen these parchment copies, all but one of which are in the Pierpont Morgan Library, but in the Library of Congress I have examined twelve copies of Tritheim's books and pamphlets—all on paper—printed before his death in 1516 as abbot of St. Jacob's in Würzburg.[9] Four of these publications are of book length. These are fine copies, showing little or no evidence of use in the way of annotations, and their paper is in excellent condition, with beauty and strength apparently unaffected by age. The other eight publications are pamphlets, most of which show evidence of some use, and several a great deal of use. The paper in five is good, in two fair, and in one (which appears to have been through a fire) quite brown and poor. Nevertheless, these books have all already more than doubled their author's expectation of life on their behalf, and seem (with the exception of the worst of the pamphlets) to have undiminished expectancy of further existence, given as good care as they have had for the past half millennium.

But the world heeded the warnings of the abbot no more than it had those of the emperor, and in retrospect the world appears to have been in the right. Certainly, there is no evidence of massive losses of paper records, manuscript or printed, from the fifteenth and earlier centuries from causes that would not have been equally destructive to parchment. On the contrary, hundreds of thousands of printed books from the second half of the fifteenth century and of manuscripts from the fifteenth and earlier centuries survive intact in libraries and archives. It may be conjectured that the combination of sound raw materials (clean linen rags), little or no chemical treatment (limited principally to retting the flax and bleaching the linen in sunshine), the minimum of mechanical treatment (pounding in stamping-mills which left the fibres long), and an abundance of clean water account for the resulting stable product. If to these conditions is added that the hard water likely to have been found in the mountain streams often sought by papermakers to operate their stamping-mills undoubtedly often contributed the mineral salts which account for the alkalinity of many early papers, then the principal elements conducive to permanence, as we now know them, would have been present.

Accustomed as we are to a preponderance of less-than-fine book-papers, how do we account for the many fine papers of the fifteenth century? Without going into the organizational complexities of medieval trade and industry, it may simply be suggested that until about 1500 book-paper was competitive with parchment and was measured by the standard set by that product. By 1500, however, paper had won the day, and only a few nostalgics like Tritheim (who had, besides, a vested interest in an obsolete practice) still insisted on the more expensive material.

Note to Introduction

1 The Council on Library Resources (CLR) was merged with the Commission on Preservation and Access in 1997 to become the Council on Library and Information Resources (CLIR). Like Barrow's, Clapp's reputation has undergone revision over time; see William J. Crowe, "Verner W. Clapp as Opinion Leader and Change Agent in the Preservation of Library Materials" (PhD diss., Indiana University, 1986).

Notes to Reading

1 '*Permanence* is the degree to which paper resists chemical action which may result from impurities in the paper itself or agents from the surrounding air. *Durability* is the degree to which a paper retains its original qualities under continual usage'

(Report of the [Technical Association of the Pulp and Paper Industry] Committee on Permanence and Durability of Paper, *Paper Trade Journal* 97:4:33–6, 27 July 1933).

2 J.-L.-A. Huillard Bréholles: *Historia diplomatica Frederici II . . . imperatoris* (Paris: Plon, 1852–9), II, 440. This is a latin translation made by the magistrates of Reggio when reconfirming the charter in 1252. The expression for 'paper' used by the magistrates was 'carta papiri'; it would be interesting to know which of several available terms was used in the greek in which the charter was originally written.

3 Roccho Pirri: *Sicilia sacra* (Panormi [i.e., Palermo]: Bua, 1633–1733) 4 (1649) 91–3. Again, this is a latin translation of the greek original, by Pirri, and we do not know the greek term for paper. The latin phrase used by Pirri, 'carta cuttunea,' means paper made from cotton according to C. du F. Ducange (Glossarium mediae et infimae latinitatis [Paris: Librairie des sciences et des arts, 1937–8], s.v. Charta cuttunea), but paper made from flax or linen according to Santifeller, 1953 (cited by G. Piccard: 'Carta bombycina, carta papyri, pergamena graeca,' *Archivalische Zeitschrift* 61 [1965] 46–74, p. 47). But would Pirri have distinguished in 1649?

4 'Volumus etiam & sancimus, ut praedicta instrumenta publica, & aliae similes cautiones, nounisi in pergaminis in posterum conscribantur. Cum enim eorum fides multis futuris temporibus duratura speretur, justum esse decernimus, ut ex vetustate forsitan, destructionis periculo non succumbant.' *Constitutiones regum regni utriusque Siciliae mandante Friderico II irnperatore per Petrum de Vinea . . . concinnatae . . . cum graeca earundem versione* (Neapoli: Ex Regia Typographia, 1786), Tit. LXXX (p. 81–2). The latin terms for paper in this title are 'charta papyri' and 'charta bombycina'; the corresponding term in the greek text is (see original document for Greek text).

5 [Curt Bühler (1905–1985) was curator of printed books at the Morgan Library & Museum for fifty-two years.—Ed.]

6 Curt F. Bühler: *The fifteenth century book* (Philadelphia: University of Pennsylvania Press, c. 1960), p. 15 ff.

7 Johannes Trithemius: *De laude scriptorum pulcherrimus tractatus* (Mainz: Peter von Friedberg, 1494). [. . .]

8 Frederick R. Goff: *Incunabula in American libraries* (New York: Bibliographical Society of America, 1964), items T-433, 436, 439, 443, 444, 449, 451, and 457 (2 copies).

9 Ibid., items T-433, 434, 437, 438, 441, 446, 448, 452, 455, 456. Also: *De purissima et immaculata conceptione Virginis Mariae* (Strassburg: Matthias Hupfuff, 1406 [!]), 2 copies; and *Compendium sive breviarium primi voluminis Annalium* (Mainz: Peter Schoffer, 1515).

Reading 2

Thomas Churchyard

A Sparke of Frendship and Warme Goodwill, That Shewes the Effect of True Affection and Unfoldes the Finenesse of This World (1588)

Thomas Churchyard (1520?–1604) was an English author, poet, and soldier of fortune whose writings rarely lacked a political agenda. This eighteen-page poem, dedicated to Sir Walter Raleigh, pays homage to Queen Elizabeth, whom the hapless Churchyard had managed to offend, and a German papermaker named Spillman (John Spielman), nicknamed "Help-man" in the poem because of the many social benefits of his papermaking mill established in 1588 in Kent. Echoing Clapp, Churchyard proclaims that although paper is less durable than parchment, its ubiquity guarantees its triumph: "parchment comes in few mens hands" while paper "from man to man doth go," an argument often promulgated today on behalf of electronic media. Paper's metaphysical powers, according to Churchyard, are mirrored in the papermaking process itself; just as dirty and useless rags are transformed into pristine paper, humanity is redeemed by the ideas that paper transmits: "Of drosse and rags, that serues no other meane, and fowle had shreds, comes Paper white and cleane. And euen so, the baddest people may, become good folke."

[. . .]
I prayse the man, that first did paper make,
the onely thing that sets all vertues forth:
It shoes newe bookes, and keepes old workes awake,
much more of price, than all this world is worth:
It witnesse beares of frendship, time and troth,
and is the tromp of vice and vertue both,

From Thomas Churchyard, *A Sparke of Frendship and Warme Goodwill, That Shewes the Effect of True Affection and Unfoldes the Finenesse of This World* (London, 1588), reprinted in *Paper in Printing History IX: A Celebration of Milestones in the Graphic Arts*, ed. Leonard Schlosser (New York: Lindenmeyr Paper Corporation, 1979), n.p.

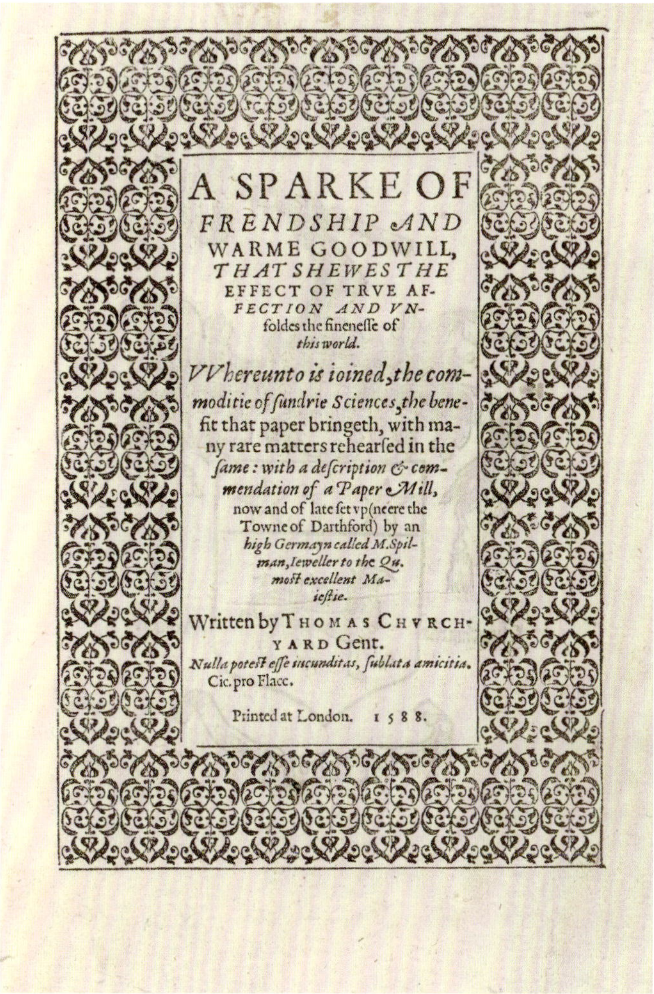

A SPARKE OF
FRENDSHIP AND
WARME GOODWILL,
THAT SHEWES THE
EFFECT OF TRVE AF-
FECTION AND VN-
foldes the finenesse of
this world.

VVhereunto is ioined, the com-
moditie of sundrie Sciences, the bene-
fit that paper bringeth, with ma-
ny rare matters rehearsed in the
same: with a description & com-
mendation of a Paper Mill,
now and of late set vp (neere the
Towne of Darthford) by an
high Germayn called M. Spil-
man, Ieweller to the Qu.
most excellent Ma-
iestie.

Written by THOMAS CHVRCH-
YARD Gent.
Nulla potest esse iucunditas, sublata amicitia.
Cic. pro Flacc.

Printed at London. 1588.

Figure 1
A Sparke of Frendship . . . , title page. Print Collection, Miriam and Ira D. Wallach Division of Art, Prints and Photographs, The New York Public Library, Astor, Lenox and Tilden Foundations.

Without whose helpe, no hap nor wealth is won,
and by whose ayde, great workes and deedes are done.

It flies from friend and foe in letter wise,
and serues a state, and kingdome sundry wayes,
It makes great winde, where neuer dust doth rise,
and bréedes some stormes, in smoothest sommer dayes.
It telles of warre, and peace as things fall out,
and brings by time, ten thousand things about.
For schollars fit, and merchants all alike,
for plowe men good, that digs and delues the dike,

For good Deuines, and lawyers not amisse,
for Saylers too, and those that trauell farre,
For Students best, that knowes what learning is,
for pleaders meete, for men of peace or warre,
For all degrees that are of manly kinde,
a right good meane, that may expresse the minde,
A néedefull thing, that no good wit may want,
a thing moste vsde, yet neuer will be skant.

What man, or sex, or shape of worthy molde,
can paper lacke, but buies it lesse or more?
Things present are, in paper long enrolde,
so things to come, and things long past before.
Though [parchment] duer, a greater time and space,
yet can it not, put paper out of place:
For paper still from man to man doth go,
when parchment comes in few mens hands, you knowe.

If paper be, so precious and so pure,
so fitte for man, and serues so many wayes,
So good for vse, and wil so well endure,
so rare a thing, and is so much in prayes:
Than he that made, for vs a paper mill,
is worthy well, of laue and worldes good will.
And though his name, be *Spillman* by degrée,
yet *Help-man* nowe, he shall be calde by mée.

Sixe hundred men, are set a worke by him,
that else might starue, or seeke abroad their bread.
Who nowe liues well, and goes full braue and trim,
and who may boast, they are with paper fed.
Straunge is that foode, yet straunger made the same,
Spillman, Help-man, so rightly call the same:
For greater help, I gesse he cannot giue,
than by his help, to make poore folke to liue.

Fewe helps these dayes, to bring vs any wealth,
some sundry wayes, doe still more harme then good:
So such as help, doe breede good bloud and health:
and in best part, ought well be vnderstoode.
If paper mill, helps poore and harmes no rich,
the gayne is great, and the inuention much,
The worke not small, the labor worth the viewe,
because old Art, is nowe reuiude a newe.

One *Thirlby*[1] went, Embassador farre from hence,
to Charles the fift, an Emperour of great fame,
And at returne, did bring with him from thence,
a learned man, *Remigius* by name.
Who *Thirlby* loude, and made by his deuise,
a paper mill, but not so much in price,
As this that nowe, néere *Darthford* standeth well,
Where *Spillman* may, himselfe and houshold dwell.

Well this is he, that first héere profite brought,
first triall made, of thinges not héere well knowne:
First framde the forme, that sundry paper wrought,
first tooke in hand, by charges of his owne
A doubtfull worke, that others earst begun,
Who spent thereon more wealth then well they won,
This man alone, the substance shewes so right,
that all the rest, were Lampes that gaue no light.

The mill it selfe, is sure right rare to sée,
the framing is, so queint and finely done,
Built all of wood, and hollowe trunkes of tree,
that makes the streames, at point deuice to runne,
Nowe vp, now downe, now sideward by a sleight,
nowe forward fast, then spouting vp on height,
As Conduits colde, coulde force so great a heate,
that fire shoulde flame, where thumping hammers beat.

The Hammers thump, and make as lowde a noyse,
as Fuller doth, that beates his wollen cloth,
In open shewe, then sundry seceete toyes,
makes rotten ragges, to yéelde a thickned froth:
Then is it stampt, and washt as white as snowe,
then flong on frame, and hangd to dry I trow:
Thus Paper streight, it is to write vpon,
as it were rubde, and smoothde with slicking stone.

Through many handes, this Paper passeth there,
before full forme, and perfect shape it takes,
Yet in short time, this Paper yucke will beare,
whereon in haste, the workeman profit makes.
A wonder sure, to see such ragges and shreads,
passe dayly through, so many hands and heads,
And Water too, that Papers enmy is,
yet Paper must, take forme and shape from this.

This Water doth, not onely driue the mill,
but giues it grace, and makes it fine and fayre,
Is cause and ground, to giue it fashion still,
for it is made, with Water, winde and ayre.
And takes his forme of compoundes mixed well,
wherein there doth, a secreate nature dwell:
A heauenly power, that earth and ayre hath knit,
by cunning Art, and worke of humane wit.

For cloth and silke, and mettalles fine or bace,
are wrought of thinges, that haue a substance great,
This findeth forme, and stampe in straunger cace,
as Water mill, made rags and shreds to sweate.
Of whose thick froth, a creame or crudde should rise,
that shoulde take shape, and strength by breath of skyes:
Though sure a meane, there is to worke the same,
some secrete cause, brings Paper first in frame.

As corne is sowne, and there must rotte in grounde,
before it blade, or takes good roote or strength,
Then reapt and thrasht, and to the myll full rounde,
is sent to grinde, and made good dow at length,
Then kneaded well, then bakte and made good bread,
so paper sure, through many a hand and head
Doth passe like drosse, that of it selfe is nought,
till it be tried, by skill and throughly wrought.

From drosse commes gold, when fier hath searcht it well,
so all thinges haue, their worth from some great cause:
The pearle some say, is fetcht from oysters shel,
thus each thing yéeldes, to Art and natures lawes.
As fier from slint, through stroake of stéele we finde,
so world may see, what wonders worketh kinde.
Glasse was at first, as straunge to make or vewe,
as Paper nowe, that is deuisde of newe.

Of newe I meane, in England saue one man,
that had great wealth, and might much treasure spare.
Who with some charge, a Paper mill began,
and after built, a stately worke moste rare.
The Royall exchaunge, but got by that more gayne,
than he indéede, did loose by former payne.
But neither he, nor none before his dayes,
made Paper mill, that merits so much prayse.

As this that nowe, is not full farre from hence,
where Water ranne, in waste and vaine a way,
Nowe profit yéeldes, and brings in poundes and pence,
that quittes the cost, and doth the charge defray.
This had not bene, if Prince had not retained,
the straunger héere, by whome these giftes are gayned:
Her highnesse than, sawe in her déepe foresight,
what famous worke, this man coulde bring to light.

The glory then, and honor of this deede,
is hers, and ours, shall be the gayne therein.
We reape the Corne, whoeuer sowde the seede,
who ere haue lost, we shall be sure to winne.
This mill remaynes, a sampler to the rest,
that after comes, to shewe whose worke is best,
No doubt but some, this course will followe on,
a straunger left, this worke to looke vpon.

That many moe, by this may builded be,
and many heads, and handes may thriue thereby,
He merits much, that first plants fruitfull trée,
they purchase prayer, that first doth practises try.
They ought not reap, that neuer ment to sowe,
they winne great hap, that can through hazards go,
They lose no time, that toyles for publike state,
they glory gayne, that first a conquest gate.
Nowe gallant witts, that ioyes in doing well,
ply pen a pace, whiles learning may be had,
Now stripplings yong, but late come out of shell,
to Schoole good boyes, to make your parents glad.
Now Printers presse, that sets foorth many a booke,
besturre the stampe, that worlde for newes may laake,
Now [illegible] Stationers, that worketh all the yeere,
sell bookes good cheape, for Paper is not déere.

Nowe writers graue, that studies heauenly things,
your workes shall shine, by meane of Papers grace,
Nowe Marchant wise, that home great profit brings,
send letters out, abroad to euery place.
For Paper doth, present it selfe to those,
in common weale, that writeth verse and prose,
The merry myll, nowe grindes and goes so braue,
that worlde at will, shall alwayes Paper haue.

When Paper was, not throwly knowne of men,
they wrote in stones, and barks of trees for shift,
But loe long since, the Paper and the Pen,
by deepe deuice, found out a finer drift.
And most to prayse, because of trifling toyes,
so great a wealth, our worthy world enioyes,
Of drosse and rags, that serues no other meane,
and fowle had shreds, comes Paper white and cleane.

And euen so, the baddest people may,
became good folke, if they will bide the stampe,
Which people first, with many a worthy way,
must be well wrought, like oyle that burnes in Lampe.
For Oyle is tried, and pur'd eare it be solde,
and searcht throw out, as fyre tries out the golde,
And when the oyle, is fit to blaze or burne,
it is applied, at néede to serue our turne.

So man is méete, to serue his natiue soyle,
when thumping worlde, abroad hath tried him throwe,
Or heere at home, his life hath scaped foyle,
and he thereby, may shew a blotlesse browe.
But this must be, as Paper passeth mill,
mans doubtfull dayes, must passe through perilles still,
And though great blowes, do beat him backe a space,
hee bides the brunt, to get the greater grace.
And yéeldes to thumps, and thwartes as yce to thawe,
as frost to fier, will soft and gentle waxe:
Or as stiffe neckes, will stoupe and yéelde to lawe,
compelde by flame, to yéelde to fire like flaxe.
For nothing more, becomes a noble minde,
than bide the blastes, and puffe of euery winde,
Whose bell owes blowes, to hinder well wonne fame,
when doe well shall, in spight possesse good name.

If ship passe storme, and tries the surging seas,
comes quiet home, to harber in a roade,
Man must of force, through torment purchace ease,
and must beleeue, great burthen is no load.
And so bace ragges, whereof is no account,
through straining hard, past tenter hookes may mount,
And bodie cost, and tumbled vp and downe,
may come to rest, and reap right rare renowne.

Mans secrete faultes, and foule defects of minde,
must be reformde, like ragges in Paper mill,
When hammers help, hath changde his cankered kinde,
and [illegible] clensde the heart, from spots and former ill.
A second shape, and forme full fresh and new
he doth receiue, in nature grace and hiew.
When Water streames hath washt him ouer quite,
than man becomes, like paper faire and white.

If Water were as scant as deerest wine,
how should this world mainteine each science heere?
In Water thin there dwels a power diuine,
where face is seene, as in a Christall cleere.
An Element that euery creature needes,
wherein full oft both fowle and fishes breedes:
Whereby a world of people daily liue,
and God to man doth manie a blessing giue.

What earth or soyle can flourish where it wants?
colde Water sweete doth coole the scalded brest,
The drops whereof doth comfort hearbes and plants,
and graces great by Water is possest.
Then muse not man, if Water thee reforme,
that art but earth, and foode for scraling worme,
A bladder pufte with winde and ayer full thinne,
that can not bide the push of baggage pinne.

Our finest coyne of siluer or of golde
in grossest sort is handled as ye knowe,
And beaten long, and thumped treble folde
before it doth for currant money goe.
The wollen cloth that from the walkemill comes,
at first must passe, through manie handes and thums:
Yea washt and walkt with Water where it goth,
ere it do take his breadth and thicknes both.

What linnen, lawne, or cambricke can be white,
if Water do not throughly wash the same:
It scowres that cleane, that is as dunne as Kyte,
and brings fowle cloth in perfect forme and frame.
All slubbred things must needes be washt anue,
fowle things are nought, if prouerbe olde be true.
Thus prooue I plaine, by course of Water mill,
and hammering world, mens manners changeth still.

Though some do say, in *France* and other place,
are Paper mils, as fayre and straunge as this,
Whats that to vs, this giues our Country grace,
and to all Kent, a double honor is.
That in the soyle, was borne our worthy Quéene,
by straungers meane, so straunge a worke is séene:
And straungers are, so glad with straunge deuice,
to serue and please, our Prince of péerelesse price.

In other Realmes, there milles are not in woorth,
scarce halfe so good, the proofe may well be founde,
This is so fine, with workmanship set foorth,
so surely built, and planted in the ground.
That it doth séeme, a house of some estate,
a Mill moste rare, a worke deuisde of late,
Whose goodnesse great, excéedes the outward showe,
and from whose stampe, shall publike profite flowe.

And troth to tell, the mill is blacke and white,
and Water doth worke all the Paper there,
The sight thereof, shall bréede more rare delight,
than man with eye, beholdes in many a where.
This somewhat more, may moue a maruell héere,
no profite may, be reapt in many a yéere,
The author than, of this newe Paper Mill,
bestowes great charge, and gaynes but worldes goodwill.

Death may preuent, his hope and purpose too,
death cuts off all, from him if it so hap,
If losse so fall, what then shall *Spilman* doe,
but so receiue, the losses in his lap.
This daunger great, deserueth some regard,
or of the worlde, doth merit some reward,
Giue him good speech, as reason doth require,
yéelde duety, so the labror hath his hire.

An high *Germaine* he is as may be prooude,
In *Lyndoam Bodenze* borne and bred:
And for this Mill may héere be truly loude
and praysed too, for déepe deuice of head.
But if the hope of gaine quit not the cost,
the world will iudge his labour is but lost.
To hazards hap he doth commit the same,
and séemes as yet to care for naught but Fame.

Wealth, wit and time, with toyle and trauaile great,
he plyes a pace, and spareth for no charge,
The Mill goes round, the workmen moyie and sweate,
the streame goes straight, that earst ranne all at large,
The whéeles conueyes the Water diuers wayes,
the Hammers thump, the stamp but seldome stayes:
The ragges and clowts, becomes as white as snowe,
and all these knackes, the master néedes must knowe.

Whose purse, whose paynes, and purpose is not small,
whose plot, points out, a peece of worke right fayre,
To hinder none, but made to please vs all,
to which braue Mill, do thousandes still repayre.
To sée what things, are wrought by cunning skill,
to Gods great prayse, and Princes honor still,
And to the place, and soyle where it doth stand,
a goodly grace, and Paper neere at hand.

Loe heere how man to paper is comparde,
that readie is to take both stampe and print
Though triall great, and manie a passage hard,
more stiffe than steele that strikes out fire from flint:
But though most hard the path and passage be,
in the right way it sets man franke and free:
That hath bene brought in bondage from his birth,
and makes him seeme a little God on earth.

Full fraught with wit, with Art and science great,
with learned lore, with skill and knowledge deepe,
With giftes diuine, that feares not fortunes threat,
with quick deuice, that can both get and keepe,
With reason such, as rules each other thing,
of Beast and Fowle, the onely Lord and King,
A Prince of all the earthly pleasures héere,
found out with paine, and bought with trauayle déere.

Note to Reading

1 [Thomas Thirlby, ambassador to the Low Countries in 1545–48, brought back to England a papermaker named Remiger or Remigius, who appears to have established a papermaking mill near Cambridge, which by 1585 had failed. "Early English Paper Mills," *Paper* 13, no. 7 (October 29, 1913): 18.—Ed.]

Reading 3

ROBERT BELL

Observations Relative to the Manufactures of Paper and Printed Books in the Province of Pennsylvania (1773)

Robert Bell (1732?–1784) was a radical Scottish publisher, printer, and bookseller who settled in Philadelphia in 1766 and published several editions of Thomas Paine's Common Sense. *The work from which this excerpt is taken, though a simple broadside in appearance, delivers a subversive message: Bell's enraptured description of the magical transformation of worn-out linen rags, or holland, into tender billets-doux barely disguised his political proposition that a reliable source of domestically produced paper would ensure economic security and, by extension, eventual independence for the struggling American colonies.*

[. . .]

It is pleasant enough to consider the changes that a linen fragment undergoes, by passing thro' the several hands above-mentioned. The finest pieces of holland, when worn to tatters, assume a new whiteness more beautiful than their first, and often return in the shape of letters to their native country. A lady's shift may be metamorphosed into billets-doux, and come into her possession a second time. A beau may peruse his cravat after it is worn out, with greater pleasure and advantage than ever he did in a glass. In a word, a piece of cloth, after having officiated for some years as a towel or a napkin, may by this means be raised from a dunghill, and become the most valuable piece of furniture in a prince's cabinet.

From ROBERT BELL, *Observations Relative to the Manufactures of Paper and Printed Books in the Province of Pennsylvania* (Philadelphia: Broadside, 1773), 1.

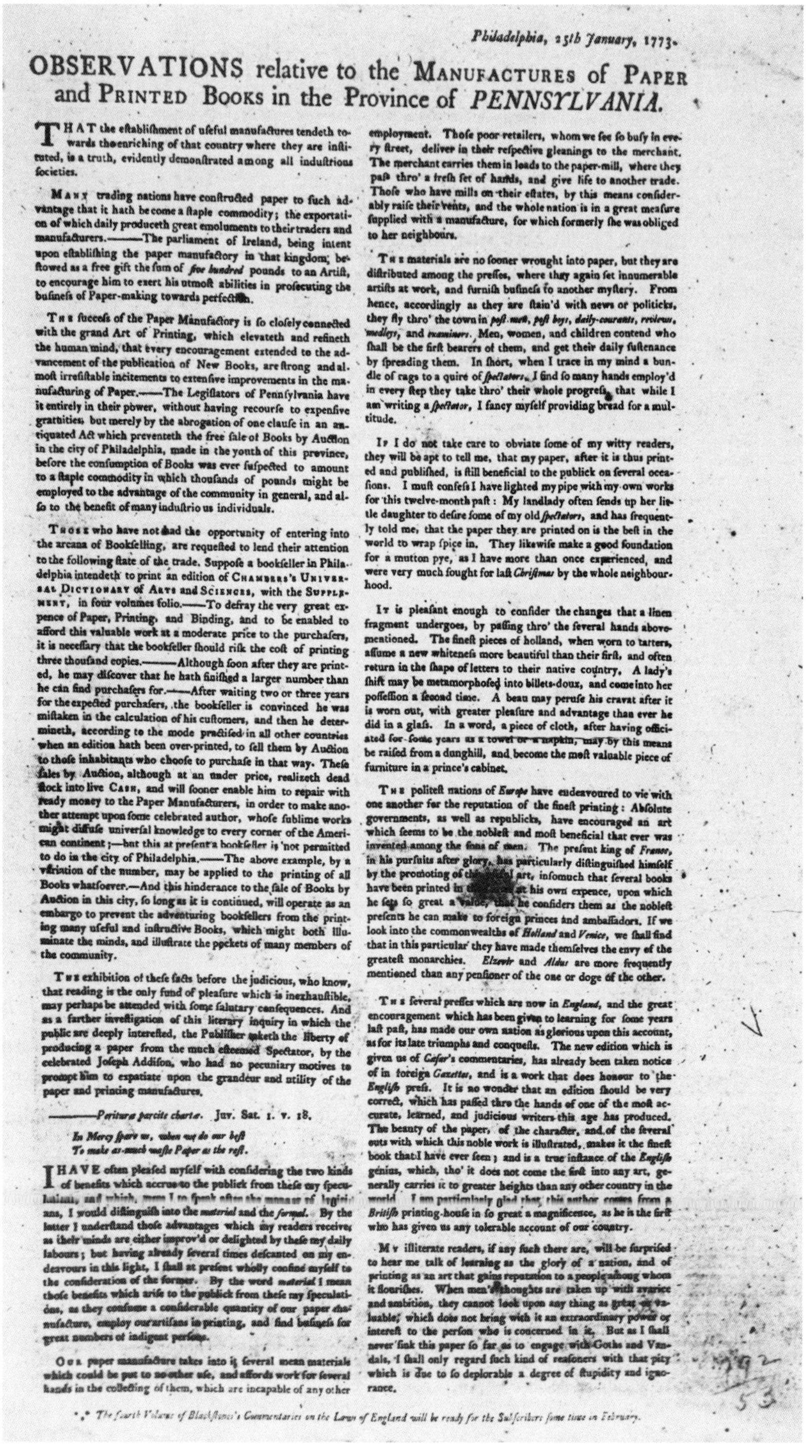

Philadelphia, 25th January, 1773.

OBSERVATIONS relative to the MANUFACTURES of PAPER and PRINTED BOOKS in the Province of *PENNSYLVANIA.*

THAT the establishment of useful manufactures tendeth towards the enriching of that country where they are instituted, is a truth, evidently demonstrated among all industrious societies.

MANY trading nations have constructed paper to such advantage that it hath become a staple commodity; the exportation of which daily produceth great emoluments to their traders and manufacturers.——The parliament of Ireland, being intent upon establishing the paper manufactory in that kingdom; bestowed as a free gift the sum of *five hundred* pounds to an Artist, to encourage him to exert his utmost abilities in prosecuting the business of Paper-making towards perfection.

THE success of the Paper Manufactory is so closely connected with the grand Art of Printing, which elevateth and refineth the human mind, that every encouragement extended to the advancement of the publication of New Books, are strong and almost irresistable incitements to extensive improvements in the manufacturing of Paper.——The Legislators of Pennsylvania have it entirely in their power, without having recourse to expensive gratuities, but merely by the abrogation of one clause in an antiquated Act which preventeth the free sale of Books by Auction in the city of Philadelphia, made in the youth of this province, before the consumption of Books was ever suspected to amount to a staple commodity in which thousands of pounds might be employed to the advantage of the community in general, and also to the benefit of many industrious individuals.

THOSE who have not had the opportunity of entering into the arcana of Bookselling, are requested to lend their attention to the following state of the trade. Suppose a bookseller in Philadelphia intendeth to print an edition of CHAMBERS's UNIVERSAL DICTIONARY of ARTS and SCIENCES, with the SUPPLEMENT, in four volumes folio.——To defray the very great expence of Paper, Printing, and Binding, and to be enabled to afford this valuable work at a moderate price to the purchasers, it is necessary that the bookseller should risk the cost of printing three thousand copies.——Although soon after they are printed, he may discover that he hath finished a larger number than he can find purchasers for.——After waiting two or three years for the expected purchasers, the bookseller is convinced he was mistaken in the calculation of his customers, and then he determineth, according to the mode practised in all other countries when an edition hath been over-printed, to sell them by Auction to those inhabitants who choose to purchase in that way. These sales by Auction, although at an under price, realizeth dead stock into live CASH, and will sooner enable him to repair with ready money to the Paper Manufacturers, in order to make another attempt upon some celebrated author, whose sublime works might diffuse universal knowledge to every corner of the American continent;—but this at present a bookseller is not permitted to do in the city of Philadelphia.——The above example, by a variation of the number, may be applied to the printing of all Books whatsoever.—And this hinderance to the sale of Books by Auction in this city, so long as it is continued, will operate as an embargo to prevent the adventuring booksellers from the printing many useful and instructive Books, which might both illuminate the minds, and illustrate the pockets of many members of the community.

THE exhibition of these facts before the judicious, who know, that reading is the only fund of pleasure which is inexhaustible, may perhaps be attended with some salutary consequences. And as a farther investigation of this literary inquiry in which the public are deeply interested, the Publisher taketh the liberty of producing a paper from the much esteemed Spectator, by the celebrated Joseph Addison, who had no pecuniary motives to prompt him to expatiate upon the grandeur and utility of the paper and printing manufactures.

——*Periturae parcite chartae.* Juv. Sat. 1. v. 18.

In Mercy spare us, when we do our best
To make as much waste Paper as the rest.

I HAVE often pleased myself with considering the two kinds of benefits which accrue to the publick from these my speculations, and which, were I to speak after the manner of logicians, I would distinguish into the *material* and the *formal*. By the latter I understand those advantages which my readers receive, as their minds are either improv'd or delighted by these my daily labours; but having already several times descanted on my endeavours in this light, I shall at present wholly confine myself to the consideration of the former. By the word *material* I mean those benefits which arise to the publick from these my speculations, as they consume a considerable quantity of our paper manufacture, employ our artisans in printing, and find business for great numbers of indigent persons.

OUR paper manufacture takes into it several mean materials which could be put to no other use, and affords work for several hands in the collecting of them, which are incapable of any other employment. Those poor retailers, whom we see so busy in every street, deliver in their respective gleanings to the merchant. The merchant carries them in loads to the paper-mill, where they pass thro' a fresh set of hands, and give life to another trade. Those who have mills on their estates, by this means considerably raise their rents, and the whole nation is in a great measure supplied with a manufacture, for which formerly she was obliged to her neighbours.

THE materials are no sooner wrought into paper, but they are distributed among the presses, where they again set innumerable artists at work, and furnish business to another mystery. From hence, accordingly as they are stain'd with news or politicks, they fly thro' the town in *post-men*, *post boys*, *daily-courants*, *reviews*, *medleys*, and *examiners*. Men, women, and children contend who shall be the first bearers of them, and get their daily sustenance by spreading them. In short, when I trace in my mind a bundle of rags to a quire of *spectators*, I find so many hands employ'd in every step they take thro' their whole progress, that while I am writing a *spectator*, I fancy myself providing bread for a multitude.

IF I do not take care to obviate some of my witty readers, they will be apt to tell me, that my paper, after it is thus printed and published, is still beneficial to the publick on several occasions. I must confess I have lighted my pipe with my own works for this twelve-month past: My landlady often sends up her little daughter to desire some of my old *spectators*, and has frequently told me, that the paper they are printed on is the best in the world to wrap spice in. They likewise make a good foundation for a mutton pye, as I have more than once experienced, and were very much sought for last *Christmas* by the whole neighbourhood.

IT is pleasant enough to consider the changes that a linen fragment undergoes, by passing thro' the several hands abovementioned. The finest pieces of holland, when worn to tatters, assume a new whiteness more beautiful than their first, and often return in the shape of letters to their native country. A lady's shift may be metamorphosed into billets-doux, and come into her possession a second time. A beau may peruse his cravat after it is worn out, with greater pleasure and advantage than ever he did in a glass. In a word, a piece of cloth, after having officiated for some years as a towel or a napkin, may by this means be raised from a dunghill, and become the most valuable piece of furniture in a prince's cabinet.

THE politest nations of *Europe* have endeavoured to vie with one another for the reputation of the finest printing: Absolute governments, as well as republicks, have encouraged an art which seems to be the noblest and most beneficial that ever was invented among the sons of men. The present king of *France*, in his pursuits after glory, has particularly distinguished himself by the promoting of [illegible] art, insomuch that several books have been printed in [illegible] at his own expence, upon which he sets so great a value, that he considers them as the noblest presents he can make to foreign princes and ambassadors. If we look into the commonwealths of *Holland* and *Venice*, we shall find that in this particular they have made themselves the envy of the greatest monarchies. *Elzevir* and *Aldus* are more frequently mentioned than any pensioner of the one or doge of the other.

THE several presses which are now in *England*, and the great encouragement which has been given to learning for some years last past, has made our own nation as glorious upon this account, as for its late triumphs and conquests. The new edition which is given us of *Caesar's* commentaries, has already been taken notice of in foreign *Gazettes*, and is a work that does honour to the *English* press. It is no wonder that an edition should be very correct, which has passed thro the hands of one of the most accurate, learned, and judicious writers this age has produced. The beauty of the paper, of the character, and of the several cuts with which this noble work is illustrated, makes it the finest book that I have ever seen; and is a true instance of the *English* genius, which, tho' it does not come the first into any art, generally carries it to greater heights than any other country in the world. I am particularly glad that this author comes from a *British* printing-house in so great a magnificence, as he is the first who has given us any tolerable account of our country.

MY illiterate readers, if any such there are, will be surprised to hear me talk of learning as the glory of a nation, and of printing as an art that gains reputation to a people among whom it flourishes. When men's thoughts are taken up with avarice and ambition, they cannot look upon any thing as great or valuable, which does not bring with it an extraordinary power or interest to the person who is concerned in it. But as I shall never sink this paper so far as to engage with Goths and Vandals, I shall only regard such kind of reasoners with that pity which is due to so deplorable a degree of stupidity and ignorance.

, *The fourth Volume of Blackstone's Commentaries on the Laws of England will be ready for the Subscribers some time in February.*

Figure 1
Observations relative to the . . . , reproduction of the broadside. Library of Congress, Rare Book and Special Collections Division.

Reading 4

RICHARD HERRING

Paper & Paper Making, Ancient and Modern (1856)

As a young man, Richard Herring (1829–?) delivered a series of lectures at the London Institution, founded in 1806 for the "diffusion of Science, Literature and the Arts." Thus Herring's message appropriately focused on the transformative power of paper to bring about social change through the diffusion of ideas. In this lecture from the series one would be hard-pressed, however, to decide which invention noted by Herring—the papermaking machine or the printing press—guaranteed the success of the medium of print.

[. . .]

[L]et us now proceed to trace out, in some measure, the history and progress of that more perfect and ingenious invention, MODERN PAPER; and in so doing, I can hardly forbear making some allusion to the incalculable advantages which have resulted to mankind from the introduction of so ingenious and extraordinary a discovery. It certainly would appear very remarkable, that not only amongst mankind generally, but even with those intimately associated with that branch of commerce, so little interest should be found to exist in an acquaintance with its origin and advancement, beyond the bare knowledge which directly concerns them. It is true that with them, no less than with people in general, the very indispensableness of the material renders familiarity at once an unconscious stumbling-block, to any conception of the grandeur of its importance, or its vastly interesting and varied associations. Yet what infinite trouble and labour, what fruitless consumption of time, has not been saved by the invention of paper. How many toilsome and dangerous experiments have not philosophical projectors been spared. What laborious investigations and study have not thus been abridged, by the facts of oth-

---

From RICHARD HERRING, *Paper & Paper Making, Ancient and Modern* (London: Longman, Brown, Green, and Longmans, 1856), 28–32.

ers' researches being so conveyed to posterity—knowledge, more than any one man could have attained to in a thousand years, though born with faculties in maturity. To enumerate all the advantages which the invention of paper has afforded mankind, it were, indeed, useless to attempt; for, whether we look at the traveller, traversing sea and land, without the knowledge of geography and navigation, without those beautiful charts of the ocean, by which he is now enabled to proceed with safety, and even to predict with certainty his arrival at the most distant ports; or, whether we look at the man of science, who being neither artist nor manufacturer, is thus enabled to communicate his plan and projects with accuracy and ease, for mechanics afterwards to improve and perfect; or, indeed, whether we view the growing youth, educated with such facility in the principles of their duty, backward even to barbarous states, softened and enlightened by means of the discovery; its value, in the applicability of its purposes, stands out alike in each, declaring it distinctly, above all other inventions, as truly the most wonderful, useful, and important, which has ever yet transpired in any age of the world; inasmuch as without it, every other discovery must necessarily have continued comparatively useless to society. For, be it remembered, that in contrasting the results of this invention with the productions of former periods, we are, in fact, arraying in our train the mighty arm of the press against the feeble efforts of an unwieldy style, or the tedious and uncertain process of the slow-paced pen, which, prior to an acquaintance with the art of printing, were the only means mankind possessed for spreading the influence and advantages of learning amongst their fellow-creatures. And, again, how highly interesting is it, to observe the prodigious advancement resulting from an ingenious and successful application of machinery in the one case, serving at the same time to develop to our wonder and amazement the extraordinary capabilities of production which have since been revealed by the Printing Machine. Truly may we now pronounce—

> "The Press! the venerated Press!
> Freedom's impenetrable shield—
> The sword that wins her best success,
> The only sword that man should wield."

It is stated that the daily aggregate printed surface of the *Times* alone, actually exceeds that of thirty acres; and the *Illustrated London News*, on one occasion, sent forth no less than 500,000 double numbers, or one million sheets. In fact, 2000 reams, exceeding seventy tons in weight.

The manufacture of four or five hundred square feet of paper per minute, and 12,000 impressions per hour, are now matters of everyday occurrence, although it should be borne in mind, that without the *paper machine*, pouring forth its miles of web, these corresponding advantages in printing could not have been developed.

Reading 5

Henri Pourrat

The Devil in the Paper Mill: A Tale (1948)

Henri Pourrat (1887–1959), French writer and anthropologist, spent a lifetime recording the folklore of rural nineteenth-century Auvergne. This tale was written by Pourrat for the members of La Feuille Blanche, the French Association of Friends of Paper and the Graphic Arts and Industries, which he founded. It conveys the desperate lengths to which Damian, a young apprentice, or feuille blanche *boy, would go to relieve his miserable existence in a paper mill.*[1] *The term* feuille blanche *colloquially means "blank page." Much like the phrase "clean slate," it alludes to paper's magical power of transformation and redemption. By the end of Pourrat's tale, the* feuille blanche *boy has become the master of the mill, married to the foreman's pretty daughter.*

The backbreaking labor endured by Damian in south central France was similar to the scene described by the Scotsman Charles Cowan (1801–1889) after touring a paper mill in Kent, England, in 1819. Like Herring, who equated the significance of the invention of the papermaking machine to that of the printing press, Cowan recognized the machine's positive role in the spread of social ideas, as well as in alleviating the toil of papermakers. He observes, "The labour imposed upon the vatman and coucher, owing to their constant stooping posture aggravated by the heat of the vat and the often dense steam, is peculiarly severe. I do not consider it actually unhealthy, but I am confident that such men become prematurely old, and at fifty years of age have the appearance of having reached fully the 'threescore and ten.'"[2]

Pourrat's story underscores the excruciating conditions and incredible resilience of the clannish community of papermakers. After being cleverly thwarted in his attempt to cast the young apprentice into the flames of Hell, the Devil never again calls upon a papermaker.

From Henri Pourrat, *Le diable au moulin à papier* (Paris: La Feuille Blanche, 1948), 3–15.

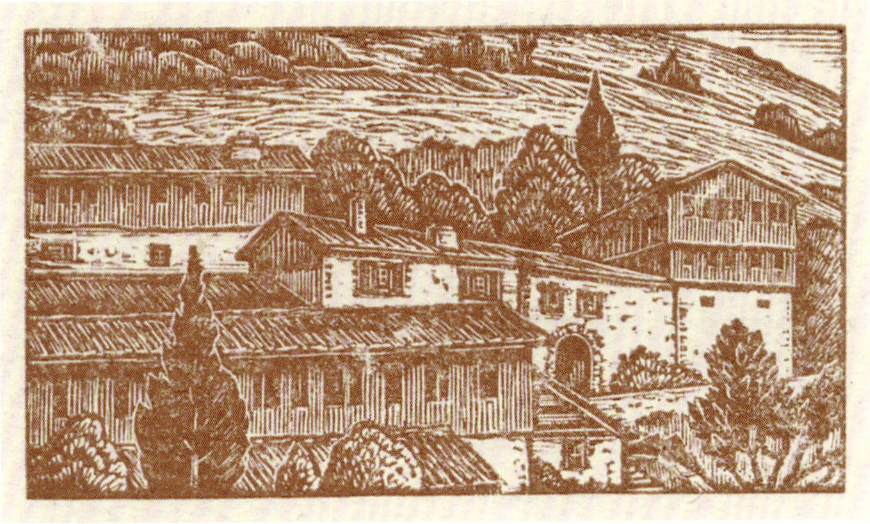

Figure 1
© La Feuille Blanche (Ambert, France).

Once upon a time, there was a boy who had lost his father. The father had been a paper-maker and that trade is passed on from father to son. As soon as the boy turned twelve, the mother apprenticed him to a paper mill.

The mills were located in the hollow of the ravine, on the raging waters, among the rocks, the leaning trees, the splashing, cascading currents. Under the flume, the wheel turned quickly, quickly, driving the pounding beaters. Everything shook, pulsated, crackled. From the railway, through the bracken, the mules brought in bundles of cloth, dirty old rags, and brought out nice white sheets. How could such a transformation take place? . . .

That's a secret. A secret that the Crusaders brought back from the Holy Land in olden days, and one that is closely guarded.

*

In the mill where Damian—who was known as Damian, the apprentice paper-maker—worked, to prevent loitering around the vats, the masters had a black dog, a big booming mastiff. The dog's name was Rapatou, which is one of the names of the Devil. A devourer who did look ready to swallow everyone. But he and Damian hit it off right away. Rapatou warmed the little boy, in the space where he slept under the stairs. The masters were big earners and hard drinkers; they gloried in their status, in their secret; this made them hard people; they did not treat the apprentice too gently. The work began in the middle of the night—so it could be done more covertly, yes, always in secrecy!—so Damian had to be up at midnight. By four in the morning, the poor boy often couldn't keep his eyes open and would drop the pile of felt or the lantern he was holding. Then, the master, the boss, the workers would rap him hard on his knuckles. And he had chilblains . . . His poor hands were already all numb, swollen, cracked and chapped! So when they were hit . . .

Those mills, moving, rolling, beating feverishly, those vat rooms and especially the beater rooms, those vaulted cellars dripping with water—their murk and din could make them seem like Hell to a child. And that is truly what it was for him.

*

One night, in tears, he tore outside, then pulled up short on the path, a few steps from the door. "My mother needs my labor to survive, so I can't just run away. Yet, I see too much of what goes on in their diabolic mill. Oh, if only someone could fix it so I don't get hit anymore . . . I'd devote myself to that someone, even if it was who I think . . . Never mind the Devil! He surely can't be as bad as the trouble the masters give me!"

And suddenly the Devil appeared before him.

There he was, near an ash tree and a parapet over the stream, looking a little like a rag picker, only stouter.

"So, little one, you've had your fill of misery?"

"Oh, yes, my fill!"

"Good. One word from me and you can live in peace. Your masters will think twice, three times even, before they so much as snap their fingers at you. How they will eat crow for you, if you so wish! But do you so wish?"

"Do I so wish? Ha! I'd pay dearly for that."

The Devil hastened to seize his advantage. "I will give you an entire year. When the year is up, I will return and give you a sign."

Only one year of peace and then to have to go with the Devil? Some bargain! But for the youngster, a year was an endless amount of time. "Let me rest for only that time, a year without raps on the hands and without misery. And later? Well, that's later . . ."

"One year?," he asks. "Wow, really!"

"Done," says the Devil. "When they want to scold you, you need only say:

O mill wheel turning around and around
Run in reverse and then break down!

You'll see what happens."

The boy went back into the mill with such assurance that he went right up to the boss.

"You guys don't frighten me anymore. Now I can defend myself. Just try me and I'll give it to you good!"

What with the din, the man only half heard. He thought it was just some childish boasting. He was all involved in maneuvering the rags and preparing the pulp, so he wasn't about to stop for anything this little rascal had to spout.

*

The next day—or rather the next night—at midnight, he went to awaken Damian, who was lingering in bed. He had already called him four or five times . . .

"Just you wait and see what I have for you here on the end of my sleeve!"

But he had barely rapped the boy's knuckles when:

O mill wheel turning around and around
Run in reverse and then break down!

In an instant, what chaos . . . Everything started to go wrong in the mill. The beaters beat from bottom to top, the machine came apart, the wheel fell to pieces, the water jumped aside and began flooding everything, the troughs tipped over and the pulp mixed with the mud, the vat opened and released its contents. In the smoother itself, the piles of paper crumbled and were ruined, while on the drying lines, the sheets slipped from the cords and flew out through the openings, like a cloud around the building.

Neither master, nor boss, nor journeyman, no one found a remedy for this. They all ran around in the mess, getting caught on their white canvas aprons, which tore like sheets of paper . . . They tried to raise the beaters, to save the pulp, to stop the water, but they could not do a thing. What a ring-around-the-roses there was that morning—ashes, ashes, and they all fell down!

There was damage on top of damage, and no one had any idea what could have happened. It took a long time to reassemble the machine, howsoever diligently they tried. Twelve or fifteen days it was idle.

Ceaselessly they rehashed it, discussing it, wondering what devilment might have taken place—and they would not have believed how right they were. But for all their talk, they said not a thing.

Then, as if enraged, they returned to beating the pulp and making the sheet. Everything had to hustle in order to make up for the fifteen lost days!

Then, the first time that Damian lolled in bed, they started to punish him. Only, it wasn't long before . . .

O mill wheel turning around and around
Run in reverse and then break down!

And presto, the feverish pitch of the mill came to a halt. Then the wheel went berserk, the master shaft broke, the beaters came off their hafts, the troughs turned upside down, the vat was upended, the cords of the drying lines snapped, the sheets flew through the air.

"Just what we needed . . . an antichrist!"

If their eyes had been daggers, Damian would not have survived safe and sound.

However, they no longer dared touch him, or even say a harsh word to him.

"Remember what I promised you!"

The boss remembered . . . With a sign, Damian could have blown out the windows!

But Damian no longer wished anyone ill. Once delivered, such ease! He could now sleep to his heart's content, and linger on his stool, warming his hands on the bottom of his porringer, before eating his cabbage soup. And no more clouts. And especially no more raps on his chilblains . . .

So he, who took everything in stride, had completely changed. Until then, his only friend had been Rapatou, the big black dog. Rapatou who followed him everywhere, licking his poor hands, cramming that goodly head against his legs. Rapatou remained his faithful companion. But Damian made friends with the master's daughter, Marinette. Washing the felt became a pass-time, as it is in the good mills. Since by tapping the felt on the washboard, you can spray water in your neighbor's eyes. They splash you in turn. You tease one another and chase one another along the banks of the stream, in the grass. These outdoor washdays turned into a real party.

Then, there were all those festivals as the seasons turn: the Feast of Saint John Porte Latine,[1] who is the patron saint of paper-makers, the Feast of Saint Peter, who is also their patron saint, the Feast of Corpus Christi, its procession where they all sing in a drone, covered with a surplice, and the fairs, and the weddings, and the trout fishing; and the celebrations when any worker passing through was hired by the master, and then paid for the wine . . .

In all of that, there were no more rebuffs. Damian had his place at the table and sang his song. As he had grown strong, increasingly cheerful, imaginative, and skilled, he was always pleasant company. And people looked kindly on him. He worked Marinette's garden, from the lilac bush at the front to the quince tree at the back, and Marinette knit mittens for him for the winter. When he went to town, he brought the girl two dozen pins or a pretty ribbon for her hair. And when she served him, she gave him the crust of the fresh bread or the heart of the lettuce, topped with a look of friendship.

He was at ease, ha, more at ease than a king. Besides, the song of the paper-makers told him:

That if the king were aware
Of the life that we share
His throne he'd give away
And make paper all the day

Damian was forgetting only one point, a mere trifle: the promise he had made to a certain so-and-so one winter's eve.

*

But one year later to the day, on an evening at that same dark and reluctant time of year, Damian came face to face on the path with someone who was carrying a large sack.

"So? The year went well? Went by fast, perhaps?"

The Devil sniggered and spoke in an affected manner.

"Be that as it may, a deal was made. I kept my promise, did I not? Now keep yours."

Like the rag pickers bringing their haul of rags to the mills, he carried his sack on his back. The sack was jiggling, for it was already filled with a good half dozen souls.

The idea of getting stuffed in the bag was causing Damian a fair amount of grief.

"What a deal! Was I ever duped! So how can I dupe him in return? . . . I was too young at the time, a bit too much of an apprentice. Now I have to act like a journeyman, and quickly, real quickly . . ."

Just at that moment, his big dog Rapatou crept into the shadows, silently, as if coming to his aid.

The Devil did not see the dog, occupied as he was with setting down his sack. As soon as he had opened it:

"Here you go," he said to Damian, "hop right on in."

Either by merit or by luck, the boy then had the idea that saved him.

"I'm a bit heavy," he said. "You'll have trouble lifting the sack onto your back again. Pick it up, bend over, and I'll climb in."

But instead of climbing in, he took his Rapatou and stuffed him in the bag.

The Devil didn't even turn around. He felt the sack get heavy—"It's true the little rascal weighs a lot!"—and with a jerk, shouldered the sack and left straight away for Hell.

He reached the end of his journey. What's more, the sack was so full, he could not have fit in a single customer more.

Racing toward his home, as the load was heavy, to give himself courage, he sang as he went:

I'm carrying a full load for the oven
And the last one in is the journeyman

The last, but the *pièce de résistance*. He was jumping for joy to bring a papermaker back to his Hell where, apparently, he did not yet have one.

As soon as he entered, he called to his imps. "Just wait until you see what you see!" He grabbed the sack, attempting to empty it onto the middle of the rotisserie. But the dog Rapatou was the first to fall out and he did not see things quite that way! Hell had nothing to do with him, the big dog who had always honestly done his doggy job . . .

Before they knew what had happened, it was as if a bomb had exploded. Barely had Rapatou felt the flames when he leapt up, yapping, howling, sending brazier, pitchforks, and horns flying. Front and back, the firebrands went flying off into sparks. The blaze seemed to jump. And from the midst of these fireworks, thrusting himself at the heads and tails of the devils, was Rapatou, biting, toppling,

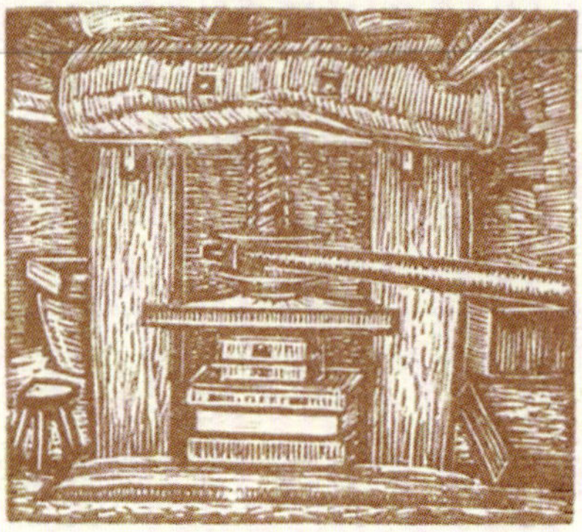

Figure 2

dispersing . . . Before the others were even aware of what was happening, he had turned Hell upside down.

"What an idea, what an idea I had, to bring one of those awful creatures here!," the great Devil, one of his horns missing, said to himself. And he had no other thought in mind but to see his paper-maker gone from there post-haste.

So, more affectedly than ever, he proclaimed:

Slide back the bolts and open the door

Of this paper-maker let me see no more!

"Yet I knew that those people have a reputation for being bad to deal with. An apprentice! What if it had been journeymen! Let them go where they will, but may they no longer come here!"

In three leaps, Rapatou was back at the mill.

He returned there. As for Damian, he had remained there.

Afterward, Damian even stayed there for good. He became its lord and master and took Marinette for his wife. And much later, in the fullness of age, when he had to leave the mill, no, it was not to go down to that rag picker with the big sack. Since that famous day when he thought he was bringing back a *feuille blanche* boy, the Devil has no longer admitted a single paper-maker into his Hell.

Notes to Introduction

1 Pourrat's booklet was printed on handmade Chiffon d'Auvergne from the Moulin Richard de Bas, which he was instrumental in preserving. See www.richarddebas.fr.

2 Charles Cowan, *Reminiscences* (n.p.: privately printed, 1878), 68.

Note to Reading

1 [Saint Jean Porte-Latine was plunged into a vat of boiling oil before the Porta Latina in Rome by order of Emperor Domitian on May 6, 95. Boiled oil is a component of printing ink. *Inland Printer* 66 (March 1921): 764.—Ed.]

Reading 6

Eden Phillpotts

Storm in a Teacup (1919)

Eden Phillpotts (1862–1960) was an English poet, dramatist, and author of a series of novels based on different trades. His unflinching description of the backbreaking toil associated with making paper by hand documents the decline of this British industry, which was already well under way in the early twentieth century. In this excerpt from Phillpotts's novel about the paper trade, the vatman Jordan Kellock, infatuated by Medora, wife of a fellow vatman, loses his "stroke" or "shake." As observed by the Scotsman Charles Cowan, "The formation of the sheet on the mould is a very delicate operation, requiring great skill and practice, particularly the 'shake.' There is nearly as much difference in the 'shake' of different workmen as in the features of their countenances."[1] The cataclysmic impact of Kellock's physical and psychological breakdown reveals not only the superstition rife among the proud brotherhood of papermakers but also the fine-tuned discipline crucial to maintaining the rhythm of the mill. Kellock's God-given gift to form sheets of pure paper is destroyed by his ambiguous moral state.

[. . .] He could only lick his smarting wounds. Then he fell into genuine grief for Medora's plight. Actual physical symptoms intruded. He found his eyes affected and strange movements in his heart and stomach. His hands shook in the morning, and he cut himself shaving—a thing that he had not done for years. He could not eat, yet suffered from a sensation of emptiness. Daylight by no means modified his sense of loss and chaos. It found him before all things desirous to see Medora; but, by the time he was up and dressed, this purpose failed him for a season, and his thoughts were occupied with Knox [a fellow vatman]. Then he turned again to Medora, and felt that life must be suspended until he could see her and break to her what had happened. It was now too late to visit Cornworthy until the day's work

---

From Eden Phillpotts, *Storm in a Teacup* (New York: Macmillan Company, 1919), 267–70.

should be done, and remembering how often work had saved a situation, solved a problem and helped him through difficult hours, Kellock proceeded to the Mill, and was thankful to be there. He felt that labour would calm his nerves, restore his balance, and assist him, before the evening came, to survey his situation in the light of this convulsion. He found himself entirely interested in what Medora would do; and he believed that he knew. His heart bled for her.

Thus absorbed, he reached the vat. He was engaged upon the largest sheets of drawing-paper at the time—work calling for more than average lifting power and muscular energy—and he was glad that now, for a while, work must take the first demand upon mind and body alike.

The vats were full, and the machinery hummed overhead; coucher and layer stood at their places, and Jordan, slipping his deckle upon the mould, grasped it with thumb on edge, and sank it into the pulp.

Elsewhere Knox, Robert Life and others had taken up their positions at the breast of the vats with their assistants about them, and the work of paper-making went on its immemorial way.

Then that happened that was long remembered—an incident of interest and concern for the many, a tragedy for the one. Kellock brought up his mould, and instead of proceeding with the rhythmic actions to right and left—those delicate operations of exquisite complexity where brain telegraphed to muscle, and motor and sensory nerves both played their part in the completion of the "stroke"—instead of the usual beautiful and harmonious gestures that drained the mould and laid a sweet, even face of paper upon it, he found forces invisible at his elbows and an enemy still more terrible within. His brain hung fire; a wave of horrible doubt and irresolution swept over him. It ran through the physical parts engaged—his arms and breast muscles and the small of his back. He stared at the mould, turned and washed off the faulty sheet he had created, and made an attempt at a jest to Harold Spry, who was watching, all eyes.

"Where are my wits, Harold?" he said. Then he took a deep breath, and dipped the mould again.

Spry and the layer watched sympathetically. To their eyes there seemed no failure as Kellock drew up his load; but he knew. A condition of tremendous tension raised his heart-beat to a gallop, and his eyes grew misty. He gasped like a drowning man, and felt the sweat beading on his forehead.

"I'll—I'll just get a breath of air and come back," he said, dropped the mould, and went out of the shop. Spry washed the mould, then he walked down the line of vats and spoke to Knox. A man came from the engine house with a message, and Ernest Trood also entered with some information for Robert Life. What he heard made him hasten out of doors to find Kellock sitting up on a form at the entrance of the vat house with his head in his hands.

"What's the matter, my son?" asked Trood, kindly enough; but a look at Jordan told him all he feared to hear.

The young man's expression had changed, and there was fear in his eyes, as though they had just mirrored some awful thing. The resolute, closely-knit Kellock seemed to have fallen to pieces. Every limb indicated the nerve storm under which he suffered. Trood was experienced, and knew the danger. He believed that Kellock had given in too soon.

"Fight—fight like hell!" he said. "Don't run away from it. Don't give it time to get into you. Come back now, lad—this minute. At your age, it's nothing—just indigestion, or a chill about you. If you let it fester, you'll go from bad to worse, and very like have to knock off for six months before you look at a mould again."

"It's no good—it's gone," said the younger man; but he obeyed, and followed Trood into the vat house.

Knox had warned the rest to ignore the sufferer, and no man took any notice of Kellock as he returned.

Spry was waiting, and greeted him cheerfully.

"You're all right again—your eyes are all right," he said.

Trood turned his back on Kellock, and everybody was at work as usual. He made a tremendous effort with himself, called up his utmost resolution, smiled and nodded to Spry, who was whistling, gripped his deckle to the mould, and then strove to think of something else, pursue his business in the usual mechanical fashion, and let his unconscious but highly trained energies pursue their road.

But it was not to be. Some link had strained, if not broken, in the complexus of brain and nerve and muscle. Perfect obedience was lacking; a rebel had crept into the organism. For once, the man's expressionless face was alive with expression; for once his steady and monotonous voice vibrated.

"It's all up," he said to Harold Spry.

Then he put down the mould.

Trood was beside him in an instant, and Knox came also. Elsewhere those who had no love for Kellock talked under their breath together. Others, who came and went, took the news.

Trood made the vatman try again; but only once. He saw in a moment that the breakdown could not be bluffed; the fault in the machine was too deep.

Jordan put on his coat, and Trood arranged to drive him to Totnes presently to see a doctor. The young man was calm, but his will power appeared suspended. He looked into the faces of his companions for any ray of comfort; and the fact that he could do so was testimony to his collapse.

He went back to "The Waterman's Arms" presently; and through the Mill like lightning flashed the news that Kellock had lost his stroke.

Note to Introduction

1 Charles Cowan, *Reminiscences* (n.p.: privately printed, 1878), 68.

Reading 7

HENRY MORRIS

Nicolas Louis Robert and His Endless Wire Papermaking Machine: With Facsimiles of the Inventor's Original Drawings of the First Paper Machine . . . (2000)

Nicolas Louis Robert (1761–1828) was employed as a bookkeeper in a paper mill charged with producing the paper used for assignats, the paper currency put into circulation during the French Revolution. In response to ever-growing production quotas due to hyperinflation and regular labor strikes, the mill owner, Saint-Léger Didot (1767–1829), was desperate to decrease the number of his employees through mechanization. Robert, a natural tinkerer, eventually produced a machine able to produce "endless" (i.e., rolls of) paper. His invention was later improved thanks to underwriting by the Fourdrinier brothers, Henry (1766–1829) and Sealy (1773–1847). This account, quoted by Morris, was recorded by the Englishman John Gamble, Didot's brother-in-law and early investor.

Louis Robert, a native of France, is the person to whom we are indebted for the papermaking machine. I had frequent opportunities of seeing him at work on his first model, in 1796, 1797 and 1798, at the Paper Manufactory at Essonnes, eight leagues from Paris, Département de Seine-et-Oise (which at that period belonged to Monsieur Leger Didot, my brother-in-law), where Louis Robert was employed as bookkeeper. His first model was no larger than a bird organ,[1] and the slips of paper he made with it were not wider than a piece of common tape, but of various lengths. Robert possessed a natural mechanical genius, and was never so happy as when employed in inventing or improving some piece of machinery, but unfortunately, the only time he could devote to his favourite pursuit was in the evening, after the counting-house business was over, and many a time I have heard him blamed and reproached by his employer, Leger Didot, for wasting so much

---

From HENRY MORRIS, *Nicolas Louis Robert and His Endless Wire Papermaking Machine: With Facsimiles of the Inventor's Original Drawings of the First Paper Machine . . .* (Newtown, Pa.: Bird & Bull Press, 2000), 21–22.

time on the invention that would never be brought to perfection. However, Robert persevered, and in about two years produced a model which performed so satisfactorily that M. Didot was, at last, induced to afford him the means of making a machine upon a larger scale, which was called the working-model. Orders were given to the carpenters, smiths, millwrights and other workmen employed at the manufactory to take directions from, and execute any work, that Robert might require for his paper-making machine. In a few months from that time a machine was completed capable of making paper of the width of Colombier (24 in.),[2] and of various lengths, being the kind of paper usually employed for paper hangings of which the consumption in France was immense. Owing to the great difficulty, at that time, of sizing and pressing sheets of paper of great length with the presses usually employed in papermills, the length was generally limited to twelve yards, but a fresh difficulty arose in the process of drying it; it was suspended on lines in the drying loft, but from its weight when in the wet state it contracted so much in the middle, during the drying, that it was rendered unfit for the use of the paper stainers; the next attempt was by laying these long sheets horizontally on the lines; and by adjusting several of the frames or trebles on a level line, the inconvenience of the partial shrinking of the paper when suspended in a vertical direction was remedied. Many of these long sheets were sold to the paper strainers of Paris. After a series of experiments and improvements, Louis Robert applied to the French Government for a patent, or brevet d'invention.

Notes to Reading

1 [A bird organ was a small, hand-cranked music box intended to teach caged birds how to sing. It was small enough to be placed on a table.—Ed.]

2 [Colombier was a standard paper size in France and England; its length varied by an inch, 34½″ and 35½″ respectively, from England to France.—Ed.]

Reading 8

JACQUES DERRIDA

Paper or Me, You Know . . . (1997)

In his "spleen de papier," *the deconstructionist philosopher Jacques Derrida (1930–2004) advances his blunt defense of paper against the assault of electronic media. Derrida cautiously accepts electronic media but only as tools to be used in the service of the written word, much as a driver needs to look at oncoming traffic and in the rearview mirror while simultaneously moving forward.*

[. . .]

And then longing or even "acting" for the bookish arts obliges no one to stick with them. Like many, I take my longing as proof and, without giving up a thing, I try—always with uneven success—to adapt my "economizing" to all paperless media. I use the computer, of course, but not e-mail, and I do not "surf" the Net, even though I use it as a theoretical topic in my teaching and elsewhere. Abstention, abstinence, but also self-protection. One of the problems is that now any public speech (and sometimes any private act, any "phenomenon") can become "globalized" in under an hour without your being able to exercise the slightest control over it. Sometimes this is terrifying (less novel still in its potential than in its power, the rhythm and the scope, the objective technicality of its phenomenality); sometimes amusing. It always involves new responsibilities, another critical archival art, in short, another "history." But why would one sacrifice one potential when inventing another? Saying "adieu" to paper today would be a bit like deciding one fine day not to talk any longer, just because you know how to write. Or no longer looking in the rearview mirror, just because the road is before us. We drive with two hands and with the feet, looking both forward and backward. We presumably cannot look *at the same time*, in a single and indivisible instant, both

---

From JACQUES DERRIDA, "Le papier ou moi, vous savez . . . ," *Pouvoirs du papier, Cahiers de médiologie* 4 (1997): 56.

backward and forward, but we do drive while switching very quickly, in the blink of an eye, from the windshield to the mirror. Otherwise, it's a choice between blindness and an accident, if you see what I mean: the end of paper isn't about to happen anytime soon.

In closing, two more comments on my "paper spleen." On the one hand, as I dream of absolute recall, so it is that when I sigh (which is my very breath) in truth for keeping everything, my imagination continues to project that archive *onto paper*. Not onto a screen, although that could happen too, but on a strip of paper. A multimedia strip, with phrases, letters, sounds and images: everything, it would retain the imprint of everything. A single original, of which copies would be made.

Reading 9

Pierre-Marc de Biasi

A Paradoxical Substance (2011)

Pierre-Marc de Biasi (1950–), director of the Institute of Texts and Modern Manuscripts, Paris, is a codicologist, paper historian, and artist specializing in the systematic revelation of physical and theoretical meaning—extending as far back as the "genetic code"—of modern literary manuscripts and works of art on paper. Parallels to conservation can be drawn from De Biasi's analogy of paper as the physical culmination of Leonardo da Vinci's cosa mentale*.*[1] *Barbara Appelbaum, describing the goal of conservation as enhancement of meaning, writes, "Ultimately, then, the optimal practice of conservation rests not in our hands but in our heads."*[2] *With this final reading in Part I, the reader has traversed the dawn of paper and its ascendancy over parchment in the fifteenth century to what some prophesy will be its dusk in the twenty-first century.*

[. . .]

It is used for everything. There is an abundance of it. But it is there without being there. It is not a material; it is an environment, a world, a sort of all-purpose genie, a partner in the most intimate moments and most official circumstances of our existence. We are so used to having it always around that we no longer see it. Omnipresent, invisible, we live with it in a relationship of coexistence as constant as it is imperceptible. Paper is a subliminal substance.

And even when considered in a vacuum, it is its capacity as a medium that stands out. Its physical reality disappears behind the symbols or images it carries. The written page is text; the printing conceals the page, the legible effaces the visible. The almost preternatural drawings of Matisse make us forget the Arches vellum[1] they are on. The photographic print charms us like an epiphany of the image

---

From Pierre-Marc de Biasi, "Une substance paradoxale," in *Le papier à l'œuvre* (Paris: Hazan Louvre Éditions, 2011), 19–24.

or the thing depicted, while the paper that reveals it to us disappears. Perhaps this spiriting away of the physical appertains only to looking at the finished work, in an after-the-fact illusion that has little to do with the reality of the creative act.

A Material That Resembles Us

For the writer shaping his sentences in his notebook, for Matisse preparing to draw on his page the pure line drawing of his mind's eye, for the wizardly hand of the photographer sweeping the light beam of the enlarger over the silvered paper, there is nothing virtual about paper. For the artist at work, paper is definitely real, central and essential, in all the intensity of that practically physical relationship with the material that is at the heart of creation. From the mind to the eye, from the eye to the hand, from the hand to the sheet of paper, the path is the opposite of sublimation. It is a nerve fiber that runs from concept to drawing, a manifestation of the *cosa mentale*. The true enthusiast will fully grasp the spirituality of the work only if he in turn becomes capable of sensing its innate materiality. As much as the artistic creation itself, he will love the color, the texture, the weight, and the contrast, and even the watermark of the paper selected by the artist.

The unobtrusiveness of paper is due to our inadequate perception of its appearance. But the fact is that, at the very least, paper has an exceptional flair for camouflage: a medium of excellence, the grander it is, the less we see it. On the contrary, it is in its deteriorated forms that its presence becomes the most compelling: when it changes into rubbish, when it burdens us, when it acts as a screen for its own evanescence . . . we crumple it, we throw it away, we burn it, we recycle it, we wish it away. And yet, miracle of art that it is, even in these most insignificant or most humiliated forms, paper can still be transformed into desired beauty [. . .]

[. . .]

Ubiquitous, paper? Yes, superabundant, but essential: just run out of it, and it's a catastrophe. Because the paper we toss away so offhandedly is also what we need for developing, communicating, and transmitting our thoughts, what we submit to defend our rights or prove our identity—and, unfortunately for them, some people know how distressing it can be to be "undocumented" or "paperless." Artists have also sought to document this issue.

As imperceptible as the water in its bowl to a goldfish, paper is our daily scenery, our human environment: vital and negligible like that crowd of strangers we pass each day without paying the slightest attention, or like that silhouette we glimpse in the mirror every now and then. That might actually be the mystery of paper: if it seems to us so near and yet so far, so familiar and yet so strange, it is because it is made in our image. Fragile and tough, trivial and serious, flexible and stiff, sensual and abstract, it ignites for no reason, it is easily damaged, it endures, it remembers, it becomes wrinkled over time: we are like two peas in a pod.

So it's no surprise that, as early as the Renaissance, this paradoxical material became the preferred medium of artists. To work with paper is to manipulate a substance that is humanized through and through, to find in the very support of the artistic creation that measure of all things that is the human body. But there is more. Paper is not only a metaphor for our flesh, it is also a symbol of the human adventure. A sheet of cellulose is a temporal sedimentation, a sort of symbolic material that has collected the memory of its own travels around the planet. Unassuming hero of the most dazzling adventures of civilization, paper is a bit like the Hegelian concept of the Spirit: each nation passing the torch in order to advance it from East to West, it is an incarnation of the future that carries in it the entire history of the world.

The Entire Memory of the World

Born in the Middle Kingdom more than two millennia ago, paper remained the monopoly of the Buddhist world for a thousand years. The armies of the Prophet seized it at Samarkand in the 8th century to place it in the service of the Muslim empire and the new monotheism. Islam in turn kept the secret of its manufacture for five hundred years, until the fall of the Abbasid Dynasty.

After circling the planet from east to west, and after serving as a medium for painters and literati from the Far and Near East, paper arrived at the shores of Italy in the 13th century, giving birth to a new Christian artifact. European paper, which in due time acquired preindustrial manufacturing processes, enabling it to become the essential instrument of the Western Renaissance two centuries later, for both literature and the figurative arts.

In a few decades, the printing press made it the universal medium of Modern Times: books flourished; the concept of invariable text took shape; wood and then metal engravings proliferated. Scholars, engineers, artists, and architects discovered in rag paper an ideal support for sketching and imagining forms of modernity. Well beyond the limitations imposed by the skimpy size and slow artisanal preparation of parchment, the creative act opened itself to huge tangible images, increasingly well suited to the artists' ambitions, in the form of large, solid sheets produced daily by the thousands in the mechanized mills of the Old World.

Since then, paper has been adapted, refined, and industrialized, ultimately becoming accessible to all, to the extent that it became the leading product of mass consumption in the 20th century. With cheap wood pulp, the high-speed rotary press, and the popular press, it became the very substance of our culture, the soul of our democracy of opinions, and the symbol of the rule of law. Material of memory, constituting the foundation for language and nationhood, paper is authentic, it rules, and it is the basis for our dialogue relationship with past and future generations.

The medium of the visual artist's most diverse drawings and creations, paper is not only the physical support of the creative message but also its design space, its secret laboratory: it plays an active and a passive role, through its pentimenti as much as through its metamorphoses and its stigmata. [. . .]

Like the composer's score or the writer's draft, the sketch of the painter, sculptor, or architect makes paper the fragile support of the essential, the miracle of a guardian medium that preserves the traces of creativity in a nascent state. Paper is the sanctuary of genesis.

Communicating or Transmitting

It is this great adventure that is now perhaps coming to an end. In the space of a few years, for the first time in half a millennium, the digital revolution has challenged the unchecked power that paper had acquired as the practically exclusive medium for the still image and the written culture.

A master of metamorphosis, the cellulose sheet immediately fell into line with the new inventions by showing—as in the early hours of metal stamping, woodcutting, printing, engraving, lithography, photography, rotary presses, and photocopiers—that it was the essential partner in the great transformation that was commencing. The challenge is now being fulfilled. With nearly 400 million metric tons per year, the world has never consumed as much paper as it has since it entered the silicon era.

But make no mistake, paper has become a backup for the screen; it no longer rules. Everyone feels, more or less confusedly, that paper is already what links us to the past, to tradition, to the world of yesterday or the day before, and that, in a world of intangible media, the future is already inventing other media resources for circulating texts and images at the speed of light.

New visual art forms are being born in its wake: digital art that recycles the resources of digital photography, 3D simulations, interactivity, and augmented virtual reality. This fledgling Net Art, which is betting on silicon against cellulose, is tying its destiny to a belief: faith in the all-communicationful. Our times are witness to competition building between two perceptions of text and image: one expeditious, for reference, information, indexing, and communication—that of the screen and the Internet; the other slow, for acquisition, reflection, deliberation, reverie, enjoyment, or pleasure—that of paper.

The machines with which we calculate, write, and draw have no weaknesses. As primitive as they may still be, they already no longer have anything in common with the rhythms of our thoughts and our bodies. They need only a tenth of a second to read *Madame Bovary* in its entirety, while it takes me at least twenty hours. So I read a million times more slowly than the machine . . . or even slower, if I take the time to read for pleasure, savoring each sentence. As for writing *Madame Bovary* . . . for that Flaubert devoted five years of his life.

It's like this: our eyes perceive what's real at only some twenty frames per second. There is, in our relationship with paper, a luxurious slowness that matches our biorhythms and our insights. It takes time to reflect, judge, learn, appreciate, and understand. And it takes much more to make an enduring work. But what does it mean *to endure*? Perhaps the opposite of to communicate.

Communicating instantly through space is, admittedly, a useful and remarkable feat. Generating a buzz that ensures worldwide notoriety in just a few days is a miracle artists have always dreamed of, at the risk of dragging their work through the limitation of fashion and short cycles. Quickly famous, quickly forgotten: time always wreaks vengeance on what we try to do without.

But taking the time to develop a message or artistic work worthy of being transmitted through time, over the long term, is that not the intention of what we call art? The long cycle is not necessarily that of monumentality; it can even take the paradoxical path of fragility. [. . .]

[. . .]

Tomorrow, Paper?

We cannot rule out the more or less long-range possibility that paper is destined to lose most of its privileges, one by one, as the traditional medium for communication and transmission of texts and images. But, even if that were the case, it is clear that we are certainly not moving toward a paperless world. On the contrary, demand should double in the next twenty years, to the point of making paper a rare and expensive commodity, as valuable as it was in the 18th century. Nor are we moving toward a world where the creative people forget paper. On the contrary, by freeing itself from the monumentality of print, paper will only deepen its original involvement in that initial moment of creative thinking that gives birth to art. From the emerging idea to the first graphic representation, the shortest path is the one that goes from the eye to the hand, and from the hand to the sheet of paper. It is the shortest nerve pathway and also the freest—free from any standard and any prior constraint. In a time of 3D simulations and computer-assisted design software, why does the architect continue to use the blueprint to devise his plan? Because the paper blueprint is a design space where he can sketch his vision, allowing his hand to act and think without scale or yardstick, in that interval that is both physical and absolute, which the *thought-design* needs in order to express itself.

At the threshold of a world dominated by virtuality and intangibility, perhaps we are inexorably headed for an era of art paper or sensation paper—a passion for its visual and tactile density, an "Asian-style" culture of paper, where the ancient and subtle substance of the graphosphere could advantageously metamorphose into a pleasure material and become the universal medium of a new aesthetic relationship to the world. Paper resembles us. It is its age-old intimacy and complicity with the human body that will make it the irreplaceable material of our future. If it pleases the eye, the mind, and the hand, it is because it contains a secret: freedom of movement freed from all technical servitude, and the magic formula of a projective space that offers itself, as if naturally, to our desire for figuration, to make each of us a creator and artist.

Notes to Introduction

1 *Cosa mentale*, roughly translated as "mental occupation," refers to Leonardo's belief that painting is a mental, not a manual, process.

2 Barbara Appelbaum, *Conservation Treatment Methodology* (Oxford: Elsevier, 2007), xxix.

Note to Reading

1 ["Arches vellum" is a heavy wove paper.—Ed.]

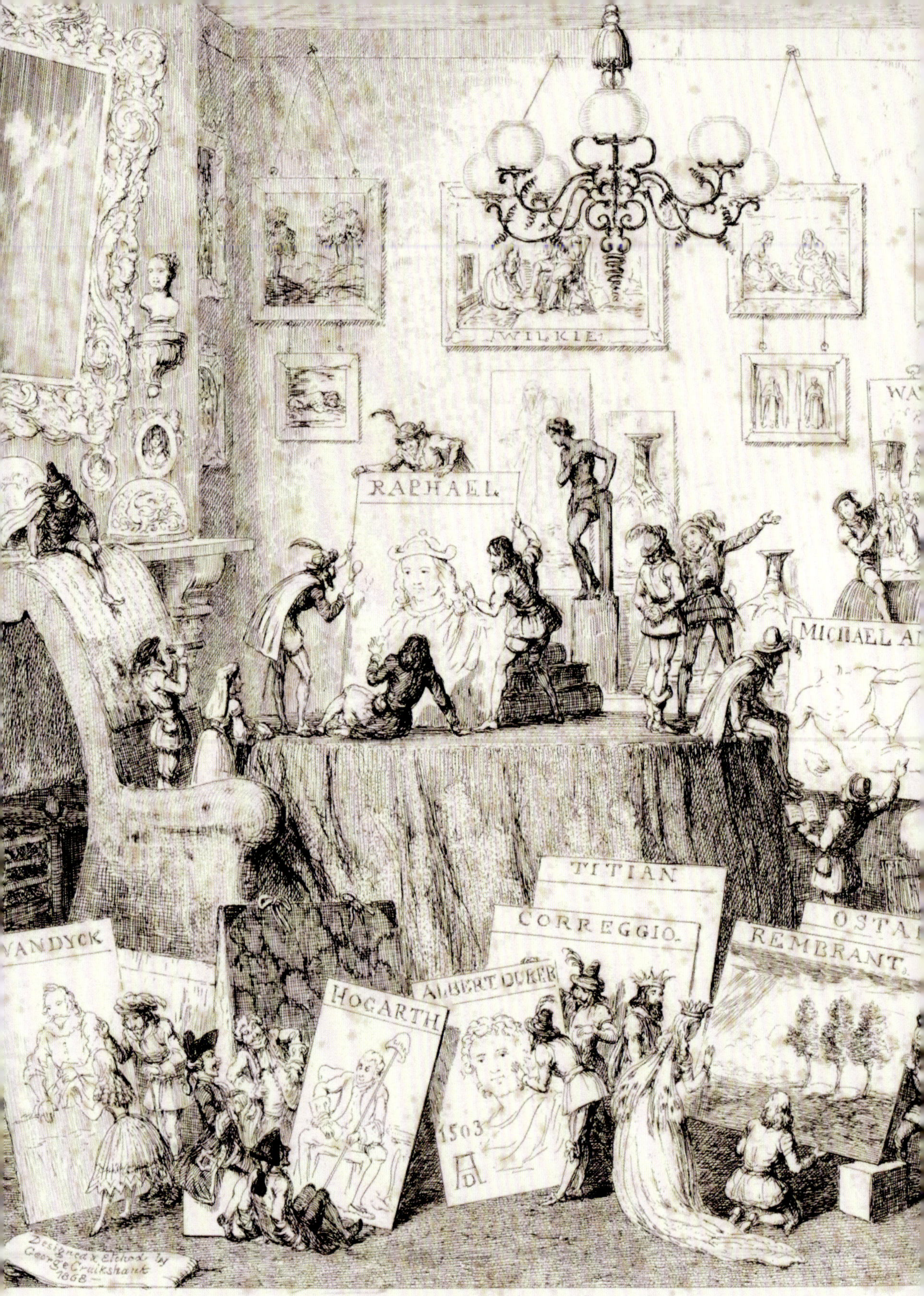

Fairy Connoisseurs Inspecting Mr. Frederick Locker's Collection of Drawings &c. &c. &c.

PART II

The Mastery of Drawing

Part II emphasizes the importance of accurately identifying and describing the materials and methods used in the creation of drawings. The part title is the freely translated title of Winslow Ames's translation of Joseph Meder's classic *Die Handzeichnung: Ihre Technik und Entwinklung* of 1919, which refers not to the mastery of how to draw but how to systematically look at drawings in order to "see" and understand their meaning. Conservators are quite familiar with the initial looking phase, during which the data recorded by their eyes are often augmented by the data revealed by magnification, different types of visible light or radiation sources, imaging techniques, or instrumental analysis. In designing treatments, assessing condition, identifying former function, or determining date and authenticity, conservators routinely characterize media and techniques with great precision.[1]

This type of information gathering is now considered a standard methodology in technical art history and forms the foundation on which other, less tangible assessments can be built. In the words of David Bomford (1946–), director of conservation at the Museum of Fine Arts Houston, "Technical art history concerns itself with all the processes for making art, and the technical and documentary means by which we throw light on those processes. It is principally concerned with the physical materials and structures of works of art and how they are prepared, used, combined and manipulated. But—and this is what makes technical art history so intellectually satisfying—it also interests itself in how an artist arrived at the finished, or, indeed unfinished, work. It charts the stages of invention, development, realization, elaboration and revision: in short, it is a route into, and our access to, the heart of the artist's intentions and changing ambitions."[2] Bomford's "intellectually satisfying" stage is an interdisciplinary endeavor. The thought of scientists, art historians, social historians, and artists, especially if they are still living, is essential. Their expertise, combined with an objective description of what one observes, results in deeper insight into a drawing's artistic or historic significance

George Cruikshank, *Fairy Connoisseurs Inspecting Mr. Frederick Locker's Collection of Drawings*, 1868.

and its multiple social meanings. By acknowledging a drawing's cultural value, conservators are more acutely aware of the potential for loss of intangible information during unsuitable treatments.

As chronicled by Letizia Montalbano in reading 11, research specifically into the materials and techniques of drawings, and, by extension, its subsequent influence on conclusions made concerning their structure and appearance, has been sporadic since Meder's publication. Montalbano credits the emergence of the modern restoration theories of Cesare Brandi (1906–1988) and Umberto Baldini (1921–2006) for the current legitimacy of a phenomenological approach by which "matter becomes ostensibly the 'vehicle for the epiphany of the image.'"[3]

Marks, whether drawn or printed, as will be seen in Part III, can be considered simple substantive matter or intangible instruments of social change. In order to preserve their physical and psychic powers, these marks must be intensely examined. Devoted enthusiasts of drawings recognize that the exercise of looking exclusively at media, as distinct from deciphering what the picture depicts, is indispensable; it cannot be skipped over in a rush to arrive at an attribution or to announce a new discovery, a lesson hard-won by many less experienced observers.

Meder and Marjorie Cohn insist on the examination of original drawings, as many as possible, front and back; this approach forms the backbone of traditional connoisseurship. The various schools of connoisseurship, as practiced since the late seventeenth century, cannot be considered in all their intricate and controversial incarnations in this volume.[4] However, close scrutiny of original materials and techniques remains a principal mandate throughout connoisseurship's history. One early English connoisseur, Sir Jonathan Richardson Sr. (1665–1745), noted, "But 'tis on the Works Themselves we must Chiefly, and Ultimately depend."[5] And in keeping with the Enlightenment philosophy of his time, Richardson further believes that through diligent study anyone, regardless of class or education, can learn to truly "see" paintings. Thus even the youthful Johann Peter Eckermann (reading 17), despite his peasant roots, can fall under the sway of Flemish painting and dedicate himself to art.

The voices of the artist, art historian, teacher, curator, conservator, engineer, chemist, inventor, manufacturer, and purveyor are heard in Part II. All have a vested interest in the accurate identification, classification, and characterization of drawing media, whether for immediate pragmatic reasons or for purely intellectual ones.

The reader is encouraged to note the authors' tendency to evoke other human senses when describing works on paper. For example, Konrad Oberhuber (reading 13) observes that an artist's style of drawing is as idiosyncratic as his writing, an observation originating with the Florentine artist and writer Giorgio Vasari (1511–1574). By extension, then, the lines of a drawing can be heard to speak in a regional dialect. Or, later, the color chemist A. H. Munsell (reading 38) draws an analogy between reading musical notation and then hearing music when distinguishing colors. Some paper conservators only half-jokingly recommend using all five senses in the examination process.[6] Indeed, most conservators are well acquainted with the look, sound, smell, and feel of works on paper but, perhaps, not their taste.[7]

Notes

1 Numerous publications demonstrate the valuable information derived from instrumental analysis, specialized imaging techniques, and intense observation. See Janet Ambers, Catherine Higgitt, and David Saunders, eds., *Italian Renaissance Drawings: Technical Examination and Analysis* (London: British Museum, 2010); Lorenza Melli, *I disegni italiani del Quattrocento nel Kupferstich-Kabinett di Dresda* (Florence: Centro Di, 2006); and Carlo James, *Visual Identification and Analysis of Old Master Drawing Techniques* (Florence: Olschki, 2010).

2 David Bomford, "Forbes Prize Lecture," *Studies in Conservation* 53 (2008): 199.

3 Casare Brandi, "Theory of Restoration I," in *Historical and Philosophical Issues in the Conservation of Cultural Heritage*, ed. Nicholas Stanley Price, M. Kirby Talley Jr., and Alessandra Melucco Vaccaro (Los Angeles: Getty Conservation Institute, 1996), 232.

4 For a discussion of the history of connoisseurship, see M. Kirby Talley Jr., "The Eye's Caress: Looking, Appreciation, and Connoisseurship," in Price, Talley, and Melucco Vaccaro, *Historical and Philosophical Issues in the Conservation of Cultural Heritage*, 2–41. A particularly interesting account of the history of Rembrandt drawing connoisseurship can be found in Introductions I, II, and III in the online *Catalogue of Drawings by Rembrandt and His School at the British Museum*, by Martin Royalton-Kisch: www.britishmuseum.org/research/publications/online_research_catalogues/rembrandt_drawings/drawings_by_rembrandt.aspx.

5 Sir Jonathan Richardson Sr., *An Essay on the Whole Art of Criticism as it relates to Painting and an Argument on behalf of the Science of a Connoisseur* (London: W. Churchill, 1719), 110.

6 Elizabeth Lunning and Roy Perkinson, *The Print Council of America Paper Sample Book: A Practical Guide to the Description of Paper* (n.p.: Print Council of America, 1996), 4.

7 See Fontenelle, Murray, and Balzac, readings 40, 48, and 49, respectively, for instances of tasting paper. Printers traditionally touched paper with their tongues to detect alum, an acrid compound used to harden gelatin sizing.

Reading 10

WINSLOW AMES

Introduction to *The Mastery of Drawing*, by Joseph Meder (1978)

Joseph Meder (1857–1934) was curator of prints and drawings at the Albertina, Vienna, from 1909 to 1923. Chiding art historians that an unfamiliarity with the working properties of drawing materials led to their misidentification and ultimately to an imperfect understanding of the works themselves, Meder was among the first to replicate historical recipes and visually compare the results with original drawings. His book, Die Handzeichnung: Ihre Technik und Entwicklung *(Drawing: Its Technique and Development), published in 1919, served as a primary source on media identification for generations of art historians and conservators, Marjorie Cohn among them (see readings 21, 65, 93, and 94). This important work was finally translated into English (1978) by the art historian Winslow Ames (1917–1990) under the title* The Mastery of Drawing. *Meder's contribution to the discipline of technical art history was his practice of enhanced looking, or hypervision, today a standard methodology. Having an abiding interest in paper restoration, including the controversial practice of paper splitting, Meder occasionally carried out treatments himself.*[1]

[. . .]

Beyond the historical development of styles and the personal development of artists, this book concerns itself with the widely varying techniques of drawing. The author was interested not only in checking old recipes; he also knew from his daily handling of originals in every ancient technique that there was a glaring gap in the study material. Whoever has seriously investigated drawings realizes from his own experience how little the average reference book has to say about drawing techniques. Many present-day beliefs about specific media and processes rest only

From WINSLOW AMES, trans. and rev., introduction to Joseph Meder, *The Mastery of Drawing* (New York: Abaris Books, 1978), 7. Originally published as *Die Handzeichnung: Ihre Technik und Entwicklung* (Vienna: Anton Schroll, 1919). Reproduced courtesy of the publisher.

on traditions, which, as usual, are liable to error. There is also the remarkable lack of actual personal experience in drawing on the part of most art historians; sympathy is no substitute for participation. We are speaking here not of the blunders that come of mere heedlessness, but of statements to which the weight of learned reputations gives credence. What should we say to a passage in the biography (1903) of a Florentine fifteenth-century master, "The artist of that period never used charcoal, which first appeared in the nineteenth century as a new technique," or "The small design is executed almost impressionistically in white and brown carbon inks on reddish paper"?

One may object that charcoal or chalk, pen or silverpoint, blue paper or white, are irrelevant for the understanding of an artist, and that the study of his forms and their development is the historian's task. In a limited sense this is so, but it is not so where the object is to make a comprehensive and exact study. What seems to a modern writer a scrap of paper was once a necessity to the artist. Every artist by the grace of God chooses his own medium, and the means affect the end. Today it is recognized that this has always been true in the formation of styles. The revolution in form that followed the introduction of oil painting is evident in both Netherlandish and Italian art.

A too cursory treatment of media has led many a specialist astray. The perpetuation of false traditions could have been avoided by better technical knowledge. Even in Thausing's day the famous "Portrait of Wohlgemuth"[1] was reproduced as a drawing by Dürer.[2] Had it been determined that the medium was graphite, the anachronism would have been evident and the sheet recognized as foreign to Dürer. Critical observation of the form alone did not necessarily protect the critic from error. And when a "Saint Christopher" in red chalk (Uffizi, Florence) is given to an anonymous fourteenth-century artist, the attribution errs by at least a century on technical grounds. Likewise, a silverpoint is not to be taken away from a master simply because its color—a mere matter of degree of oxidation—differs from that of another silverpoint. Such oversubtle critical observations are based upon that little learning which is a dangerous thing. A marked instance was that of Hercules Seghers' etching of the "Lamentation,"[3] which in 1902 figured in a Cologne auction catalogue as a drawing, and, what is more, as a drawing by Mantegna.[4] One is obliged to ask why, if a technical description is given at all, it has any right to be incorrect. Particularly common is the misuse of the word *sepia* for brownish colors of all centuries; true sepia appears late in the eighteenth century and has never been common. "Sepia wash" on a Gozzoli[5] drawing sounds truly strange.

Handicraft, or technique, means not just the professional execution of commissions as the middle ages expected guildsmen to execute them, not just the narrow mechanical operation of a master's school with ready cartoons, patterns, and manikins, such as might have been found in the shops of the aging Perugino or of Lucas Cranach.[6] No, handicraft includes everything within the concept *pratica*, *praxis*, or *usus*—exercise, experience, skill of hand, preparation—which artists today must still learn if they are to be worth their salt.

Note to Introduction

1 Petra Helm, "Notizen aus dem Nachlass von Joseph Meder," *Maltechnik Restauro* 85, no. 3 (July 1979): 211–15.

Notes to Reading

1 There is a copy in the Albertina. [. . .]
2 [Mori(t)z Thausing (1835–1884), Austrian art historian specializing in the work of Albrecht Dürer (1471–1528). As a teenager, Dürer was apprenticed to Michael Wohlgemut(h) (1434–1519), who oversaw a busy multimedia workshop. Dürer's 1516 oil on panel portrait of his master is now in the Germanisches Nationalmuseum, Nuremberg. The drawing attributed to Dürer by Thausing was probably copied from this painting.—Ed.]
3 Jaro Springer, ed., *Die Radierungen des Herkules Seghers* (Berlin, 1910–12), plate 1.
4 [Hercules Seghers (1589?–1638?), Dutch painter and printmaker. Mantegna (1431–1506), Italian painter and printmaker.—Ed.]
5 [Benozzo Gozzoli (1421?–1497), Florentine painter.—Ed.]
6 [Perugino (1446–1524), Italian painter. Lucas Cranach (the Elder) (1472–1533), German painter and printmaker.—Ed.]

Reading 11

Letizia Montalbano

Drawings Revealed: Analyses of Graphic Techniques, Challenges, Directions and Points of View (2008)

In this reading, Letizia Montalbano (1959–), paper conservator in the Opificio delle Pietre Dure in Florence, traces the gradual emergence of the study of the materials and techniques of drawings and establishes its legitimacy by building on the modern restoration theories of Cesare Brandi and Umberto Baldini.[1]

"We cannot be conservators if we are not familiar with and knowledgeable about, to the greatest possible extent, the objects that are to be preserved, knowing them for what they are and for what they are worth; we cannot treat them if we rely only on our esthetic tastes, no matter how refined they may be, or upon our technical ability alone, however elevated it may be":[1] with those words, Umberto Baldini emphasized, in his introduction to the book *Teoria del restauro e unità di metodologia*, just how important it is for those who are working in the field of art conservation to have a thorough knowledge of any given artwork. On the other hand, it is precisely when we undertake a restoration that we are given the opportunity, to a greater extent than anyone else, to examine, from all points of view, "the material of an artwork,"[2] allowing us to attain an unusual and noteworthy understanding of that object.

[. . .]

Unquestionably, the *forma mentis* of our professional expertise induces us to immediately overlook the stylistic aspect of an image in order to reconstruct the artist's creative process and delve into the qualitative issues of materials and the

From Letizia Montalbano, "I disegni rivelati: Analisi delle tecniche grafiche, problematiche, ricerche e punti di vista," in *Leonardo e Raffaello, per esempio . . . : Disegni e studi d'artista*, ed. Cecilia Frosinini with Letizia Montalbano and Michela Piccolo (Florence: Mandragora, 2008), 25–26.

effects wrought by the passage of time and understood as a set of human, physical and chemical, and environmental factors.

Analyzing materials, in fact, does not only mean identifying their constituent elements: if we can identify the technique involved in the execution of the piece, we can thereby understand the mechanisms involved in the deterioration, historical changes, and certain "pathologies" that the artwork suffers. At the same time, other clues can serve to tell us how and where it has been saved, whether it has undergone manipulations or modifications, and, of course, whether it has been subjected to restoration.

The technical analysis of an object is often positioned on an intermediate plane between the strictly historical and artistic realm and the scientific domain, serving as a *trait d'union* between the endeavors of both fields of study, until recently so sharply distanced from one another. The challenge of immersing oneself in their reciprocal languages, however, remains quite unmistakable: there are enduring reluctances and doubts about the introduction of research procedures that are, more properly speaking, scientific because of a failure to grasp their actual validity; in contrast, any examination of the critical and artistic aspects alone may seem like a literary exercise to someone who is accustomed to mathematical synthesis. Likewise, the language of restoration is anything but easy, and one runs the risk of stultifying any reader with minute descriptions of materials used and methods of their application, aspects that seem to add little to comprehending the artwork and its intrinsic significance.

In any case, the conversation among the various parties has by now oriented itself primarily in the context of the restoration of pictorial materials, and the understanding of many technical topics has likewise spread. The analysis of the artwork, from this point of view as well, is by now widely and rightly considered to be a crucial aspect of art historical studies.

Acknowledgment of the importance of a more technical and scientific approach to the field of study of drawings, on the other hand, is a very recent achievement, despite the interest that the enormous wealth of graphic works in our possession has always aroused among experts.

Even now the benchmark for a more systematic kind of research is Joseph Meder's book, published in 1919, a point of departure for the various publications on graphic techniques that were subsequently published over the course of the twentieth century.[3] Grassi called it one of those "highly useful books that were not later emulated," but immediately afterward he dismissed it with the notation, "its critical approach to drawing and drawings is inadequate."[4] This book, however, should be considered a forerunner of a more modern approach that attempts to reassemble historical threads and changes of the technical parameters as well, complementary to its theoretical and philological counterparts. Meder not only undertook an enormous effort in terms of reconstucting and interpreting ancient sources but also was one of the first to venture into the realm of laboratory experimentation, so to speak, in an effort to understand the physical and chemical changes of materials in order to attain a deeper appreciation of technology. One pioneering attempt that is still famous today was his effort to establish the rate and manner in which 7marks left by metal point oxidize (expressly fabricated by him and and similar to knitting needles!) on paper samples that he had prepared.[5] Subsequently translated and published by Ames,[6] it allowed scholars to establish precise historical baselines, minute descriptions of technical procedures, and above all a vision of ancient graphic techniques that was not narrowly Italian, ranging from the texts of Alcherio and Cennini to those by Vasari, Pacheco, and Mayerne, by way of the writings of Leonardo and Dürer, to name just a few of the most widely used sources.[7]

This important publication was followed by a number of handbooks on drawing techniques, such as those by Tolnay, Lavallée, and Watrous,[8] but their approach was less in-depth than Meder's.

A major change in the way that the study of artworks was undertaken was unquestionably introduced by the birth of a full-fledged theory of restoration, dating from the sixties, with first Cesare Brandi and shortly thereafter Umberto Baldini. What had up to then been confined to periodic collateral publications associated with exhibitions of restored artworks began, thanks to them, to free itself from the narrow specialized context of laboratories, proceeding to establish a full-fledged critical theory that, by focusing attention on "material as an epiphany of the image,"[9] also pushed the historical and artistic world to analyze its twofold status of structure and appearance, terms that were opportunely introduced by the two scholars. In any case, it has been a slow and by no means linear process, and in fact we have had to wait the past thirty years to see a few texts of a slightly more specialized nature published, registering a further change in course in the way that drawings are studied.

Note to Introduction

1 Translations of the texts by Brandi and Baldini may be found in Nicholas Stanley Price, M. Kirby Talley Jr., and Alessandra Melucco Vaccaro, eds., *Historical and Philosophical Issues in the Conservation of Cultural Heritage* (Los Angeles: Getty Conservation Institute, 1996), 232.

Notes to Reading

1 U. Baldini, *Teoria del restauro e unità di metodologia*, Florence [1978], p. 5.

2 C. Brandi, *Teoria del restauro*, lectures collected by L. Vlad Borrelli, J. Raspi Serra, G. Urbani, Rome 1963, published in Turin 1977, pp. 9–12.

3 J. Meder, *Die Handzeichnung. Ihre Technik und Entwicklung*, Vienna 1919.

4 L. Grassi, *Il disegno italiano dal Trecento al Seicento*, Rome 1956, p. 49.

5 Meder 1919, op. cit., pp. 85–86.

6 See J. Meder, *The Mastery of Drawing*, translated and revised by W. Ames, 2 volumes, New York 1978.

7 Giovanni Alcherio, *De coloribus* (1398–1411; published in *Original Treatises on the Arts of Painting*, edited by Mary Philadelphia Merrifield, New York 1967 [originally published in London 1849], pp. 259–291); Cennino Cennini, *Il Libro dell'Arte* (late fourteenth century; see the edition edited by F. Brunello, Vicenza 1971); Giorgio Vasari, *Le Vite de' più eccellenti architetti, pittori et scultori italiani da Cimabue insino a' tempi nostri*, in Fiorenza, appresso Lorenzo Torrentino, 1550; idem, *Le Vite de' più eccellenti pittori, scultori, e architettori*, . . . riviste et ampliate, 3 vols., in Fiorenza, appresso i Giunti, 1568; Leonardo da Vinci, *Il trattato della pittura* (late fifteenth century–early sixteenth century; see the edition edited by G. Manzi, Rome 1817); Albrecht Dürer, *Speis der Malerknaben* (late fifteenth century–early sixteenth century; published in *Dürers schriftlicher Nachlass: auf Grund der Originalhandschriften und theilweise neu entdeckter alter Abschriften*, edited by K. Lange and F. Fuhse, Halle 1893); Théodore Turquet de Mayerne, *Pictoria, Sculptoria, Tinctoria, et quae subalternarum atrium spectantia* (1620–1646; translated into Italian with the title *Pittura, scultura e delle arti minori*, 1620–1646: London, British Library, MS Sloane 2052, edited by S. Rinaldi, Anzio 1995); Francisco Pacheco, *Arte de la pintura* (Seville, Simon Faxardo, 1649, published posthumously; published by B. Bassegoda i Hugas, Madrid 1990); Filippo Baldinucci, *Vocabolario toscano dell'arte del disegno*, in Florence, Santi Franchi, 1681.

8 Charles de Tolnay, *History and Technique of Old Master Drawings: A Handbook*, New York 1943; P. Lavallée, *Les techniques du dessin. Leur évolution dans les différentes écoles de l'Europe*, second edition, revised and corrected, Paris 1949; J. Watrous, *The Craft of Old-Master Drawings*, Madison, WI 1957.

9 Brandi 1963, 1977, op. cit., p. 9.

Reading 12

Walter Koschatzky

Technique as Artistic Expression (1987)

Walter Koschatzky (1921–2003) was the director of the Albertina, Vienna, from 1962 to 1986. His approach embodies the methodology established by Meder: material-based study is the critical first step in building a comprehensive understanding of drawings. As with the discussions of the role of paper in Part I, Koschatzky views mark making as a simultaneous act of preservation and communication—a physical "residue of the intangible," or, as more famously said by Henri Matisse (1869–1954), "Drawing is like making an expressive gesture with the advantage of permanence."[1] Koschatzky equates the various techniques of drawing to languages, which suggests that the sense of hearing enhances one's observations. As conservators will attest, a paper's "rattle," the sound that a sheet makes when it is shaken, often corresponds to its condition; for example, gelatin-sized papers stored in damp conditions are prone to losing their rattle.

The technique of drawing is not in fact something "technical." It is not a question of technology or a progression from a state of lesser ability to one of greater ability in accomplishment, but something quite different. Neither is it about different materials and their applications, nor their qualities. In essence, it concerns the variety of drawings as they evolved over the course of history. When drawing changed, it was not for technical reasons. The technique of drawing is, in fact, a visual expression of the perpetually changing perception of the world and of the place of man in that world. The artist thereby takes on the role of a seismograph that records changing times and their challenges. The following discussion attempts to demonstrate how a drawing goes even further than this.

From Walter Koschatzky, "Technique as Artistic Expression," in *Drawings Defined*, ed. Walter Strauss and Tracie Felker (New York: Abaris Books, 1987), 105. Reproduced courtesy of the publisher.

To ask about the technique of drawing is to ask about the meaning of art. It is certain that nothing in the ever changing course of the history of mankind can provide an answer with as much sensitivity and depth as the artist's drawing. The famous philosopher Georg Wilhelm Hegel noted the great importance of the art of drawing as a demonstration of the spirit miraculously transformed in a visual image by the dexterity of the hand.

What is it, then, that makes drawing so different, so multi-faceted? Is it the preparatory function of a drawing, or is it its freedom as an independent work, or is it determined by an era, by the surroundings in which it was created, or the individuality of the artist? In fact it is *all* of these elements that govern the will or intent, and therefore the style and technique of the drawing.

Therefore, in speaking about the technique of drawing, one is speaking of nothing less than the momentary expression of a personality, of its uncertainty, of its wish and yearning within the framework of an era. Unlike any other creation produced by an artist, a drawing is the spontaneous residue of the intangible, made visible through a personality.

[. . .]

The artist chooses the medium for his message. He chooses it according to the needs of what he wishes to express. The technique must comply with his internal vision, which is why we in turn presume to analyze his choice. The selection of the medium, that is, the technique, is a most decisive part of the creative process and the visual expression of its intangible content.

Note to Introduction

1 Henri Matisse, "Notes of a Painter on His Drawings," quoted in John Elderfield, *The Paintings of Henri Matisse* (London: Thames and Hudson, 1984), 4. In reading 76 in this volume, Arthur David Baynes-Cope describes a drawing as "comparable only to a song composed for the joy of singing."

Reading 13

Konrad Oberhuber

Introduction to *Sixteenth Century Italian Drawings from the Collection of Janos Scholz* (1973)

Konrad Oberhuber (1935–2007), director of the Albertina, Vienna, from 1987 to 2000, was a preeminent authority on the drawings of Raphael (1483–1520). He recognized the difficulty of reconciling visual evidence with textual information when attributing drawings—attribution once being the ultimate goal in traditional connoisseurship—and likens the differences in materials and techniques to those in regional dialects. Thus, like Koschatzky before him, Oberhuber called on the sense of hearing to recognize the individual voices of drawings.

Oberhuber was not the first to use geographic differences in style as a tool in the characterization of Italian drawings. A generation earlier, the German art historian Bernhard Degenhart (1907–1999) proposed that the way an artist unconsciously arranged lines on paper, die Strichbildung, *was characteristic of a specific place (central vs. northern Italy) and period (Gothic vs. Mannerist).[1] At the time, Degenhart's graphological analysis could be applied only to drawings since, as he lamented, the artist's "handwriting" is obliterated in paintings. Thanks to the use of infrared reflectography to detect underdrawings in paintings, this is no longer the case.*

[. . .]

Up to a point, a drawing can be likened to a sample of handwriting. Beyond the meaning of the images and words, both can tell the informed student the historical period of the sheet, the country and region in which it originated, and, once these norms are recognized, the character of the individual hand. Connoisseurs of Italian drawing are able to state with certainty whether a work is Venetian rather

From Konrad Oberhuber, "Introduction," in *Sixteenth Century Italian Drawings from the Collection of Janos Scholz* by Konrad Oberhuber and Dean Walker (Washington, D.C.: National Gallery of Art; New York: Pierpont Morgan Library, 1973), p. ix. Reprinted by permission of the National Gallery of Art, Washington.

than Florentine or Lombard rather than Roman, just as Italians can distinguish with ease the dialects spoken in these different cities.

As every traveler to Italy knows, cities there still preserve a striking individuality, often immediately noticeable in accents and building style, irrespective of distance and centuries of separate cultural development. For the art historian, the drawing styles of these places have the distinct characteristics of a regional accent which by no means negates an individual or influential voice.

Note to Introduction

1 Bernhard Degenhart, "Zur Graphologie der Handzeichnung: Die Strichbildung als stetige Erscheinung innerhalb der italienischen Kunstkreise," *Kunstgeschichtliches Jahrbuch der Biblioteca Hertziana*, 1 (Leipzig, 1937), 223–343.

Reading 14

Max Klinger

Painting and Drawing (1891)

Max Klinger (1857–1920) was a German Symbolist painter, sculptor, and printmaker. He assigned a different "spirit" to each medium, for to overlook its significance or to thoughtlessly substitute one medium for another was to alter the meaning of a work, much like changing the key in which a piece of music is played.[1] *Where Oberhuber saw a comparison between regional dialects and drawing styles, Klinger viewed drawings as having their own harmony.*

In the late nineteenth century, Klinger, along with Wilhelm Ostwald (readings 19 and 69), was a member of Leipzig's intellectual elite. The contributions of scholars in both Leipzig and Vienna—here represented by Meder, Koschatzky, and Oberhuber—were fundamental to the rapidly growing discipline of drawing and print connoisseurship.

[. . .]

[T]here is a unique spirit and unique poetry in every material, owing to its appearance and the way it can be worked, which with artistic treatment enhances the character of the depiction and cannot be replaced by anything else: roughly in the same way that the character of a piece of music is based on its established tonality, and is lost if transposed into another one. If the spirit of the material is not taken into account and incorporated in the conception and execution—whether owing to material shortage or caprice, willfully or as a result of outside pressure, one material is substituted for another—the artistic integrity of the impression[1] is destroyed even before one begins. Our thinking, easily distracted in any case, turns away from the depiction itself to the medium employed, and to the difficulty of the captured effect. The too artificial technician and the dilettante have equal success; in both one loses sight of the intent because of the material.

From Max Klinger, *Malerei und Zeichnung* (Leipzig: Verlag von Arthur Georgi, 1891), 11.

Note to Introduction

1 I am grateful to Amy Kelly Hamlin and Felix Billeter for their help tracking down this passage.

Note to Reading

1 [*Eindruckes* in the original; translated here as "impression," in the sense of a unique printed impression rather than a mental response.—Ed.]

Reading 15

Henri Zerner

What Gave Connoisseurship Its Bad Name? (1987)

Like Oberhuber, the art historian Henri Zerner (1939–) struggles to have faith in what his eyes see in a drawing regardless of what he reads in a book. "Ours is a logocentric culture," he writes. "We trust the written document much more readily than our visual understanding of an image." Through enhanced looking, Zerner believes, drawings can be read as objectively as texts. Acknowledging the importance of material identification, Zerner declares that true connoisseurship of drawings "is in its infancy." At the time of Zerner's writing, the antagonism between traditional Morellian connoisseurs, epitomized by Bernard Berenson (1865–1959), and the historical positivists, Zerner's "logocentrics," was raging.[1] Conservators are likewise exhorted to rely on the visual evidence at hand by the chemist Arthur Baynes-Cope (readings 53 and 76), who quotes the philosopher Maimonides (1135–1204): "Only a fool regards something as proved merely because it is written in a book."[2]

The antagonism between art history and connoisseurship, or perhaps more exactly, between connoisseurs and art historians, is nothing new. Here is how the great Morelli characterized it in his typically pugnacious way: "It has been said, sarcastically, that the art connoisseur is distinguished from the art historian by knowing something about the art of the past. If he happens to be of the better sort he abstains from writing on the subject. On the other hand, the art historian, although writing much upon art, really knows nothing about it; whilst the painters who boast of their technical knowledge are neither competent critics nor competent historians."[1]

From Henri Zerner, "What Gave Connoisseurship Its Bad Name?," in *Drawings Defined*, ed. Walter Strauss and Tracie Felker (New York: Abaris Books, 1987), 289–90. Reproduced courtesy of the publisher.

The painters have mostly dropped out of the game, even if Frank Stella did recently give the Norton lectures on Caravaggio and Rubens.[2] It looks like they might be replaced by the conservators. These days, the offensive is mostly on the art historian's side, and it is the connoisseurs that are under attack. In the present effort to rejuvenate art history, the whole enterprise of connoisseurship has been more or less discredited.

There are two main reasons for this. The first and most obvious is venality. Connoisseurship appears inextricably linked to the commerce of art. The specific case of Berenson[3] looms large, but he is only the most visible example. The art trade depends on the authority of experts to authenticate the objects in circulation. There tends to be a blanket disapproval of this commercial involvement. There is no reason to be so righteous. Scholarship does have its ethics, which do not have to be incompatible with commerce. There are honest picture experts, just as there are honest bankers and grocers. Of course there are crooked ones too. What was unfortunate in the case of Berenson was not that he was working for dealers, or even that his deals were occasionally of a dubious nature, but that he was so ashamed of the connection.

The real question is to what extent scholarship has been directed and shaped by the marketplace, and this brings me to connoisseurship's second major blemish. Connoisseurship is synonymous with attribution, with the decision as to when and where objects were made, and in most cases, in assigning names. It devotes its energy to establishing masters and masterpieces greater, lesser, or miniscule. That this has happened under the pressures of commercial interests is obvious enough, but what is important is the intellectual result of this development rather than some possible moral stain. The situation is made all the worse because the work and knowledge of the connoisseur is often badly expressed, if verbalized at all, so that it appears as a kind of divination or arbitrary authority. The notion of quality, so often invoked by those who practice the art of connoisseurship, as Berenson called it, is such an elusive and mystical one that it comes under easy attack for those who would condemn the whole enterprise as bogus. All in all, connoisseurship may well have deserved its bad name. But I also believe this is extremely unfortunate.

What is at stake is the interpretation of art and particularly the use of art as historical evidence. I will surprise no one if I say that the tendency of recent studies has been towards a contextual and historical understanding of art, moving away from a more traditional aesthetic approach. The danger of this trend is to impose upon art, to read into it what one has learned elsewhere, mainly from literary or documentary evidence, and, I am afraid, in the case of many art historians, from secondary work by historians or anthropologists. I consider Antal's[4] study of fifteenth-century painting a case in point, despite his apologists.

Of course, what we should retrieve from art is precisely what we cannot find elsewhere, what it alone is capable of telling us. In order to do this, we need a serious examination of the visual evidence itself. To look and to see—to make fine visual distinctions, to identify specific visual features, to correlate them to one another through notions of rhythm, recurrence, and relative irregularity—is not an easy matter. To be attentive to all the indications we can obtain from the examination of artifacts, from the slightest inflections of the maker's hand to the largest configurations, to make this study disciplined and coherent, this should be the task of connoisseurship. It can tell us an extraordinary amount about how things are made, to what purpose, and about their later destiny and use.

It is ironic that connoisseurship has concentrated mostly on the products of literate cultures, when there are so many cultural areas where the only evidence is visual. This is true of prehistoric societies, of course, but also of certain aspects of culture, like popular culture, where the verbal material is often deficient. What can visual material tell us about the cultural disruptions and disparities of a complex society? It is in answering this kind of question that connoissership could test its powers.

If I had to define connoisseurship, I would say it is the articulation and symptomatic examination of the visual evidence. If we understand it in this sense, it is hard not to agree with Morelli that it has to be the necessary foundation of art history.

I would not like to end on such an all too obvious note, but by raising a major problem that art history has never quite faced. How can we weigh the visual evidence against the verbal or documentary evidence? Ours is a logocentric culture. We trust the written document much more readily than our visual understanding of an image. This must be changed and we must attend to visual clues if we want to get something out of our visual legacy. But this is not easy. Connoisseurship is in its infancy.

Notes to Introduction

1 See Gary Schwartz, "Connoisseurship: The Penalty of Ahistoricism," and David Ebitz, "Connoisseurship as Practice," *Artibus et Historiae* 9, no. 18 (1988): 201–6 and 213–15, respectively.

2 Arthur Baynes-Cope, "Ethics in Paper Conservation," *Tidens Tand* 1 (1982): 152. Maimonides, or Moses ben Maimon, was a rabbi, physician, and philosopher.

Notes to Reading

1 Giovanni Morelli, *Italian Painters: Critical Studies of Their Works*, vol. 1 (London, 1892), 4. [Morelli (1816–1891), an Italian physician and art critic, developed a method of identifying an artist's hand by its habitual, often-unconscious rendering of inconspicuous details, the most famous example being the folds of the ear.—Ed.]

2 [The Charles Eliot Norton Lectures at Harvard University, established in 1925, are presented by an invited artist, poet, writer, architect, or musician whose work has been accepted into the broader humanities curriculum. The American painter Frank Stella (1936–) served as the Charles Eliot Norton Professor in 1982–84.—Ed.]

3 [Bernard Berenson (1865–1959), American art historian and disciple of Morelli. His reputation has suffered due to questions regarding the extent to which he profited financially from his attributions.—Ed.]

4 [Frederick Antal (1887–1954), Hungarian art historian whose approach was heavily sociological.—Ed.]

Reading 16

J. D. Harding

Elementary Art; or, The Use of the Lead Pencil Advocated and Explained (1838)

In his many popular how-to books, the English painter James Duffield Harding (1797?–1863) promulgated his instructions on mastering proper technique and composition through intensive practice in drawing.[1] *Conveniently, the artists' supply company Winsor & Newton was able to readily provide amateur and professional artists with Harding-endorsed graphite pencils in four different hardnesses and Imperial-sized papers stamped "J.D.H." As proof of his persuasiveness, Harding's pencils and papers were still available from Rowney and Co. over fifty years later. In that Rowney catalogue, echoing Meder's mantra* practica, praxis, *and* usus, *is an eloquent exhortation to practice:*

> *Thy pencil (blest companion) magic plays,*
> *In cadence, here, with Nature's song of praise.*
> *And still 'tis thine, when Summer's charms are gone,*
> *When Autumn brings the fields of rip'ning corn,*
> *To watch the transient hues the trees unfold,*
> *As steals the first chill wind across the wold.*
> *Nor deem as lost, bleak Winter's darkest day:*
> *Beside the fire thy pencil then should play*
> *An active part: the good old maxim thine,*
> *To "let no day go by without a line."*[2]

[. . .]

There is one peculiarity connected with Art, however, in which it may be said to differ from Language, inasmuch as the latter has its established characters, while those of the former are the productions of the Artist's own ingenuity,

---

From J. D. Harding, *Elementary Art; or, The Use of the Lead Pencil Advocated and Explained* (London: Charles Tilt, 1838), 3–4.

invented to convey more distinctly his own ideas, and to exhibit to others his views of Nature. It is this individuality of character, as it may be called, that gives occasion to the varieties observable in the works of the different Painters, ancient and modern, and which constitutes the originality and distinguishes the manner of each. Hence every original Painter has found it necessary to invent the characters in which he has communicated his thoughts: this has rendered the careful and attentive study of Nature imperative, as well as the best works of his predecessors or contemporaries, from which he might derive such aid as he could most usefully apply, to assist his own observation; thus making their thoughts and practice conduce towards the attainment of a method which he considers strictly true and original. [. . .]

Notes to Introduction

1 I am grateful to John Krill for bringing to my attention the writings of J. D. Harding.
2 Leonidas Clint Miles, *A Guide to Sketching from Nature in Water Colours*, 4th ed. (London: George Rowney and Co., ca. 1890), 85.

Reading 17

Johann Peter Eckermann

Conversations with Goethe in the Last Years of His Life (1836–1848)

Johann Peter Eckermann (1792–1854), German poet and author, is best known for his contribution to Goethe scholarship.[1] *In* Conversations, *Eckermann describes his early fascination with drawing. He recounts a vivid memory of copying a horse from a printed tobacco wrapper, brought home by his itinerant laborer father. Any aspirations of becoming an artist, however, were promptly squelched by his peasant stock parents, who "thought of a painter only as one who paints doors and houses." "They, therefore, advised me earnestly against it," he writes, "saying it was not only a very dirty, but very dangerous trade, and that those who worked at it, especially in Hamburg, where houses are seven stories high, were constantly in danger of breaking their necks."*[2] *Eckermann joined the military in the period of intermittent campaigns against the Napoleonic empire, eventually reaching Flanders. It was here, as related in this reading, that the twenty-two-year-old discovered the "great pictures of the Netherlands."*[3] *Eckermann continued to practice his drawing by copying prints by the popular painter and illustrator Johann Heinrich Ramberg (1763–1840). Desperate for formal instruction, the naive artist determined to seek out the well-known Ramberg in Hanover: "So I knotted up my bundle, and took, in the winter of 1815, my walk of almost forty leagues, quite alone, over the heath and through the deep snow."*[4] *Ramberg took him in, but Eckermann quickly became acquainted with an apprentice's drudgery, working until his trembling hands "could no longer hold a pencil."*[5] *Although he gave up his "devotion to Art"*[6] *in favor of a more dependable livelihood and the less taxing art of poetry, Eckermann's transformation was nonetheless complete.*

From Johann Peter Eckermann, *Gespräche mit Goethe in den letzten Jahren seines Lebens* (Leipzig: Brockhaus, 1925), 15–16.

[. . .]

Here, before the great paintings of the Netherlanders, a new world opened up to me; I spent entire days in churches and museums. These were essentially the first paintings I had ever seen. I now saw what it would mean to be a painter; I saw pupils' successful progress, and I could have wept that I was prevented from taking a similar path. Yet I made up my mind on the spot: In Tournai I made the acquaintance of a young artist. I bought myself black chalk and a large sheet of drawing paper and promptly sat down in front of a picture to copy it. My great eagerness in the matter made up for what I lacked in practice and direction, and so I successfully captured the outlines of the figures; I also began, working from the left side, to add shading to the whole, when a marching order interrupted such pleasant activity. I hastened to indicate the shading with separate letters, in the hope that in this way I might complete it at my leisure. I rolled up my picture and placed it in a tube, which I carried on my back along with my rifle on the long walk from Tournai to Hameln[1] . . . I got back home . . . I immediately began to continue with my drawing; I first finished the picture I had brought from Brabant, and since I had no more suitable patterns, I worked from the small Ramberg engravings that I executed in a larger format in black chalk.

Notes to Introduction

1 Thanks are extended to Irene Brückle for sharing the story of Eckermann's discovery of drawing with me.
2 Johann Peter Eckermann, *Gespräche mit Goethe in den letzten Jahren seines Lebens* (Leipzig: Brockhaus, 1925), 11.
3 Ibid., 13.
4 Ibid., 14.
5 Ibid., 15.
6 Ibid.

Note to Reading

1 [The distance from Tournai to Hameln is approximately 336 miles.—Ed.]

Reading 18

Henry Petroski

The Pencil: A History of Design and Circumstance (1990)

Henry Petroski (1942–) is a civil engineer and professor of history at Duke University who has written on the social significance of unremarkable things, including the toothpick, the cup holder, and the graphite pencil. His opening assessment of the pencil as having "little intrinsic, permanent, or special value" may also be applied to expendable paper-based works, whose reevaluation as cultural artifacts is evolving.

Petroski endows the humble pencil with the psychic energy of creativity. This view stands in contrast to Wilhelm Ostwald (reading 19), who carefully delineates its physical properties, especially its erasability. It is this erasability that Petroski sees as graphite's advantage, as well as the cause of its second-class status.

Petroski, however, is not the first to champion an underappreciated drawing and writing medium or to recognize its power to effect change. The American inventor, manufacturer, and indefatigable purveyor of writing inks Thaddeus Davids (reading 70) was incensed that various inks had never been defined or given consistent nomenclature, writing, "Ink is history, in the common acceptation of the word; for, what is generally denominated history—is ink diffused on paper in certain definite lines. Yet ink has no history written or composed hitherto."[1] *One suspects, however, that Davids had a vested financial interest in setting the record straight, since he peddled indelible ink.*

Davids's frustrated attempts to trace the etymology of historical inks and Meder's distrust of the term sepia *prompt a closer look at Petroski's reference to Henri de Toulouse-Lautrec's quote, "I am a pencil," and the persistent problem of the*

From Henry Petroski, *The Pencil: A History of Design and Circumstance* (New York: Alfred A. Knopf, 1990), 6–7.

inconsistent terminology used to describe drawing media. Given Toulouse-Lautrec's intense involvement with lithography, it is likely that he was referring to a wax-based lithographic crayon and not a graphite pencil: crayon *traditionally means any drawing medium in stick form regardless of the nature of its binder.*[2]

[. . .]

These stories of absence are interesting not so much because of what they say about the lowly status of the wood-cased pencil as an artifact as because of what they say about our awareness of and our attitudes toward common things, processes, events, or even ideas that appear to have little intrinsic, permanent, or special value. An object like the pencil is generally considered unremarkable, and it is taken for granted. It is taken for granted because it is abundant, inexpensive, and as familiar as speech.

Yet the pencil need be no cliché. It can be as powerful a metaphor as the pen, as rich a symbol as the flag. Artists have long counted the pencil among the tools of their trade, and have even identified with the drawing medium. Andrew Wyeth described his pencil as a fencer's foil; Toulouse-Lautrec said of himself, "I am a pencil"; and the Moscow-born Paris illustrator and caricaturist Emmanuel Poiré took his pseudonym from the Russian word for pencil, *karandash*. In turn, the Swiss pencil-making firm of Caran d'Ache was named after this artist, and a stylized version of his signature is now used as a company logo.

The pencil, the tool of doodlers, stands for thinking and creativity, but at the same time, as the toy of children, it symbolizes spontaneity and immaturity. Yet the pencil's graphite is also the ephemeral medium of thinkers, planners, drafters, architects, and engineers, the medium to be erased, revised, smudged, obliterated, lost—or inked over. Ink, on the other hand, whether in a book or on plans or on a contract, signifies finality and supersedes the pencil drafts and sketches. If early pencilings interest collectors, it is often because of their association with the permanent success written or drawn in ink. Unlike graphite, to which paper is like sandpaper, ink flows smoothly and fills in the nooks and crannies of creation. Ink is the cosmetic that ideas will wear when they go out in public. Graphite is their dirty truth.

A glance at the index to any book of familiar quotations will corroborate the fact that there are scores of quotations extolling the pen for every one, if that, mentioning the pencil. Yet, while the conventional wisdom may be that the pen is mightier than the sword, the pencil has come to be the weapon of choice of those wishing to make better pens as well as better swords. It is often said that "everything begins with a pencil," and indeed it is the preferred medium of designers. In one recent study of the nature of the design process, engineers balked when they were asked to record their thought processes with a pen. While the directors of the study did not want the subjects to be able to erase their false starts or alter their records of creativity, the engineers did not feel comfortable or natural without a pencil in their hands when asked to comment on designing a new bridge or a better mousetrap.

Notes to Introduction

1 Thaddeus Davids, *The History of Ink including Its Etymology, Chemistry, and Bibliography* (New York: T. Davids and Co., 1860), 5.
2 Margaret Holben Ellis and Brigitte Yeh, "The History, Use and Characteristics of Wax-Based Drawing Media," *Paper Conservator* 22 (1998): 48–55.

Reading 19

W. Ostwald

Letters to a Painter on the Theory and Practice of Painting (1904)

Wilhelm Ostwald (1853–1932), winner of the 1909 Nobel Prize in Chemistry, was a renowned professor at the University of Leipzig, where he taught Svante Arrhenius (1859–1927). In the tradition of Giorgio Vasari, Cennino Cennini, and Charles Eastlake, Letters to a Painter on the Theory and Practice of Painting, *originally published in 1904 as* Malerbriefe . . . , *was among the earliest endeavors to inform artists about the science of their materials. Though Ostwald's work was followed by other popular manuals—by Arthur Laurie (1861–1949),*[1] *Joseph Meder (1857–1934) (see reading 10), Max Doerner (1870–1939),*[2] *and Ralph Mayer (1895–1979),*[3] *among others—Ostwald's explanations are still, despite the book's age, mostly accurate, as well as charming in their unusual format, a series of personal letters to a curious and somewhat petulant artist.*

Ostwald meticulously describes the difference in particle sizes between the dry drawing media that accounts for their appearance and variable tractability on paper. He was the first to observe the subtle decrease in shininess in "fixed" graphite drawing, caused by the reorientation of their microscopic platelike particles—a phenomenon also observed by conservators who debate the effects of washing such drawings or even spraying them for flattening.

Like Clapp (reading 1), Ostwald recognizes the critical role played by paper as the carrier of information and recommends one of the first do-it-yourself light-aging tests to assess paper permanence.

From W. Ostwald, *Letters to a Painter on the Theory and Practice of Painting*, trans. H. W. Morse (Boston: Ginn & Company, 1907), 3–7, 14–15; originally published as W. Ostwald, *Malerbriefe: Beiträge zur Theorie und Praxis der Malerei* (Leipzig: S. Hirzel, 1904).

[. . .]

To begin with, all *drawings*, in the narrower sense of the word, are made without the use of any chemical medium. They are produced by putting in the desired lines and surfaces with sticks or crayons of pigment. All transitions between the color of the ground and that of the pigment can be brought out by lighter and stronger strokes of the crayon and by a single or repeated treatment of the surface. In order that such a process may be carried out the ground must be rough. Its irregularities act like a file on the crayon, taking off its material in the form of powder, which remains lying on the places over which the crayon has passed. Whether or not a really stable drawing is produced in this way depends on the fineness of the colored powder which the ground takes from the crayon. If this is very fine, each little particle finds a resting place in the roughness of the ground and is not easily disturbed by shaking, tapping, or even touching the drawing. Graphite, of which ordinary lead pencils are made,[1] acts in this way; but since the grains simply lie freely in crannies of the ground and are not held fast by any binding material, they can be removed from their position by strong or repeated mechanical disturbances. Lead-pencil drawings are therefore subject to smudging and lead-pencil lines can, for the same reason, be removed with a rubber eraser. Soft erasers (those of dark color) act because of their somewhat adhesive properties, by virtue of which they pick the grains of color out of their resting places. The harder sorts contain, in addition, a sharp powder[2] which grinds away the ground on which the drawing is made and so removes the grains of color together with a part of their surroundings. If a drawing of this kind is to be kept from smudging, each grain must be fastened in some way to the spot where it lies, and this is done by flowing or spraying with a gummy substance. Only a very small amount of this needs to be applied (very dilute solutions of one to five per cent), since the amount of pigment which is to be held in place is very small. The coarser the colored powder, the smoother the ground, and the thicker the coat of pigment the greater will be the danger of smudging and the necessity for fixing the drawing. Black or red chalk is much less fine in grain than graphite, and the particles of charcoal are the coarsest of all. This latter substance consists of the carbonized remains of wood cells and is not a powder at all until it is broken up by the filing action of the ground. Paper is almost exclusively the substance used as ground, and it has a double role to play in a crayon drawing. It must have, in the first place, a certain degree of roughness; the smoother the surface, the less the amount of powder which it can file off, and the less the chance of the powder being tightly held. Strong strokes can then be produced only by a correspondingly strong pressure, by which both the filing and the binding actions are increased. It is therefore possible to use the smoothest papers (when other reasons make it desirable) for the finest powder (lead pencil), while papers of increasing roughness must be used for chalk and charcoal.

The other duty of the ground is to serve as carrier for the finished work of art, and for this it should have chemical and mechanical powers of resistance in the highest possible degree. The chemical power of resistance is to be chosen especially with reference to inalterability under the oxidizing action of the oxygen of the air. This is an influence which cannot be excluded except by very special means, and which must therefore always be taken into account, so that in what follows we shall continually refer to it. The pure vegetable fiber (cellulose), of which good paper consists, is extremely durable, as is the glue with which such paper is usually hardened. Proof of this we have everywhere in papers preserved in libraries and archives, many of which are hundreds of years old. But wood pulp, added nowadays to papers of the cheaper sort which are in enormous demand, is unstable, and so much the more so when it has had but slight chemical treatment. Changes show themselves in the yellow or brownish color which such papers take on in the light (most rapidly, of course, in direct sunlight), and also in a decrease in toughness. Both of these processes, turning yellow and turning brittle, take place without the aid of light, but very much more slowly. The artist who desires the greatest possible durability for his creations must therefore carefully avoid such papers. They can usually be recognized by laying a strip of the paper to be tested half in and half out of a book, like a bookmark, and subjecting the exposed half to the action of direct sunlight. A few hours' time is usually sufficient to show a change in the exposed half of the suspected paper.

A sufficient idea of the mechanical strength of a paper is obtained by tearing it. Sheets which are to be moved about, such for example as are kept in portfolios, are best fastened to a larger sheet of stiff paper, which bears the brunt of all mechanical wear. In order that the backing may be changed if necessary, the sheet should only be fastened to it firmly enough to avoid folds during handling, while permitting of easy removal without damage when this is desirable.

[. . .]

Let us turn our attention for the time being to the problem of the lead-pencil marks. The graphite of the pencil may be seen by microscopic examination to consist of scales with smooth (and therefore reflecting) surfaces. When the lines are drawn the particles are laid down parallel with the surface of the paper, like the scales of a fish, forming a smooth surface, which reflects a great deal of surface light when held in certain positions and which is therefore glossy. The particles of black chalk and charcoal are, however, wholly irregular fragments, which do not form a smooth surface and are therefore not glossy.

You also mention in your letter that lead-pencil drawings show much less gloss after being fixed. The paper swells during fixing; the individual fibers move and take up other positions, moving the particles of graphite from their regular parallel arrangement, and the result is a corresponding decrease in the surface reflection. It follows directly that chalk and charcoal drawings show no change in appearance on being fixed, for here there is no parallel arrangement of particles to disturb.

Notes to Introduction

1 Arthur Laurie, *The Painter's Methods and Materials* (Philadelphia: J. B. Lippincott, 1926).
2 Max Doerner, *Malmaterial und seine Verwendung im Bilde* (Munich: F. Schmidt, 1921).
3 Ralph Mayer, *The Artists' Handbook of Materials and Techniques* (New York: Viking Press, 1940).

Notes to Reading

1 ["Lead" pencil refers to the physical construction of the implement: a rod or "lead" of graphite is encased in wood. Although etymologically related, the metal lead is not present.—Ed.]
2 [The "sharp powder" is abrasive pumice.—Ed.]

Reading 20

Joseph Farington

The Farington Diary (1795)

This excerpt from the sixteen-volume diary of the Englishman Joseph Farington (1747–1821), currently in the Royal Library at Windsor Castle, continues the theme of the education of artists.[1] *Apparently due to the exorbitant cost of natural rubber erasers in 1795, Royal Academy students needed to make do with bread crumbs for correcting their drawn copies of plaster casts. Joseph Priestley described the introduction of a new eraser in 1770: "Since this Work was printed off, I have seen a substance excellently adapted to the purpose of wiping from paper the marks of a black-lead-pencil. It must, therefore, be of singular use to those who practice drawing. It is sold by Mr. NAIRNE, Mathematical Instrument Maker, opposite the Royal-Exchange. He sells a cubical piece, of about half an inch, for three shillings; and he says it will last several years."*[2] *Ostwald's erasers (see reading 19) were most likely made of vulcanized rubber, which was discovered by Charles Goodyear in 1839.*

Mr Wilton (Keeper of the Academy) having mentioned to me that the students in the Plaister Academy continue to behave very rudely; and that they have the practise of throwing the bread, allowed them by the Academy for rubbing out, at each other, so as to waste so much that the Bill for bread sometimes amounts to Sixteen Shillings a week; and this relation to Mr Wilton being collaborated by Mr Reynolds,[1] I moved that "in future no bread be allowed the Students," this was unanimously agreed to. —Mr. West[2] said independant of every other consideration it would be productive of much good to the Students to deprive them of the use of bread; as they would be induced to pay more attention to their outlines; and would learn to draw more correct, when they had not the perpetual resource of rubbing out.

From Joseph Farington, *The Farington Diary*, 3rd ed., ed. James Greig, vol. 1, *July 13, 1793–August 24, 1802* (London: Hutchison and Co., 1923), 129–30.

Notes to Introduction

1 John Krill brought this fascinating resource to my attention.
2 Joseph Priestley, *A Familiar Introduction to the Theory and Practice of Perspective* (London: J. Johnson and J. Payne, 1770), xv.

Notes to Reading

1 ["Mr Reynolds" is probably Sir Joshua Reynolds (1723–1792), first president of the Royal Academy of Arts.—Ed.]
2 ["Mr. West" is probably Benjamin West (1738–1820), who, in 1795, was the second president of the Royal Academy.—Ed.]

Reading 21

Marjorie B. Cohn

A Note on Media and Methods (1970)

Marjorie Cohn (1939–) is former paper conservator and curator of prints at the Fogg Art Museum, Harvard University. Her 1970 technical essay on the materials and techniques of Giambattista Tiepolo (1696–1770) and his son Domenico (1727–1804) was among the first to be included in an exhibition catalogue, starting a trend in museum publications and on websites that underscores the public's fascination with the materials and techniques of drawing. Antoinette King (1932–2014), former paper conservator at the Museum of Modern Art, was also an early contributor; her 1977 essay on the paper cut-outs of Henri Matisse (1869–1954) deserves mention.[1]

In an effort to differentiate between the brown inks found in Tiepolo drawings, Cohn replicated historical recipes for bistre. Outpacing Meder, she not only compared them visually but also assessed the unique working properties of bistre, identifying a silty mixture, as well as a clear golden solution that penetrates to the verso of the paper.[2] *Cohn recommends that one examine original drawings—including their versos—in order to fine-tune the ability to see the nuances of their materials and techniques.*[3]

Cohn points to the importance of all senses, not only sight, as an aid in connoisseurship, commenting, for example, on the fragrance of freshly made bistre and the tactile quality of its grittiness. Bernard Berenson, discussed in the introduction to reading 15, was also an advocate for the sense of touch. Even more impressive is Cohn's translation of the information gleaned from the senses into clear and concise descriptive prose—a difficult exercise explored further in Part IV.

From Marjorie B. Cohn, "A Note on Media and Methods," in *Tiepolo: A Bicentenary Exhibition, 1770–1970* (Cambridge, Mass.: Fogg Art Museum, Harvard University, 1970), 211, 214–21.

No student of Tiepolo drawings remains indifferent to their sheer virtuosity. Giambattista, in his wash drawings above all, achieved a new abstraction of illusionism through physical means that were direct, immaculate,[1] and apparently effortless. His most gifted son, Domenico, developed another distinctive wash style, more forgiving of uncertainty of hand, perhaps, but it too enchants our senses responsive to the physical expressiveness of paper and ink. And both men's chalk drawings are accomplished, if not so dazzling.

These artists and their followers made no original technical contribution. They popularized no new format nor paper nor pigment, but their vitalization of old methods under the leadership of Giambattista is so essential a part of their accomplishment that it deserves some special analysis. [. . .]

[. . .]

It is the Tiepolo wash drawings which present the most complicated interaction of style, technique, and materials. A particular problem lies with the pigments themselves, for we are not even sure of what brown inks were used. As George Knox has indicated in an earlier Tiepolo drawing catalogue, in a short but very accurate and perceptive analysis of Giambattista's drawing materials, bistre and iron gall ink, both available to the artist, can be differentiated in a few sheets with no trouble. However, ". . . between the two extremes there remains a considerable group for which a definite decision seems difficult to make."[2] He notes the known characteristic of iron gall ink of being originally a black ink which has since discolored to brown[3] and then advances a tentative conclusion: ". . . all now brown pen and wash drawings, with perhaps the exception of a few bistre drawings, were originally black in color."[4] The current writer's own observations lead her to a similar conclusion: that there must have been more black ink and wash drawings among sheets now appearing brown. She believes, however, that bistre was used to a predominant degree, so that a majority of Tiepolo wash drawings by both father and son are in this medium.[5] The following is advanced in support of this opinion, although with some hesitation, since the argument depends on color effects which cannot, of course, be reproduced in a catalogue printed in black and white. Indeed, they are so subtle that there is no substitute for the close examination of original drawings by the dozen, even by the hundred, and so the patience of the reader may be greatly tried.

Confusion over brown inks is not restricted to the study of Tiepolo drawings alone. In the individual passages in *Die Handzeichnung* which discuss various brown pigments, Meder remarks in a resigned tone on the habit of cataloguers who call all brown drawings, bistre[6] or, worse, sepia. Sepia, a pigment derived from the ink sac of the cuttlefish, was not employed until the eighteenth century: ". . . yet its name still ornaments—mistakenly—books and catalogues for every kind of brown ink of an unusual richness. With the greatest ingenuousness, Pisanello, Raphael, Leonardo, and Rembrandt are considered to have used it."[7] Perhaps it will be easier for us to evaluate its possible use by the Tiepolos after examining the potential of the other two known inks.

Iron gall ink is formed by a chemical reaction between oak galls (containing tannin) and ferrous sulphate cooked in water. It is a corrosive, intensely saturated ink, originally purplish black, which is very difficult to dilute to a strength suitable for modulated transparent washes. There are very few Tiepolo drawings which show the lacy effect so familiar in Cambiaso and Guercino drawings,[8] for example, where a very rich brown, highly corrosive pigment, certainly iron gall ink, has eaten away the supporting page. One of the few is *Moses' Address to the People* (Shickman, No. 49) of this exhibition, which, significantly, was not washed over but was executed in pen alone.

Bistre is an extract in water of soot (not ashes) collected from chimneys where hard wood has been burned.[9] Unlike iron gall ink, it is comparatively unsaturated and only slightly corrosive, even at its highest concentration. It does not seem to change color with time and can be any shade of brown and, more astonishingly, *many* shades of brown. There are a number of variables, the first being the obvious one of the kinds of soot used."[10] In the author's experience, oak makes a yellower and maple a redder ink. The soot commonly available to Venetian manufacturers is not known, although oak and beech would seem likely possibilities.

A second cause for color variation is more complicated. It depends on the nature of bistre, which has two components: one, a more or less golden brown liquid which is an extract from the sooty tars, and the other, a silt of black particles. Meder notes the existence of the second component, saying that it can be removed by filtering,[11] but in practice this is difficult to accomplish completely. Thus, depending on the care used in manufacture, bistre can be more or less blackish and grainy. Also, the very finest particles seem always to remain, settling to the bottom of the ink pot to be stirred up by the artist depending on the amount in the bottle and the depth to which he dips his brush.

It is probably the difference in color between the solid and the liquid components of bistre which causes its most interesting functional characteristic: ink from the same bottle will vary tremendously in color on different sheets of paper depending on the paper's sizing. Hard-surfaced paper will tint golden while a relatively more absorbent sheet will show more grey. It is conjectured that on the well-sized paper all the wash remains on the surface, while on the more absorbent sheet the yellow liquid soaks in, leaving only black particles on the surface. This guess is confirmed by the appearance of the verso of many of the Domenico Tiepolo drawings in the Metropolitan Museum, those from the Oriental and Centaur series in particular, where the ink color varies from grey to golden on the same sheet. In areas where the artist has attempted to attain a very dark value he has loaded his brush heavily and puddled the washes. These darkest areas are very blackish and granular. On the verso in these same areas, however, the ink which has soaked through is golden; the paper fibers have acted as a filter. This effect is most noticeable on

Abraham Visited by the Angels (Metropolitan 37.165.5) with a further variation. The darkest areas of wash are blackish and granular on the recto and golden on the verso; lighter but still intensely colored areas on the recto have remained golden. Why did the latter washes not sink through the paper, too, leaving a lighter grey silty patch? Probably these lighter areas, less heavily reworked by the artist, dried faster, before the paper's sizing could soften. Also, very small heavily washed areas tend to be warmer in tone than larger ones; such an area would likewise tend to dry faster and thus remain on the surface. A Giambattista drawing, *The Annunciation* (Morgan, IV 134), provides a clear example in the leg at the bottom of the sheet, where a narrow strip of wash, unreworked and unconnected to the larger masses of wash, is the reddest area on the page.

Yet another, even more subtle effect of the two color elements in bistre can be seen in a few of Domenico's drawings such as the *Assumption of the Virgin* (Metropolitan 37.165.65) and the *Cherubs in Clouds* (Museum of Fine Arts 95.1400). In these sheets some of the more firmly drawn pen lines within the siltier washed areas have very fine *lighter* outlines, lighter than the ink, of course, but also lighter than the surrounding wash (Fig. 1). Presumably the blunt pen point was applied with such pressure that it threw up a minute ridge of paper on each side of the stroke. When the wash was laid over it the blackish silt would run off these microscopic mountains into the valleys on each side, leaving only the more golden fluid to dry on their peaks.

One may well ask to what purpose such a ridiculously fine technical observation is made; yet while it is true that the effect is almost below the level of perception, still in sum it gives the lines of some of Domenico's drawings a strange radiance in their darkest shadows, an effect peculiar to his work and not shared by Giambattista's. Obviously no one can positively declare the artist to have been aware and appreciative of this effect, but in his choice and continued use of materials which would produce it he has expressed a preference.

Accepting the silty darks and transparent lighter washes and this potential for color modulation by a single wash within a single drawing as characteristic of bistre, and also the possibility of many shades of bistre on many sheets, one will more willingly conclude that most of the brown ink and wash drawings by the Tiepolos are in fact bistre. Of the few that are not so modulated, some, as has been stated, are obviously iron gall ink and originally appeared as black ink drawings. Indeed, the Tiepolos were not unaccustomed to using what we call "India" ink, a black pigment triturated from various forms of carbon black which retains its hue better than iron gall ink. It is most frequently seen in the work of Domenico, as in two drawings of greyhounds in a landscape owned by Mrs. Charles Slatkin; but it is also occasionally encountered in the work of Giambattista, as in the two caricatures also owned by Mrs. Slatkin (Nos. 87 and 88).

A few drawings by both Tiepolos, while not the modulated gold or olive brown of bistre, are also not the flat rusty brown of iron gall ink. They are instead a sort of heavy medium brown which gives a curiously opaque impression even in

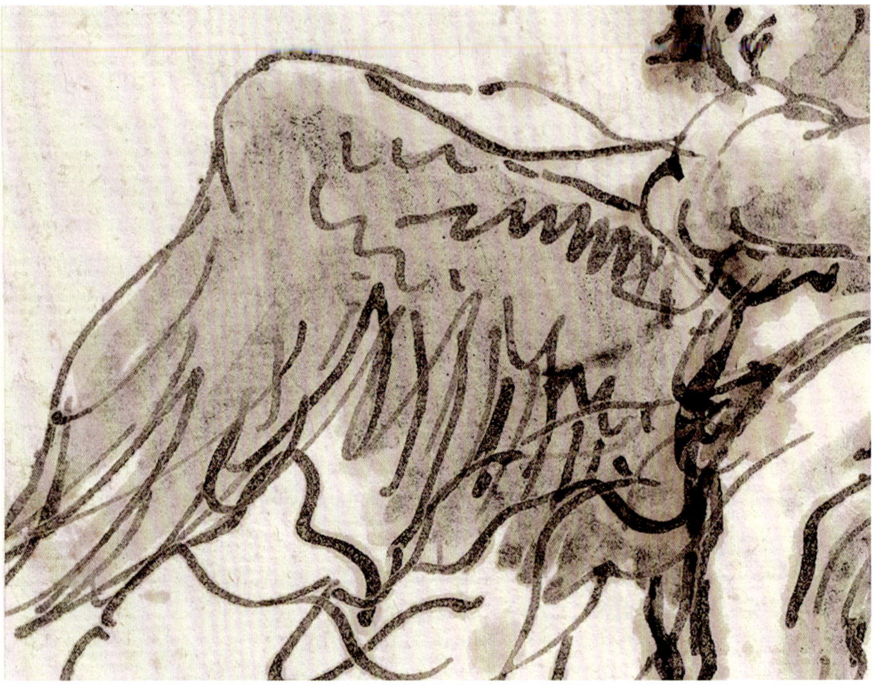

Figure 1
Domenico Tiepolo, *Flying Angels and Cherubs with the Instruments of the Passion*, enlarged detail. Museum of Fine Arts, Boston. Gift of Edward Habich, 95.1400. Photograph © 2014 Museum of Fine Arts, Boston. Here and elsewhere are seen the fine light outlines around the heaviest pen strokes, which are perhaps explained by the silty ink falling away from ridges made by the pen's pressure. They are especially visible at the intersection of the lines. Broader lighter outlines around pen strokes, also visible in this drawing, are perhaps better explained by the selective application of dark wash.

the lighter washes. Here, perhaps, we detect the use of sepia. Meder cites its first use since ancient times as occurring in Rome in the late eighteenth century,[12] but by this time when it had reached the notice of an encyclopaedist, it may well have been employed for several years and in other Italian cities. Meder provides a color sample[13] in original wash which is an intense pure brown, "flatter" than the adjoining real bistre sample. At a guess, drawings in the Morgan Library which might be in sepia include a group comprising *The Massacre of the Innocents* (Morgan, IV 149), *Christ and Martha* (Morgan, IV 151), and *Saint Kneeling at Wayside Shrine* (Morgan, IV 151a) and also the *Crowd of Soldiers* (Morgan, IV 151e). There is another sort of ink, an unmodulated mouse brown seen in the *Apotheosis of an Aged Warrior* (Morgan, IV 104), which might be a more dilute sepia.[14] The author has not been able to make sepia herself, cuttlefish being in short supply, and so has not had the chance to experiment on differently sized papers with different saturations of ink. Thus any conclusions regarding sepia are more conjectural than those on iron gall ink and bistre presented here.

Again, the reader may well ask the purpose of these attempts at exact identification of individual pigments, especially when it is evident at first glance that there is none of the consistency or uniqueness of type useful in the detection of forgeries. The author admits to the predictable pedantic urge toward exact description, but there are also more instructive reasons.

Every drawing by Giambattista sharing the virtues of his genius at their best—an immaculate brilliance, a lively surface, a breadth of touch that makes each irrevocable stroke seem inevitable—draws some of its quality from the positive values of its medium, usually bistre. Also, the relationship of Domenico's wash drawing style to that of his father can be better understood in terms of their handling of bistre. To analyze this further, let us first examine their actual methods in making a wash drawing.

As we have indicated, there is usually an underdrawing. Over this Giambattista drew a pen contour as concise as the underdrawing was prolix. Occasionally pen would be employed without a preparatory sketch, and here, as in the group at the Morgan Library already mentioned, the pen line is a little more discursive. This is also true in cases where the wash is omitted and the pen line stands alone, as in *Truth* (Princeton 48.863, No. 72) of this exhibition. Giambattista seems always to have used a quill pen. Not using a reed, he could not exert such pressure as to deform the paper surface as seen in Domenico's pen strokes discussed above; and in any case his touch was lighter and more fluid. Rarely, he would substitute his brush tip for the quill, as he may have done in *Zephyrs with the Horse of Apollo* (Morgan, IV 120, No. 19). Such virtuosity and exquisite control are amazing but hardly surprising considering the touch evidenced in his washes.

Over the pen lines would be laid washes. Usually the line was allowed to dry first, although in the same group of drawings at the Morgan Library which may be described as impulsive in that they are executed without the conventional underdrawing, the wash was applied immediately over the wet pen lines [fig. omitted]. This wiped the ink lines from the darkened areas, creating an optically illusionistic effect analogous to that found in laying one wet wash over another, which is discussed more fully below. Here, a mechanical technique simulates the visual experience of seeing contours in bright light with perfect definition but finding them only suggested in shadow. A most persuasive exploitation of this effect may be seen in *The Virgin and Child and the Young St. John* (Kaplan, No. 89), also.

In most of Giambattista's mature and characteristic wash drawings[15] the wash is not continuously modulated from light to dark but is applied in two discrete values, one medium and one dark, often darker than the pen line itself. Bistre is of low saturation and when drawn out in a fluid line it may thin to a medium tone; whereas an area laid in with a brush can be loaded with pigment (often silty) in a small spot.

The washes were applied in this sequence: the ink design was accented at points with a fully loaded brush and then amplified and unified with a more expansive wash of medium value. There are many drawings such as the *Flight into Egypt* (Private Collection, New York, No. 14) and a *Drunken Punchinello* (Thaw, No. 86) of this exhibition where the dark accents can be seen to have been only partially dry before they were swept over again with the brush. Their edges, which would have dried first, are sharply defined and dark; but their centers are lighter, often wiped clean of the darker tone and retaining only the depth of color of the lighter wash (Fig. 2). Occasionally streamers of dark pigment can be seen flowing from the darker accents into the paler tone. If the sequence of wash application had been the reverse and the lighter wash applied first, and if this were allowed to dry before the darker accents were laid in, then the darker areas would always be uniformly toned and one would never see them running into the lighter wash. If the darker accents were laid in before the lighter wash was dry, their contours would not be precise but blurred, and their edges would be lighter than their centers. The sequence—dark, then light—is not inflexible; one can sometimes guess that a lighter wash was intensified, as in the veil behind the shoulders of Ariadne in *Bacchus and Ariadne* (Davidson, No. 36); but it never appears to have been completely reversed.

The sequence used by Giambattista is obviously the best adapted to his rapid working methods, as the small accents dried more quickly than broader washes would, and he could proceed to the completion of his drawing that much more rapidly. Yet this order probably appealed to him for a more essential artistic reason: It could give even the darkest accents, with their "wiped" centers, a striking luminosity. The natural actions of the materials approximated the visual effect to the eye upon moving from brilliance to darkness, when the edge of shadows seems blacker than their depths. Also, the sequence from contour to accent to general modeling may have appealed to him as a gradual approach to the desired effect much as a similar procedure did to his son in his chalk drawings.

Domenico's personal wash drawing style, which he developed early in his career,[16] was a sort of elaboration of his father's, with everything—contour, sequence, color, accents—made more complicated. In fact, there are no perfectly defined sequences; pen (quill, reed, or both) overlays wash, and one depth of wash overlays another until a constantly modulating tangle of lines and tones is shaped into a composition.

Giambattista seems always to have used a single ink in his washes, light and dark, although because of the color-modulating effect of bistre they appear varied and lively. Occasionally he would use a different ink for his pen line, probably not for coloristic reasons but only to achieve a darker contour than was possible with bistre. Domenico, however, seems often to have depended on several inks in the same drawing, all in subtly differing tones of grey and brown; and he was supremely sensitive to the two-color potential of bistre. One is constantly perplexed in his more complicated, layered wash drawings: are grey washes added or is the varying absorbency of worked and reworked paper simply exploited to its utmost? This

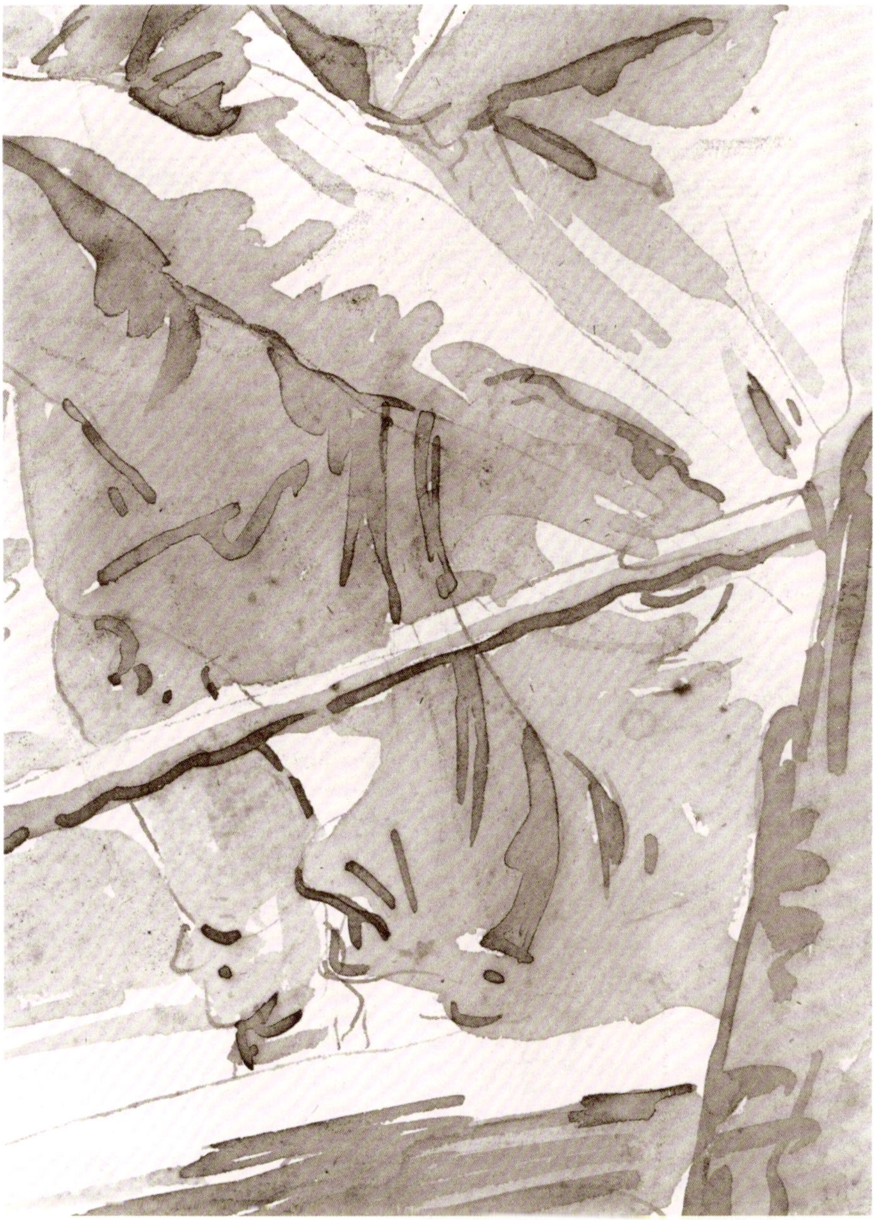

Figure 2

Giambattista Tiepolo, *The Flight into Egypt*, enlarged detail (No. 14). Private Collection.
This area shows clearly the "wiping" effect of laying in medium wash over areas of darker wash which were not completely dry. The larger areas in particular have their centers wiped clean of the darker tone; they would have dried more slowly than the smaller areas, which are comparatively dark throughout. The edges of the darker areas, which dried more rapidly, are hard and dark. At the very bottom left are some medium dark areas which seem not to have been affected by the darker wash. They may have been added later, after the lighter wash had dried. This detail also shows, in the upper left, the tendency of the bistre wash to be silty in its darkest areas. Note, too, at the upper center and right, the extreme transparency of the short strokes of the black chalk underdrawing, where the paper grain shows throughout.

sensitivity was perhaps the natural development of an artist whose youthful talent lay in grisaille painting.[17]

Sometimes the use of two inks is very obvious, of course, although even here the contrast is not usually between grey and brown but between two shades of brown. Thus in *The Milliners* (Museum of Fine Arts 47.2, No. 102) the foreground lady's costume is picked out in a golden brown while all the other tones are in a more olive brown ink. Giambattista's undeviating concentration on matter only as it is shaped by light has been transformed here by Domenico's more reportorial interest in local color and detail. The change is made explicit by their technique. One is reminded of Giambattista's pen lines in shadow wiped away by his washes and Domenico's, by contrast, intensified by a halo of light—both effects being achieved by varying manipulations of the same ink. However different their artistic intentions, both men were equally served by their identical materials. Their appreciation and exploitation of the subtle coloristic and tonal potential of bistre in particular contributed greatly to their larger achievement.

Notes to Introduction

1 Jack Cowart et al., *Henri Matisse: Paper Cut-Outs* (St. Louis: St. Louis Art Museum, 1977), 272–77.
2 The conservator Carlo James has also re-created many historical recipes for bistre in order to differentiate it visually from iron gall ink. See Carlo James, "A Closer Look at Bister Inks," in *Dear Print Fan*, ed. Craigen Bowen, Susan Dackerman, and Elizabeth Mansfield (Cambridge, Mass.: Harvard University Art Museums, 2001), 161–66; also, Carlo James, *Visual Identification and Analysis of Old Master Drawing Techniques* (Florence: Olschki, 2010), 19.
3 Not surprisingly, one of Cohn's graduate students organized a small exhibition on what can be discovered on versos: see James Harper, *Verso: The Flip Side of Master Drawings* (Cambridge, Mass.: Harvard University Art Museums, 2001) (excerpted in reading 42).

Notes to Reading

1 Because of their style a real appreciation of Tiepolo drawings, even more than of drawings by other artists, depends on their being in good condition. Fortunately, most of them are very well preserved indeed, for most have been kept protected in albums until this century. Many of these entered museum collections immediately upon the albums' dispersal and have since been stored in favorable environments.
2 George Knox, *Catalogue of the Tiepolo Drawings in the Victoria and Albert Museum*, London, 1960, p. 3 f.
3 Joseph Meder, *Die Handzeichnung, ihre Technik und Entwicklung* (Vienna: Anton Schroll, 1919) p. 62 f.
4 Knox, op. cit., p. 4.
5 This catalogue makes no attempt to identify the pigment of each drawing in it because the author of this note has been unable to see them in the original. Also, some would undoubtedly resist visual identification even after the most careful

examination. There is a chemical test for iron gall ink, given by Meder, but it is destructive (it turns the drawing blue!); and it is obviously impractical for drawings of quality. Modern spectroscopic methods using sensors rather than sample analyzers may come to our aid some day, but till now such equipment is too rare and expensive to have been widely applied.

6 Meder, op. cit., p. 67.

7 Ibid., p. 69.

8 [Luca Cambiaso (1527–1585), Genovese painter; Guercino (1591–1666), Venetian painter.—Ed.]

9 In the author's experience a bistre of good quality is by far the easiest artist's pigment to make "from scratch." Considering the availability of its ingredients, wood soot and water, the artists may well have made it up themselves. Also, it has a delightful and unforgettable fragrance. For its exact recipe and for those of other inks, see: James Watrous, *The Craft of Old Master Drawings*, Madison (Wisconsin), 1957, pp. 66–88. Watrous gives a short but suggestive summary of bistre's color potential, p. 78.

10 Meder, op. cit., p. 68.

11 Ibid.

12 Ibid., p. 70.

13 Ibid., plate 51.

14 [These drawings belonging to the Morgan Library & Museum do not share characteristics associated with sepia as made from historical recipes. My thanks to Mary Broadway for her observations.—Ed.]

15 Knox, op. cit., p. 10, where it is noted that Giambattista's wash technique was fully developed before the age of thirty.

16 James Byam Shaw, *The Drawings of Domenico Tiepolo*, London, 1962, p. 29 f. He cites *St. Ambrose* (Fogg 1965.415, No. 55) of this exhibition as an exceptionally early example of Domenico's wash style, done while the artist was collaborating with his father.

17 Ibid., p. 13 f., p. 61, and elsewhere, where sensitive analyses of Domenico's technique are found.

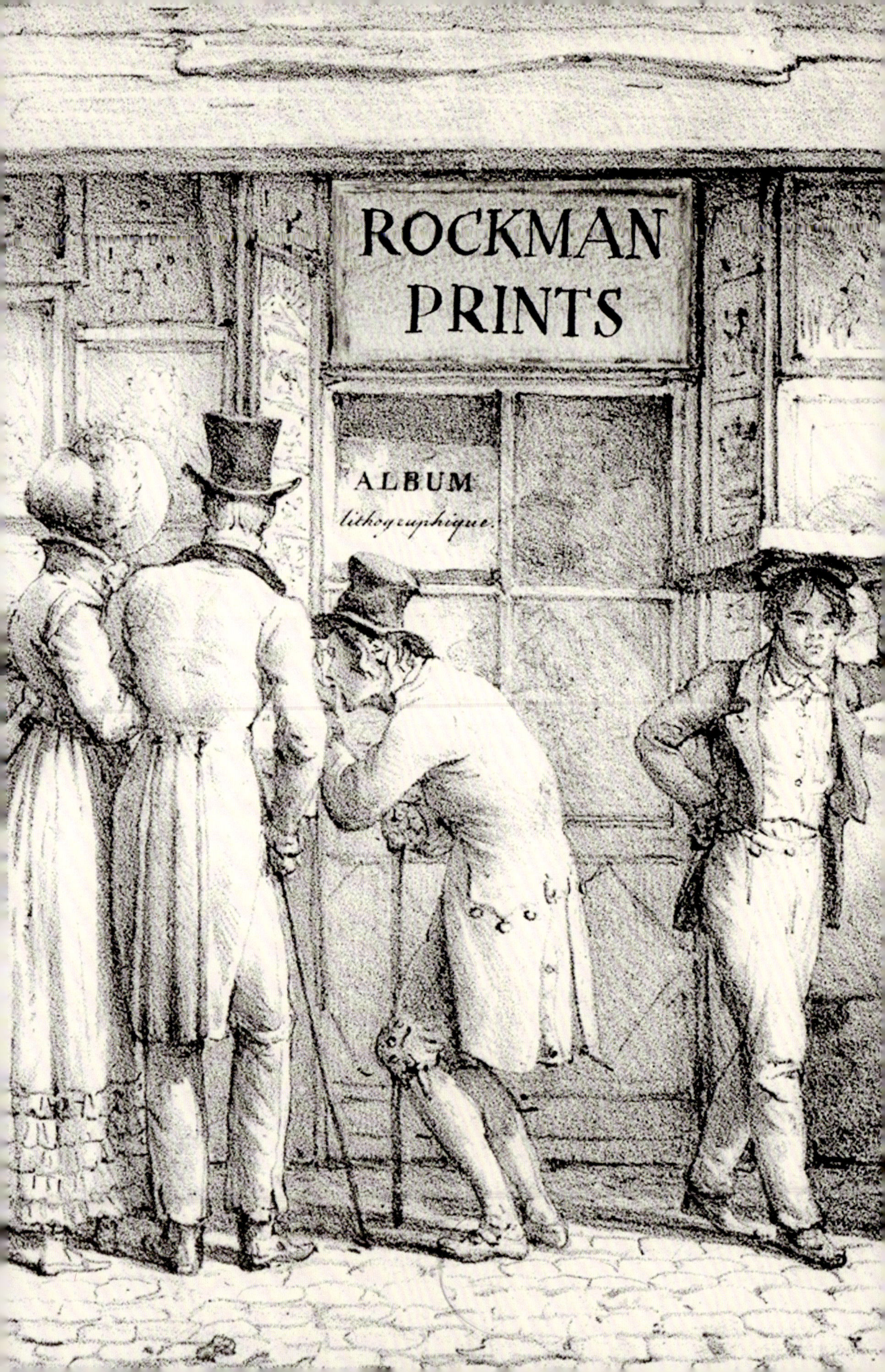
ROCKMAN
PRINTS
ALBUM
lithographique.

PART III

Printmaking: Multiple Originals

The accurate identification of printing techniques is a critical skill for conservators of works on paper and, with the rapid introduction of new methods of digital reproduction, an increasingly difficult one to master. But as early as 1897 one frustrated print curator lamented, "There are even now numerous artists—and by artists I mean painters—who confuse etching with pen-and-ink drawing and who are unable to tell one style of engraving from another. Even a specialist newspaper belonging to the printing trade quite recently spoke of Hokusai's delightful coloured prints as 'engravings on copper.'"[1] And as recently as 1965 Carl Zigrosser observed a "growing indifference to means and techniques" (reading 33).[2]

Unlike the various classification systems for drawing media—dry versus wet, oldest to newest, broad versus fine—printing techniques have traditionally been categorized by the appearance of the printed lines on paper; they are either raised above it, or pressed into it, or as one with the paper surface.[3] Or, as more passionately described, they are the traces of "an embrace, one body pressed against the other."[4] The conservator, therefore, needs to scrutinize the intimate relationship between paper and ink in order to classify the print as one of three types, relief, intaglio, or planographic, each category having unique visual and physical properties and, by extension, its own conservation issues.

A thorough knowledge of printing techniques and their associated paper and ink requirements has both theoretical and practical advantages. Along with a familiarity with the unique physical properties of prints comes an appreciation for their subtle aesthetic effects, both of which influence treatment decisions. During the preliminary examination of a print, the conservator considers how the presence of certain attributes might be altered by treatment. Questions might include the following:

- Will the depth of a platemark be changed by washing and flattening?
- Does the presence of a vulnerable *chine collé* preclude immersion washing?

Charles Vernet (1785–1836), *François Delpech's Print Shop*, ca. 1820 (detail). Lithograph with 20th-century lettering addition ("Rockman Prints"); publisher: F. Delpech, Paris. Image: 6 5/8 × 9 1/2 in. (16.8 × 24.1 cm), sheet: 8 5/8 × 11 9/16 in. (21.9 × 29.3 cm). The Elisha Whittelsey Collection, The Elisha Whittelsey Fund, 1977 (1977.614.25). The Metropolitan Museum of Art, New York, NY, U.S.A. Image © The Metropolitan Museum of Art. Image source: Art Resource, NY.

- Can desiccated nineteenth-century engraving inks be safely washed and deacidified?
- How evenly will an aged plate paper absorb water?
- Will aniline dyes in German lithographic inks bleed?
- Will thick stratified silk screen inks interfere with flattening?
- Will the reduction of yellowed oil migration from the ink alter the integrity of the history of the materials used?
- When is resizing soft-sized plate paper appropriate?

Simultaneously, the appearance of the print is evaluated, often with attention to quality. The overall quality of a printed impression is governed by the nuanced relationship between its paper—color, texture, and absorbency—and its ink—velvety, transparent, pale, or crisp. The topography of the lines and any platemarks are assessed; in Japanese *ukiyo-e* prints, for example, the verso is examined for shiny striations left by the barren. The precise points that add up to "quality," a heretofore subjective assessment under the umbrella of connoisseurship, are difficult to quantify, although paper conservators have made great strides in devising more objective methods for measuring it.

As with the choice of drawing media, an artist's preferred printing technique can be considered significant art historically. The French art critic André Mellerio (1862–1943) in his defense of color lithography (reading 32) asserted in 1898 that "any method or process which an artist develops to express himself is, for that very reason, legitimate."[5] Felix Brunner (1929–) goes so far as to correlate the choice of technique with the personality of the artist: "The woodcuts and the dry points of the German expressionists give vent to a certain aggressiveness aimed at the existing order, which they wanted to change."[6] And the master printmaker Jules Heller (1919–2007) claimed that the very materiality of each printmaking technique communicates a message: "The print is employed as a medium in its own right. It is utilized by the printmaker for what it alone can accomplish in serving his particular needs. This precious sheet of paper bears the autographic trace of the printmaker on its surface; in his own 'handwriting,' then, we read the record of his dreams, his hopes, aspirations, play, loves, and fears."[7]

Occurring periodically throughout the history of printmaking and appearing with regularity in Part III are questions concerning the definition of an "original" print. Bickering about the originality of prints has been sporadic but continuous, starting as early as the sixteenth century (reading 22). Usually following on the heels of a technological advancement, elucidating the exact difference between a mass-produced and a fine art print may become litigious, as occurred in 1890 when Sir John Herkomer (1849–1914) was accused of passing off photogravures as original etchings[8] and in 1896 when Walter Sickert (1860–1942) claimed that transfer lithography, as practiced by Joseph Pennell (1857–1926) and James McNeill Whistler (1834–1903), is not a genuine artistic process.[9]

Issues of originality are included in Part III because paper conservators, in particular, are routinely confronted by the consequences of the shifting definition of originality and the value judgments that result. As written in the Preamble of the *Code of Ethics* of the American Institute for Conservation of Historic and Artistic Works (AIC), to which professional conservators subscribe, "Cultural property . . . is material which has significance that may be artistic, historical, scientific, religious, or social, and it is an invaluable and irreplaceable legacy that must be preserved for future generations."[10] It is interesting to note that one requirement, if not a driving force, for preservation is not specified—that of financial worth. Our society places high value on a work's originality and uniqueness. A print's financial value increases substantially if it is declared original, whereas an unsigned twin from the same edition is worth less. In the case of prints specifically, financial value often determines their preservation fate. Original prints are more likely to be accorded a higher level of institutional protection, influence funding priorities, or benefit from the ministrations of a paper conservator. At the same time, the AIC *Guidelines for Practice* caution, "Professional fees must relate directly to the provision of service, and must not be based on the value of the cultural property."[11] To paraphrase Bamber Gasgoigne, the printed "picture of milk cascading onto cornflakes . . . is susceptible to exactly the same conservation treatment as its commercial predecessors or as a great printed work of art, such as an etching by Rembrandt."[12] As every paper conservator knows too well, it requires at least the same amount of skill, judgment, and time to treat an offset color lithograph from the nineteenth century as an etching by Rembrandt. Therefore, many individual or institutional owners will simply not expend the money required for the conservation of nonoriginal prints.

As noted in Part I, works on paper are increasingly appreciated for their social or use, in addition to their artistic, value.[13] This has resulted in broader acceptance of the intellectual capital contained in nonoriginal prints, even old newspapers and has increased the incentive to preserve them. At the same time, the cost-effectiveness of broad-based preventive conservation programs has been recognized.[14] As defined by the AIC, preventive conservation is the mitigation of deterioration and damage to cultural property through the formulation and implementation of policies and procedures for appropriate environmental conditions; handling and maintenance procedures for storage, exhibition, packing, transport, and use; integrated pest management; emergency preparedness and response; and reformatting or duplication.[15] Preventive conservation has become an established principle in funding policies, so that it now has precedence over more expensive single-item treatment. It is through participation in conservation assessment surveys and other activities emphasizing the importance of proper housing and storage that paper conservators can spread the mantle of preservation over prints not deemed to be original—at least not yet.

Notes

1 Hans W. Singer and William Strang, *Etching, Engraving and the Other Methods of Printing Pictures* (London: Kegan Paul, Trench, Trübner and Co., 1897), 2.
2 Carl Zigrosser, "The Historical Background of Originality," in Carl Zigrosser and Christa M. Gaehde, *A Guide to the Collecting and Care of Original Prints* (New York: Crown, 1965), 34.
3 A second and confusing system was to differentiate between linear and tonal techniques, as seen in Roy Vernon Sowers, *Enjoy Your Museum II: Prints* (Pasadena, Calif.: ESTO Publishing, 1934), no pagination; corresponds to p. 2.
4 Michael Rothenstein, in Ruth Weisberg, "The Syntax of the Print: In Search of an Aesthetic Context," *Tamarind Papers* 9, no. 2 (Fall 1986): 58.
5 André Mellerio, "La lithographie originale en couleurs" (Paris: L'Estampe et L'Affiche, 1898), trans. Margaret Needham, in Phillip Dennis Cate and Sinclair Hamilton Hitchings, *The Color Revolution: Color Lithography in France, 1890–1900* (New Brunswick, N.J.: Rutgers University Art Gallery, 1978), 80–81.
6 Felix Brunner, *A Handbook of Graphic Reproduction Processes*, 6th rev. ed. (New York: Hastings House, Visual Communications Books, 1984), 31. One is also reminded of the fictional German Expressionist artist, Professor Kuchen, who cried, "Art is a fight to the finish between black charcoal and white paper" (Günter Grass, *The Tin Drum* [New York: Random House, 1990], 462).
7 Jules Heller, *Printmaking Today: An Introduction to the Graphic Arts* (New York: Holt, Rinehart and Winston, 1958), xiv.
8 The pictures in question appeared in a souvenir publication for Herkomer's dramatic work, *An Idyll*; *Musical Times* 31, no. 566 (April 1, 1890): 210.
9 See E. R. Pennell and J. Pennell, *The Life of James McNeill Whistler* (Philadelphia: Lippincott, 1911), 342.
10 AIC, *Code of Ethics*, www.conservation-us.org, accessed October 1, 2011.
11 AIC, *Guidelines for Practice*, www.conservation-us.org, accessed October 1, 2011.
12 Bamber Gascoigne, *How to Identify Prints* (New York: Thames and Hudson, 1986), 7.
13 See Irene Brückle, "Aqueous Treatment in Context," in *Paper and Water: A Guide for Conservators* (Oxford: Elsevier, 2011), 419–36.
14 The reader is encouraged to consult Sarah Staniford, ed., *Historical Perspectives on Preventive Conservation* (Los Angeles: Getty Conservation Institute, 2013).
15 AIC, *Commentaries to the Guidelines for Practice*, www.conservation-us.org, accessed August 27, 2012.

Reading 22

Ugo da Carpi

Petition to the Venetian Senate (1516)

Today marks on paper that are deemed original works of art are considered to be of higher financial and, by implication, cultural worth than copies or other derivations of them. Originality is typically determined by physical evidence of the artist's hand or, if at some remove, by proof of the artist's unambiguous involvement in the conception, direction, or fabrication of the work. In the case of prints, the determination of originality can be problematic. The final printed product, due to the mediating print matrix or the intervention of a skilled printer, traditionally a different person and later a machine, is often completely divorced from the artist's direct touch.[1] *This reading from the sixteenth century poses the question, how can visually and physically identical products be considered exclusive and original works of art and hence eligible for protection? While today few will disagree with the concept of intellectual property, debate over what constitutes originality surfaces with great regularity in the world of printmaking, especially as imaging technologies proliferate. To some extent, the signing and numbering of limited-edition prints, along with the destruction of the printing matrix, artificially (and ironically) solves this problem.*[2] *Technological progress also serves to expand the boundaries of originality; to wit, after the invention of photography, many reproductive engravings gained ground, albeit slowly, in both financial value and curatorial recognition, eventually earning the hard-won imprimatur "original."*[3]

The designation of originality, with its associated value judgment, can influence institutional preservation policies and to a large extent dictate which prints arrive on a paper conservator's bench in the first place. Of course, to a conservator each print, even the supposedly identical ones by Ugo da Carpi (1480?–1520/23), is physically and

---

From Jan Johnson, "I chiaroscuro di Ugo da Carpi / Ugo da Carpi's Chiaroscuro Woodcuts," *Il conoscitore di stampe* 3–4, no. 57–58 (1982): 12 n. 1. Original document in the Archivo di Stato di Venezia, Notario di Collegio Re 18, 1515–20, c. 38v.

visually unique. Differences in paper manufacture, inking, and press pressure and the vicissitudes of function and time, not to mention the whims of collectors, have resulted in each print's individual response to conservation treatment. As noted by Ruth Weisberg (reading 36), when artists act as their own printers, the probability of nonidentical impressions is even greater. For Ugo da Carpi, the issue was not only the uniqueness of his designs, including those still residing in his imagination, but also, as he claimed, his invention of the chiaroscuro process.[4] *His petition to the Venetian senate introduces the concept of multiple originals.*[5]

24 July 1516

Most Serene Prince and Exellent Senate,

Whereas I, Ugo da Carpi, engraver of figures in wood, have long been established in your illustrious city of Venice, and I have spent my youth there and having lived to old age, and wishing as I do to continue living from the practice of my customary craft: and since I have discovered a new way of printing in chiaroscuro, a new thing that has never before been done, and a fine thing, and useful to many who find pleasure in drawing: and moreover, since I have engraved and plan to engrave things that have never before been done, nor even dreamed of by others: of these efforts of mine, I humbly ask, beg, and entreat by the grace of your illustrious lordships, so worthy in your clemency, to grant me, that no one be allowed, nor dare, to copy or counterfeit any design or engraving of mine, and I speak with the greatest reverence, that has not been taken in violation of the wishes of others; only by my own device and wit, which I shall prove to be necessary and useful, and those who venture and presume to counterfeit against my wishes, and desiring to print here, or away from here, cannot then sell them in the places that are subject to your illustrious dominion, under penalty of losing those illustrations, and for every illustration, ten ducats, divided into three parts. One part goes to charity, the second part to the dominion in which the complaint has been made, and the third part to the accuser. I humbly entrust myself to the benevolence of your illustrious excellencies.

Translated by Antony Shugaar

Notes to Introduction

1 This relative distance forms the basis for the argument that "an etching was really an artist's direct drawing on the copper-plate, instead of on paper. An etching was thus equivalent to a drawing, which an ordinary engraving was not." Attributed to Mr. Fortnum, *Journal of the Society of Arts* 32, no. 1, 593 (June 1, 1883): 772.

2 There is an inherent contradiction in limiting a medium whose original purpose was to disseminate images to the widest audience possible. The practice began in the late nineteenth century.

3 A reproductive engraving, as opposed to a reproduction, is a copy of another work, typically a painting done without the aid of photography.

4 I am grateful to Irene Brückle for bringing this selection to my attention. See Rebecca Zorach and Elizabeth Rodini, *Paper Museums: The Reproductive Print in Europe, 1500–1800* (Chicago: David and Alfred Smart Museum of Art, 2005), 46; Linda Stiber Morenus, "The Chiaroscuro Woodcut Printmaking of Ugo da Carpi, Niccolò Vincentino and Antonio da Trento: Technique in Relation to Artistic Style," in *Printing Colour in the Hand-Press Period: History, Techniques, Function and Reception, ca. 1400–1800* (London: Brill, 2014).

5 Jules Heller coined the phrase "multiple originals" in his *Printmaking Today: An Introduction to the Graphic Arts* (New York: Holt, 1958), xiv.

Reading 23

William M. Ivins Jr.

How Prints Look: Photographs with a Commentary (1943)

William M. Ivins Jr. (1881–1961) was the first curator of prints at the Metropolitan Museum of Art, New York (1916–46). Largely self-taught as a print connoisseur, Ivins was among the first to use photographic enlargements to illustrate the distinct visual characteristics of various printing techniques, which were categorized according to the topography of the surfaces. Felix Brunner and Bamber Gascoigne (see Further Reading) developed Ivins's methodology into more intuitive flowcharts.[1] *Today the differentiation of printing techniques can also be done on the Internet. Those who have compared the two methods—scrutinizing printed details in a book or pixels on a monitor—seem to favor the former paper-based approach, though this may be a generational preference. One must keep in mind that the illustrations of the various printing techniques in books are usually created using offset lithography, further complicating precise identification by the interference of halftone patterns. According to Ivins, however, our eyes automatically filter out the "implicit distortions" from the contemporary imaging technology to which we are accustomed.*

Ivins believed that "seeing is doing" and tried his hand at printmaking to complement the direct comparison of originals. The latter requisite activity is underscored in this quotation from the philosopher Plotinus (205–270) on Ivins's title page: "Remember that there are parts of what it most concerns you to know which I cannot describe to you; you must come with me and see for yourselves. The vision is for him who will see it."

Ivins was convinced that the ability to produce exactly repeatable images was of enormous social significance, as was his successor, A. Hyatt Mayer (1901–1980), author of Prints and People: A Social History of Printed Pictures *(see Further*

From William M. Ivins Jr., *How Prints Look: Photographs with a Commentary* (New York: Metropolitan Museum of Art, 1943), 142–43.

Reading). While this topic has grown into a stand-alone academic discipline within media studies and many more types of prints since Ivins's time have been anointed as "originals," for conservators the precise identification of printing techniques remains crucial for characterizing ink and paper and their potentially alterable physical and visual characteristics.

In recognition of its enduring usefulness, How Prints Look *was reissued in 1987 under the editorial guidance of Marjorie Cohn, who chose to overlook Ivins's period prejudices: "Yet his crotchets are as nothing to his perceptions, and every attempt has been made to respect his conviction that pictures, and looking hard, are more essential to an understanding of prints than words."*[2] *The strength of Cohn's assertion is reflected by the regular interjection of Ivins's voice throughout the readings in Part III.*

On the Social Importance of the Graphic Techniques

A print may be defined as an image made by a process that is capable of producing a number of exact duplicates. Such an image cannot be made directly by hand, for no copy is like the original from which it is copied, and no two copies are alike. This means that no two handmade pictures can tell the same things about the same object.

It can be reasonably argued that the great event of the fifteenth century was not the rediscovery of Greek or any other sort of ancient learning, but the discovery of mechanical ways to make pictorial records in duplicate, exactly, cheaply, and in vast quantities. This discovery made it possible for the first time to make pictorial statements and records in exact duplicate and to distribute them in invariant visual form simultaneously to many different people in many different places. The importance of this to human society can hardly be overstated. It has possibly had greater effects than any mechanical discovery since the invention of writing, as it is basic not only to our knowledge of the past and the present but to a very large number of our modern technological and scientific developments.

Since sometime in the 1400's men of the western European civilizations have derived from duplicate pictures most of their knowledge of the shapes and textures of things not immediately at hand.

Only the very smallest proportion of these duplicate pictures was made for specifically artistic as distinct from informative reasons. So-called fine prints are only an infinitesimal and indefinable portion of the total number of printed pictures. Fashions in collecting and the blindness of students of "art" to all things not baptized as "art" have led to an almost complete misunderstanding and even ignorance of the actual functions and developments of the duplicate picture.

Each generation has had its particular and favorite form of duplicate picture. First there were woodcuts, then engravings, then etchings, after which there came in order across the centuries, mezzotints, aquatints, wood engravings, and lithographs. Finally in the 1830's came the beginnings of photography, out of which in turn came the photomechanical processes which produce halftones, line blocks, and all the other modern means of illustrating books, magazines, and newspapers. With the exception of lithography each of the pre-photographic processes requires special techniques of draughtsmanship that are based upon the nature of its peculiar tools and materials. Lithography, having no specially shaped drawing tool, uses those of ordinary men, crayon or pencil, pen, and brush. Whatever tools and materials are used, only certain very restricted kinds of statements can be made with any combination.

Seeing the world through its duplicate pictures, each generation has unwittingly read the qualities and defects of its duplicate pictures into its knowledge of the world, just as the man who wears green glasses sees the world all green. The last great change in the conditioning of human vision happened when people began to see in terms of photographs. The recent development of color photography and its reproduction in printer's ink has already begun to have its marked effect upon the public's sense of color. [. . .]

In view of all this the importance of being able to recognize the technique or process by which a printed or otherwise precisely duplicable image was made becomes obvious—for this knowledge enables us to discount or make allowances for the limitations, the blind spots, the distortions, implicitly and unknowingly introduced by techniques and processes into duplicate images and their testimony about the world. These implicit distortions are a most important part of the unconscious, unphrased, common assumptions of any human society, which basically determine its ideas and action. Very few people ever realize the extent to which "objective facts" as known by us are actually no more than peculiarities of our instruments of observation and record.

From a purely artistic point of view, however, the recognition of graphic techniques has little reward for anyone who is not himself a practicing graphic technician. The fact that a particular print is in one or another process has no more bearing upon its artistic merit than has the fact that it was printed in the basement or on the top floor.

Notes to Introduction

1 A more expensive but superior method for the identification of printing techniques is to insert original examples into instructional manuals; one such example is Sir Frank Short, *Etchings and Engravings: What They Are and What They Are Not* (London: Royal Society of Painter-Etchers and Engravers, 1912).

2 William M. Ivins Jr., *How Prints Look: Photographs with a Commentary*, rev. Marjorie B. Cohn (Boston: Beacon Press, 1987), viii.

Reading 24

Seymour Haden

About Etching (1878–1879)

Francis Seymour Haden (1818–1910), a British printmaker who also trained as a surgeon, had little regard for reproductive engravings, which dominated commercial printmaking in the nineteenth century.[1] Instead of these "translations" of painted, printed, or sculpted works into hyperstylized dots, dashes, and diamonds, he favored what he considered to be the more expressive lines of etching, as did his brother-in-law, the American expatriate James Abbott McNeill Whistler (1834–1903), and spearheaded the revival of etching in England. Here Haden vividly describes the differences between etched and engraved lines according to their characteristic movements—lively and spontaneous for the former, stilted and constrained for the latter.[2] As Ivins would later agree, "Just as on the street it is impossible to recognize a still figure a long way off but easy to do so the moment it begins to move, so with practice in looking it will be discovered that the lines produced by each graphic process have a characteristic gait or motion of progress that is frequently recognizable even when the lines are too small to be seen clearly or singly."[3] Many conservators will attest that stepping back and circling a print while viewing its surface can serve to diagnose the "lifelessness" of a photomechanical reproduction.

Haden championed the use of antique papers for their subtle colors and receptive printing surfaces, which added a romantic and atmospheric aura to his etchings. This practice was taken up by a number of British printmakers, among them Samuel Palmer (1805–1881) and James McBey (1883–1959).[4] In America, Whistler's use of aged papers was adopted by a group of enthusiasts that included Frederick Childe Hassam (1859–1935), Joseph Pennell (1857–1926), and John Taylor Arms (1887–1953), among many others.

From Seymour Haden, *About Etching*, 3rd ed. (London: John Strangeways, Fine Arts Society, 1878–79), 10–13.

Because the use of vintage papers, even sporadically in the same edition, was so popular in the late nineteenth century and well into the twentieth, conservators cannot assume that stains, tidelines, discolored edges, foxing, stray marks and sewing, and worm holes constitute damage. Further, because aged paper often absorbs water quickly and unevenly due to the breakdown of its gelatin sizing, conservators must recognize that it may not respond as would be expected by the date of the print.

NOTE IV.

The Engraver's Art.

The way being so far cleared, it may not be amiss that the next note should be on the subject of Engraving generally, and as to how, when it fell into the hands of the painter, it came to be merged into Etching.

As the motive of this note, and, indeed, for nearly all I shall have to say, will be found in the three following paragraphs, I think it best to give them first, as a sort of text:—

> 'Could we chain down the original artist to the engraver's stool and get him to carry his work as far as he could with intelligent execution, we should have a "finished" engraving far more valuable than any etching.'—'*V.*'[1]

> 'L'eau-forte était le délassement et le caprice favori des maîtres. Ils jetaient sur la planche l'idée ou la conception que le pinceau aurait ralentie; ils dessinaient sur le vernis comme sur le papier. Leurs plus intimes confidences sont écrites sous cette forme libre et rapide; car l'eau-forte d'un grand peintre suppose toujours un éclair de verve qu'il n'a pas voulu laisser refroidir.'—*Paul de St. Victor.*[2]

> 'Etching, "a blundering art."'—*J. Ruskin.*[3]

If we take the great roll of engravers on copper from the invention of the art in the fifteenth century to its decline in the nineteenth, we find that, with the exception of a few men who interposed and for a time practised a novel method, it consists entirely of workers with the burin or graver. Secondly, that these burinists divide themselves into two classes,—*those who were original artists and engraved their own works, and those who copied or translated the works of others*. The first of these groups, which may be designated that of the Painter-engravers, and which represents the use of the burin in its simplest and purest form, begins with Martin Schoen[4] and has its perfect type in Dürer. The other, that of the Line-engravers commonly so called, takes its impulse from Goltzius, Cornelius Cort, and Agostino Caracci,[5] and brings their innovations, with but slight modifications, down to our time. The Etchers and the Mezzotinters appear to have come in somewhat, as it were, *en contumace*, the first to protest with the early engravers in favour of greater liberty and a more natural treatment—the last with the Mechanical Engravers for a more painter-like quality. We find, as might be expected, little bond of union between these two main groups beyond the plate they worked upon. The early painters with their imperfect chemistry appear to have been content to use the graver as the only tool known to them capable of ploughing the copper, and some of these, being great masters, so far triumphed over the instrument as to produce with it work which excites our wonder and admiration to this day; but the moment the possibility of acting upon the plate by an implement used like a pencil was shown to them, the burin fell from their hands and they became Etchers; while the graver descended at once to a class of men who thenceforth undertook, by a slow and laborious process, to which the instrument was not ill-adapted, to reproduce the works of others; and no sooner had they obtained possession of it than they seemed bent on showing to what extravagance its use might be carried, and how independent it might be made of the painter's art. Ridiculing the attempts of Marc Antonio[6] to make it effective as an agent of expression, they proceeded at once to show how tones might be rendered by lozenge-shaped intervals with or without a dot in the centre of each—the bursting rain-cloud by an arrangement of concentric curves not unlike the engine-turning at the back of a modern watch—atmospheric backgrounds by a sort of tooling having something of the effect of watered silk, and skies by a machine. *There is not, therefore, and there never was, anything that could be called a rivalry between the Etchers and the later Engravers. They represent distinct classes—the class of artist and the class of copyist,*—a distinction to be firmly insisted upon, partly because it is a radical distinction, but principally because we shall have to show that the adoption of the tool, (except in the case of the painter-engraver, who, it is to be remembered, was an original artist,) implies the practice of a secondary art.

NOTE V.

Etching Compared with Engraving—The Etched Line with the Engraved Line.

The comparison of the etching-needle with the burin is the comparison of the pen with the plough. In one case a finely-pointed style obedient to every movement of the sentient hand—in the other a tool driven by the elbow against an object brought to meet it half way; in one case suppleness, liberty, rapidity and directness of utterance, and the faculty of keeping pace with ideas as they are formed—in the other, the combined action of two hands and the active opposition of two forces—that of the instrument against the plate, and of the plate against the instrument. What wonder that the line described by one should be free, expressive, full of vivacity—by the other, cold, constrained, and uninteresting; that one should be personal as the handwriting—the other without identity; that the difference between the two should be, in point of fact, as the difference between any two things of which mechanical effort is the parent of one, and brain impulsion of the other. The able writers who contend that more painter-like qualities are to be obtained from the burin than from the point, have only to reflect for a moment upon the fashion of the two instruments and on the way in which they are used; on the mobility and independence of one, and the limitations and restraints which its very form imposes upon the other. What infinite memoranda may not be made by the point while the burin is delving its conventional and inexpressive furrow!

Notes to Introduction

1 Reproductive engravings are differentiated from photogravures by the absence of photographic methods used to transfer or fix images onto the print matrix.
2 See Emma Chambers, *An Indolent and Blundering Art? The Etching Revival and the Redefinition of Etching in England, 1838–1892* (Farnham, Surrey: Ashgate, 1999).
3 William M. Ivins Jr., *How Prints Look* (New York: Metropolitan Museum of Art, 1943), 3. See reading 23.
4 See Deborah Carton and M. Brigitte Yeh, "Invoking the Past: John Taylor Arms' Use of Antique Papers," in *Looking at Paper—Evidence and Interpretation: Symposium Proceedings, Toronto, 1999* (Ottawa: Canadian Conservation Institute, 2001), 17–25; Merrily Smith, "Hunting for Old Paper with James McNeill Whistler," *Book and Paper Group Annual* 19 (1997): 90–91; Harriet K. Stratis, "Whistler's Papers: Their Appearance, Selection and Use," in *The Lithographs of James McNeill Whistler*, vol. 2, *Correspondence and Technical Studies*, ed. M. Tedeschi (Chicago: Art Institute of Chicago, 1998), 298–304.

Notes to Reading

1 ["V." remains unidentified. It seems that Haden is overreacting to a mediocre review, "Mr. Haden's Etchings," that appeared in the *Spectator* 39 (February 3, 1866), 132.—Ed.]
2 ["Etching was the favorite relaxation and caprice of the masters. They were throwing down on the plate the idea or concept that the brush would have slowed; they drew on varnish as on paper. Their most intimate confidences are written in this free and rapid form; for the etching of a great painter always supposes a flash of verve that he didn't want to let grow cold." Paul de Saint-Victor (1819–1894) was a conservative art critic for the Parisian newspaper *Le Presse*.—Ed.]
3 [John Ruskin (1819–1900) described etching as physically "blundering" and "indolent," or morally questionable, in *Ariadne Florentina: Six Lectures on Wood and Metal Engraving* (New York: John Wiley and Sons, 1887), 149.—Ed.]
4 [Martin Schoen is the German engraver Martin Schongauer (1498?–1491).—Ed.]
5 [Hendrik Goltzius (1558–1617), Dutch engraver and painter. Cornelius Cort (1553–1578), Dutch engraver best known for his copies of Titian. Agostino Caracci (1557–1602), Italian reproductive engraver.—Ed.]
6 [Marc Antonio is Marcantonio Raimondi (1480?–1534?), an Italian engraver who copied the prints and even the monogram of Dürer without the artist's permission

(see reading 22) and the paintings of Raphael with the artist's blessing (see Alexandra M. Korey, "Creativity, Authenticity, and the Copy in Early Print Culture," in *Paper Museums: The Reproductive Print in Europe, 1500–1800*, ed. Rebecca Zorach and Elizabeth Rodini (Chicago: David and Alfred Smart Museum of Art, 2005), 31–50; and Lisa Pon, *Raphael, Dürer, and Marcantonio Raimondi: Copying and the Italian Renaissance Print* (New Haven, Conn.: Yale University Press, 2004). It seems likely that Haden is referring to Raimondi's early copies of Raphael in which the engraver more freely improvised in copying secondary passages. Raimondi also sought innovative ways to reproduce Raphael's tones. As seen in the engraving *The Judgment of Paris* (1510–20), he roughened the surface of the plate with pumice. The resulting shallow scratches and pits would later be filled with ink, presaging the mezzotint technique (www.metmuseum.org/collection/the-collection-online/search/337058).—Ed.]

Reading 25

Joseph Pennell

Etchers and Etching: Chapters in the History of the Art Together with Technical Explanations of Modern Artistic Methods (1925)

While the American printmaker Joseph Pennell (1857–1926) had little use for the emerging processes that rapidly came to dominate the printing industry in the late nineteenth century, he enthusiastically embraced lithography later in life, probably due to the influence of his good friend and fellow expatriate James Abbott McNeill Whistler. This poignant addition to the preface of Pennell's Etchers and Etching *addresses the far-reaching devastation of war that touches all aspects of life, including the arts. The popularity of vintage papers for etchings persisted well after World War I, with newly discovered caches periodically announced with fanfare by suppliers.*

Postscript to Preface

The publication of this book was stopped for four years owing to the war, and instead of being issued in London by Mr. Fisher Unwin first, it has been made by Messrs. Macmillan in New York. This however is but a small matter in comparison to what art in a big way has suffered. Many artists have given their lives; more have been ruined; and a few have found opportunities—subjects—in the horrors and miseries of the war. Galleries in Europe ceased to acquire the few contemporary works that were made, unless these were commanded by the state. Some of the galleries are reported even to be destroyed. Exhibitions, save for the raising of war funds, mostly ceased, especially in Europe. And even as peace dawns, art still flies away. The belief in some quarters that a new art, or a new inspiration, would come from the war has not been realized. No one who knew anything ever thought it would, save in the case of those who recorded the war.

From Joseph Pennell, *Etchers and Etching: Chapters in the History of the Art Together with Technical Explanations of Modern Artistic Methods*, 3rd ed. (New York: Macmillan, 1925), xi–xii.

But art will never die, it is everlasting, eternal; and though artists have suffered more than the members of any other profession, they will come into their own again.

Precious records have vanished. For a while in Europe even those etchers who had the opportunity to work, unless in the government service, could obtain neither copper nor tools nor acids to carry on with. Paper mills in Italy have been burned and bombed. Old paper has disappeared. Technical schools have closed. Dealers were unable to obtain prints. Collectors had no time to collect. That such a state of things should come to pass was incredible. Yet it happened in our day and generation.

Tradition in art, too, was in danger of being forgotten. It is with a view, then, of recording what I have seen and studied and experienced and practised, with a view of trying to carry on tradition and recording facts, I am glad to have had the volume written and ready to issue in the first year of the war—published now that I hope it is ended, now that I hope there may be no more war. And if the world really cared for art and literature and the arts of peace, there would be no more war or rumours of war. We have relapsed into vandalism and vulgarity. The world is now made up mostly of prigs and prohibitionists; they will venture on the suppression of art as they have ventured on the suppression of wine and song and brought about the unsexing of women.[1] But art will arise again, and laziness, hypocrisy, and sentiment, which crush and cumber the earth, again will be swept away. I shall not see the new earth, but it will come forth.

Note to Reading

1 [Prohibition was in effect at the time of this publication, as was the fashion for women's dropped-waist tubular dresses, creating a boyish silhouette.—Ed.]

Reading 26

W. Biscombe Gardner

Wood Engraving as Compared with Other Reproductive Arts, and Its Future as a Fine Art (1896)

The British illustrator W. Biscombe Gardner (1847–1919) fought tirelessly but fruitlessly against the relentless march of technology in the illustration industry, specifically objecting to "process" work that used photographic methods to transfer and fix images. Despite Gardner's advocacy of wood engraving as the truest reproductive technique for capturing tonal values and interpreting the idiosyncratic touch of the artist, images created by hand in boxwood were doomed. As Ivins caustically commented, "Having thus lost their general economic justification and stabilization most of the old techniques became languages used for their own sweet sakes rather than for their power and ability. They left the work and sweat of life and took their places at the vicar's tea table, where their accents and manners were of more account than what they thought and said."[1] *In the discussion following Gardner's presentation to the Society of Arts, of which this reading is a transcription, one gentleman observes that Gardner's engraved copy of a portrait of Lord Alfred Tennyson painted by George Frederick Watts (1817–1904), propitiously on display for the event, was a perfect example of the superiority of wood engraving in more accurately transforming heavily charged brushstrokes than was possible using a photographic technique. This work justifies Gardner's claim that reproductive printmakers, as innovative translators of paint, sculpture, or architecture into ink, put their unique stamp of originality on the printed product—an observation eventually endorsed by art historians.*[2]

We should note that Gardner is describing the booming illustration trade in England and Europe. Many of the pseudoluxury products that he condemns are today entering conservation studios and presenting their own set of complications, such as heavily coated or gilt-edged papers, the chine collé *process typical of* livres de luxe, *elaborate large-format* Prachtausgaben, *heavily engraved picture plates whose inks*

From W. Biscombe Gardner, "Wood Engraving as Compared with Other Reproductive Arts, and Its Future as a Fine Art," *Journal of the Society of Arts* 44, no. 2, 272 (June 5, 1896): 611–18.

are "well bronzed by time,"[3] *and stencil techniques including* pochoir *for the application of local colors and the natural gums used for glazing them.*

In the short paper I am about to read, I wish to place before you—first, the many advantages wood engraving possesses over all other reproductive arts; secondly, what I think ought to, and what I think will, be the future fields of this art; and thirdly, to follow up with a suggestion as to a way to arrive at that result. Whether my suggestion will be taken up I must leave the future to determine.

The great power wood engraving possesses beyond all other reproductive materials, I hope to prove, by bringing before your notice the variety of methods of handling the graver which has been resorted to in the production of the specimens that are on view to-night.

Wood engraving, we are told by some people, is dying out, or will be one of the past arts before many years, and process or mechanical work[1] by means of photography will take its place as a medium for illustrating books, magazines, and papers.

This is what we hear on all sides, and if we look at the illustrations at present produced in most of the publications, it undoubtedly does appear at the first glance as if there were something in this almost universal opinion. But what is this so-called popular opinion? I think it only comes from readers who confine themselves to the sensational side of literature, and not having their imaginations fully developed, want something beyond the letter-press to help them to realise what they are reading; *then* cheap art comes in to help the cheap publications, which are on the increase to such an extent that process illustration looks as if it were taking everything before it.

But by dipping a little deeper into the question we find a very different state of things. We see that wood engraving was never more alive as a fine art, or in a higher state of perfection than it is at the present period; and it is still capable, in the hands of right good earnest workers, of being lifted to a much higher position.

We may grant that wood engraving has been superseded for a time, for purposes of illustration, by process work, on account of the latter being so much more rapid, and also much—very much—cheaper. I will not say very much inferior, for it is unquestionably better than bad wood engraving. And here I may say, I do not mean essentially bad work, but work that has to be done in a rush, such as war subjects, which sometimes have to be engraved in a few hours.

Yet so long as the public are blind enough (blindness which makes one wonder, considering the art culture that is going on all over the country) to take bad or rather, as some put it, "soft and delicate" kinds of illustration, instead of good art, both delicate and beautiful, so long will some publishers give them bad rather than good work, charge the same for the publication, and pocket the difference, and why not, if the public will be so foolish? Every opportunity is afforded them to distinguish the good from the bad; but they, or the majority of them, judging from the publications that pay best, prefer the bad to the good. Then, what happens? So long as this blindness or apathy lasts so long will wood engraving suffer in this particular department of publication.

But we have a much brighter side to look at. Let us leave behind the lovers of cheapness and speak to those who are lovers of art for art's sake. To them I say that wood engraving, as a reproductive fine art, never can be touched or ever will be touched by any process yet invented. I am now speaking of all the mechanical processes at present being used for illustration. With regard to the higher or fine art processes I hope to show further on that wood engraving is far and away above any of them in its marvellous versatility of *technique*, which enables the engraver to translate not only the values, but the very individual touch of each artist, from whose picture he may be engraving.

All processes dependent upon photography are bound to go wrong in the rendering of values, since photography has not yet been brought to such a state of perfection as to master the difficulties of exact colour translation. In fact photography is utterly inadequate in the most simple wash drawings in black and white. For instance, a drawing executed in black, no matter what black, when Chinese white[2] is used mixed with the black, will come out, as a whole, entirely untrue as to values in the process block, because of the brown in the black and the blue in the white. The slightest difference in the admixture of the two colours will be entirely mistranslated, although the drawing may look perfectly right as to the values to the eye.

Even in a pure wash drawing, without the use of Chinese white, the paper drawn upon being more or less yellow, the high lights that are left are lower in tone in consequence.

Take any process, and if you have an intimate knowledge of the original painting or drawing from which the process plate has been done, you will see at once how very far short it falls of the original. All the spirit has departed from it—it is weak and insipid. Indeed, although photography has made great strides in late years, it has not yet been able to get over its main difficulty, values, and I must say that I do not see at present how it can get over it. No process plate that I have ever seen, either from a painting or a black and white wash drawing, has had a pure black, or a properly gradated white, which is as much to say, it cannot carry out the most simple rules of translation of values. But, I admit with pleasure, that a pen-and-ink drawing in pure black ink could hardly be reproduced in a better manner than by the best process. I say the best, as I only now and again see a very fine line reproduced thoroughly well by process. As a rule it is either broken or thicker than the original, and thus loses its value. But beyond this, I most emphatically assert that nothing does or ever can compare with the work done through the sensitive medium of the eye and hand of man. In fact, I consider wood engraving far better than any, or all the reproductive arts, as it stands quite alone in its wonderful adaptability; for any variety of texture one likes can be produced on the boxwood block. This cannot be said for either etching, mezzotint, steel, or copper, each having its own methods, great as the masters have been who have worked upon one or the other of these materials; and as far as I have seen, they never have departed from the original mode of treatment, peculiar to the material they worked upon. Thus, they have necessarily missed the individuality of the painter they translated. The reason, of course, being that the material will not lend itself to variety of treatment.

Now, with regard to wood engraving on boxwood, any texture you like can be engraved upon this material, it is so adaptable. But it requires an artistic engraver, not a mechanic, to arrive at anything like good results. Happily, at the present time, we have plenty of right good artistic engravers, not only here in London, but all over the world.

The crowning advantage enjoyed by wood engraving, through which it obtains its immense superiority over all other methods, is that the engraver is enabled to work in both black and white line. This, in the hands of an artistic engraver, one who knows his work, and, beyond that, loves it, is a most valuable power.

Nothing is out of the range of imitation possible to wood engraving. The differences of textures of flesh, silk, satin, cloth, wool, steel, glass, the grain of wood, marble, weather-worn stone, furs and skins of animals, atmospheric effects, foliage of all kinds; all these it can represent, and beyond everything, it can render the differences between oil and water colour, and can accurately transcribe the old master's work with all its cracks and blemishes from damp and shrinkage. It has been said by some that the old masters cannot be rendered in this way except by photography.

[. . .]

All things, I have said, are possible on wood; and this is secured by variation of line, admixture of dots, or round or lozenge-shaped spots or strokes of the graver, cut in different directions across one another. In fact, one can hardly lay down hard-and-fast rules with regard to the exact methods to be used. All depends upon the subject to be engraved and the feeling which it engenders. This, again, depends upon the thoroughly artistic sensibility and training of the engraver. He must first be an artist, in feeling at least, well grounded in his anatomy, daily using his pencil or brush in drawing or painting; in fact, an art student. This is most essential—that is to say, if he wishes to do good work. Then he must be well up in all the branches of art. He must take a pleasure in his work and turn out his best. Such an engraver will not feel inclined to do, what we will call "manufactured engravings," but rather will feel a fresh delight in every subject that comes before him. No matter who the master may be from whose work he may be engraving, each one is a new study or problem. He has to reflect how he can best render that particular painter's feeling and individuality.

[. . .]

That these specimens somewhat demonstrate the wonderful versatility of wood engraving, is to me absolutely conclusive; while as regards the means employed in producing any of the blocks here introduced, engravers, or lovers of engraving, ought to see at a glance how they are done, and what advantages wood possesses over other methods.

But will all wood engravers take the trouble to arrive at these results? I sometimes doubt it. So many of them, I am afraid, think too much of the money to be got out of their work, and will not sacrifice the time or give the thought requisite to arrive at what they know to be their best. "That is good enough," is sometimes, if not often, the quiet cry with some. They, perhaps, have not sufficient feeling or love for their art, grand as it is. "Publishers," they say, "will not pay more than a certain price, so why should I sacrifice myself for their sakes?" Hence the stereotyped cry, "That's good enough; it will pass the art editor." But publishers will pay for good work when they want it; I only wish, for the good of wood engravers, they wanted it a little more often.

Something has latterly been said about the public not appreciating good art in the publications—that they are indifferent; and on that account we are also told that the publishers are not to blame in the matter of the bad art produced in a number of the current magazines and papers.

Perhaps. But, as the artist or engraver is powerless, if the publisher will not take of his best, we must look to the public for the encouragement so much required. What the public insist upon having, the publisher will be bound to give, if he wishes to prosper. In fact, you see the very best plates in some publications are either etchings, wood engravings, or photogravure. This is done simply to keep well with the public, which shows that they like a good thing when they can get it.

A very large section of the public, I am glad to hear, are crying out for better illustrations, and are longing for more knowledge of illustrative art. Many publishers see that want, and, with a few exceptions, they take undue advantage of it. They give anything that will pass for the week or month, and, I have no doubt, judge, from the rise or fall of their publications, whether to give better art or leave bad alone.

Now, let us take each of the reproductive fine arts in turn, look into their different methods of work, see the results, and compare these with what I have previously shown that wood can produce. How very far short they are in reproducing the *technique* of the painter! To begin with, I will take etching, as it is the first and highest of those methods, though not one of the true reproductive arts.

[. . .]

[. . .] I look upon etching as a painter's art to delineate the feeling of the artist himself in that particular medium; in fact, an original art. Decidedly, in my way of thinking, it never was, and never could have been, thought of as simply an imitative art. The more you look into it, the more you recognise that every line conveys the feeling of the painter-etcher, not of the copyist. And yet, at the present period, etching is being used regularly as a reproductive art, and the public think they cannot have anything much better than a proof-etching[3] of any well known picture; they do not see that the etcher can only convey to the eye a very general idea of the original. That is to say, this art gives you the subject—the true values of the picture—but what is wanting is the individuality or *technique* of the painter.

In a good etching from a picture we have the etcher, not the painter. We have a beautiful art, and yet, if I may be allowed to say so, one which misses the actual art in this particular department. As to the question, who is answerable for this misapplication of an art that is so personally interwoven with the fibre of the creative artist and not the copyist, I leave others to answer.

Next, take mezzotint. This is a delicately beautiful art, but I think what I have said about etching will apply in some ways to this also. It misses the one essential of a reproductive art—the power of preserving the individuality of the painter. We feel that want intensely in looking at the result. His touch is not there. It is the mezzotinter's, and his alone, that we find in the engraving.

Then again, take steel and copper, I cite them together as they are pure and simple line and dot work. Everything depends upon the beauty of the line, the true translation of the subject in the outline and values. Beyond that aim they fail utterly. As I said before the material on which they work does not lend itself to the handling or *technique* of the painter. To understand what I mean by this take a picture full of impasto and another very smooth and highly finished. Both will, when engraved on steel or copper, come out equal in value. The impasto has gone, but the finish remains.

With regard to photogravure, this process does give you somewhat the handling and touch of the artist; but for the very reason that its foundation is due to photography it does not come out quite true in its values.

I now approach the subject of the future of wood engraving and the likelihood of its permanence.

This I must say greatly depends upon the desire on the part of the public, or perhaps, I should say, to their gradually being led up to a stronger wish for a better class of illustration or reproduction than they get at the present time. In fact they ought now to get a far higher class of work given them considering the enormous profits made out of nearly every class of publication that is issued. But as it is they get cartloads of rubbish, and are as much as told that it is art and good enough for them.

I sometimes think that the British public must be greatly wanting in feeling for what is good in art, or otherwise they would not put up with so much that is given them. Anyhow, the question here comes to the front, How long will they go on putting up with this bad work? when they could have better if they would only strike for it, or insist strongly enough upon having it. If they would only do this, they would get the higher and more beautiful work, for the publisher would be bound to let them have their heart's desire. This would mean a higher class of work all round, and wood engraving would prosper accordingly.

But this is not all that wood engraving is destined to accomplish. It will be in the future, and not very distant future either, if worked by the engravers with their whole heart and strength, the leading reproductive fine art. I know there will be a big fight to go through, but what matters, I feel we must win.

I have only very faintly shown you what is being done at the present, in methods of handling, and with other methods added by earnest students, there is no end to what could be done with the material. It must come to the front. I do not say this idly, or without a knowledge of what I see before me, as to its capabilities beyond other reproductive arts; but all depends upon the earnestness of the engravers, they must throw themselves heart and soul into their work, and be ready to make almost any sacrifice for it.

Then what encouragement there is in the fact, that a strong taste for the beauty of the art is growing, and manifesting itself in the eager way in which artists' proofs are being bought, not only in this country, but in America, France, and, I think, Germany also.

This taste will naturally spread, it cannot help doing so; and I ask all wood engravers to work for this grand opening for the output of their work. I would advise the purchase of copyrights of good pictures, selecting subjects that would be likely to take with the public. Engrave and publish them in the same way that all other reproductive plates are published, that is to say, a limited number of artists' proofs with prints after letters at half the price to follow.

This would place upon the market a good and comparatively cheap art for the general public to indulge in, as the wood engraver could sell his artists' proofs at a lower rate than that at which the other fine art plates are published, simply because the printing would cost him less; and I feel confident that the public would largely respond.

If the engraver could not see his way to purchase the copyright, I would suggest a partnership between the artist and engraver. The engraver to put his work against the value of the artists' copyright, each to contribute an equal share towards the outlay for printing, paper, and other expenses necessary to the publication of a plate, and divide the profits. This would work to the benefit of both.

The wood engraver in this way would free himself to some extent, and show what could be done, independently of magazines or papers. Not that he would necessarily cut himself adrift from these sources of work, but would be helping himself onward to a higher line, in which he would not be so much ground down to help to build up the fortunes of others, as is too often the case at the present time; but thank goodness all publishers are not alike in that respect.

It has been said, and with truth, that the individual workers must die out—very true, and that fresh blood will be wanted to take the place of those defunct members of the art. I believe one way to accomplish this would be through the action of the Royal Academy of Arts, who are our pioneers in most of the fine arts. They should recognise wood engraving as a fine art, as well as steel and copper. They should elect members of this art as associates, as well as open their schools to students of the same, under the same rules as those which apply to students of painting, sculpture, and architecture, whose ultimate career might be that of the wood engraver. These students would then be able to have the good training both in drawing and painting, which knowledge is most essential to anyone wishing to be thoroughly able to reproduce engravings from the paintings of others. For the eye must be trained in feeling and colour, as well as in drawing, if the engraver is to render with any approach to truth the relative values of any painting he may have to reproduce.

One sometimes wonders what becomes of all the Royal Academy students, of whom a very small proportion, in the end, come to the front. What happens to the remainder, who intended at least to try and make a great name? Perhaps it was that they were wanting in creative power. Then if not painters, sculptors, or architects, they would, perhaps, make first-rate engravers. And why? Simply because of the splendid training they would have gone through in the antique, life, and painting schools.

Just think once more, if they only had had the chance of studying one or other of the reproductive arts at the same time that they were studying to be painters or sculptors, what a splendid band of workers in the reproductive arts we should have to go on with in the future. The cry of decay in this or that special art would not be heard; it would fade away to be repeated no more. These arts would revive in every way, and there would be a splendid future for them all.

I feel most strongly on this point, that it is to the members of the Royal Academy of Arts we must look to for some of the help that is required to prolong the life of this beautiful art. It ought not to be left to the dogged determination of this or that individual, to force his art to the front without aid or encouragement.

These reforms would, I feel sure, open up one way to give the impetus to wood engraving that it now so much requires. The art would be more widely recognised as a fine art; a more general market would be opened up for the collections of fine art wood engraving plates; and the public would be enabled to get good art at a price they could readily afford.

It is principally to this fine art outlet that I look forward to as the great and future field for wood engraving.

Notes to Introduction

1 William M. Ivins Jr., *How Prints Look* (New York: Metropolitan Museum of Art, 1943), 145.

2 I am grateful to John Krill for directing my attention to Timothy Clayton, *The English Print, 1688–1802* (New Haven, Conn.: Yale University Press, 1997), xi. And see Stephen Bann, *Parallel Lines: Printmakers, Painters and Photographers in Nineteenth-Century France* (New Haven, Conn.: Yale University Press, 2001); Anthony Dyson, *Pictures to Print: The Nineteenth Century Engraving Trade* (London: Farrand Press, 1984); Rebecca Zorach and Elizabeth Rodini, eds., *Paper Museums: The Reproductive Print in Europe, 1500–1800* (Chicago: David and Alfred Smart Museum of Art, 2005).

3 George Cumberland, *An Essay on the Utility of Collecting the Best Works of the Ancient Engravers of the Italian School* (London: W. Nicol, 1827), 23. See reading 90.

Notes to Reading

1 [In the proceedings of the House of Commons in 1899, the definition of process work was proposed as follows: "The expression 'Process Work' shall mean any work such as photogravure or half-tone block, of which the design or subject is executed on the printing surface, whether in whole or in part, by photographic, electric, or mechanical means, and shall include any such work executed on any material whence by any process, whether immediately or mediately, prints may be taken and multiplied, and any prints so taken, but shall not include an engraving nor a photograph." *Report from the Select Committee of the House of Lords on the Copyright Bill [H.L.] and the Copyright (Artistic) Bill [H.L.]; together with the Proceedings of the Committee, Minutes of Evidence and Appendix* (London: Wyman and Sons, Ltd, August 9, 1899), 242.—Ed.]

2 ["Chinese white" = zinc white.—Ed.]

3 [By "proof-etching," Gardner is referring to the limited number of prints available for purchase prior to the addition of "letters," which could include title, date, artist, printer, and copyright and publisher information.—Ed.]

Reading 27

Hans W. Singer and William Strang

Etching, Engraving and the Other Methods of Printing Pictures (1897)

This reading describes the limitations of photographic or photogravure reproductions of paintings. Any suggestion that they possess even a spark of originality is extinguished by Hans W. Singer (1867–1959), head of the Print Department of the Dresden Picture Gallery and author of a catalogue raisonné of Max Klinger's (see reading 14) strangely obsessive etchings, and his coauthor, William Strang (1859–1921), a Scottish painter and printmaker. As a protest against the inclusion in its exhibitions of photomechanical reproductions of paintings, Strang resigned from the Royal Society of Painter-Etchers, which was founded by Seymour Haden (reading 24).

[. . .] To call a photo-mechanical picture producer a fine art publisher is applying a misnomer. Art of any kind is creation, and photography of any kind is description. The writer of a gallery catalogue describes a painting verbally, the photographer does it pictorially; neither of them is an artist. In fact, the catalogue writer is even the nearer the artist of the two, for he at least performs an act of mental creation in the putting together of words. The photographer's business is merely to remove or reduce all obstructions in the way of the process, nothing more.

Anybody who claims that a photograph or a photogravure gives him any artistic pleasure, is his own dupe. It may help to recall the pleasure that he experienced once upon a time in face of the original painting, and thus cause him to rehearse it mentally, but that is all. By virtue of our powers of association it may even enable him to experience such a mental pleasure if the photograph be after a picture which he has never seen in the original. But the photograph itself is only a cue to enjoyment, not a source of it. It ranks in this respect with the type of a common penny paper,[1] that happens to reprint a fine poem.

From Hans W. Singer and William Strang, *Etching, Engraving and the Other Methods of Printing Pictures* (London: Kegan Paul, Trench, Trübner and Co., 1897), 186–87.

As soon as the public will be persuaded to eschew the pretentious photogravures and befriend itself with true graphic art, even if it be only the simple, original woodcut, we may hail an improvement of taste. For all photo-mechanical processes should be banished from our walls or portfolios, where they strut with borrowed feathers, and relegated to that province where they belong—the purpose of study and the illustration of books.

Note to Reading

1 [A "common penny paper" was a sensational tabloid.—Ed.]

Reading 28

Arthur Hayden

Chats on Old Prints (1906)

A prolific commentator on collectibles, Arthur Hayden (1868–1946) wrote on a diversity of odds and ends, such as old silver, old furniture, old English china, cottage and farmhouse furniture, Sheffield plates, and old clocks and watches. His conversational Chats *reflects the prevailing tastes of the educated classes; hence it is not surprising that Hayden bemoans the passing of "traditional" steel and wood engraving as championed by Gardner (reading 26). Here Hayden introduces the inescapable issue of print connoisseurship in an oblique reference to the characteristic cut lines of engraving, a forlorn reflection that the wood engraver "with his sandbag, his boxwood block, and his graver has gone to that most permanent of all furrows, the grave itself, which Time has cut enduringly." He seems to be paraphrasing an earlier quotation from Seymour Haden's archrival, John Ruskin, in* Ariadne Florentina: *"That engraving means, primarily, making a permanent cut or furrow in something. The central syllable of the word has become a sorrowful one meaning the most permanent of furrows."*[1]

Even a leading art critic like Ruskin had difficulty finding words to describe the carved grooves of engraving, which he extolled. Enthusiasts assigned metaphorical adjectives such as "harmonious" and "lyrical" to the crisp and controlled furrows of engravings, whereas detractors described them as "mechanical" and "contrived." As was the case with drawing media, printed lines were often compared to poems, music, or dance, a practice that infuriated Ivins (reading 23), since none of these terms communicates the actual appearance of the lines.

Preface

The study of prints begins in the nursery in the contemplation of toy-books with pictures, unaccompanied by any uneasy qualms as to "states."

From Arthur Hayden, *Chats on Old Prints* (New York: Frederick A. Stokes Company, 1906), 7–8.

It is a matter open to question whether the younger units of the present generation, who have grown up in the environment of the pictorial magazine and journal and the thousand-and-one forms of illustration by modern process, quite realise the departure from the older methods of engraving. Finely wrought steel engraving is a lost art, and the wood engraver of middle-Victorian days with his sandbag, his boxwood block, and his graver has gone to that most permanent of all furrows, the grave itself, which Time has cut enduringly.

In order therefore to approach the study of prints, the beginner has to look back to accustom his eye to engraver's work of a time immediately preceding our own. To-day photo-mechanical processes have taken the place of the graver and the skilful hand and trained eye of the craftsman behind it. In the course of the volume it will be shown what sins of commission and omission have almost led to the extinction of both steel and wood engraving, and it may be here briefly stated that time and the cost of production are the greatest factors governing this result.

It is not my intention to cast a stone at modern process work, which has done so much to extend the knowledge of the fine arts. The engraver's interpretation of a picture or his rendering of a design was not always too faithful, and by the aid of the camera, line-drawing in particular can be rendered line for line in facsimile. It would be as logical to quarrel with Caxton's printing press[1] because it supplanted the manuscripts of ecclesiastical writers with their wealth of illumination.

Note to Introduction

1 John Ruskin, *Ariadne Florentina: Six Lectures on Wood and Metal Engraving* (New York: John Wiley and Sons, 1887), 1.

Note to Reading

1 [William Caxton (1415(22)?–1492?) established the first printing press in England in 1475; Johannes Gutenberg (1398?–1468?) established his press in Mainz, Germany, in 1440. Handwritten versions of printed books are known; see Curt F. Bühler, *The Fifteenth Century Book* (Philadelphia: University of Pennsylvania Press, 1960), 15–39.—Ed.]

Reading 29

Roy Vernon Sowers and Pauline Sowers

Enjoy Your Museum II: Prints (1934)

Although they share Ivins's passion and opinions, Roy Vernon Sowers (1897–1970) and Pauline Sowers have a far different approach to the aesthetic appreciation of prints, as evidenced in this reading describing a Rembrandt etching and a Dürer engraving. The Sowerses come close to convincing their readers that color and sound really do exist in black and white, an old-fashioned epithet favored by print aficionados. The perception of various colors in black and white was perpetuated in printing manuals. The English caricaturist Henry Alken (1785–1851), for example, gave directions for adjusting the "color" of the lines in his soft-ground etchings, even though they were invariably hand-colored in brilliant watercolor washes.[1] *For the conservator, it is crucial to differentiate between color prints and colored prints (or a combination of the two techniques) when designing treatments or establishing exhibition restrictions.*

III

Aesthetic Appreciation

[. . .]

Let us say that we are looking at an etched landscape by Rembrandt, that master of the masters of etching. What do we see in it? A clear and frank representation of a cottage, a stream, some trees; in the distance, the spires of Amsterdam. But isn't there more to it? Is there not some hint of other colors than black and white? That crinkly, brittle edge to the trees, does it not shine with the golden yellow of leaves in sunlight? The sturdy and sparing strokes that go to make the trunks give us the feeling of the roughness of the bark, the delicate, sensitive touches

From Roy Vernon Sowers and Pauline Sowers, *Enjoy Your Museum II: Prints* (Pasadena, Calif.: ESTO Publishing Company, 1934).

on the dimly seen city are vaporous and blue with distance, while the cottage stands before us in all its whites and browns. This it is to see color in a black and white, and an experienced eye can see it even in the worst of prints and always in the great ones.

[. . .]

Now let us look at an engraving by Durer, one of his lovely Madonnas. Here again we see color; light on rich damask and satin, or the humbler but no less beautiful colors of wool, dominant green in the luscious grass that grows at the feet of the gracious figure, the red brown of the tree trunk, and in the distance, so carefully and thoughtfully drawn, one sees the gray of a city's walls and the red roofs of the houses so snugly guarded. Yes, there is plenty of color in Durer's engravings, but it is different from that of Rembrandt's etchings. It is more forceful, more a statement of act, less poetic.

[. . .]

Look well at these two prints; study them and their kind and you will begin to realize what every print lover knows, that there is something in a black and white that no painting can ever give. He who has not stood before a "singing" impression of an etching by Rembrandt with feelings compounded of awe and a desire to dance, has missed one of life's major aesthetic experiences.

Note to Introduction

1 Henry Alken, *The Art and Practice of Etching* (London: S. and J. Fuller, 1849), 59.

Reading 30

William M. Ivins Jr.

How Prints Look: Photographs with a Commentary (1943)

William M. Ivins Jr. had little patience for "experts" who use irrelevant flowery language when describing a printed impression, stating that a "second-hand verbal description" of obscure "describable marks" would not lead to a meaningful assessment of quality. Ivins observed that print connoisseurship is impossible in an absence of technical knowledge pertaining to ink formulations, printing papers, and the mechanics of print production, neatly forecasting trends in technical art history. In the case of early Italian prints, this lack was subsequently addressed by David Landau and Peter Parshall in The Renaissance Print, 1470–1550.[1] *Their research into contemporary printing practices formed the foundation for further technical studies by conservators, who used specialized imaging techniques to document sequence and impression quality.*[2] *For example, after hundreds of hours spent scrutinizing prints by Andrea Mantegna (1431?–1506), the paper conservator Shelley Fletcher (1943–) pulled the rug out from under even the most confident connoisseur of early engravings, suggesting that our "definition of quality based on 'impression' be flexible."*[3] *Similar studies, also with confounding results, have been carried out on the prints of Antonio del Pollaiuolo (1431–1498)*[4] *and Albrecht Dürer (1471–1528).*[5] *Ivins also made an offhand reference to the use of watermarks in print connoisseurship. In this respect, too, conservators have contributed to print connoisseurship studies by reassessing the information provided in watermarks and their interpretation.*[6] *Indeed, even papers lacking watermarks can yield a great deal of evidence through their structural and appearance properties, as discussed in Part IV.*

From William M. Ivins Jr., *How Prints Look: Photographs with a Commentary* (New York: Metropolitan Museum of Art, 1943), 150–52.

Of Quality of Impression

"Quality," unfortunately, has nothing to do with the major artistic virtues such as imagination, composition, or draughtsmanship, but is a matter solely of the printer's skill and the condition of his printing surface. Whether or not a particular impression is a good or a fine impression is a matter only to be decided by direct comparison with other impressions of the same print or by memory of such other impressions. Only a very small number of people have dependable memories of this kind of thing, and no one has an infallible memory. If other impressions are not available then one has to take what he can get—and if anyone likes what he can get, it is a good impression for him. What is fine quality for one print has no bearing on what is fine quality for another by another artist. Any good artist has a quality of his own that is just as idiosyncratic as his drawing. When two artists have very nearly the same quality one or the other (and probably each!) is of very little artistic importance.

It is important to remember that as yet no one seems to know how the early engravers and etchers printed their plates, in what kind of a press, or under what pressure. The early impressions of early engravings, especially the large Italian ones, are often very irregularly printed, as though the plates had been very carelessly wiped after being inked in soft ink or oil paint that was carelessly rubbed into the lines, and as though the pressure in printing had not been the same all over the plate. While some lines have a lot of ink in them, others have little or none. The early impressions not infrequently seem to have slipped in passing through the press or its equivalent, so that some of the lines are smeared or doubled. There is reason to believe that the hard, even, late impressions of many of the early plates were printed long after the plates were made and by a press or other device of a kind not used in printing the early soft impressions.

There is rarely any need for having a modern print in poor condition. Fifteenth-century prints in fine condition are apt to be also rather late and poor impressions. Really fine early impressions of them are apt to be in rather poor condition. Almost the rarest things in the world are very old prints that are fine in both quality and condition.

Too great an emphasis on condition or quality of impression is apt to wind up by turning prints into precious stones or postage stamps instead of the drawings and pictures they actually are and removes them into some category quite other than that of works of imaginative art. It is said that according to modern fashionable standards many of the prints which Dürer proudly gave to his great friend and patron the King of Denmark are mediocre impressions.[1] A print which is only worth having in a very fine impression is doubtfully worth having in any impression whatever. The poorest impression of a really great print, no matter how shabby it is, is still a great work of art.

It is useful to remember that print makers, print collectors, and print users of the past never dreamed of us of today or of our tastes and prejudices in regard to quality and condition. The feeling that prints should not be colored, like the feeling that they are precious things to be kept in mats and handled as little as possible, is comparatively modern. Most of the very great old collections were pasted rather carelessly in scrapbooks, and big prints were kept loose in portfolios or drawers. Just as most primitive woodcuts were daubed up with crude color, so early Italian engravings and etchings were frequently either colored or heightened in effect by more or less discreet use of pen or brush. For a long time prints were regarded simply as pictures of which there were many duplicates (much as photographs are today), and not as things of a special kind with special values and special rituals. They were bought by artists and craftsmen for use and were part of the customary working equipment of the studio or shop. On occasion they were pricked all over or squared off as the easiest way of taking off their designs for use.

Of States and Watermarks

The beginner should make most careful differentiation between state, condition, and quality. They are quite different and in many instances are independent variables. Every change in the work on a printing surface as shown by comparison of the prints pulled from it indicates a change in state. These changes are sometimes made by the artist in furtherance of his pictorial idea, sometimes they amount to no more than a change in the punctuation of the name or the address of the publisher or dealer, sometimes they are made by any skilled hand that is available to cover up or disguise the wear that is inevitably undergone by the printing surface in the course of its use. Mere wear, by itself, however, even though it result in the complete disappearance of lines and the altering of values, does not make a difference in state. Condition has no relationship to either state or quality, but is a matter solely of the physical condition of the piece of paper on which the print was printed. Prints frequently occur that are late in state and horrid in quality but in superb condition.

At the very beginning of his study of prints the student is faced by one of the basic problems in that branch of philosophy known as epistemology. Most of the vast accumulation of learning about states, watermarks, and so forth, arises from the understandable but fruitless attempt to find describable marks such that knowledge of them through second-hand verbal description will take the place of first-hand sensuous recognition of quality of impression. For many print collectors these marks play the role played by the wine label for the man who tells whether a wine is good or bad by reading in a book and not by tasting in a glass. It is important always to remember that the condition, state, or watermark of a print carries no guaranty of its being a good impression, let alone a fine one. Most of the talk about quality of impression goes back not to first-hand judgment but to memory of verbal representations made in view of sale—*i.e.*, "mere campaign talk," which should be recognized as such. It is worth noting how few statements about quality are to be found on receipted bills.

Notes to Introduction

1 David Landau and Peter Parshall, *The Renaissance Print, 1470–1550* (New Haven, Conn.: Yale University Press, 1994).

2 Digital imaging has resulted in the ability to compare details of printed impressions simultaneously, for example, of Rembrandt's *Three Crosses*; see http//82nd-and-fifth.metmuseum.org/rembrandt-rembrandt-van-rijn-dutch-christ-crucified-between-the-two-thieves-the-three-crosses-41.1.

3 Shelley Fletcher, "Qualities of Impressions in Mantegna's Engravings," in *Dear Print Fan*, ed. Craigen Bowen, Susan Dackerman, and Elizabeth Mansfield (Cambridge, Mass.: Harvard University Art Museums, 2001), 123. See also Shelley Fletcher, "A Closer Look at Mantegna's Prints," *Print Quarterly* 18, no. 1 (March 2001): 3–40.

4 See Moyna Stanton, *Discover the Battle of the Nudes*, www.clevelandart.org/exhibcef/battle/html/4286621.html.

5 Angela Campbell, "A Fine Line: Microscopic Plate Wear in Albrecht Dürer's Meisterstiche Impressions," paper presented at the Metropolitan Museum of Art, Fellows Colloquia, April 2010; Angela Campbell and Andrew Raftery, "Remaking Dürer: Investigating the Master Engravings by Masterful Engraving," *Art in Print* 2, no. 4 (November–December 2012).

6 Nancy Ash and Shelley Fletcher, *Watermarks in Rembrandt's Prints* (Washington, D.C.: National Gallery of Art, 1998).

Note to Reading

1 [Albrecht Dürer (1471–1528), the German painter and printmaker, recorded in July 1521, "I presented to the King of Denmark the best pieces of my prints, they are worth five florins." Roger Fry, ed., *Records of Journeys to Venice and the Low Countries by Albrecht Dürer* (Boston: Merrymount Press, 1913), 97. These prints are now in the Statens Museum in Copenhagen.—Ed.]

Reading 31

Rudolph Ackermann

Some Account of the Art of Lithography, or Process for Taking Fac-Simile Impressions of Drawings from Stone (1817)

The Repository of Arts, Literature, Fashions, Manufactures, &c. was a print shop established in 1795 on the Strand in London and a popular and influential journal (1809–29), both of which were established by Rudolph Ackermann (1764–1834), renowned print dealer, publisher, and shameless self-promoter.[1] In this reading, Ackermann distracts his readers with Quatremère de Quincy's discourse on the origins of intaglio and relief printmaking to disingenuously imply that lithography has an ancient and thus distinguished pedigree and to optimistically speculate on its potential. Lithography rapidly gained traction in France, Germany, and England, where it was energetically spurred on by Ackermann. Despite Ackermann's avowal that lithographic prints were not "the copy of a design, but the design itself" and "the original work of the draftsman," controversy soon arose about the technique's fine art status, probably due to its swift commercialization and overnight ubiquity in tabloid illustration. For Walter Benjamin (reading 34) it was the contemporaneous nature of lithography—the very speed with which lithography "put its products on the market, not only in large numbers as hitherto, but in daily changing forms"—that denied it any claim to originality.[2]

By the mid-nineteenth century the artistic reputation of lithographs was in extremis. An American reporter at the 1854 New-York Exhibition observed that while lithography "is intrinsically a most beautiful art, and one in which artistic delicacy of touch may be made to tell most effectively," it had sunk, especially in America, to new depths in the pursuit of profit, such that "taste and decency are so often outraged

---

From Rudolph Ackermann, "Some Account of the Art of Lithography, or Process for Taking Fac-Simile Impressions of Drawings from Stone," *Repository of Arts, Literature, Fashions, Manufactures, &c.* 3, no. 13 (January 1, 1817): 222–25.

by forlorn caricatures, flagrant daubs, ghostly portraits, bald and blotched maps, city views through some monster bird's eye, and sooty peeps into fogland scenery. . . . There is no effective restraint on the most desperate competition, hence style is sacrificed to cheapness until a depth of artistic turpitude is reached, too profound to tempt purchasers even among the most benighted Calibans."[3]

It should be noted that these once-disdained products of consumerism exist by the thousands in historical collections and routinely come under the care of conservators, who must be alert to the challenges of their brittle wood pulp papers and unpredictable printing inks.

The lithographic art resembles that of engraving only in the property of indefinitely multiplying the same design. Though it does not seem capable of ever rivaling engraving upon copper in the charm and harmony of the strokes, and in the gradation and shades of the chiaroscuro, it has, however, this advantage over it—of not being the copy of a design, but the design itself; of being the original work of the draftsman repeated in every impression that is taken off.

Aloysius Senefelder,[1] an indifferent singer at the theatre of Münich, was the first that noticed the property possessed by calcareous stones of retaining marks traced upon them with a greasy ink, and transmitting fac-similes of them to paper applied by strong pressure to their surface. He discovered, moreover, that it was possible to repeat the same effect by damping the stone, and charging the same marks afresh with ink. In 1800 he obtained of the King of Bavaria an exclusive privilege for the practice of his process for thirteen years; and in conjunction with Baron Aretin, he formed a lithographic establishment at Münich, where music and collections of copies of different kinds are still printed. In the sequel, several similar establishments sprung up in the Bavarian capital; but none of them contributed so essentially to the progress of the lithographic art as that founded by Baron Aretin and M. Manlich, whence issued the beautiful collection of copies from tile productions of the great masters, that adorn the cabinet of his Bavarian majesty.

Notwithstanding the success that has attended the cultivation of this art at Münich, it is but little known and practised in any other part of Europe. The failure of the different attempts made some years since to introduce it into this country, has hitherto operated as an effectual check to new experiments. In France, the Count de Lasteyrie,[2] convinced of the advantages of this process, paid several visits to Münich, and endeavoured to form a lithographic establishment at Paris. He has even composed a treatise, containing all the details of this art; but neither this work, nor the result of his essays, has ever been made public: so that artists would still be in the dark on this subject had not M. Engelmann,[3] of Mühlhausen, after practising the art at one of the extremities of France, surmounted various difficulties to introduce it into the capital.

Our readers will not expect a full and particular description of all the details of the lithographic process; but we shall endeavour to give them such a general idea of it as all may comprehend.

The effects produced by a stroke made upon stone with a fatty or resinous matter, are the very simple results of affinities, the applications of which had not been previously remarked.

1. It is certain that a mark made with a crayon, or an oily ink, adheres so strongly to the stone, that mechanical means must be employed to remove it.

2. All the parts of the stone not covered with an oily coat, receive, retain, and absorb water.

3. If upon the stone thus prepared there be laid a coat of greasy and coloured matter, it will adhere only to the marks made with the oily ink, and be repelled on the other hand by the wetted parts.

In short, the lithographic process depends upon this circumstance, that the stone damped with water refuses the ink, and that the same stone, when greased, repels water and takes the ink. If, then, a sheet of paper be laid and pressed upon the stone, the oily, resinous, and coloured strokes alone will be transferred to the paper, and produce on it a counter-proof of what they represented upon the stone. Fac-similes may likewise be obtained, looking the same way as the original, by transposing upon the stone a drawing made upon prepared paper. We may hence conclude, that certain lithographic processes are totally different from those of engraving; and as they depend on affinities and repulsions produced by substances of different natures, it is possible, by varying them, we may obtain effects which we are not yet aware of.

All stones susceptible of taking in a greasy substance, and of imbibing water with facility, are suitable for lithography, provided they be compact, capable of receiving a fine polish, and of a clear and uniform colour. All these qualities are found in certain calcareous stones furnished in abundance by the quarries of Solenhofen, near Pappenheim, in Bavaria. They consist of a nearly pure carbonate of lime, and are met with in various parts of England. When the stone is prepared and polished, the artist may, without any further preparation, sketch his design in any manner whatever, and finish it either with the crayon,[4] the pen, or the pencil—the grain of the stone being smoother and finer than the finest and best-strained drawing-paper for the two latter.

The learned Quatremère de Quincy,[5] in a paper lately published on the subject of this art, has thrown out some ingenious observations on the probability that some such process was known to the ancients.

"It has always been matter of astonishment," says he, "that the ancients, who executed a great number of works cut out, or in relief, upon wood, metals, and hard stones, and having of course taken impressions of all the objects thus engraved, should not have found out either the art of printing or that of line-engraving. We shall not, however, be so much surprised at this, if we consider that necessity is the primary cause of almost all inventions. The ancients wrote and read much less than the moderns. Their way of life, their political occupations, public exercises, games, and spectacles, kept the men almost always from home. Neither were there the same communications of commerce, taste, and science, between all parts of the ancient world as at present exist. The multiplication of writings, by means of manuscript copies, was sufficient for all their wants.

"The state of things was very different in Europe at the time of the revival of the arts and sciences. When the zeal of the learned recovered the writings of the ancients, it was not merely a few scholars, but whole nations who wished to benefit by their discoveries. Civil and religious studies had produced a prodigious multiplication of books and readers. An economical method of copying could not fail to result from the extraordinary demand for copyists; and as the price of the latter kept continually rising, necessity caused that to be discovered, or, in other words, to be observed, which had hitherto escaped observation.

"No sooner were books multiplied by means of printing, than it became necessary also to multiply the images which illustrated them, and were formerly produced only by the slow and expensive processes of painting and drawing by hand; and then it was that Mazzo Finiguerra[6] applied to this purpose the method which he employed for obtaining impressions of his works. Hence arose copper-plate engraving.

"I have always been of the opinion that the ancients employed some similar process, especially at Rome, when the taste of amateurs had called for the multiplication of images in libraries and of family portraits, in the collections that were formed of them. Atticus[7] had already formed a considerable collection of portraits in a volume, when Varro's[8] iconographic collection of illustrious persons amounted to seven hundred. But are we to suppose, that what was termed Varro's invention—an invention which Pliny so highly extols—was confined to the collecting of portraits, coloured or uncoloured? Atticus, we see, was a collector before him. Augmentation of number is not an invention. Can it be said to have consisted in this, that Varro, instead of possessing a single set, multiplied his portraits by copies? Neither would this be any invention, for in all ages works in every branch of the arts had been multiplied by copying. Could Pliny have characterized so vulgar a method as 'an invention, of which the very gods were jealous, which enables great men to triumph over death and time, and not only gives them immortality, but, by the general diffusion of their images, causes the whole world to enjoy their presence in the collections which contain them?' Most certainly the idea of sending volumes into different countries could not constitute an invention either. It is, therefore, extremely probable, that some method of multiplying images, drawn or coloured, formed the groundwork of this invention of Varro's; and this process, which might be very different from our engraving, has not been transmitted to us. Who knows, in fact, how many equivalents for engraving may exist? and who could have supposed that an art would ever be invented for multiplying designs themselves by mechanical means? How much yet remains to be discovered! Necessity alone, as it comes to be felt, will unfold new methods of multiplying works of art."

The lithographic art was brought over to this country in 1901 by M. Andrée,[9] of Offenbach, in its rude and original state. He published some specimens of different artists, but no improvement has since been made in it here. The admirable productions, however, which have of late appeared at Münich, consisting as well of the works of modern artists as of imitations of ancient masters, for which lithography is peculiarly adapted, have excited a spirit of emulation in Mr. Ackermann, who is determined to use his best endeavours to rival the professors of this art on the continent.

Notes to Introduction

1 I thank John Krill for bringing this account to my attention.
2 Walter Benjamin, "The Work of Art in the Age of Mechnical Reproduction," in *Illuminations*, ed. Hannah Arendt, trans. Harry Zohn (New York: Schocken Books, 1969), 218.
3 B. Silliman and C. R. Goodrich, eds., *The World of Science, Arts and Industry Illustrated from Examples in the New-York Exhibition, 1853–1854* (New York: G. P. Putnam and Co., 1854), 132. Caliban is a character in Shakespeare's *The Tempest* who is an evil but ultimately pitiful beast.

Notes to Reading

1 [Alois Senefelder (1771–1834) discovered the process of lithography in 1796 and was granted exclusive rights to it in 1799. Ackermann translated and published Senefelder's original treatise in 1819.—Ed.]
2 [Count Adrian Jules de Lasteyrie (1759–1849) established the first lithographic press in Paris on the rue du Bac in 1815. He published *Receuil de différents genres d'impressions lithographiques* in 1817.—Ed.]
3 [Godefroy Engelmann (1788–1839) was an early advocate of lithography, establishing a rival workshop to Lasteyrie's in Paris in 1816. He went on to perfect the process of chromolithography, for which he received an English patent in 1837.—Ed.]
4 For crayons, it is necessary that a grain or tooth be made upon the stone with fine sand.
5 [Antoine-Chrysostome Quatremère de Quincy (1755–1849) was secretary of the Académie Royal des Beaux-Arts.—Ed.]
6 [Mazzo (Tommaso) Finiguerra (1426–1464), a goldsmith, was known for his work in *niello*, a process of engraved and pigmented designs in metal. He reputedly preserved his intricate patterns by pressing and transferring them onto paper, a natural precursor to engravings.—Ed.]
7 [Titus Pomponius Atticus (110–32 B.C.) was a close friend and relative through marriage of the Roman writer Marcus Cicero.—Ed.]
8 [Marcus Terentius Varro (116–27 B.C.) was a Roman scholar and author who, according to Pliny the Elder (23–79), invented a method of immortalizing great men using portraiture: "[Varro] had the idea of inserting in his numerous books not only

the names, but, by aid of a certain invention, the images of seven hundred illustrious persons." *Natural History*, bk. 35, chap. 11; translation from Theodore L. De Vinne, *The Invention of Printing* (New York: Francis Hart and Co., 1978), 111 n. 1.—Ed.]

9 [Johann Anton André (1775–1842), a German music publisher, was Sennefelder's first partner.—Ed.]

Reading 32

André Mellerio

Original Color Lithography (1898)

The French art critic André Mellerio (1862–1943) fought strenuously for the admittance of color prints into the "art of the Black and White" from which it had been excluded, that is, at least as an entrant in the annual Salon of the Society of French Artists in 1898. Defending its prohibition against color prints of any ilk, Henri Lefort (1846–1912), president of the society's section on engraving and lithography, stated, "By its essential principles, its origins and its traditions, the art of the print is unquestionably the art of Black and White. This is the traditional classification to which it is attributed."[1]

To twenty-first-century eyes, color in prints is taken for granted, but before the ability to efficiently register the successive printing matrices required for each color, multicolor printing remained out of reach for most fine art printmakers until the late nineteenth century.[2] *The complex skills and equipment required to produce color lithographs, or chromolithographs as they came to be called, precluded many artists from producing them, hence the sullied reputation of commercial chromos as cheap and tasteless. Mellerio proposed that a name change to "original color lithographs" would serve to distinguish the prints he championed from chromos, the same public relations strategy used by silk screen and ink-jet printers in the twentieth century. He claimed further that original color lithographs would serve to "purify" public taste. In rebuke, one commercial printer protested, "This same commercial spirit, however, has inspired and encouraged a charming variety of effect both in colour and design, and lithography of to-day, in almost every form of its manifestation, is infinitely*

From André Mellerio, *La lithographie originale en couleurs* (Paris: L'Estampe et l'affiche, 1898), trans. Margaret Needham, in Dennis Cate and Sinclair Hamilton Hitchings, *The Color Revolution: Color Lithography in France, 1890–1900* (New Brunswick, N.J.: Rutgers University Art Gallery, 1978), 79–81, 94–97. Reproduced courtesy of Zimmerli Art Museum at Rutgers University.

more attractive and capable of considerably more expressive power than could ever have been hoped for before commerical utility and value demanded a full recognition. Pleasing and harmonious effects, which are almost invariably sought after in lithography, need not be inartistic; and it is quite possible for the technique of the lithographic draughtsman to translate original work without a serious depreciation of its pictorial and artistic value."[3] *It should be noted that brightly colored chromolithographs such as trade and greeting cards, postcards, and baseball cards are highly collectible today and constitute the bulk of many important collections of ephemera.*

In the meantime, faced with the relentless onslaught of technology, Black and Whites struggled to maintain their artistic purity. The prejudice against color in prints persisted for decades, like the argument against color in moving pictures and still photography (color did not appear on the front page of the New York Times *until 1997). Like so many innovations in printing technology, color needed to enter the fine arts domain through the back door of commercial illustration.*

[. . .]

What has been the role of color in the history of lithography?[1] It must be said that it has existed almost since the origins of lithography, which go back only to the beginning of this century. The first efforts in color, still timid, were simply destined to serve as documents in the study of archaeology. It was only later, as improvements were made, that works of artists were reproduced. But it was necessary to reckon with the inking, the registering and the printing—all very complicated techniques. Furthermore, the work of the printer resulted in further modifications, which intervened between the sketch and the final work. These developments led to chromolithography, with all that that word implies of hard work and satisfactory technical success, but also—it is necessary to add—mitigated by an indelibly inartistic quality. In reality, no matter how perfect it was, it was simply a method of reproduction by a more or less skillful technique, with a process more or less developed, of an original work, conceived often from an entirely different point of view.

[. . .]

Should we be asked, given the present movement, how chromolithography crossed the boundary line into the domain of original art, we would not hesitate to state that it was in the work of Chéret.[2] The renovator of the poster, or rather its true creator, from the modern point of view, he not only achieved a personal statement, but he also exerted a large and resounding influence. The elegant variegated patterns, which brightened our city walls, led to the creation of original forms at the same time that they brought a sparkling and luminous range of gay colors. The print—in its ardent search for regeneration—was naturally affected. Chéret's posters opened up a new path—a path which the print happily followed.

[. . .]

How did the modern print differ from the old and ordinary chromolithograph to such a degree that it is necessary to look today for a name which clearly indicates this break? Precisely in the way Chéret freed his work and made it a creative form. The modern print was no longer a *facsimile* reproduction of just any original work in color, but a personal conception, something realized for its own sake. An artistic inspiration joined in advance with a technique expressed itself directly in the chosen process of execution. This principle, applied victoriously by Chéret to the poster, whose nature and function made it special, was to be applied by others to the print, whose characteristics differed in certain ways. And so original color lithography was born, and a simple sheet of paper for which mechanical means procured the advantage of unlimited copies, attained a real value as an art form.

But it is important to deal immediately with a previous question, the importance of which is evident, and upon which rests the matter of whether this effort will appear corrupt at its very base. We want to ask, what is the legitimacy of the color print? Should it be considered simply as an encroaching and diminishing incursion into the domain of painting? Or, on the contrary, does it have an intrinsic essence and its own particular range and scope?[3]

We lean resolutely toward the latter affirmation. And we leave to quibblers the task of distinguishing the more or less real preeminence of one genre over another. But the right of the color print to exist comes directly from the principle which we consider an axiom: any method or process which an artist develops to express himself is, for that very reason, legitimate. Now, people cannot deny that this is so for color lithography. And for the present period we will go further: not only has it been happily used by a number of artists, but it seems that some of them can assert even more forcefully that it is their principle if not their unique mode of expression.

[. . .]

The color lithograph is a print, and so the general laws of the latter must be applied to it. What are they? Logically and necessarily we find them in the essence of the print itself: a simple piece of paper, on which by mechanical printing a drawing is reproduced numerous times. The picture surface is neither as solid or sturdy as that for oil painting, just as the process does not have the same deep and bountiful richness. Neither does it have the texture or the brilliance of pastel, the penetrating light of the watercolor. Nor can it claim to have the imperceptible delicacies of an original drawing. It is a fact that mechanical printing, as perfect as it may be, takes away the tiny refinements of the brush strokes or of the artist's touch by which he transmits his sensation directly. It seems then that for these reasons prints should avoid ambitious effects, and that a simplicity of means demanding a less active part from the printer should work in its favor.

Moreover, we have to examine the special process of color lithography. It has its own advantages and its own pitfalls. Color permits different approaches and effects from simple black and white. However, successive printings do not mix colors thoroughly, but rather superimpose them. It thus seems wise to avoid excessive mixtures ending in the pretentious insipidities of chromolithographs. It is to the advantage of the modern multi-color print to keep the color areas separated, with a simplicity of tone, aiming more for the harmony of the whole than for complicated shadings. If one can say that posters are prints in large format, then original color lithography by its very origins remains still very close to its big sister, the poster.[4] It has, however, a different goal and special points of view: made to be held in the hand and looked at from close range, it naturally involves a greater detail and a greater refinement. With these possibilities the original print enjoys a harmonizing play of colors and of line that might even be described as decorative. The range of tones can be light, gay, very lively, but not, however, the boisterous burst of color called for by the utilitarian goal of the poster. We will be attracted to a varied play of color harmonies over which the eye can travel easily. Lines that are too complicated, tangled composition, or elusive delicacies of atmosphere do not seem to fall within its competence. It would be vain, as we have said, for the original print to compete with the varied and flexible means which oil painting has to achieve, perfect and delicate modeling, or with the sparkle of the pastel or the washes of the watercolor.

But on the other hand, what holds a special charm is the way the mat impression inks the paper without seeming too heavy. We can really appreciate this when, like the Japanese, the artists delight our eyes with original tonal harmonies united with elegantly decorative lines. But it must be said that this simplicity of means requires true artists. Yet many different resources are available. For example, look at the way Lautrec uses a little of everything in his prints: flat tints, outlines, bright spots, spatter, fine strokes of the pen, etc. But however original the working, it always remains a true color print which we will define as follows: a piece of paper decorated with colors and line which are part of it without hiding it or weighing it down. It is neither a *facsimile* of, nor a substitute for painting; it is another process, lacking certain elements, but with its own charm, of equal artistic value and the appreciable advantage of a printing with numerous copies.

Let us note in passing that among the modern color processes, lithography seems to have produced more varied results in the contemporary movement, and to have presented richer possibilities. Woodcuts and etchings have a rather similar appearance in some works; it seems sometimes that they can be used almost indifferently, without a marked character of their own, by using flat colors and outlines. Lithography, with less bite in the lines, is warmer, has more variety and depth, and lends itself perhaps better to nuances while remaining very much a print.[5]

But here, just as in our earlier statements, we do not intend to set absolute limits. This is a broad overview, not an unchanging code designed to limit originality. Each artist, with his own feelings and his way of thinking, must choose and adapt a particular technique to his personality, just as he does with the particular effects which he wishes to produce. Furthermore, ingenuity, technical discoveries, all the advances, can and must continually modify the application of the processes.

But there is another question of a different order which we feel it is necessary to consider. What role can the color print, particularly the easier and more economical lithograph, aspire to play in our present social order with its new developments? We live in a time described as democratic. A very big word in which each person hears what he wants! If we look for equality at a common level, it is the very death of art, because while we can lower the highest, we will never raise the lowest to the highest level. It would be the triumph of mediocrity. But, there is another mode of democratization worthy of attracting reasonable and generous minds. It does not degrade the personality of the artist, thanks to means that science gives him, to be able to put for very modest prices true works of art within the reach of ever larger groups of people. That is, we think, all that can and should be done in our desire to further purify public taste. The continual improvement in the reproduction of museum masterpieces, which have now become accessible to all, is a beautiful and eternal means of instruction[6]—as long as we understand, however, that its essential teaching is to enlighten the mind and not to enslave it. Now it is not less true that we live in our own time, that we share its impulses, even its mistakes, if you will, and its preconceptions, but that is the very substance of our lives, not an artificial varnish made from the residue of dead civilizations. Antiquity lives in our modern times only because of that part of humanity which is common to it and to us. And so it is that in comparison to the works of ancient art which, no matter what one says, require an introduction, the general public will find in the representation of its daily feelings a more direct nourishment, easier to digest, and which will lead it to the past.

Now it has happened that paintings, statues, and all works that have a value as unique objects by their rarity, have become costly collector's items, accessible only to a limited number. What remains? Prints and original prints. Their artistic value is indisputable. The number of prints that can be pulled, and the reasonableness of price, put them within reach of the general public. And it cannot be denied that besides the big collectors, who sometimes are lovers of art, but often are nothing more than wranglers over expensive objects, either because of vanity, or for hidden speculation, there exists here and now numerous representatives of the middle class who dedicate their leisure time, a portion of their intelligence and their money, to nosing about, to being interested in, and to buying. There is a potential enthusiasm, that could easily be triggered among ordinary people for this art, that is not yet perfect, but is already more refined. They are beginning to recognize the chromolithograph and to abhor it, preferring the original print. Now, while possession instructs, it also makes the interest more passionate and the search more active. Little by little, cultivated by efforts of artists, publishers, and magazines, and by the growing culture of a more educated public, this group will become numerous, with clearer ideas, asking only to see itself encouraged, enlightened, and helped.

But, it may be asked, why color prints rather than black and white? Can it be claimed that the former will kill the latter? Certainly not! And don't we see an old master, Fantin-Latour,[7] a painter with such a gift for color, remaining obstinately faithful to the black and white print, in spite of all the enticements and direct entreaties. In our personal opinion, one black and white lithograph by Odilon Redon,[8] just to take modern artists, achieves a summit of art. And if we look back to past masters, who can ever prevail over a Dürer or a Rembrandt? We will even say that there is in this kind of print where one can be a supreme colorist in the true sense of the word, just as in a pure drawing, a certain something which is more essential, more austere, which will always be a challenge for the strongest artists.

But there is a question of fact. From the standpoint of artistic diffusion, it is the general public, which turning to the print, finds in color a more accessible, more direct, and more engaging quality. We must understand that black and white involves an abstraction, relying only on certain qualities of the concrete. Thus the public's interest will contribute even more to the production of color prints, to which, culturally, artists have already been led by the various influences which we have indicated.

And even science itself, which in our day is concerned with improving public health, seems to encourage the print. After having opened up the wide boulevards and avenues, science wants light flooding into our homes, and it even advocates gaiety, that health of the soul, as a physical remedy. Now, isn't color the embodiment of light and gaiety? Furthermore, there is the everyday decoration of our interiors, which is pursued so rightly in its aim, but so erroneously and clumsily in its applications. Certainly, color lithography, without usurping anything from painting, holds its own in our decorative schemes, fitting in with simple furniture and natural woods from which gold leaf is excluded. One might almost say of Chéret's poster and Rivière's[9] prints that they are the frescos, if not of the poor man, at least of the crowd.

Notes to Introduction

1 Dennis Cate and Sinclair Hamilton Hitchings, *The Color Revolution: Color Lithography in France, 1890–1900* (New Brunswick, N.J.: Rutgers University Art Gallery, 1978), 1.

2 Some argue that there still exists a bias against color in the academic study of early modern prints. It would appear that brightly colored inks were not uncommon and added commercial value, didactic meaning, and visual emphasis to subjects as diverse as anatomy, art, astronomy, biology, cartography, medicine, militaria, and polemics in both single-sheet prints and books. Conference, "Impressions of Colour: Rediscovering Colour in Early Modern Printmaking, ca. 1400–1700," December 8–9, 2011, Cambridge University.

3 Alfred Seymour, *Practical Lithography* (New York: D. Van Nostrand Co., 1903), v–vi.

Notes to Reading

1 *La Lithographie*, by Herni Bouchot, librarian at the Cabinet des estampes at the Bibliothèque nationale, Paris, Libraries-Imprimeries réunies.
This book, naturally brief, because it aims to be part of the Fine Arts Teaching Library, gives a clear view of the fluctuations of lithography from its birth to our day. In it, one finds useful documentary information. As for the new movement, it is studied, if not with indulgent affect on, at least with a very keen curiosity.

2 [Jules Chéret (1836–1932), French painter and lithographer.—Ed.]

3 The debate is now taking place actively and passionately on all sides. Mr. Henri Lefort, president of the print and lithography section of the Society of French Artists, presented to the Committee of 90, last January 24, a report stating the reason for the non-admissibility of color prints to the Salon. In it is stated that "by its essential principles, its origins and its traditions, the art of the print is, unquestionably, the art of *Black and White*. This is the traditional classification to which it is attributed." Mr. Charles Maurin, painter-printmaker, counters by a virulent article in *Le Journal des artistes*, on March 20, 1898: "It is perhaps important to protest against Mr. Lefort's claims. In doing so, we are sure that we represent numerous printmakers, who for the last ten years have given a new scope to the original French print, which the new president classes under the disdainful and childish heading: 'colored pictures'."

Mr. Thiébault-Sisson seems to share Mr. Lefort's opinion: "The painter-lithographers in my humble opinion . . . waste too much time searching for effects which are not in the domain of lithography; in particular, they are almost all struck by a manic desire to translate into color their elucubrations, and I find that mania deplorable . . . it is impossible to make color lithography produce anything but very contrasting effects, which are marvelous in posters, but which bombard and irritate the eye in a carefully done print, made for a collector's portfolio or for an album" (*Le Temps*, November 5, 1897).

As a counter argument comes Mr. Roger Marx's article "*Pour la gravure en coulers*" in the *Voltaire* of March 30, 1898: "the contemporary school contains too many eminent colorists to be fairly judged without them; furthermore this ostracism is in formal contradiction with the preferences so many times expressed by historians and collectors. . . .

The works of the masters teach that in the recent and in the distant past, wood, stone and metal have never refused to accept a multicolored inking; between the materials and the process no incompatibility exists."

4 It is to be noted, moreover, that many of the artists of whom we have spoken have attained success not just in color prints, but concurrently in posters. We should mention among others, Toulouse-Lautrec, with *Jeanne Avril*, *Le Divan Japonais*, his remarkable *Babylone d'Allemagne*, Bonnard: *France-Champagne*, *La Revue blanche*, *L'Estampe et l'affiche*. From Ibels, *Mévisto*, *Irène Henry*, *Le Champ de Foire*, etc. From Maurice Denis: *La Dépêche de Toulouse*.

5 However, the objection can be raised, and we will raise it ourselves, praising the previously mentioned Japanese prints which are woodcuts. Furthermore, in our own time, Mary Cassatt has produced remarkable color etchings. Raffaelli is perhaps even more typical in that respect: his numerous plates for the same print are all cut by him and generally rest on the principle of different inks penetrating the incision, which is the basis of all engraving, rather than using tints laid flat on copper.

Our colleague, Mr. Geroge Lecomte, is now preparing a work on color printing: etching, dry pot, etc., that will more completely elucidate these interesting points.

6 Let us remember in reference to this the excellent article by Charles Saunier, *La Chalcographie du Louvre* (*l'Estampe et l'affiche*, November 15, 1897, issue). "Must one really be so rich to own works which do not dash with basic good taste, which divert the eyes and impose on the mind an idea of beauty. We are going to prove that it is not so. . . . People of taste and of modest means desiring to own beautiful things can have their attention called to a gallery of masterpieces of which it will be easy to take advantage.

"Their purchase will sometimes cost less than a franc, more often from two to five francs, very rarely a little more. For a modest sum they will have extremely important works. . . ."

This treasure trove is in the Louvre's reproductive department.

7 [Henri Fantin-Latour (1836–1904).—Ed.]

8 [Odilon Redon (1840–1916), French painter and printmaker.—Ed.]

9 [Henri Rivière (1864–1951), French illustrator.—Ed.]

Reading 33

Carl Zigrosser

The Historical Background of Originality (1965)

More than a century after the development of photogravure and the transvaluation of prints from a technology designed to produce exactly repeatable images into an independent art form, Carl Zigrosser (1891–1975) continued the struggle to define what constitutes an "original" print, revising here his 1961 essay "What Is an Original Print," published by the Print Council of America.[1] *According to Zigrosser, eliminated from consideration are prints that rely on photography as a means of producing and fixing the image to be printed, that is, photomechanical reproductions of any kind. He reluctantly designates prephotography reproductive engravings "translations" and reserves the honorific "original" for prints whose design and execution belong solely to the artist. Probably catalyzing but certainly influencing Zigrosser's essay were contemporary publishing practices in France, such as the release of "limited" editions of signed and numbered lithographs by Joan Miró (1893–1983) and Marc Chagall (1887–1985) that had appeared originally in large runs of* Verve *magazine. One sentence, diplomatically excised from this 1965 version of his essay, complains, "If there are fools, chiefly from America, eager in their ignorance, to pay high prices for reproductions, in the belief that they are original prints, who is to disillusion them?"*[2] *In 1961 other forms of prints were vying for the cachet of originality, such as silk screen prints, which had shed their commercial taint by a name change to serigraphs.*[3] *Given Zigrosser's grudging acceptance of certain offset lithographs of Jean Charlot (1898–1979) as original, one wonders if he would embrace the born-digital giclée prints of today. The term* giclée *was coined in 1991 specifically to designate limited-edition "fine art" IRIS ink-jet prints. Over time, however, it became associated with commercial art reproductions and fell out of usage entirely.*[4]

From Carl Zigrosser, "The Historical Background of Originality," in Carl Zigrosser and Christa M. Gaehde, *A Guide to the Collecting and Care of Original Prints* (New York: Crown Publishers, 1965), 19–24, 32–35. Reproduced with permission of the Print Council of America.

Felix Brunner (1929–) (see Further Reading), who was involved with the commercial printing industry, has a more sanguine outlook when distinguishing mass-produced prints from fine art prints: "Only the genuine will last and retain its value whereas the spurious will sooner or later be recognized for what it is."[5] *And the multi-talented Bamber Gascoigne (1935–) (see Further Reading and www.historyworld.net and www.timesearch.info) refuses to assign any measurement of artistic merit to printing techniques: "We all live surrounded by prints. From billboards and brochures, from books and magazines, printed images compete for our attention. There can hardly be a house in the developed world, which does not have a print on the kitchen table, decorating a packet of cereal or biscuits or tea. The picture of milk cascading onto cornflakes is just as much a print as any other. As with so many prints of the past, it is the modern printer's best available answer to a commercial need. It is susceptible to exactly the same processes of detection as its commercial predecessors (say a sheet music cover of the 1850s) or as a great printed work of art, such as an etching by Rembrandt."*[6]

The paper conservator Christa Gaehde's (1922–2002) contribution to the Guide *is notable for being among the first efforts to delineate the conservation needs of prints. Previous publications, aimed at collectors, usually confined their advice to finding a trustworthy restorer. Constituting about one-fifth of the* Guide, *mostly in illustrations, it reflects the adolescence of the discipline of paper conservation in 1965.*

[. . .]

Today we are much more conscious of originality in all senses of the word than our forefathers were. In the late Middle Ages when prints began to be made in Europe, the idea of originality did not exist; there were traditional themes and traditional modes of depicting them that were transmitted from artist to artist and generation to generation. Artists copied and recopied each other's work without any sense of guilt. In the medieval, and to a large extent in the Oriental conception of art, the artist's personality was submerged in his work. Pictures were not signed. The earliest signatures on prints were marks or monograms such as E S or M S (Master E S or Martin Schongauer); and it has been suggested that these marks—following the practice of goldsmiths—were hallmarks or guarantees of honest and masterly workmanship rather than signatures in the modern sense of the word. Gradually, however, as prints and easel paintings became transportable, and therefore acquired use and value as personal property, the artist's name became a valuable asset; and his production, issued under his own trademark, became almost a special brand of merchandise. Beginning at the time of the Renaissance, anonymity was replaced by the emphasis and exploitation of the artist's individual personality. The concept of plagiarism and forgery came about very gradually as a controversial issue. When Dürer went to Venice in 1505 to protest Marcantonio Raimondi's wholesale plagiarism of his *Life of the Virgin* series and other prints, the only satisfaction he could obtain from the authorities was that Raimondi was enjoined from using Dürer's monogram. In the seventeenth century artists occasionally received protection against fraudulent copying as a special favor from ruling monarchs. On certain prints published by Rubens, for example, are engraved the words *cum privilegiis regis* . . . (with the privileges or protection of the king). The first general copyright law was passed by the British Parliament in 1735 upon petition of Hogarth and others who had suffered from plagiarism and piracy. Thereafter, Hogarth's engravings—the series *Rake's Progress*, for example—bear the line *Published according to Act of Parliament*. Since then, the artist's rights in his own design have been fairly well established in most countries, in principle at least.

Again, today, we are more conscious of execution, the artist's personal touch, than were earlier print amateurs. They were more apt to value the print not for its own sake but as a surrogate of a drawing or painting. They were more concerned with a generalized outline of the composition as suggestive of sublime and noble design. They accepted the reproductive limitation of the print and did not demand the personal touch of the designer's hand. It must be remembered that the chief function of printmaking throughout its early history was reproductive. The "original" print, as we value it today, by Rembrandt, Goya, Degas, Mantegna, for example, was the exception rather than the rule. A striking example of this attitude may be seen in Van Dyck's *Iconography*. Of the hundred-odd designs that Van Dyck made for his gallery of famous men, only five of the eighteen he actually etched remained intact. The other thirteen were "finished," and all the rest completely engraved by professional craftsmen after his drawings. He had intended to do the whole set himself, but had abandoned the idea because his own presentation was unpopular. Today we are extravagant in our appraisal of his original etchings, in comparison with the rest of the *Iconography*.

It was in the nineteenth century that the concept of the original print began to emerge in tangible form. The invention of photography early in the century was a critical point in the history of printmaking, but its full impact was not realized until the end of the century, when its applications to photomechanical reproduction were perfected. The effect was revolutionary and far-reaching. As was said in *Six Centuries of Fine Prints* (New York, 1937): "Through the development of photoengraving, the line cut and the halftone, it [photography] stripped regular printmaking completely of its reproductive function. . . . The artist who now makes prints speaks not as a copyist but as a creative artist working directly in a graphic medium. This has necessitated a new orientation, a new justification for prints. They must stand or fall as an independent art."[1]

There were also active spokesmen on behalf of the original print from about the middle of the nineteenth century onward. Whistler preached the gospel by precept and example. Seymour Haden wrote a pamphlet in 1883, *The Relative Claims of Etching and Engraving to Rank as Fine Arts*, in which he used the phrase *painter-etchers* and *painter-engravers* as opposed to reproductive craftsmen. For the purpose of his argument he classified all the creative virtues under the heading of etching and all the dull mechanical practices under the head of engraving: "The essential differences between etching and engraving may, therefore, be described as of two kinds—differences of principle, and differences of technique—and these again be expressed, not inaptly, by some such formula as the following: 'Etching, depending on brain impulse, is personal; and the creative faculty being chiefly engaged in it, invention, sensibility, and the various attributes which make up the sum of genius, belong to it and constitute it an art. Engraving being without personality—except such as may be supposed to be involved in the act of copying or translating the work of another—originality, and all the attributes which attend the exercise of the creative faculty, are absent from it, and constitute it a *métier*.'"[2] The question of originals versus photomechanical reproductions also came up later in the nineteenth century. Sir Hubert Herkomer was sharply criticized in the British press by Walter Sickert and Joseph Pennell for selling photogravures of his paintings as original etchings.[3] The influence of Whistler and Haden bore fruit in England and America in the high regard placed upon original etching at the beginning of the twentieth century. In spite of the fact that this appreciation was limited to etching (and, as it has turned out, often to etchings by artists of mediocre potential) it was a step toward the recognition of printmaking as a major medium. In France, although many of their great artists have made original prints in one form or another, there is still a large body of opinion that has no high regard for printmaking as a creative medium, and considers it a reproductive process for the luxury trade. Even after the photomechanical reproductive processes were fully perfected, "deluxe" publications were issued containing reproductions of paintings etched by mediocre artists or professional printers, designed to have a luxury or snob appeal (including such eye-catchers as Japan vellum paper, marginal *remarques*,[4] limited editions, and fancy bindings), although in reality these "handmade" productions were inferior—as far as fidelity to the original paintings was concerned—to regular process prints. This fact and the presence of highly skilled craftsmen in printing and color work have brought about some of the questionable practices in vogue today in France.

Printmakers in America, more than in any other country today, feel an obligation to perform every step in the production of a print from the preparation of the plates, blocks, or stones to the printing of the finished impressions. This may be due in part to a dearth of skilled professional printers, who might relieve the artist of part of the burden, and in part to a sense of dedication on the part of the artist to what he considers a major creative medium, and which impels him to participate in every step of it.

In our era, then, the graphic artists—particularly Hayter and the Americans—tend to view printmaking as a major medium; and this point of view, which has also spread among critics, museum people, and the buying public, has tended to glorify originality and creation. It may be that too high a premium is being placed upon these values in the light of graphic tradition. There is an originality of design (which can still appear, although diluted, in reproductions), and there is an originality of execution (upon which the modern artists set great store). Perhaps the distinction can be made clearer by reference to another art, literature in translation. There are numerous works—often very great masterpieces—which do not suffer much by translation into a foreign tongue. Their outstanding qualities—the originality of invention, the grandeur and universality of theme, the powerful articulation of structure, the profound implications of meaning—can be carried over into another medium without too much loss of emotive power. Again, there are other works, some poems, for example, that are so idiosyncratic in expression that they can never be translated. So it is with prints. One might consider the reproductive engravings of old as translations of pictures into another medium. A translation implies a translator, who may be either pedestrian or inspired (as was Raimondi at his best). An artist can sometimes translate his own work, as in the example of Manet's *Olympia*. When Emile Zola wrote his pamphlet in defense of Manet's much criticized painting *Olympia*, and it seemed desirable to include an illustration, Manet himself made an etching of it. It was not a reproduction of the painting, though it served as one. It was actually a translation of the subject into another medium, a variant of the artist's conception. We are grateful for the occasion that induced Manet to make an original etching, designed and executed by himself. Today such a pamphlet would probably be illustrated with a process color reproduction.

If engravings are translations, then photomechanical prints might be called reproductions, pure and simple. The personality of the translator, with all its advantages and disadvantages, does not intervene. The original artists' prints of today might be called poems that cannot be translated into a foreign tongue.

[. . .]

In conclusion, one may summarize the problem of reproductive versus original prints somewhat as follows: Owing to the impact of photography and photomechanical processes, a new attitude toward printmaking has developed that stresses the original, the creative factor. In general, one may say that handwork is bound up with art and original execution, as opposed to automation and mechanical processes.

In past print history, the invention per se, the design and the message, were what the public looked for and prized. To be sure, the original artist of old also was concerned with the execution and with the effort to clothe his conception in the most perfect form possible. But that concern was his private affair, related to his artistic conscience, and of interest, possibly, only to his fellow practitioners, but certainly not to the layman in general. The modern artists (and through them now the public at large) tend to be conscious not only of what they say but also of how they say it. Indeed, they almost make the latter the prime creative motivation. Thus, when they make a print, they maintain that it is a complete aesthetic entity, a perfect fusion of concept and form, a work of art that could not exist in any other shape or form and that is fully the equal in validity and impact of an oil painting, irrespective of whether it exists in one or more impressions. But in any estimate of rank between major and minor art, one must remember that printmaking really cannot count on the still potent asset of uniqueness, as can painting, drawing, and, to a lesser extent, plastic art (which also has its problem with casts as multi-originals). The transvaluation of printmaking from minor into major, however, is in the spirit of the age, and must be reckoned with.

It is imperative above all that we come to terms with certain attitudes—holdovers from the past—which, being less scrupulous in discrimination between original and reproductive, are causing confusion and misunderstanding. These notes are designed to review objectively the conflicting standards of business morality held by some artists and some dealers. It may turn out that the problem is one of semantics rather than ethics. We must make the issue widespread and make clear just what the difference is between original and reproductive. These notes also aim to place this very modern problem into some sort of historical perspective in order to serve as a corrective of any uninformed criticism from the modern point of view of various practices in the past. These procedures might be deemed questionable today, whereas they were quite legitimate in the framework of their time. One can be rigid in applying certain standards to the works of our time, but one may not to those of the past.

So much for the original print in the past. What about its status in the future? There is of course no absolute assurance that the criteria of the past or even of the present will continue to be accepted unchanged and unchallenged in the future. There are already some faint signs that a transvaluation of values may be in the making. The present concept of the original print represents the climax and apotheosis of individuality. The ideal of the established artists of our day, and particularly of the printmakers, is idiosyncrasy of expression, the artist's personal touch. The artist has his own message, which he expresses in his own inimitable way, performing each step toward the ultimate end with his own hand and eye. A public, having grown to like and admire his productions, has placed a cachet of value upon his expression because of its personal and technical idiosyncrasy. But the continued ascendancy of such a concept is by no means secure. Young artists and new movements are emerging, and will continue to emerge, to vie with the older artists and win their own following. The names of the movements are irrelevant—they succeed each other in bewildering profusion, and here is not the place to present a history of modern "isms"—the movements are significant only insofar as they indicate a trend, a growing indifference to means and techniques.

Many of the artists of the *avant-garde* are inclined to reject traditions of the past, particularly the tradition of handwork in the production of prints. They tend to concentrate their efforts on the invention and perfection of their personal "gimmick," which distinguishes them from other competing artists. When they seek a wider audience through the production and distribution of prints, they are apt to be impatient of the "drudgery" involved, and fall back upon the impersonality of the photomechanical processes, which interpose the least possible barrier between the original and its multiplication. Why use cumbersome and antiquated methods, when easy modern methods are at hand? In many cases the artists may not even know that other methods exist, for they make a point of ignoring the past. Nor would the public that collects their work by fad or inclination care one way or another how their prints were made: it will accept anything from a successful artist.

As a matter of fact, the public is becoming more and more accustomed to impersonal mass production in every branch of life. "Mass-Man" would not balk at mass-produced images. It could happen, therefore, that the word *original* might lose its special meaning, and the distinction between original and reproduction, or between art (as we know it now) and non-art might lose its force. No one would care about the ethics or semantics involved. Such suppositions are of course pure speculation. Nonetheless they remain a possibility. If the tradition of the original print and its production were forgotten, let us hope that some new word would be found for the new form that would take its place, and that the old words *original prints* would still be reserved to describe the masterpieces of our own age.

Notes to Introduction

1 Carl Zigrosser, "What Is an Original Print?," ed. Joshua Binion Cahn, in *What Is an Original Print?* (New York: Print Council of America, 1961).

2 Ibid., 20.

3 In the popular textbook *Silk Screen Techniques* by J. I. Biegeleisen and Max A. Cohn (New York: Dover Publications, 1958), Zigrosser is credited with having approved serigraphs as "fine art" (13).

4 See "The True Story of Giclée," www.dpandi.com/giclee/giclee.html; and Martin Jürgens, *The Digital Print: Identification and Preservation* (Los Angeles: Getty Conservation Institute, 2009), 24–25.

5 Felix Brunner, *A Handbook of Graphic Reproduction Processes*, 6th rev. ed. (New York: Hastings House, Visual Communications Books, 1984), 10.

6 Bamber Gascoigne, *How to Identify Prints* (New York: Thames and Hudson, 1986), 7.

Notes to Reading

1 [Carl Zigrosser, *Six Centuries of Fine Prints* (New York: Garden City Publishing Co., 1939), 163. Only the first half of the quote could be found.—Ed.]

2 [Seymour Haden, "The Relative Claims of Etching and Engraving to Rank as Fine Arts," *Journal of the Society of Arts* 32, no. 1, 593 (June 1, 1883): 717.—Ed.]

3 [Sir Hubert von Herkomer (1849–1914) was an English playwright and artist. The incident occurred in 1890. Walter Richard Sickert (1860–1942) was associated with the Camden Town Group of painters in London, and Joseph Pennell (1857–1926) was an American illustrator and printmaker.—Ed.]

4 [A *rémarque* is a whimsical printed or drawn sketch added in proximity to the artist's inscribed name or signature.—Ed.]

Reading 34

Walter Benjamin

The Work of Art in the Age of Mechanical Reproduction (1936)

No discussion of originality would be complete without homage paid to the seminal essay "The Work of Art in the Age of Mechanical Reproduction," *first published in 1936, by the provocative German cultural commentator Walter Benjamin (1892–1940). For today's conservators, Benjamin's narrow definition of originality, wherein most prints and all photographs would be denied the label "auratic original," seems proscriptive. Put into historical context, however, his argument can be seen as a natural response to the continuum of technological forces that buffeted the status of prints as works of art, or at least as items having significant cultural worth, since their beginnings.*

Numerous art and social historians have strongly disagreed with Benjamin's disturbing conclusion that photography, preceded by lithography, would ultimately lead to the "liquidation of the traditional value of cultural heritage." The passage of time has allowed for the assimilation of photographs and other underappreciated printed matter, including reproductive engravings, ephemera, and offset lithographs (see reading 35), into museum collections and, by extension, has afforded them the protection provided by institutional preservation policies. Perhaps the time has come when the question of originality is no longer the fulcrum on which the appreciation, both financial and sentimental, and, hence, the survival of a print balance. Conservators, whom Benjamin indirectly credits for their role in the physical and chemical analysis of "authentic" works of art, will continue to regard prints, including those born digital, as multiple originals.

From Walter Benjamin, "The Work of Art in the Age of Mechanical Reproduction" (1936), in *Illuminations*, ed. Hannah Arendt, trans. Harry Zohn (New York: Schocken Books, 1969), 218–22. Reproduced with permission of Houghton Mifflin Harcourt.

[. . .]

I

In principle a work of art has always been reproducible. Man-made artifacts could always be imitated by men. Replicas were made by pupils in practice of their craft, by masters for diffusing their works, and, finally, by third parties in the pursuit of gain. Mechanical reproduction of a work of art, however, represents something new. Historically, it advanced intermittently and in leaps at long intervals, but with accelerated intensity. The Greeks knew only two procedures of technically reproducing works of art: founding and stamping. Bronzes, terra cottas, and coins were the only art works which they could produce in quantity. All others were unique and could not be mechanically reproduced. With the woodcut graphic art became mechanically reproducible for the first time, long before script became reproducible by print. The enormous changes which printing, the mechanical reproduction of writing, has brought about in literature are a familiar story. However, within the phenomenon which we are here examining from the perspective of world history, print is merely a special, though particularly important, case. During the Middle Ages engraving and etching were added to the woodcut; at the beginning of the nineteenth century lithography made its appearance. With lithography the technique of reproduction reached an essentially new stage. This much more direct process was distinguished by the tracing of the design on a stone rather than its incision on a block of wood or its etching on a copperplate and permitted graphic art for the first time to put its products on the market, not only in large numbers as hitherto, but also in daily changing forms. Lithography enabled graphic art to illustrate everyday life, and it began to keep pace with printing. But only a few decades after its invention, lithography was surpassed by photography. For the first time in the process of pictorial reproduction, photography freed the hand of the most important artistic functions which henceforth devolved only upon the eye looking into a lens. Since the eye perceives more swiftly than the hand can draw, the process of pictorial reproduction was accelerated so enormously that it could keep pace with speech. A film operator shooting a scene in the studio captures the images at the speed of an actor's speech. Just as lithography virtually implied the illustrated newspaper, so did photography foreshadow the sound film. The technical reproduction of sound was tackled at the end of the last century. These convergent endeavors made predictable a situation which Paul Valery pointed up in this sentence: "Just as water, gas, and electricity are brought into our houses from far off to satisfy our needs in response to a minimal effort, so we shall be supplied with visual or auditory images, which will appear and disappear at a simple movement of the hand, hardly more than a sign" [. . .].[1] Around 1900 technical reproduction had reached a standard that not only permitted it to reproduce all transmitted works of art and thus to cause the most profound change in their impact upon the public; it also had captured a place of its own among the artistic processes. For the study of this standard nothing is more revealing than the nature of the repercussions that these two different manifestations—the reproduction of works of art and the art of the film—have had on art in its traditional form.

II

Even the most perfect reproduction of a work of art is lacking in one element: its presence in time and space, its unique existence at the place where it happens to be. This unique existence of the work of art determined the history to which it was subject throughout the time of its existence. This includes the changes which it may have suffered in physical condition over the years as well as the various changes in its ownership.[2] The traces of the first can be revealed only by chemical or physical analyses which it is impossible to perform on a reproduction; changes of ownership are subject to a tradition which must be traced from the situation of the original.

The presence of the original is the prerequisite to the concept of authenticity. Chemical analyses of the patina of a bronze can help to establish this, as does the proof that a given manuscript of the Middle Ages stems from an archive of the fifteenth century. The whole sphere of authenticity is outside technical—and, of course, not only technical—reproducibility.[3] Confronted with its manual reproduction, which was usually branded as a forgery, the original preserved all its authority; not so vis-à-vis technical reproduction. The reason is twofold. First, process reproduction is more independent of the original than manual reproduction. For example, in photography, process reproduction[4] can bring out those aspects of the original that are unattainable to the naked eye yet accessible to the lens, which is adjustable and chooses its angle at will. And photographic reproduction, with the aid of certain processes, such as enlargement or slow motion, can capture images which escape natural vision. Secondly, technical reproduction can put the copy of the original into situations which would be out of reach for the original itself. Above all, it enables the original to meet the beholder halfway, be it in the form of a photograph or a phonograph record. The cathedral leaves its locale to be received in the studio of a lover of art; the choral production, performed in an auditorium or in the open air, resounds in the drawing room.

The situations into which the product of mechanical reproduction can be brought may not touch the actual work of art, yet the quality of its presence is always depreciated. This holds not only for the art work but also, for instance, for a landscape which passes in review before the spectator in a movie. In the case of the art object, a most sensitive nucleus—namely, its authenticity—is interfered with whereas no natural object is vulnerable on that score. The authenticity of a thing is the essence of all that is transmissible from its beginning, ranging from its substantive duration to its testimony to the history which it has experienced. Since the historical testimony rests on the authenticity, the former, too, is jeopardized by reproduction when substantive duration ceases to matter. And what is really jeopardized when the historical testimony is affected is the authority of the object.[5]

One might subsume the eliminated element in the term "aura" and go on to say: that which withers in the age of mechanical reproduction is the aura of the work of art. This is a symptomatic process whose significance points beyond the realm of art. One might generalize by saying: the technique of reproduction detaches the reproduced object from the domain of tradition. By making many reproductions it substitutes a plurality of copies for a unique existence. And in permitting the reproduction to meet the beholder or listener in his own particular situation, it reactivates the object reproduced. These two processes lead to a tremendous shattering of tradition which is the obverse of the contemporary crisis and renewal of mankind. Both processes are intimately connected with the contemporary mass movements. Their most powerful agent is the film. Its social significance, particularly in its most positive form, is inconceivable without its destructive, cathartic aspect, that is, the liquidation of the traditional value of the cultural heritage. This phenomenon is most palpable in the great historical films. It extends to ever new positions. In 1927 Abel Gance exclaimed enthusiastically: "Shakespeare, Rembrandt, Beethoven will make films . . . all legends, all mythologies and all myths, all founders of religion, and the very religions . . . await their exposed resurrection, and the heroes crowd each other at the gate."[6] Presumably without intending it, he issued an invitation to a far-reaching liquidation.

Notes to Reading

[Preface to the essay has been elided.—Ed.]

1 [Quoted from Paul Valéry, "The Conquest of Ubiquity," in *Collected Works of Paul Valéry*, vol. 13: *Aesthetics* (Bollingen Series, vol. 45), trans. Ralph Manheim (New York: Pantheon Books, 1964), 226. Ambroise-Paul-Toussaint-Jules Valéry (1871–1945) was a French poet and philosopher.—Ed.]

2 Of course, the history of a work of art encompasses more than this. This history of the "Mona Lisa," for instance, encompasses the kind and number of its copies made in the 17th, 18th, and 19th centuries.

3 Precisely because authenticity is not reproducible, the intensive penetration of certain (mechanical) processes of reproduction was instrumental in differentiating and grading authenticity. To develop such differentiations was an important function of the trade in works of art. The invention of the woodcut may be said to have struck at the root of the quality of authenticity even before its late flowering. To be sure, at the time of its origin a medieval picture of the Madonna could not yet be said to be "authentic." It became "authentic" only during the succeeding centuries and perhaps most strikingly so during the last one.

4 [Process reproduction refers to commercial methods of photomechanical reproduction; since the late nineteenth century, Germany had excelled at high-quality photogravure and collotype reproductions of works of art.—ED.]

5 The poorest provincial staging of *Faust* is superior to a Faust film in that, ideally, it competes with the first performance at Weimar. Before the screen it is unprofitable to remember traditional contents which might come to mind before the stage—for instance, that Goethe's friend Johann Heinrich Merck is hidden in Mephisto, and the like.

6 Abel Gance, "Le temps de l'image est venu," *L'Art Cinématographique*, Vol. 2, pp. 94 f. Paris, 1927. [Abel Gance (1889–1981) was a French filmmaker famous for his monumental film of the life of Napoleon that he began in 1927 and continued to work on throughout his life.—ED.]

Reading 35

Johanna Drucker

Offset: The Work of Mechanical Art in the Age of Electronic (Re)production (1998)

It is significant that in her 1998 essay, Johanna Drucker (1952–) neglects to mention Walter Benjamin's seminal essay, "The Work of Art in the Age of Mechanical Reproduction" (reading 34), despite her repeated references to its title and her use of his jargon, especially "auratic original" and "democratic multiple." Her assumption that Benjamin's vocabulary has completely infiltrated the language of art criticism and visual theory is, by and large, true.

As mentioned frequently throughout Part III, the lack of concrete evidence of the artist's touch in the production of prints has always been of concern in the question of originality. Drucker emphasizes that for offset lithographs this proof is often difficult if not impossible to detect. In the course of their examination of prints, conservators are often asked to decipher the various methods used by the artist and to untangle the order of their application, for example, in determining the sequence of printed states. This evaluation contributes to an understanding of the work's appearance and acknowledges the widespread fascination with the creative process. How or to what extent an artist manipulated the making of a plate, its inking and printing, however, can present a challenge to even the most experienced eye. As a prime example, consider the 1946 print Cinq Personnages, *by Stanley William Hayter (1901–1988),[1] variously described as a soft-ground etching, a silk screen, and an engraving. It is only the artist who can ultimately unravel the many techniques employed, one of which, color viscosity printing, was invented by Hayter himself. Fortunately Hayter published a detailed account of his procedures, highlighting the importance of documentation of processes, especially unusual ones, for the future.[2]*

From Johanna Drucker, "Offset: The Work of Mechanical Art in the Age of Electronic (Re)production," in *Figuring the Word: Essays on Books, Writing, and Visual Poetics* (New York: Granary Books, 1998), 184–93. Reproduced courtesy of the author.

Offset printing was designed for commercial purposes to facilitate the high-speed, high-volume production of multiples through photomechanical processes. [. . .]

Offset's Identity

Offset printing is situated at the intersection of individual creative expression and industrial production. The theoretical premises on which the identity of offset as a creative medium can be legitimated have yet to be fully articulated. In addition, practical considerations, such as the difficulty of learning the process, have kept the number of offset artists low. However, with the widespread availability of the electronic darkroom on personal computers the pre-press aspects of offset production are becoming part of the common currency of information processing. But the essential characteristics of offset as a printing process remain unchanged: the photomechanical transfer of visual information from film to a planographic aluminum plate to a soft blanket and then to paper on a high speed press. The most recent technology extends the electronic reach for Heidelberg's newest press[1]: the plates are made from disks right on the press. But offset printing itself is still very much a mechanical art; its technology is based in nineteenth century process and machinery, in the mechanization of older craft practices which were slow, labor intensive, and dependent on skilled handwork. Offset is the high end output for electronic media, the most efficient, accurate, and versatile mode of rendering the digital, rasterized,[2] processed data of the electronic environment into material form as print.

Offset is not an outmoded technology, like stone lithography or handset letterpress, acquiring status by virtue of its obsolescence. Offset art, instead, is comparable to video: it is an artistic use of a viable industrial mode. Not surprisingly, video has also been slow to obtain recognition—or, even more importantly—identity as an art form. As a creative medium in an age of electronic (re)production, offset can only be understood in relation to its continuing functionality as an industrial process. As a mechanical mode of production, offset can only be assessed aesthetically in relation to the eroding status of modernist claims about the autonomy of fine art. And the prosaic familiarity of offset printed sheets can only be theorized in terms of their specific cultural identity if the invisible aspects of skilled labor and production processes on which it depends are taken into account—and the ordinary character of offset defamiliarized. The purpose of this essay is to at least suggest some of the parameters within which such an assessment might be framed.

History and Technology

[. . .]

Designed to perfect the rapid reproduction of images and texts through a combination of photographic and lithographic printing processes, offset generally hides its complexity under the appearance of efficiency. In fact, offset is a process which requires elaborate and highly skilled work in every aspect of its production. In the pre-press stage, hours in the darkroom, at the drawing board, or in the design studio all contribute to the eventual making of a piece of film. Whether recording hand processes—drawing, mark-making, fingerprints—or using photographic methods—electronic or traditional—offset plates have to have their own photosensitive surfaces exposed and developed before they can be used in printing. At the stripping and platemaking stages, negatives, mylar, positive plates, stencils, opaquing fluid, elaborate burning and registration—any number of manipulations—provide tools for composing the visual image ultimately fixed in the photographic emulsion. Once made, the plate provides the means for transferring ink to the soft blanket[3] and onto the paper. In printing, a plate is continually dampened to keep the oil base ink off the non-image area, and the variables of water, pressure, speed and inking require continual vigilance. When used as an artform and creative medium, offset sometimes calls attention to and disturbs the transparency of the reproductive process. It is possible to transform offset from a reproductive to a productive medium through interference at any point in the process. And yet, the very photomechanical character of offset is such that it will always, inevitably, conceal the labor involved in its operation.

Auratic Originals vs. Democratic (?) Multiples

The beginning of the modern period in visual art corresponds with the development of mass production modes in industry. The technology for such mass production of images (including offset) made it necessary for fine art to distinguish itself from other forms of image production. Modern fine art had two identifiable characteristics: it was autonomous, separate and distinct as a commodity form (an object among others and an image "free" from reference) and it was comprised of unique, original works—or at the very least, those in highly limited editions.

The modern concept of an autonomous fine art depended on the idea that art was somehow distinct from mass culture in its imagery, style, and method of production—and that it was capable of serving a function which could be served by no other cultural activity. As theorized by twentieth century critics, especially Clement Greenberg,[4] the function of art was to preserve the values of individual expression in the face of the powerful instruments of information, control, and seduction of both totalitarian and capitalist ideologies. Art production was defined as a practice of individual expression that found its social and aesthetic identity in the institutions of museums, galleries, and critical discussion. The myth of the auratic original allowed status to accrue to the tradable commodity of the oil painting, while the use of certain print media was perceived to be too intimately intertwined with commercial, advertising, and industrial aims and methods to retain its aesthetic autonomy. Simply stated: lithographic prints too closely resembled the products of industry to pass as fine art. Needless to say, the more that machinery and less that hand work was involved in its making, the lower a work was on the hierarchical spectrum defined by "art" at one extreme and "mass media images" at the other. The mass produced images of high speed media have only been permitted to enter the fine art realm through imitation, iconic reference, or in parodic, limited ways which quickly recuperate them as fine art (Cubist collage, Dada assemblage, Pop art, Fluxus, recent appropriation work). Offset, after all, is real industrial production—not just a simulation. The grounds on which it could be distinguished from commercial production cannot be sustained on material terms, and thus, either have to be made on aesethetic terms (iconography of personal expression, laying bare of the mechanical devices, or the artificial limiting of editions) or social terms (through permitting offset to aspire to the condition of fine art through museum shows, gallery sales, critical review). Offset challenges the very premises of modern fine art (already severely eroded by postmodern rhetoric, if not practices) by its serious connection to industrial production, countering the premises and claims of the auratic original.

Ironically, offset is equally unable to sustain the mythic terms of the democratic multiple. One of offset's unique capabilities (the basis of its success as an industrial process) is the relation between production cost and unit cost. The high volume capacity of the printing machine is what renders the end-product affordable—if the equation is used conventionally. But offset equipment is extremely expensive. Joe Ruther's[5] salvage techniques may have permitted him to keep a few old clunkers running, but even that was at a cost beyond the average computer or letterpress shop. (If not always true in monetary terms, this is the case in terms of the mechanical skill and patience to keep the old presses running.) Price tags for high end machines creep into the six-figure bracket pretty fast—they are complicated machines with thousands of moving parts functioning at extremely tight tolerances. Offset artists, typically, make small editions which are nonetheless high volume by art standards. Where a stone lithographer might push an edition to fifty or a hundred copies in the editioning process (and consider it excessive if the run were to include more than three or four colors) in offset printing that many sheets go by in two or three minutes during the "set-up" time on the press. For the artist, the greatest investment of time and effort is in the set-up, in the pre-press, darkroom, platemaking, stripping, and working the plate up on the press to print cleanly, densely and correctly. The costs in all of these pre-press processes are as high for ten (good) copies as they are for ten thousand or more. The only change in cost factor comes with costs of paper. An affordable, democratic, accessible form of production? Not exactly . . . and yet, its artistic identity is bound up with both making claims to be so (the case with Fluxus, Conceptual art, where an artists' books sensibility became associated with cheap quick print services, and the smeary, greyed-out, fuzz-focus image) and also, with the desire not to be condemned to this category (where offset art is considered too unlimited to claim a place in the market of the fine art print).

Production vs. Reproduction

Paradoxes, paradoxes. The term *reproduction* is also fraught with contradictions. Offset is often a means of reproducing an image from another medium (art masterpiece post-cards)—but the concept of reproduction is always qualified. Transformation and mutation are effects of reproduction—the original is not only not present in its reproduced form, but is seriously altered by the photographic and printing processes—ink on paper is not, for example, oil paint, continuous tone photo, or water color on absorbent surface—it is a simulation of the same which references those visual codes by which it can be recognized that the original was such or such. But this concept of reproduction has little relevance to the offset artist/printer who is using the medium as a creative tool, since in most cases, these artists are intent on exploring the potential of the multifaceted process to make images which have no other original, do not exist outside of their production through the offset process.

To maximize the reproductive potential of the offset medium, then, one plays down, represses, the interventionary urges. To use offset as a means of producing images which have no pre-existence, no other corporeal life, the full range of manipulative operations are exploited. The simulation of sameness becomes a moot point in offset works which mutate in various stages of the printing process—except where editioned pieces follow printmaking protocol in spite of their use of mechanical means. Consistency, standardization, sameness—these are all readily achieved on an offset press, but there are offset artists for whom the mutability of process on press is part of the attraction of the medium. For others the darkroom remains the site of most interventions and manipulations, while for some the computer is the mutating instrument of choice.

[. . .]

Invisible Labor

For the moment, offset remains a mechanical process dependent on highly skilled and time-consuming labor. The seamless product tends to render the workedness of the production invisible. Labor, after all, does not disappear in a so-called post-industrial society, but it is rendered invisible, made to seem a natural function of the appearance of a product, rather than being a thing in itself. While it would be facile simply to equate art with work (however rarified) since it is clearly a much more complex activity than that, there is a sense in which visual art gains its auratic status by the way it represents specialized work as a category. One of the issues in the reassessment of offset has to be this calling to attention of the worked character of the product—not by counting runs, or ink layers, or by listing the hours of production—but by articulating the aspect of productivity which is constituted as labor, insisting on the value of production as it is, not by forcing it into some perverse imitation of fine art (limited editions or auratic originals). For offset to qualify as high art (according to modern or traditional criteria), does it have to conceal its having been made by machine? To force the offset artist into deconstructive imagery or printing techniques simply to call attention to production is artificial, though it is interesting that many of these artists do this. Todd Walker's simulations of industry style color separation (done in the darkroom, offset, and by collotype) serve as a good example here—calling attention to the medium in classic modernist terms, deconstructing it in classic postmodern method.[6] The laying bare of devices so central to the modernist critique of capitalist production, advocated by Viktor Shklovsky in literature,[7] Bertolt Brecht in theater,[8] Lev Kuleshov in cinema,[9] and Marcel Duchamp in Conceptual art,[10] finds expression in this offset work—but the use of mainstream industrial materials and production processes more aptly participates in the postmodern sensibility, one which profoundly challenges the autonomous status of art, at least in rhetorical terms.

Trying to see the history of the present requires a certain defamiliarization. Offset printing is the taken for granted stock in trade of daily print media, and yet the processes of its production are as unknown, unfamiliar, and unrecognized as those of microchip manufacturing, television broadcasting, or ultrasonic imaging. In short, offset requires professional skills and industrial equipment. In the trade one works for years to progress from apprentice to journeyperson to master. The distinctions which render offset productions creative productions rather than commercial ones are the same as those of any other artistic process—individual vision, subjective expression, and creative use of the medium. Since the terms on which work gains status as art are dependent largely on the perceived identity of the pieces, their producers, and their place in the culture, the status of offset work seems to have finally edged towards mature recognition as an artform. Because of the difficulties involved in its production, the complexity of the process, the skill required for running a press, it will probably remain the province of a limited number of artist practitioners. But this work seems ready for critical recognition and appreciation—whether it is offset art in the form of prints, or artists' books. The development of electronic media, oddly enough, has become the fulcrum point for critical reflection upon the value of and character of mechanical work—the very act of dematerializing information has made the act of rendering it back into form that much more self-conscious a process. The history of any technology is always under change—in relation to this moment in time offset has become newly defined. Offset no longer has an identity only as a means of industrial, mechanical reproduction, but is clearly capable of sustaining an identity as artistic production as well.

Notes to Introduction

1 I am grateful to Judith Walsh for bringing this essay by Hayter to my attention. *Art in Print* 2, no. 3 (September–October 2012), contains five articles on this innovative printmaker.

2 S. W. Hayter, *New Ways of Gravure* (New York: Pantheon Books, 1949), 155–61.

Notes to Reading

1 [Heidelberger Druckmaschinen AG is a leading manufacturer of offset presses.—Ed.]

2 [Today "rasterized" refers to an image that has been translated into pixels. *Raster* is the German term for the halftone screen formerly used in offset lithography to translate tones into patterns.—Ed.]

3 [The "soft blanket" is the thick rubber mat wrapped around the rotating cylinder in an offset press.—Ed.]

4 [The American art critic Clement Greenberg (1909–1994) glorified the direct, often messy manipulation of materials characteristic of Abstract Expressionism and championed the work of Jackson Pollock.—Ed.]

5 [Joseph W. Ruther (1932–) is an offset printer and artist.—Ed.]

6 [Todd Walker (1917–1998), American photographer and printmaker, is known for his image manipulation.—Ed.]

7 [Viktor Borisovich Shklovsky (1893–1984) was a Russian writer best known for his practice of defamiliarization, whereby clichés are introduced into literature, forcing readers to confront them in new ways.—Ed.]

8 [The German poet and playwright Bertolt Brecht (1898–1956) is known for his use of the defamiliarization effect in drama, that is, taking the banal out of its usual context.—Ed.]

9 [Lev Vladimirovich Kuleshov (1899–1970) was a Soviet cinematographer who advocated the highly deliberate and surgical editing of films; the process was often made obvious to viewers.—Ed.]

10 [Marcel Duchamp (1887–1968) was a French Surrealist and Dada artist.—Ed.]

RUTH WEISBERG

The Syntax of the Print: In Search of an Aesthetic Context (1986)

In this reading the American artist Ruth Weisberg (1942–) elaborates on the ramifications of the artist-printer collaboration in printmaking, opting in the end to accept the possibility of a technically inferior print in lieu of a perfect one. Since the artist and/or the printer may take creative or remedial measures in the course of their collaboration, conservators must be alert to the possible resulting treatment complications. For example, hard-to-detect corrections to spottily printed lithographs are often made using soluble felt-tipped markers. The assumption that prints from the same edition are physically identical is a risky one.

Process

[. . .]

Collaboration has almost always been a key aspect of the graphic arts. In the Orient and to a great degree in the West, printmaking has been a group effort, with tasks divided among artisans and artists. One need only think of Japanese *Ukiy-oe* to realize that collaboration need not compromise aesthetic accomplishment. Such artists as Rembrandt and Whistler, working alone, are really anomalies. While it is true that Whistler's *Nocturne: Palaces* from the Second Venice Set would be nothing but vague wisps without the artist's masterful wiping of the plate, even Whistler relied on others when making lithographs.

The consequences of collaboration seem to be twofold. It leaves the process vulnerable to the lowest common denominator. The artist's efforts can be compromised because he or she does not have total control. Because control or mastery by the individual is the preferred model of modern Western culture, contemporary

From RUTH WEISBERG, "The Syntax of the Print: In Search of an Aesthetic Context," *The Tamarind Papers* 9, no. 2 (Fall 1986): 59. Reproduced courtesy of Ruth Weisberg.

collaboration demands special social skills and sensitivity on the part of team members. Leonard Lehrer provides a most insightful examination of this aspect of printmaking, while defining some of the crucial desiderata for the non-artist members of the team: "Printers are a special breed, a breed which combines immense skill with diplomacy and endurance, patience with knowledge; they set the tone of the project, maintain its rhythm, and are expected to have answers for everything. . . . It is a unique relationship and a unique component of the art world."[1]

How much visible effect collaboration should have on the finished print is a controversial matter. Tamarind Lithography Workshop espoused a noninterventionist position. The printer should "detect the true spirit of the work and give it life, while at the same time avoiding any act which might tend to impose his own aesthetic upon that of the artist."[2] In contrast, Ken Tyler believes the relationship between the artist and printer to be symbiotic: "You can't separate it after a while and you don't know whether the suggestion came from the printer on the press or that it was the artist's idea. But you know that something's going on there and if it works, it's magic."[3] The question that should be asked is what the effect of interventionist and non-interventionist modes of collaboration may be upon the quality of the work and the aesthetic choices of the artist. Another way to put this question is to ask whether the art is more important than the artist, or whether the artist's development is more significant than the success of a particular print.

My education and formation as an artist tend to put me in the noninterventionist camp. [. . .] Authenticity, originality, and the artistic growth of the artist are important values for me, so while individual prints may benefit from the intercession of a printer or publisher, I feel the artist is impoverished in the long run. The workshop can become not just a technical resource but an aesthetic crutch.

Notes to Reading

1 Leonard Lehrer, "Artist and Printer: Some Matches Are Made in Heaven and Others . . . ," *TTP* 8 (1985): 49. [Lehrer is an American painter and printmaker.—Ed.]
2 Lucinda Gedeon, *Tamarind: From Los Angeles to Albuquerque* (Los Angeles: Grunwald Center for the Graphic Arts, UCLA, 1985): 13. [The Tamarind Lithography Workshop (TLW) was founded in Los Angeles in 1960 to "rescue" the dying art of lithography. Today it is affiliated with the University of New Mexico. http://tamarind.unm.edu.—Ed.]
3 Pat Gilmour, *Ken Tyler—Master Printer, and the American Print Renaissance* (New York: Hudson Hills Press in association with the Australian National Gallery, 1986), 32. [Ken Tyler (1931–) is an innovative American printer and publisher who actively collaborated with leading artists in the revival of limited-edition fine art prints in the mid- to late twentieth century.—Ed.]

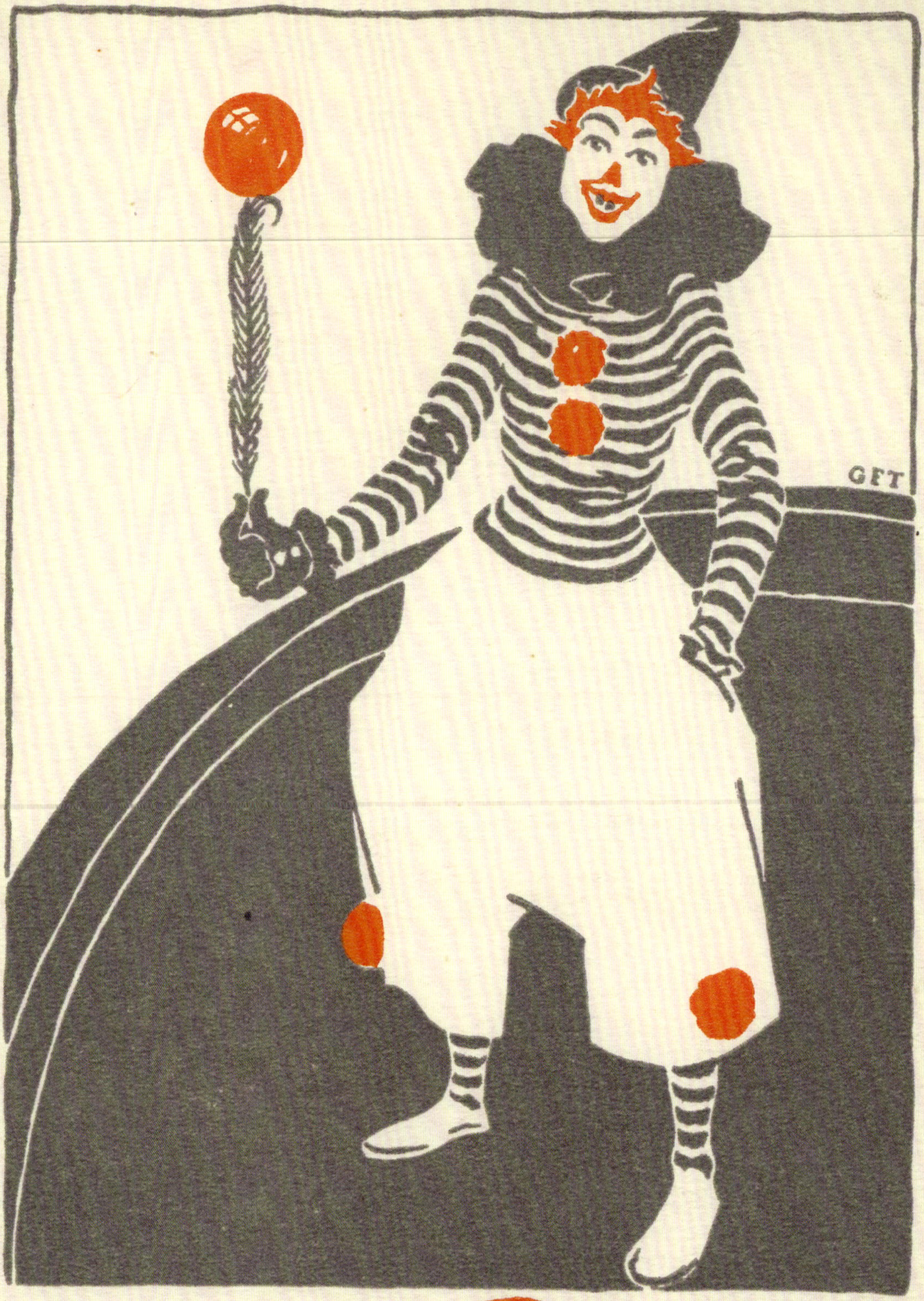

Paper is
part of the picture

PART IV

"Paper Is Part of the Picture"

As seen in Parts II and III, an indispensable skill for the paper conservator is the precise characterization of written, drawn, and printed marks. Marks transmit meaning not only through the interpretation of the words or images they depict but also through their physical substance and visual appearance. It has been recognized that an assessment of the materiality and facture of works on paper enhances their understanding, as evidenced by the increase in technical essays about specific media and techniques and an overall improvement in the accuracy and detail of gallery wall labels and exhibition catalogues.

Part IV considers the paper on which marks appear as a critical component of their materiality, one that, through its substance and appearance, also conveys meaning. What is to be gained by a fuller description of the physical and visual characteristics of the papers used by artists over the centuries? Searching for the answer is frustrating, since few truly evocative and informative descriptions of paper are found in the literature of art. Not only are observations about the paper substrate of prints and drawings rarely recorded, but there is a curious lack of language to describe paper; many of the terms that are employed are vague, misleading, or outdated. Yet most readers will agree that the substance and the appearance of paper vary enormously and play a critical aesthetic role. Several factors have contributed to the paucity of information on what paper—when used as a carrier of "artistic" marks—acts and looks like.

First, as Corte (reading 37) and Munsell (reading 38) demonstrate, quantifying and qualifying the optical or appearance properties of paper—transparency, opacity, brightness, color, and gloss—are fraught with difficulties. The measuring device, whether inanimate instrument or human eye, can result in data that are incomprehensible to the layperson, as in the former case of scientific color notations, or sensations that are unreliable, as in the latter case of eyesight. When relying on the eye, appearances are famously fickle:

"Paper is Part of the Picture," print by George F. Trenholm. From *Strathmore Munsell Covers: A Demonstration Showing the Simplicity with Which They Represent a Perfect Background and Bring Color into Printing*, 1925. Reproduced courtesy of NYU IFA Conservation Center Special Collections.

"Yes!" I answered you last night;
"No!" this morning, Sir, I say!
Colours, seen by candle-light,
Will not look the same by Day.[1]

Second, those with a vested interest in the accurate characterization of paper are generally not curators, art historians, and conservators. Rather, they are the paper manufacturers and sellers at one end of the spectrum and the purchasers and consumers (artists, print publishers) at the other. Corte addresses his remarks to "the buyer of paper"; Labarre (reading 39), to the "paper dealer" and "exporter"; and de Fontenelle (reading 40), to the "stationer and printer." Since members of the museum community usually do not make paper or intentionally manipulate it to fabricate paper-based products, their physical interaction with it is minimal. The determination of grain direction or degree of whiteness or absorbency—critical to the creation of well-functioning books, luminous watercolors, or velvety mezzotints—is simply not an everyday concern. Hence familiarity with the terminology and testing used to describe and measure such properties is not well established in the literature of art.[2] By becoming fluent in the language of the paper as used by industry and related trades, more precise and consistent, and thus more meaningful, vocabulary can be used to communicate the properties of paper.

Other successful approaches for describing paper are found in allied academic disciplines. For example, paper historians, musicologists, and scholars of cartography have made steady advances in the study of paper as physical evidence in dating and authentication studies. At the forefront, however, are our bibliophile colleagues who have made great strides in developing a taxonomy of paper, perhaps because of their closer acquaintance with paper as a commodity in the publishing industry. Bibliographers specializing in descriptive bibliography, the precise description of the physical materials and structure of books, have excelled in developing systems and templates for recording the papers used in book production.[3] It was they who pointed out the importance of identifying the wire or felt side of paper, of always comparing watermarks from the same side, and of locating the watermark relative to the rest of the sheet in order to extrapolate original sheet size. Their approach demonstrates that "the study of paper need not be a haphazard pursuit, that it can be practiced systematically on the basis of scientific principles, and that it can lead to reliable results."[4] Alas, as the paper historian and bibliographer Paul Needham (reading 44) observes, drawing and print enthusiasts have been slow to adopt this approach.

Needham's disappointment would appear to be well founded, perhaps because in academic art history formal stylistic analysis has historically been favored over attention to materiality and facture. As seen in Parts II and III, the physical characteristics of media and the idiosyncratic ways they are used have gradually entered the domain of art history and connoisseurship studies. When it comes to paper, however, most catalogue entries for prints and drawings make no mention of it, except for the inclusion of watermarks, often inadequately described in the opinion of Needham, or if the paper is colored or unusual, such as the Douglass Howell[5] papers or Kurt Schwitters's (1887–1948) Dada ephemera.[6] Hence the role of paper as an indicator of date, region, function, or artistic choice has been generally underrepresented in art-related publications; few of its physical and visual characteristics have been systematically recorded or consistently presented.[7] The most successful studies on artists' paper (readings 45–47) have been conducted by comparing as many sheets as possible and systematically identifying them using well-defined terminology understood and accepted by all stakeholders.

Notes

1 *The Lady's Yes*, by Elizabeth Barrett Browning (1806 –1861). A spermaceti candle has a color temperature of 1925° K; average noon daylight is 5500° K. See Leonard T. Troland, "Report of the Committee on Colorimetry for 1920–1921," *Journal of the Optical Society of America and Review of Scientific Instruments* 6, no. 6 (August 1922): table 9.

2 Standard methods for measuring paper properties have been established in the United States by the American Society for Testing and Materials (ASTM) and the Technical Association of the Pulp and Paper Industry (TAPPI) and internationally by the International Standards Organization (ISO).

3 G. T. Tanselle, "The Bibliographical Description of Paper," *Studies in Bibliography: Bibliographical Society of the University of Virginia* 24 (1971): 27–67. The seminar syllabus for Tanselle's course Introduction to Bibliography (Charlottesville: University of Virginia, 2002) is periodically updated and available online at www.rarebookschool.org.

4 John Bidwell, "The Study of Paper as Evidence, Artefact, and Commodity," in *The Book Encompassed*, ed. Peter Davison (New Castle, Del.: Oak Knoll Press, 1998), 73.

5 Douglass Morse Howell (1906–1994) was an artist and papermaker who almost single-handedly revived the craft of hand-papermaking in the United States.

6 Two early publications by Andrew Robison do consider the materiality of paper (see Further Reading), wherein he focuses on the surface, color, and dimensions of paper in prints. Still, he is hindered by a limited vocabulary for describing the "proper" tones of white, resorting to highly subjective adjectives such as *cold* or *dead*.

7 One recent exhibition at the Louvre, which focuses on artists and the papers they choose, is a notable exception: *Le papier à l'œuvre*, ed. Natalie Coural (Paris: Hazan Louvre Éditions, 2011).

The tendency to include descriptions of media and techniques but not those of paper is reflected in two books published in conjunction with an important exhibition at the British Museum in 2010, Hugo Chapman and Marzia Faietti, *Fra Angelico to Leonardo: Italian Renaissance Drawings*, and Janet Ambers, Catherine Higgitt, and David Saunders, eds., *Italian Renaissance Drawings: Technical Examination and Analysis* (see Further Reading). In the former, media and technique are described in exacting detail, along with insightful interpretation of their significance; any mention of paper, including watermarks, is reserved for the companion "technical" catalogue. The format used in the latter publication to record paper properties serves as an excellent model.

Reading 37

HEINZ CORTE

Perception of the Optical Properties of Paper (1976)

Heinz Corte, a paper chemist, designed an objective mathematical model for the structure of paper, a material that, in a curious state of disambiguation, he later characterizes as "the life blood of civilization and the enemy of oblivion."[1]

In order to describe paper, scientists divide its properties into two categories: mechanical and optical. Mechanical, also called structural, properties include basis weight or grammage, thickness or caliper, formation (the uniformity of distribution of the fibers within the sheet), directionality or grain (the predominant orientation of fibers), two-sidedness (felt side vs. wire side),[2] *surface texture or finish, and porosity. Conservators recognize that mechanical properties not only affect the appearance of the paper but, more important, influence its behavior, especially its interaction with water during treatments such as washing.*[3] *Therefore, a thorough assessment of a paper's mechanical properties is a critical component of the pretreatment examination. For example, the degree to which a paper will expand as it absorbs water—its hygroexpansivity—will inform strategies to be used in its subsequent drying and flattening. In examining works of art on paper, as opposed to disposable test sheets, many mechanical properties, for example, thickness, two-sidedness, and directionality, can be safely determined without the need to tear or fold. Spot testing, used to estimate absorbency, can also be performed, albeit on a microlevel.*

Optical, or appearance, properties include transparency, opacity, brightness, color, and gloss.[4] *Fluorescence resulting from the introduction of "optical brightening agents" (OBAs), especially into printing papers, is now considered as an additional*

From HEINZ CORTE, "Perception of the Optical Properties of Paper," in *The Fundamental Properties of Paper Related to Its Uses: Transactions of the Symposium Held at Cambridge: September 1973*, 2 vols., ed. Francis Bolam (London: Technical Division, British Paper and Board Industry Federation, 1976), 2:627–28, 632–33, 644–45, 653–55.

optical property.[5] *As Corte takes pains to point out, our eyes are not optical instruments; they tell us only what paper looks like, not what it measurably is, as does a spectrophotometer*[6] *or a colorimeter.*[7] *Our eyes transmit sensations to the brain, which, in turn, compares and judges. Therefore, when assessing the appearance of paper, we are concerned no longer only with physics but also with physiology and psychology. Since humans have fundamental difficulty assessing the optical properties of paper, several systems and standards have been developed for use in industry and design.*

Introduction

[. . .]

Now, when we test the optical properties of paper such as the brightness, the colour or the gloss, the one practical usage property that such measurements are expected to represent is for the paper to look right. In nine out of ten cases, when we speak of optical properties, we mean the appearance of the paper. [. . .]

It is the purpose of this paper to demonstrate that there is no reason to assume that our perception of the appearance of paper and the readings of optical test instruments are governed by the same physical laws. This is because our eyes are not optical instruments designed to measure. They are extensions of the brain; they do not measure, but transmit signals by which we judge. The signal produced by electromagnetic radiation on the retina is only one step of the process. It would be meaningless if there were not a brain behind it to compare the signal with stored information and thus pass judgment against the background of our experience and expectation. Everybody knows that we often do not notice things, although they are there, because they are of no importance to us at the time. The information is stored nevertheless and may be recalled later quite unexpectedly—*déjà vu*. In addition, we often see things that are not physically there. 'If, in the middle of the night, you put your head under the blanket, close your eyes and press your finger into the corner of one eye against the root of the nose, you see *bright yellow light*. The words in italics express sensations.' (After R. W. Pohl [. . .].)[1] As soon as you are occupied with measuring *light* (photometry) or with investigating such things as *colour, brightness* or *gloss*, you have left the realm of physics and have entered the realms of physiology and psychology. Physical optics is not concerned with the sensations of *light, colour, gloss,* etc., but with [the] nature of the radiation that causes these sensations. The buyer of paper on the other hand is very much concerned with the sensations he has when looking at a sheet of paper. The chief concern of the paper physicist in this area should be to explore the nature of the interaction between electromagnetic radiation and paper that causes those sensations. Possibly, some 'instrumentisation' of visual perception is unavoidable, if only as a basis for communication, but it is important to bear in mind that, when a sheet of paper causes a reading on the scale of an optical instrument, we are not necessarily measuring what we see. Our visual perception is controlled (or accompanied?) by electrochemical processes in the neural network of the retina. They cause a number of contrast phenomena and these in turn can cause a confusion between the 'subject knowing' and the 'object known'. They can make us believe we see things that are not there and, on the other hand, prevent us from seeing things that are there. Our sense of vision tells us only what a thing looks like, what it appears to be; not necessarily what it is in measurable physical terms.

[. . .]

Retardation Time

The two illuminated areas are projected on to a screen with the dividing line being vertical. The two areas are intersected by the shadow of a horizontal rod extending over both fields (Fig. 1). When the rod is at rest, the shadow is straight; when the rod is moved up and down, there is a break at the boundary between the two fields, because our perception of the shadow in the less bright area is retarded compared with that of the shadow in the brighter area. Again, the brightness of one of the fields is varied until the shadow cast by the moving rod remains straight. The two surfaces can thus be *defined* to be equally bright.

Formation

[. . .]

Fig. 2 demonstrates the effect of eye movement. The crossroads of the white roads between the black blocks appear shadowy, except the one on which the eyes are focused. As the gaze moves across the pattern, so do the shadowy spots in a fleeting manner. This is caused, in very simple terms, by the desensitisation of peripheral retina cells that are not involved in the image-forming process plus the fact that the roads between the blocks appear brighter than the corners. The perception is quite clear, but I cannot think of an instrument that would indicate the shadows, because they do not physically exist.

Concluding Remarks

Any confusion of physiology with physics that may arise when discussing optical properties of paper stems from two roots. Firstly, we use the word *light* both for the visual sensation and for the physical radiation that causes it. (We even speak of ultra-violet light!) This is because physics is an empirical science in the sense that it originates from our sensory perception of the environment and would not

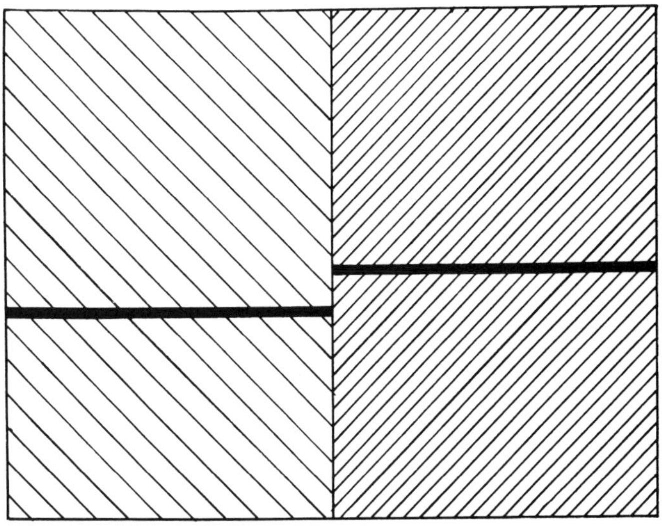

Figure 1
Method to define equal brightness as equal retardation time.

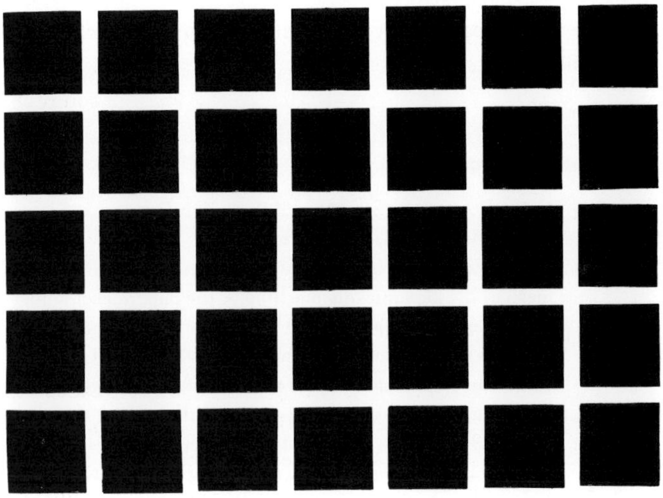

Figure 2
Perception of 'fleeting shadows' by eye movement.

exist without it. We perceive the outside world most of all with our eyes and the sense of vision is perhaps the most important one of the five, because 'light is the fastest, longest ranging carrier of information'.[2] Secondly, it is one thing to read '5·2 kg' off the scale of a tensile tester or '80 per cent' off the scale of an opacity meter, another to view a sheet of paper at a shallow angle to see how glossy it is. In the first two instances, our eyes serve to convey neutral information and everybody else will see the same; in the last instance, our eyes are the judges so to speak and different persons will generally differ. Put in another way, Fig. 1 [. . .] , for example, *convey[s]* messages, Fig. 2 *is* the message.

Finally, this paper is not about the instrumental quantification of visual perception, certainly not about instruments for quality control purposes. Yet it is worth mentioning that our knowledge of perceptive processes is far more advanced than is perhaps generally appreciated, certainly more than I knew when I started writing this paper. Several mathematical models have been devised to simulate the interplay between the various components of our visual apparatus. Many of them are virtually on/off devices and it would seem to be possible to simulate electronically at least some of the processes that control our vision. Here is a challenge and a potentially rewarding area of activity for the imaginative instrument designer. There is no danger that we shall lose our sense of modesty or a healthy respect for nature. Even over 100 years ago, Hermann von Helmholtz in his classic *Handbuch der Physiologischen Optik*[3] said that he could easily design a much better optical instrument than the eye is, but not a better eye; this is still true.

We have been given eyes to survive, that is, to observe and to judge and, let us add, to please and to be pleased. The prospects of building an instrument, however sophisticated, that really sees a piece of paper or the world at large as our eyes do are just about as bright as those of building one that recognises and appreciates beauty.

Notes to Introduction

1 Heinz Corte, "The Structure of Paper," in *Handbook of Paper Science*, ed. H. F. Rance (Amsterdam: Elsevier, 1982), 277.

2 Felt side, that part of a sheet which has not come into direct contact with the wire mesh of the papermaking machine or the papermaker's mold in the case of handmade paper; wire side, that part of a sheet which has come into contact with the wire mesh of the paper machine or, in the case of handmade paper, the papermaker's mold. Each side has a characteristic texture and, in the case of internally colored papers, may have a different color due to the settling out of the colorant.

3 For an in-depth discussion of the interaction of paper and water, see Gerhard Banik and Irene Brückle, eds., *Paper and Water: A Guide for Conservators* (Oxford: Elsevier, 2011).

4 While Corte classifies formation as an optical rather than a mechanical property, formation is typically included under mechanical properties, since it influences thickness. Corte's figure 15 (not reproduced here) is used as an illustration of the difficulty of assessing formation visually.

5 Martin A. Hubbe, Joel J. Pawlak, and Alexander A. Koukoulas, "Paper's Appearance: A Review," *BioResources* 3, no. 2 (2008): 632–33.

6 Spectrophotometry is the measurement of the relative amounts of electromagnetic power reflected or transmitted by a substance at individual wavelengths of the spectrum.

7 Colorimetry assigns unique identifiers to colors according to three independent variables.

Notes to Reading

1 [R. W. Pohl, *Einführung in die Physik*, vol. 3: *Optik und Atomphysik* (Munich: Springer-Verlag, 1948). Robert Wichard Pohl (1884–1976) was a German physicist.—Ed.]

2 Clarke, A. C., *The Lost Worlds of 2001* (Sidgwick & Jackson, London, 1972), 121.

3 von Helmholtz, H., *Handbuch der Physiologischen Optik* (Leopold Voss, Leipzig, 1867). [Hermann von Helmholtz (1821–1894) was a German physicist and philosopher.—Ed.]

Reading 38

A. H. Munsell

A Color Notation System (1905)

Albert Henry Munsell (1858–1918), American painter and art teacher, proposed his system for understanding, visualizing, measuring, and matching colors in 1905. Munsell hoped that by comparing colors with carefully calibrated color chips and coding them according to three measurements—hue, value, and chroma—his color notations could be decoded and "seen," much like musical scores, in which symbols representing pitch, intensity and duration can be decoded and "heard."[1] Immediately popular in industry and education,[2] the Munsell System's potential application also appealed to conservators and curators who wished to document colors in artworks.[3] Nonetheless, the system failed to gain broad acceptance in the museum community, especially for describing the color of paper. Several factors contributed to its unpopularity in this regard. The color reference guide Munsell Nearly Neutrals, *which most closely corresponds to the color of most papers, is very expensive.[4] A smaller, less expensive set,* Munsell Nearly Whites, *can be used only to designate the color of papers in good condition.[5] In addition, comparison with a known paper depends on a standardized light source, and matching can vary with the sample's gloss and texture. Finally, human color perception varies widely. Of seventeen experienced color matchers who were asked to arrange paper samples from whitest to darkest, only six agreed on the whitest and eight on the darkest.[6]*

Above and beyond these factors, it would seem to be basic human nature to describe the color of paper by comparison to familiar substances, for example, buff, ivory, or cream. Despite the imprecise and anachronistic nature of such terms,[7] they are still preferred over mystifying combinations of numbers and letters: a designation of 5Y 9.25/0.5 hardly brings to the mind's eye the color of eggshell, at least to those not fluent in Munsell-speak. When it comes to visualizing colors, the same can be said

---

From A. H. Munsell, *A Color Notation System* (Boston: George Ellis and Co., 1905), 7–10.

for the tristimulus data (L*a*b*; X, Y, Z; *or R, G, B) obtained by colorimeters or the % reflectance spectra produced by spectrophotometers.*[8]

An alternate option, intended to be used alongside the Munsell System, was first recommended to the arts and museum population in 1965.[9] *Introduced in* Museum News, *the Universal Color Language demonstrated how colors could be described using accurate, convenient, and easily understood terminology. The methodology had been published earlier, in 1939, and revised in 1955 as the* ISCC-NBS Method of Designating Color Names and a Dictionary of Color Names.[10] *It was followed in 1976 by a more user-friendly publication,* Color: Universal Language and Dictionary of Names.[11] *Using the Universal Color Language, any Munsell notation, as well as exotic commercial color names and historical terms, such as Corinthian Pink, can be "translated" into a Universal Color Name and visualized. Rather than comparison to familiar substances, or worse, fleeting emotions, basic color names are assigned modifiers such as pale, moderate, brilliant, and strong. Hence the Universal Color Name for 5Y 9.25/0.5 would be pale yellow, not eggshell.*

Despite being widely adopted by commerce and disseminated to the general population via Webster's International Dictionary, *the Universal Color Language failed to gain a foothold within the arts community, prompting George A. Agoston (1920–2000), a well-known engineer, painter, and arts educator, to lament that artists and designers, who were the anticipated benefactees of the system, often seemed to be unaware of its existence.*[12] *Its use in describing the color of bindings, paper boards, and dust jackets of books was also disappointing, despite having been recommended by G. Thomas Tanselle (1934–), a leading descriptive bibliographer.*[13] *Ruth Johnston-Feller (1923–2000), a well-known color chemist with close ties to the conservation community,*[14] *also endorsed the use of the Universal Color Language: "Curators and technical people in art conservation could identify a color name from this dictionary by which others could visualize an approximate color."*[15] *Despite these nationally accepted systems and standards, in the literature of art, describing color by metaphor prevails.*

Chapter I. Color Names

Writing from Samoa to Sidney Colvin[1] in London, Stevenson[2] says: "Perhaps in the same way it might amuse you to send us any pattern of wall paper that might strike you as cheap, pretty, and suitable for a room in a hot and extremely bright climate. It should be borne in mind that our climate can be extremely dark, too. Our sitting room is to be in varnished wood. The room I have particularly in mind is a sort of bed and sitting room, pretty large, lit on three sides, and the colour in favour of its proprietor at present is a topazy yellow. But then with what colour to relieve it? For a little work-room of my own at the back I should rather like to see some patterns of unglossy—well, I'll be hanged if I can describe this red. It's not Turkish, and it's not Roman, and it's not Indian; but it seems to partake of the last two, and yet it can't be either of them, because it ought to be able to go with vermilion. Ah, what a tangled web we weave! Anyway, with what brains you have left choose me and send me some—many—patterns of the exact shade."

(1) Where could be found a more delightful cry for some rational way to describe color? He wants "a topazy yellow" and a red that is not Turkish nor Roman nor Indian, but that "seems to partake of the last two, and yet it can't be either of them." As a cap to the climax comes his demand for "patterns of the exact shade." Thus one of the clearest and most forceful writers of English finds himself unable to describe the color he wants. And why? Simply because popular language does not clearly state a single one of the three qualities united in every color, and which must be known before one may even hope to convey his color conceptions to another.

(2) The incongruous and bizarre nature of our present color names must appear to any thoughtful person. Baby blue, peacock blue, Nile green, apple green, lemon yellow, straw yellow, rose pink, heliotrope,[3] royal purple, Magenta,[4] Solferino,[5] plum, and automobile[6] are popular terms, conveying different ideas to different persons and utterly failing to define colors. The terms used for a single hue, such as pea green, sea green, olive green, grass green, sage green, evergreen, invisible green,[7] are not to be trusted in ordering a piece of cloth. They invite mistakes and disappointment. Not only are they inaccurate: they are inappropriate.[8] Can we imagine musical tones called lark, canary, cockatoo, crow, cat, dog, or mouse, because they bear some distant resemblance to the cries of those animals?

Color needs a system.

(3) Music is equipped with a system by which it defines each sound in terms of its pitch, intensity, and duration, without dragging in loose allusions to the endlessly varying sounds of nature. So should color be supplied with an appropriate system, based on the hue, value, and chroma of our sensations, and not attempting to describe them by the indefinite and varying colors of natural objects. The system now to be considered portrays the three dimensions of color, and measures each by an appropriate scale. It does not rest upon the whim or an individual, but upon physical measurements made possible by special color apparatus. The results may be tested by anyone who comes to the problem with "a clear mind, a good eye, and a fair supply of patience."

[. . .]

Color has three dimensions.

(8) It may sound strange to say that color has three dimensions, but it is easily proved by the fact that each of them can be measured. Thus in the case of the boy's

faded cap its redness or HUE is determined by one instrument; the amount of light in the red, which is its VALUE, is found by another instrument; while still a third instrument determines the purity or CHROMA of the red.

The omission of any one of these three qualities leaves us in doubt as to the character of a color, just as truly as the character of this studio would remain undefined if the length were omitted and we described it as 22 feet wide by 14 feet high. The imagination would be free to ascribe any length it chose, from 25 to 100 feet. This length might be differently conceived by every individual who tried to supply the missing factor.

Notes to Introduction

1 In his introduction to the eleventh edition of Munsell's *A Color Notation* (Baltimore, Md.: Munsell Color Company, 1961), the former director of the Rhode Island School of Design, Royal B. Farnum, wonders why it took so long to devise a system for color notation: "No doubt the reason for the early development of musical notation is due to the fact that sound is transient, and so cannot be captured in permanent form as color. One is tempted to wonder whether music scores would be so common today if the phonographic record had come first" (9).

2 In the early 1920s, the Binney and Smith Company marketed sets of Munsell Crayola crayons in order to introduce the system to schoolchildren. Margaret Holben Ellis and Brigitte Yeh, "The History, Use and Characteristics of Wax-Based Drawing Media," *Paper Conservator* 22 (1998): 51.

3 All the pigments described in volume 1 of Robert Feller's *Artists' Pigments: A Handbook of Their History and Characteristics* (Washington, D.C.: National Gallery of Art, 1986) were described using Munsell notations and the Universal Color Language.

4 The *Munsell Nearly Neutral Book of Color* currently costs more than $1,000.

5 Munsell Nearly Whites have a constant value of 9.25 (10 is a nonexistent absolute white) for twenty hues and a chroma range of /0.5 to /2.5. The set falls short when matching papers that have darkened, i.e., increased in value.

6 R. E. Lofton, "A Measure of the Color Characteristics of White Papers," *Bureau of Standards Technical Paper No. 244* (Washington, D.C.: U.S. Department of Commerce, 1923), 675.

7 Today the terms *buff, ivory*, and *cream* are anachronistic: *buff* refers to the color of soft undyed buffalo hides, trade in elephant ivory was banned in 1989, and pasteurized fat-free milk is now preferred to cream; Lunning and Perkinson's reference to "buttery cream" (see reading 46) may be unfamiliar to many readers. Even eggshells are whiter than they used to be.

8 Measurements derived from spectrophotometers and colorimeters can be converted to Munsell notations.

9 Christian Rohlfing, Introduction to Kenneth L. Kelly, "A Universal Color Language," *Museum News Technical Supplement*, no. 8 (1965): n.p.

10 K. L. Kelly and D. B. Judd, *The ISCC-NBS Method of Designating Colors and a Dictionary of Color Names*, NBS Circular 553, November 1, 1955. U.S. National Bureau of Standards (NBS); Inter-Society Color Council (ISCC).

11 K. L. Kelly and D. B. Judd, *Color: Universal Language and Dictionary of Names*, NBS Special Publication 440, December 1976.

12 George A. Agoston, *Color Theory and Its Application in Art and Design*, 2nd ed. (Berlin: Springer-Verlag, 1987), 4.

13 G. Thomas Tanselle, "A System of Color Identification for Bibliographical Description," *Studies in Bibliography* 20 (1967): 203–34; Craig S. Abbott, "Designating Color in Descriptive Bibliography: The ISCC-NBS Method in Practice," *Papers of the Bibliographical Society of America* 84, no. 2 (1990): 119.

14 Ruth Johnston-Feller was the wife and close collaborator of the conservation scientist Robert Feller (see readings 63 and 64).

15 Ruth Johnston-Feller, *Color Science in the Examination of Museum Objects: Nondestructive Procedures* (Los Angeles: Getty Conservation Institute, 2001), 55.

Notes to Reading

1 [Sidney Colvin (1845–1927) was an English art historian, print curator, and literary commentator.—Ed.]

2 Vailima Letters, Oct. 8, 1902. [The Scottish writer Robert Louis Balfour Stevenson (1850–1894) is best known for *Treasure Island*: *The Vailima Letters* (New York: Charles Scribner's Sons, 1902). The quote is from p. 70.—Ed.]

3 [Heliotrope is a pink-purple color named after the flower. *Color: The Universal Language and Dictionary of Names* states that the color is quite variable and corresponds to a broad range of Munsell designations, depending on its commercial application in the fields of horticulture, interior decoration, or textile manufacturing.—Ed.]

4 [As capitalized here, the color Magenta refers to its origins as a dye discovered in the Italian town of Magenta, site of a famous battle in 1859. There appears to be no relationship other than timing between the Battle of Magenta and the dye's development. It is listed as a moderate purplish-red in *Color: The Universal Language and Dictionary of Names*.—Ed.]

5 ["Solferino" refers to the dye developed in Solferino, Italy, also a site of a famous battle in 1859. It is listed as a deep purplish-red in *Color: The Universal Language and Dictionary of Names*.—Ed.]

6 [The color "automobile" is difficult to identify. At the time of Munsell's writing, automobiles were predominantly black; however, gray, green, blue, and red models were introduced by the Ford Motor Company shortly thereafter, in 1908. The color is not included in *Color: The Universal Language and Dictionary of Names*.—Ed.]

7 [According to *Color: The Universal Language and Dictionary of Names*, "invisible green" is a dark grayish-green, which recedes into shrubbery. Popular for painted fences and trellises in Georgian and Regency England, the color, a mixture of yellow ocher, lead white, and black, was a favorite of the famous landscape designer Humphrey Repton. http://janeaustensworld.wordpress.com.—Ed.]

8 [George Field (1777?–1854) writes, "Pinks are sufficiently absurd names of yellow colours" (*Chromatography; or, A Treatise on Colours and Pigments and of Their Powers in Painting, etc.* [London: Charles Tilt, 1835], 84). Field, like Munsell, expressed his frustration with meaningless names of colors. According to *Color: The Universal Language and Dictionary of Names*, English Pink is equivalent to Stil de Grain Yellow, which is akin to sap green.—Ed.]

Reading 39

E. J. Labarre

Dictionary and Encyclopedia of Paper and Paper-Making (1952)

Emile Joseph Labarre (1883–1965) was a Dutch translator by day who dedicated his after-hours energies to the series Monumenta Chartae Papyraceae Historiam Illustrantia, issued by the Paper Publication Society, which he founded in 1948. It is no surprise that Labarre, a philologist, is bothered by the vague nomenclature and unclear etymologies of paper terminology common to the English, French, German, Dutch, Italian, Spanish, and Swedish languages. Reflecting the extent of his frustration, Labarre feels compelled to begin his chapter with a truncated quote from The Civil War, *book VI, by Marcus Annaeus Lucanus (39–65), in which a sorceress demands that a cadaver reveal its secrets. Her exhortation in its entirety better conveys Labarre's impatience: "Be not grudging, I pray: give events their names, their places; and provide a voice by which Fate may communicate with me."[1] Despite his best efforts, Labarre admits that some papermaking terms defy universally accepted definitions and easy communication.*

As Labarre notes, paper names based on their intended function are particularly problematic since technologies change rapidly. To cite one of his own examples, how many of his readers would recognize Cyclostyle paper as Japanese kite-making paper?[2] As the American graphic designer Egbert Jacobson (1890–1966) observed, "There is as much difference between papers as between a dish towel and a fine linen handkerchief."[3] A purely functional description of these two textiles as woven cotton or linen materials intended to absorb moisture does little to convey the enormous disparity in their appearance, not to mention their quality and cost. The same limitations apply to describing papers according to their correct technical terms. To read of the "cartridge" papers used by English artists in the eighteenth century and then by Georgia O'Keeffe (1887–1986) in the twentieth (see reading 47), or the reputed

From E. J. Labarre, *Dictionary and Encyclopedia of Paper and Paper-Making*, 2nd ed., rev. and enl. (Amsterdam: Swets and Zeitlinger, 1952), vii–viii.

"butcher" paper of Egon Schiele (1890–1918), contributes nothing to one's mental picture of these sturdy water-resistant papers since few readers can relate to wrapping gunpowder or custom cuts of meat.[4] *Conversely, technical names specifying a paper's ultimate usage, for example, newsprint, are good predictors of a paper's behavior and thus appeal to conservators. And, as in an artist's choice of a dish towel or fine linen handkerchief as a support, functional paper names provide tantalizing clues regarding intent or the circumstances surrounding an artwork's creation. A good example of this is the* Birds of America *by John James Audubon (1785–1851), which was printed on a drawing paper rather than plate paper. This knowledge increases our admiration that a copperplate printer could produce such fine images on a relatively unyielding paper and shows the attention paid to the high quality of the hand-coloring.*[5]

Introduction

. . . da nomina rebus
LUCAN VI, 773.

It has been said that we cannot distinguish smells because we have no names for them; we can give dozens of colours a name and are so much the richer in consequence.

This book began as an attempt to name and define clearly a number of material things and concepts in the paper field that I had found to be vague, ill-defined or even applied to different objects, processes or ideas. On asking paper experts for information, I had found that one and the same word was used for quite different objects in one and the same country, while words adopted from a foreign language tended to diverge widely from the original meaning and thus added to the confusion. All this debatable vocabulary has had to be dealt with and if the definition of the words and their meanings, collected here from books and journals or extracted by me from long-suffering experts, have become too lengthy, the fault lies rather, I think, with my material than with myself, though of course my zeal may well have outrun my discretion.

Nevertheless the idea of gathering this information springs from a genuine desire, to know accurately what a word means, so that it can be correctly applied and the thing or use distinguished by an increasing number of individuals.

As stated in the introduction to the first edition, this work was intended as an exercise in comparative philology, as an attempt to ascertain the origins, growth and methods of variation of the vocabulary of a craft. [. . .]

The restriction of word-entries and their foreign equivalents to terms recorded in books and periodicals as had been planned for the first edition, had, on the whole, to be relinquished. This was not due to lack of effort but to the fact that several countries whose languages are recorded here have not reached that stage of development requiring the publication of studies on the side lines of paper-making, such as the historical aspects, including watermarks, and all branches of the technical side, which are treated in great detail in Anglo-Saxon countries and the chief paper-producing countries of Europe. Such terms may, indeed, frequently be found in the leading foreign magazines of the paper trade and industry, but these are, almost necessarily, loan translations from some other language, or are given in a descriptive phrase.

As explained in the introduction to the first edition (p. 9) an attempt to compile a technical dictionary in two or more languages soon reveals the fact that no two terms in different languages, except perhaps in highly scientific language, which has a vocabulary of its own, are fully synonymous or equivalent. The habits of the public and the customs of the trade in the various countries will frequently differ to such an extent that the equivalents do not exist at all and any attempt to reproduce the concept in a single word approaches the ridiculous. Also, the manner in which one people or one language names a thing varies considerably; an object in Germany may, for instance, be called after its use, in France after its appearance and in England after certain historical associations untraceable to all but the philologist. These differences are particularly obvious in the case of sizes and qualities of paper, frequently so typical or national that many a paper dealer can tell at a glance in what country a book was published, where the note-paper was made or where a parcel originated from.

It is therefore not surprising that each language has coined its own terms for these things which cannot, without danger of error or confusion, be reproduced by what may be termed a translation loan word (Jespersen)[1] in another language.

Users of the foreign terms recorded in this book, particularly exporters and their salesmen, are warned that the qualities normally found in foreign countries and sold under names apparently corresponding to those used, for instance, in the countries of production, not only vary considerably but occasionally the name indicates a type of paper totally different as regards origin, composition and use. A number of the most glaring misnomers in the one or other language may be found in the English entries relating to: Art, Bristol, Cap, Carbon, Cartridge, Cyclostyle, Duplicator, Ivory, Manifold, Parchment, Stencil, and many others.

[. . .]

On scrutinizing the foreign equivalents, another characteristic may be noted, namely, the difference in method of expression of a concept as applied by the Germanic languages and by the Romance languages. The former can appose a term to another in order to indicate composition or other relationship, such as origin or purpose, while the latter will insert a preposition as, in French for instance, *à*, *de*, *pour*, etc. The Germanic languages are therefore more concise but, on the other hand, frequently leave the reader to guess at the relationship intended; in other words, they are more vague.

Notes to Introduction

1 *Ne parce, precor: da nomina rebus, da loca; da vocem, qua mecum fata loquantur.* Trans. Millar Maclure. Loeb Classical Library.

2 The Cyclostyle duplicating process was invented by David Gestetner (1854–1939) in the late nineteenth century in London. The method used a patented Cyclostyle, which resembled an engraver's roulette, to simultaneously trace writing and pierce the sturdy Cyclostyle paper placed below it. Ink was then forced through the perforated stencil using a roller. http://en.wikipedia.org/wiki/David_Gestetner.

3 Egbert Jacobson, "Preface," in *Specimens: A Stevens-Nelson Paper Catalogue* (New York: Stevens-Nelson Paper Corporation, 1953), n.p.

4 The notion that Schiele used butcher paper may be fed by a desire to reinforce the romantic image of a poverty-stricken artist. In fact, many of Schiele's drawings and watercolors are on an imported "japan" paper manufactured in America by Strathmore and identified by its characteristic thistle blindstamp. I am grateful to Judith Walsh for debating Schiele's paper choices.

5 I am grateful to John Krill for bringing Audubon's paper specification to my attention. The artist was also particular about the paper used for the original watercolors for *Birds*, which were executed on fine wove Whatman paper. See Marjorie Shelley, "Drawing Birds," in Roberta Olson, *Audubon's Aviary: The Original Watercolors for "Birds of America"* (New York: New-York Historical Society, Skira Rizzoli, 2012), 108–31.

Note to Reading

1 [Jens Otto Jesperson (1860–1943) was a Danish philologist; translation loanwords are modeled more or less closely on the foreign ones.—Ed.]

Reading 40

J.-S.-E. (Jean-Sébastien-Eugène) Julia de Fontenelle

The Papermakers' and Paper Merchants' Complete Manual (1828)

An irrepressible nationalist, Julia de Fontenelle (1790–1840), compiled up-to-the-minute status reports on diverse French industries, including hat making, leather tanning, and glassblowing. Preceding Labarre (reading 39) by well over a century, de Fontenelle addresses the difficulty of accurately assessing the appearance and structural properties of paper, especially those from different countries of origin, and the economic repercussions of misjudging them. When comparing the qualities of writing and printing papers, he appeals to the vested commercial interests of the stationer and the printer. In order to become a "connoisseur" of paper, he recommends a cognitive exercise involving all five senses: smell (to determine fiber processing),[1] *taste (to determine the degree or kind of sizing),*[2] *touch (to determine smoothness and weight), and hearing (to determine degree of sizing by "rattle" or de Fontenelle's "crumpling"), in addition to sight. While such forcible manipulation of prints and drawings is not advisable, curators and conservators would do well to experiment on expendable sheets to familiarize themselves with these properties of paper. The laborious process of becoming a paper connoisseur and the benefits of instinctively recognizing an artist's preferred papers are described by Judith Walsh (reading 47).*

In reviewing the historically desirable properties of paper, it is interesting to note which qualities are obsolete today. One property of critical importance for de Fontenelle and for Clapperton (reading 41) is the surface receptivity of the paper to a dip pen and ink; in their view, not only is high-quality writing paper good advertising, but it exerts less strain on the hand. A familiarity with past technical specifications for papers having highly specific functions is necessary in order to assess their current condition. For example, it should not be surprising that a mid-nineteenth-century copperplate paper is foxed and filled with grime due to its soft-sized and open, porous

---

From J.-S.-E. (Jean-Sébastien-Eugène) Julia de Fontenelle, *Manuel complet du marchand papetier et du régleur* (Paris: Roret, 1828), 8–10.

qualities. Likewise, a newspaper from 1820 has different properties than a newspaper of 1920 because of different printing technologies. A Renaissance drawing on blue paper should bring to mind a very different type of paper than a blue paper used by Joan Miró (1893–1983), and a 1760 writing paper is functionally different from one from 1860 and even more from one from 1960.[3]

Knowledge of Papers

Knowledge of papers is one of the key elements of the stationer and printer. Such knowledge is very difficult to acquire; above all, much study is necessary, along with long experience, in order to recognize:

1. Format at a glance
2. Ream weight at the touch of a sheet
3. Quality and degree of glue by crumpling or tasting
4. Type of fabrication (triturated or pulverized) by smell
5. Whether it is natural or bleached with chlorine or chlorinated lime by its whiteness
6. Its quality by the evenness and fibrousness pulp, with or without bubbles
7. Whether it is English, Belgian, French, Dutch, etc. and from what factory by the chain lines and laid lines, as well as by its overall fabrication

Just as high quality paper not only makes a good impression but adds beauty to the writing, we would also point out that, when the paper surface is more even, the pen encounters less friction; this means that, being less hindered in its advancement, it slides more easily over the paper: thus, with fine paper, the writer accomplishes more work and his hand is less tired.

Mr. Fournier[1] states that the key qualities of paper are those that constitute its regularity, such as the evenness of the sheets with respect to dimensions, thickness, and nuance. It is these three conditions that should especially be sought, since a lack thereof is the most harmful to book-binding. In fact, it is easy to see that unequal dimensions destroys the white balance and consequently decreases the outer margin. As for the thickness, it is certain that it produces an obvious difference in the pressure gradient and then in the printing. Finally, it is clear that uneven nuance is very unpleasant to the eye. Paper combines these three conditions when the pulp is pure, well ground, and evenly spread over the sheet, with no shadows, knots, folds, or wrinkles, with no straw or rust, no grit, dust, or gap, no water spots or glue or other stains, etc.[2] In addition to that, the sheets of paper should be squarely finished, free of the types of folds that form during fabrication. Such defects are worse than one might think; because of these, at the printing house, letters that are split do not disappear during a print run, which, as we see, is one of the most serious disadvantages.

Writing paper must be very well sized, whereas paper intended for printing need not be, unless one wishes to write on it after it is printed, as with paper intended for use as a draft copy for correction. Non-sized printing paper is whiter, softer, impregnates more uniformly, quickly, and evenly, satinizes better, etc.

A wide variety of paper qualities are fabricated, depending on the beauty, fineness, purity, and crushing of the rag. It is impossible to require that those of inferior quality meet most of the conditions we have just set forth, which are applicable only to superior paper quality. [. . .]

Notes to Introduction

1 Postdating Fontenelle's diagnostic, the popular smell of old books indicates the presence of lignin, which is chemically related to vanillin.

2 Sizing, typically gelatin, could be applied after sheet formation and drying, i.e., tub sizing; or mixed into the paper processing beater, i.e., engine sizing or E.S. Alum, aluminum potassium sulfate or later aluminum sulfate, could be added to harden the gelatin; its presence can be detected by its acrid taste.

3 I am grateful to John Krill for citing these examples.

Notes to Reading

1 [Henri Fournier, *Traité de la typographie* (Paris: H. Fournier, 1825), 258–59.—Ed.]

2 [M. Brun, *Manuel practique et abrégé de la typographie française* (Brussels: Lejeune Fils, 1826), 26, 27.—Ed.]

Reading 41

R. H. Clapperton

Paper and Its Relationship to Books (1934)

Echoing de Fontenelle's (reading 40) financially motivated sentiments, Robert Henderson Clapperton touts the economic benefits of high-quality writing papers at a time when the industry was suffering. His close observations regarding the surface of writing papers suggest a constructive exercise that conservators and curators can undertake. While we routinely describe the surface of paper as smooth, how many of us are attuned to the subtle yet different effects that tub-sized and internally sized paper can have on the appearance of pen-and-ink drawings? Certainly, artists encountered the drag exerted by Clapperton's "hairs" and selected their papers accordingly as they also chose which side to use (see reading 42). Today a paper's resistance to a dip pen is of little consequence; preference is given to extreme whiteness, typically enhanced by optical brighteners,[1] and the ease with which the paper will glide through high-speed printers. The distinction between commercial papers for printing and those for writing has drastically narrowed, perhaps reflecting the demise of the latter. Even in the tiny niche market of artists' papers, maximum functional versatility is the number one consumer requirement.

In my view all writing papers should be tub-sized, and no paper which is not so sized is as firm or lasting or so easy or pleasant to write upon. In a tub-sized paper we write on a smooth skin of gelatine; in the case of an engine-sized, or E.S., paper we write on the fibres themselves, or on the fibres and clay used to fill up the crevices. These papers are not so pleasant to write on; the pen picks up 'hairs,' which are small loose fibres, which in a tub-sized paper would have been glued down by the gelatine. No engine-sized paper has the character or dignity of

From R. H. Clapperton, *Paper and Its Relationship to Books* (London: J. M. Dent and Sons, 1934), 52–53.

a tub-sized paper, and it is always flabby in comparison. In order to hide its imperfections and shortcomings it is usually much more highly glazed,[1] but no amount of faking can give it the distinction of a rag, tub-sized writing paper, and it is never so resistant to ink.

In my opinion one of the best, and cheapest, forms of advertisement for any business is a really good letter-heading on a fairly thick high-grade, tub-sized, rag paper.

Note to Introduction

1 Optical brightening agents (OBAs) are daylight fluorescent dyes, which absorb ultraviolet wavelengths of radiation and reemit it as visible light. OBAs will eventually lose their ability to fluoresce, leading to a paper that appears less white. Daylight fluorescent dyes are also added to media to produce Day Glo colors. These will also lessen in brightness over time, often accompanied by fading of the colorant—separate but simultaneous mechanisms. See Margaret Holben Ellis, Christopher McGlinchey, and Esther Chao, "Daylight Fluorescent Colors as Artistic Media," in *The Broad Spectrum: Studies in the Materials, Techniques, and Conservation of Color on Paper* (London: Archetype, 2002), 160–66.

Note to Reading

1 [Glazing is a finishing process in papermaking whereby smoothness is imparted to the surface of a paper. Until the late eighteenth century this was done by the action of hammers, later by passing the paper between two high-pressure-heated metal rollers or calenders.—Ed.]

Reading 42

JAMES HARPER

Verso: The Flip Side of Master Drawings (2001)

As recommended by Meder and Cohn, examination of the versos of drawings and prints can result in unexpected discoveries.[1] *The art historian James Harper (1965–) delves deeply into the pitfalls of designating rectos of double-sided drawings and the ambiguous status of versos. The determination of what constitutes a drawing's recto, by inference the "correct" or most "valued" side, is not straightforward. While Harper discusses the ramifications of favoring one side over the other, it seems logical to ask if there is a "right" or a "wrong" side of paper at the time of its manufacture, that is, before the deposition of any marks on its surface. Paper has two sides—a wire side and a felt side—and each often is characterized by a distinct surface texture and occasionally by a different color. The side adjacent to the wire of the mold in handmade paper or the moving mesh screen in machine-made paper is called the wire side. It is not as smooth as the reverse, or felt, side. In handmade papers, the felt side may bear the imprint of individual hairs of the wool-fibered felts between which the paper was pressed. Although the paper surface is still smoother than the wire side, these fibrous traces impart liveliness to it, an important visual characteristic, especially notable in early prints.*[2]

In the case of printing, the felt side is preferred so as not to interfere with quality of impression. Also, in book production, paper needs to be consistently oriented according to sidedness and grain direction in order to ensure the book's smooth operation.

For wet media such as pen and ink or watercolor, the felt side is also preferred due to its more uniform texture; the pen's progress across the sheet is less hampered by the peaks and troughs of laid and chain lines, and watercolor washes do not pref-

From JAMES HARPER, *Verso: The Flip Side of Master Drawings* (Cambridge, Mass.: Harvard University Art Museums, 2001), 11, 21. Reproduced by permission of the Harvard Art Museums.

erentially settle into them. Wove paper, which still exhibits two-sidedness, was considered a significant technological advancement in terms of its continuously smooth surface texture. Its enthusiastic reception among watercolorists in eighteenth-century England is illustrated in these excerpts from a November 10, 1767, letter by Thomas Gainsborough (1727–1788): "I should take it as a particular favor if you could send me half a Doz^{n}. Quire of the same sort of Paper as the fifth Edition of the new Bath Guide is printed on, it being what I have long been in search of for making wash'd Drawings upon. . . . There is so little impression of the Wires, and those so very fine, that the surface is like Vellum." Two weeks later, after receiving the wrong paper, Gainsborough responds, "I . . . could cry my Eyes out to see those furrows."[3]

Artists are attuned to the nuances of surface texture. To wit, James McNeill Whistler (1834–1903), writing a century after Gainsborough, asks his printer Auguste Delâtre (1822–1907) for a paper whose distinctive grain or texture would enliven his dry chalk lines: "Enclosed I am sending you a sample of the brown paper with which you wrapped the etchings when you sent them to me—Well—so you have some of this paper or you know at least where you bought it—Now it is just the paper I need for my drawings. I am always looking for some—but although all the shops have brown paper, it is rare to find any with this fine grain—So be kind enough to buy a good packet of it for me at once—at the tobacconist's where you acquired this—but you understand the very same."[4]

A sheet of paper is a three-dimensional object, a fact that is too easy to forget in a standard gallery installation of drawings. Every sheet has two sides, and something interesting often lies on the flip side, hidden from all but curators and specialists. In the lingo of the art historian, the front of a work on paper is called the recto, while the back is called the *verso*. [. . .]

The terms *recto* and *verso* are of Italian derivation—hardly surprising, as Italian Renaissance writers led in the development of early modern art history and art theory. But while most Renaissance concepts have been vigorously challenged in more recent years, the distinction represented by these two terms persists, perhaps on account of its basic utility. [. . .]

In most cases, the paper itself gives no sign as to which side should be considered recto and which verso.[1] It is the artist—whether conscious or not of making a choice—who selects one side of a clean sheet by marking it, thus making a recto. But when artists work on both sides, they seldom bother to declare one recto and the other verso, and it remains to the collector, the dealer, the curator, or the scholar to introduce the distinction. Determining which side is which is almost always a retrospective act, for the need to classify is not inherent in artistic creation. Inevitably, declaring a recto and a verso is a selection that also involves a value judgment, since "the front" is implicitly better and more important than "the back." The designation of the recto, though often taken for granted, invariably reflects someone's preference for it as the more significant or more attractive side.

Interestingly, the terms *recto* and *verso* do not even come from the world of drawings, but from the language used to describe manuscripts and books. An examination of their etymologies is illuminating: as you look at a book open before you, the page on the right is the *folio recto*, or the "right leaf." The page on the left is the *folio verso*, or the "turned leaf," reached by turning over the previous *folio recto*. Old cultural associations that privilege right over left are certainly operative in the application of these terms to drawings. Like the English word *right*, the Latin word *rectus* (root of such forcefully positive words as rectitude and rectify) can translate as either "right" or "straight." Verso, regardless of its value-neutral root, stands as the opposite.[2]

The traditional museum or gallery display of drawings favors the recto over the verso by effectively suppressing the verso: when a sheet is exhibited in a frame on the wall, only one side can meet the viewer's eye. [. . .] When a double-sided drawing is matted and framed for display on a gallery wall, the result is a fragment, a selection, that denies the free movement that is part of the original object's essence. What makes matters worse is that collectors often trimmed a sheet to make the design on the recto fit more pleasingly or symmetrically on a mount or an album page, a practice that resulted in permanent losses to many versos.

Notes to Introduction

1 Intriguing information about Dürer's printing practices is revealed by examining the versos of his *Meisterstiche* opus (Marjorie B. Cohn and Craigen Bowen, "Dürer's *Meisterstiche*: The Other Side," in *Printed on Paper: The Techniques, History and Conservation of Printed Media*, ed. Jane Colbourne and Reba Fishman Snyder [Newcastle upon Tyne: Northumbria University, 2009], 21–26). See also Angela Campbell, "Finding Folds: Albrecht Dürer's *Meisterstiche* Papers," *Print Quarterly* 29, no. 4 (December 2012), 405–11.

2 Andrew Robison draws attention to the aesthetic effect of this felt texture in fifteenth-century prints: "An Aesthetic History of Paper in Prints," in *Paper—Art and Technology* (San Francisco: World Print Council, 1979), 62.

3 *Letters of Thomas Gainsborough*, ed. John Hayes (New Haven, Conn.: Yale University Press, 2001), 44. For the impact of wove paper on the development of watercolor painting, see Theresa Fairbanks Harris and Scott Wilcox, with Stephen Daniels, Michael Fuller, and Maureen Green, *Papermaking and the Art of Watercolor in 18th-Century Britain: Paul Sandby and the Whatman Paper Mill* (New Haven, Conn.: Yale Center for British Art in association with Yale University Press, 2006).

4 Note that by "fine," Whistler means "excellent," not "refined." James McNeill Whistler, *Letter to Auguste Delâtre*, October 1871/1874, Collection Frits Lugt, Institut Néerlandais, Paris, #1972-A779. www.whistler.arts.gla.ac.uk/correspondence.

Notes to Reading

1 While antique laid paper, and later papers as well, generally have one side that is noticeably rougher than the other, neither side is inherently preferable, as different effects may be achieved with each. For artists' exploitation of the texture of the

support, particularly for chalk and charcoal drawings, see Philip Rawson, *Drawing* (London, 1969), 53; and Edward Saywell, "Behind the Line: The Materials and Techniques of Old Master Drawings," *Harvard University Art Museums Bulletin* 6, no. 2 (1988): 7–39.

2 [The use of value-laden language continues even in the most recent and objective publications. To cite two examples: Edward Lucie-Smith, in *The Thames and Hudson Dictionary of Art Terms* (London, 1984), 196, defines verso as "the 'wrong' or reverse side of a two-sided object, e.g. a sheet of paper"; the *Grove Dictionary of Art* (London and New York, 1996), vol. 32, 376, uses similar language, defining verso as "the wrong, back or reverse side of a flat, double-sided object such as a drawing."—Ed.]

Reading 43

ALLAN H. STEVENSON

Watermarks Are Twins (1951–1952)

Our colleagues the bibliographers maintain the lead in developing a taxonomy of paper and can best provide methodological guidance for its description. Known for his intensive studies of watermarks and his high standards for their recording and interpretation, Allan Henry Stevenson (1903–1970) illustrates how nearly identical watermarks are the result the synchronized movements between the vatman and his coucher.[1] *The choreography of their dance is described best by Eden Phillpotts in* Storm in a Teacup *(for another excerpt, see reading 6). The vatman Jordan Kellock has completed dipping, lifting, and shaking his pulp-filled mold: "His stroke accomplished and the sediment leveled on the mould, Kellock brought his mould to the 'stay'—a brass-bound ledge on his left hand. He lifted the deckle from it as he did so and the full mould was drawn up the stay to the 'asp,' where his coucher stood. Then Kellock clasped the deckle on to his second mould, now returned from the coucher, and dipped again, while his assistant, taking the full mould from the asp, turned it over on to the accumulating pile of sheets rising on his plank. Then he ran the empty mould back along the bridge to Kellock's hand and drew to himself the next full mould now waiting for him on the stay."*[2] *Thus two molds are used interchangeably, each having the same but ever so slightly different watermark.*[3] *After establishing that watermarks are fraternal, not identical, twins, Stevenson goes on to delineate their minute differences with extraordinary precision.*

A paper's watermark can indicate a right or wrong side. Historically, according to Stevenson, the letters or numbers of a watermark when sewn onto the mold were reversed. Hence they could be read from left to right in proper orientation when viewed from the felt side. Although this practice appears to have been inconsistent

---

From ALLAN H. STEVENSON, "Watermarks Are Twins," *Studies in Bibliography* 4, (1951–52): 57–71, 87–90. Reprinted by permission of the Bibliographical Society of the University of Virginia.

from country to country and from century to century, this is the protocol followed today. Artists are instructed to use the paper with the watermark read correctly from the felt side; a similar habit is advised for fine writing papers.[4] It can be argued that watermarks have acquired disproportionate, to the point of ironic, significance in contemporary artists' papers. Prominently positioned and sometimes visually distracting, they mimic brazenly monogrammed luxury goods, a far cry from paper's traditionally modest supporting role. Like the skeuomorphic laid patterns and simulated deckle edges of today's "fine" stationery, many modern watermarks are the products of marketing rather than the actual papermaking process.

> In the first place, *there is no such thing as a duplicate.* Of course not. Can you print two copies on the same paper? Then, so far as the paper itself and the watermarks are subjects of enquiry, "duplicates" may be astoundingly different.
>
> —*Falconer Madan before the Bibliographical Society (1911)*[1]

Most early books are printed on a variety of papers. They contain a number of different watermarks. Though now and then a small volume has "one mark throughout," and a tall folio shows a single stock of fine paper, the majority of books are not so consistent. Often a well-printed folio, starting out with a run of grape paper (say), shifts to crown or fleur-de-lis paper by the time it reaches its second alphabet. In other folios, no less proud ones, the preliminaries alone disclose several marks, and the text a dozen or sixteen more, with alternation of marks through successive sheets. And, as for quartos, the next play one leafs through may offer an assortment of seven watermarks within eleven sheets. Those who have examined the papermarks in early books know these things are so.[2]

Indeed, mixed watermarks began with the Gutenberg Bible;[3] and they persisted through the Renaissance into the eighteenth century, as a normal feature of books printed on handmade paper. A volume printed by Wynkyn de Worde, the *Vitas Patrum* of 1495, is said to contain nearly fifty different marks.[4]

There have been explanations. Sometimes (it has been said) printers bought cheap job lots of paper.[5] Such an explanation neatly fits the bibliographically famous Shakespearian quartos of 1619. [. . .] As for the job-lot hypothesis, it remains plausible, but it has not been proved, and may not be necessary to account for the multiplicity of marks.

The moderate mixture of papers generally found in books allows a *sui generis* explanation. It lies in the manner in which paper was manufactured, packed in quires and reams, gathered by factors into bales, transported by waggon or boat, and shelved in warehouses. As a generalized procedure is hard to visualize, let us think in terms of books printed in England in the seventeenth century—though the implications extend beyond these limits of time and place. For in the seventeenth century the ordinary papers used in English books came mainly from just across the Channel, from Normandy and Brittany.[6]

The rushing streams near such towns as Vire, Sourdeval, Fougères, Morlaix turned the wheels of a host of papermills, small family *usines*, mostly, with just one vat to dip paper-moulds in, and a small output daily.[7] Though little is known of his operations, a *marchand papetier* or factor seems to have bought up (weekly, say) the paper from various mills in a district.[8] As sometimes many mills were clustered along one stream[9] and may have been served by the same maker of moulds, we find here a reasonable explanation of the diversity of marks in books and also of the similarity of design often seen in a sequence of such marks. Paper watermarked with the arms of France and Navarre may appear in a series of sheets, all of one pattern but with the labels of various papermakers attached.[10] When reams and bales of paper stood on the wharves at Caen or Morlaix or Rochelle,[11] we may suppose they had been sorted rather according to size and quality than according to make and mark. And no doubt the piling into cargo boats, the unloading at London, the stacking in the warehouse of the London paper merchant and later on the shelves of the printer he sold paper to—all this handling contributed to the mixture of papers we find in books. And much might happen in laying out tokens of paper for the presses. All in all, it is hardly surprising there were few books consistent in their paper, hardly surprising that sometimes reams of differently marked papers were used in the printing of individual edition-sheets.[12]

Yet this is not all. Within the reams themselves the paper was in a sense consistent. But a subtle variation occurred, due to the number of moulds, and occasionally the number of vats, used in its making.

Students of the history of paper know that pairs of moulds were used in its manufacture. Regularly, if not inevitably, two similar moulds were employed together, in all mills, in the making of all sizes and qualities of paper. It is so today for homemade paper.[13] The vatman, after placing the deckle,[14] plunges a mould into his tub of paper pulp, expertly takes up just enough of the paste or allows the excess to run over the farther edge, removes the deckle, and passes the mould along the stay[15] to his helper. The coucher, after allowing the sheet to dry a moment, turns it out on the felt or woolen on top of the post[16] he is building, puts the mould on the bridge,[17] and presently lays another felt for the following sheet. Meanwhile, the vatman has taken up a second mould, added the deckle, and is forming the next sheet as he formed the first. The process goes on all day, the vatman passing the mould with the newly made sheet to the coucher, and the coucher returning the companion mould empty to the vatman. A good team at the vat and felts work rhythmically and efficiently at their equal jobs, and produce up to eight reams of paper a day.[18] From time to time the layboy removes the posts of felts and paper to the paper press.[19]

[. . .]

Although some early accounts of papermaking make no mention of pairs of moulds, the standard accounts, of course, do. Thus, La Lande in his *Art de faire le papier* (Paris, 1761), the authoritative voice of the eighteenth century, remarks:

> Une seule couverture suffit pour les deux formes qu'on emploie dans le travail du papier; car . . . l'une des deux formes est toujours découverte au moment où l'autre se plonge dans la cuve avec sa *couverture*.

And later in describing the process:

> Le Plongeur, en ôtant la couverture de dessus cette première forme, la place tout de suite sur la seconde forme, qu'on lui donne pour la plonger à son tour.
>
> . . . le Coucheur releve sa forme, en commencement par la bonne rive; il la rend au Plongeur aussi nette qu'avant qu'elle eût été plongée, & il trouve sur le trapan de cuve une seconde feuille à coucher qui a été formée pendant qu'il couchoit la première, & qu'il releve en passant, avant que d'étendre le feutre. Ainsi l'on voit, qu'au moyen de deux formes qui sont toujours en mouvement, il n'y a point de temps perdu: pendant qu'une forme se plonge, l'autre se couche; quand le Plongeur passe une forme au Coucheur, il en [recoit] une autre qui est vuide, sur laquelle il pose la couverture qu'il retire de dessus la première, & il plonge de nouveau.[20]

[. . .]

Though historians of paper have thus stressed the fact of companion moulds, not much attention has been paid to the companion watermarks produced by such moulds. (Are they identical? Are they different?) Relatively few collectors have taken note of them; and bibliographers, at least in the English-speaking world, have seemed oblivious of them. A reason perhaps is that McKerrow, in the bibliographer's bible, makes no mention of pairs of moulds, though he gives a short description of papermaking.[21] Inasmuch as bibliographers in their innocence commonly assume that the papermarks in a quire or ream are invariant and indistinguishable, what follows has a *raison d'être*—and illustrations.

Watermarks result from designs made of wire and attached to the moulds. Le Clert points out that "La confection des formes était une industrie spéciale exercée par les *formiers ou formaires*."[22] That is, skilled artisans set up as makers of moulds and supplied the needs of a papermaking region. Less often they were papermakers themselves.[23] An examination of old moulds, such as those at the Dard Hunter Paper Museum, leaves little doubt that European mouldmakers came to be exceedingly clever workmen, seasoning the wood and securing the laid wires and buttressing the chain wires as they did.[24] They fashioned the watermark patterns out of brass wire and sewed them to the moulds with finer wire.[25] Though the ordinary wire form was not a work of art, particularly after a period of use, it was sufficient and not unattractive; and it was often so neatly duplicated as to deceive students into supposing that companion marks are the same marks. Thus cataloguers of incunabula or Renaissance books who state categorically "This copy has the same watermark throughout" are usually, if not invariably, wrong.

The maker of moulds can hardly have intended to deceive anyone. It was enough if a pair of moulds resembled each other so closely that the vatman would always know them for mates. What was important was that the *formier* should cut the mould frames precisely alike, so that the single deckle would fit them both neatly. But the twin watermarks might vary somewhat in height or position or details of design without affecting the certitude of their belonging together. Indeed, as we shall see, the mould-maker must sometimes have intended a distinction, as when he put the handle of a pot once on the left, once on the right. Some more patent differences may have resulted from two mould-makers working together, or from a papermaker improvising a filigree in imitation of one that was lost. Even when a skilled artisan took special pains with moulds for fine paper (as sometimes he did), the complexity of the design, differences in chains and lettering, and wear of the wire forms in use prevented one watermark from being or continuing to be a replica of the other. Standardization was not yet. With good sight, undiffused light, and a millimeter scale, the bibliographer today can train himself to distinguish the watermarks from companion moulds, except where heavy print or close binding serve to defeat him.

Though some have not pondered his words, Briquet spoke clearly on these matters, out of years of experience:

> . . . chacun sait que dans l'ancienne manière de faire le papier on se servait de deux moules ou formes que l'ouvrier plongeait alternativement dans la pâte. Le papier produit, portait donc par parts égales l'empreinte de chacune des formes employées et ce mélange se voit dans chaque rame et dans chaque main de ce papier. Ces deux variétés du filigrane sont parfois *identiques* au point qu'on ne les distingue que par la place un peu différente que chacune d'elles occupe sur la feuille de papier. Parfois les deux variétés sont *divergentes*, voy. *coupe* (n°s 4542 et 4543), *couronne* (n°s 4791 et 4792), *croix grecque* (n°s 5525 et 5526); le plus souvent elles sont *similaires* n'offrant entre elles que des différences légères de forme ou de dimension, voy. *pot* (n°s 12.893 et 12.894), *serpent* (n°s 13.801 et 13.802), *tête de boeuf* (n°s 14.388 et 14.389).[26]

If perchance the term *filigrane identique* is slightly misleading, this is the lesson of the master, and the reader would do well to examine the indicated pairs of marks before entering the forest.

It is convenient to refer to Briquet's several categories of degree as twin watermarks, a term suggesting either identical or ordinary pairs of twins. The term *companion* watermarks also has useful connotations. But for purposes of efficient analysis and description, I have found it necessary to observe certain *points of difference* which recur in twin watermarks. Among useful points of difference are these:

1) Difference of MOULD-END. Often the pairs of marks appear in different halves of their respective moulds and the sheets of paper made on them: variant *a* centered in the left halfsheet, *b* in the right halfsheet. Actually, this is mainly an

inference from the fact that some marks read *in* and some read *out*. Frequently the inference may be checked by the indentations made by the watermark or chain wires. Where these are clear, it appears that commonly, if not normally, the lettering on the mould itself was reversed, as in typography; that is, a watermark in a left halfsheet with left-to-right lettering usually was impressed by a wireform with right-to-left lettering in the right half of the mould; for the indented side or "smooth" side is obviously the mould side of the paper. Of course, there are exceptions, as when a label was turned and resewn non-reversed. In ambiguous or uncertain cases the difference may be called one of in and out.

2) Difference of CHAIN-POSITION. Sometimes a mark fits neatly between the chains while its twin does not. Or a small mark is centered in one instance but touches a chain in the other. Some pairs are centered on a chain, with the line cutting through at slightly different points. A few are sewn to a chain or between chains that do not correspond to those of the companion mould, the count being different from the end-chain or deckle, so that they appear higher or lower in a quarto fold. We may call a chain that splits a watermark its *center chain*, and those that it falls between or that cut its edges its *attendant chains* or *outer and inner chains*.

3) Difference of CHAIN-SPACE or CHAIN-PATTERN. When the attendant chains vary as much as 2 mm. in their distance from each other, this often is the easiest means of telling the twins apart. A smaller variation, though usable, may prove elusive, because of paper shrinkage or vagaries in the parallelism of the line. Differences of as much as 4 mm. occur. Actually, the chain-structure of two handmade moulds will never be precisely alike; thus remote chains are sometimes as useful as near ones, if they differ markedly in relative position or form. Occasionally an end-chain or other margin-chain has a wobble or a break that may be seen at a glance.

4) Difference of SLANT. At least one of the marks has not been sewn to the mould in an upright position and leans to right or left. Or an important part has been bent. In doubtful cases it is wise to measure upper and lower distances from the nearest chains; particularly in quartos, where one sees but part of the watermark at a time and may be uncertain which side of the paper one is looking at.

5) Difference of REVERSED PATTERN. The mould-maker has made one mark the reverse or mirror-image of the other, except for names and initials. In the Arms of France and Navarre, for instance, Navarre may be first on the left, then on the right, with respect to the lettering below. This makes a pretty distinction, which seems to have been intentional on the part of some mould-makers.

6) Difference of LABEL. One mark carries a "full name", the other mere initials or a name-abbreviation: as DVAVLEGARD and DVG. Or one may have initials, the other none. Or a name may appear in two spellings. The labels themselves may have different shapes. And occasionally one of the labels gets inverted or attached to the wrong end of a grape or shield watermark.

7) Difference of COUNTERMARK. Countermarks are of course the smaller marks that appear opposite the main marks in the other half of the mould or sheet;

they are common after 1650 or so. When they contain lettering, they vary much as labels do. Sometimes one of the countermarks has fallen out of its mould. Frequently, though the mould-maker has made the main marks "identical", he has taken less trouble with the countermark letters and has spaced them differently. And one set of countermark letters may be reversed or even inverted in respect to the main mark. Those countermarks that are letterless differ in detail or position as other marks do.

8) Difference of DISTINCTIVE DETAIL. One mark has an element of design not found in the other. A pot may have a fleur-de-lis within its base, its companion none. One bunch of grapes may have a curved stem, the other a comparatively straight one. Such differences may have been original or consequent to use. It was easy for separately attached bits of the design to break off or drop out during the process of papermaking or the brushing or cleaning of moulds after use.

9) Difference of SEWING. Particularly in early watermarks, where heavy sewing-wire was used, the sewing marks can be seen. These are what Sotheby calls *dots*.[27] Their differences of position may serve to distinguish nearly identical marks. When the wire-forms came loose, then arose differences of resewing. In such cases the dots often are plainer, especially if linen thread has been used in temporary replacement; and the sewings may appear in new places. At this same time the watermark itself may have been moved, so that it reappears between wider chains (say) or even in the other half of the mould.

10) Difference of DISTORTION. Parts of the design have been bent out of position. This ugliness is less often due to malformation than to deterioration. The wire-forms commonly went to pieces before the moulds to which they were attached.[28] They came partly unsewn, at one or more points. The result was sprung wires, bent wires, shifts in position—and increased differentiation from the twin mark. The wires might be further bent in resewing. Certain of these distortions in the life-history of watermarks will be illustrated. The awkward thing is that the shape of a mark may change in some difficult-to-describe particular while the detective pursues it from gathering to gathering of a pompous folio. Naturally, such changes limit the certainty of recognition when he meets the mark in another book.

[. . .]

It will be asked—it should and will!—what use it is to know that watermarks like wrens go in pairs. This is no time and place for a chapter on the bibliographical significance of twin watermarks: that may come hereafter. Yet it is evident at once that even the negative implications of twin watermarks are important. Though certain investigations may come to no harm through treating *filigranes identiques* as the same mark,[29] a bibliographer may bring perplexity or disaster upon himself by supposing that *filigranes divergentes* imply different reams or papermakers. He must take into consideration even the possibility that twin watermarks may differ in the names on their labels, bearing the names of partners or of a husband and wife, or of a papermaker and his factor. Certainly in the long run it simplifies matters to pair off the papermarks that belong together and that denote one consistent stock of paper. The basic equation is: Two watermarks equal one paper. It brings assurance and control into any study of mixed watermarks which seeks to answer questions of edition size, methods of press work, and time of printing. My purpose has been to rid watermarks of one besetting ambiguity.

[. . .]

Forty years ago Falconer Madan uttered his belief and warning: *"There is no such thing as a duplicate."* (Librarians have scarcely heeded him.) We see new truth in his words. In terms of mixed watermarks, twin watermarks, press variants, cancels, and the gathering of sheets, every copy of every book (above the rank of mere pamphlet) may be expected to differ from all others. Even a thin play-quarto "with one watermark throughout" yields startling arithmetical combinations. The variant-patterns soar into the millions and beyond. *Every original copy of the Pide Bull* Lear *was in all probability unique.*[30]

Notes to Introduction

1 John Bidwell points out that Karl Theodor Weiss, *Handbuch der Wasserzeichenkunde* (Leipzig: Fachbuchverlag, 1962), had previously alerted researchers that variant watermarks could be twins. John Bidwell, "The Study of Paper as Evidence, Artefact, and Commodity," in *The Book Encompassed*, ed. Peter Davison (New Castle, Del.: Oak Knoll Press, 1998), 71.

2 Eden Phillpotts, *Storm in a Teacup* (New York: Macmillan Company, 1919), 90.

3 Occasionally, the orientation of the watermark and countermark may be mirrored in the alternate mold (Peter Bower, pers. comm., 2013).

4 The Austrian botanical artist Ferdinand Bauer (1760–1826) preferred to use the felt side of hard-sized writing papers for his preparatory sketches for the *Flora Graeca*, according to Nancy Bell, "History Revealed: Looking at Ferdinand Bauer's *Flora Graeca*," in *Looking at Paper: Evidence and Interpretation; Symposium Proceedings, Toronto, 1999*, ed. John Slavin et al. (Ottawa: Canadian Conservation Institute, 2001), 29.

Notes to Reading

1 "The Duplicity of Duplicates," *Transactions of the Bibliographical Society*, XII (1911–13), 16.

2 Various writers have noted the mixture of watermarks in books. Dard Hunter says: "In examining old books a great profusion of watermarks may be noted in the paper of an individual volume, some fifteenth-century works containing a dozen or more different papermarks in a single book." [*Papermaking: the History and Technique of an Ancient Craft*, 2d ed. (1947), p. 261.] Thus William Blades speaks of "the astonishing variety of water-marks frequently found in one volume" of Caxton; and adds: ". . . there was a great intermixture of qualities [of paper], including the make of several mills. We have never yet seen one of Caxton's books in which the same water-mark runs through the whole volume, and in many cases the variety is astonishing. Thus, in a copy of the

first edition of the 'Canterbury Tales,' now in the library of Mr. Huth, there appear no less than fifteen distinct water-marks." [*The Biography and Typography of William Caxton*, 2d ed. (1882), pp. 97–98.] Recently the collector Edward Heawood has cited as examples of the "large number of different marks, often seen in a *single book*," Camden's *Brittania* (1637), Ogilby's *Atlas Chinensis* (1671), Fryer's *New Account of East India and Persia* (1698), and certain volumes printed in Italy, Holland, and Germany. [*Water-Marks Mainly of the Seventeenth and Eighteenth Centuries* (Hilversum, 1950), p. 33.]

3 The 42-line Bible (*c.* 1456) contains three quite different watermarks: Bull's Head, Grapes, and Bullock, each with two variants. See Samuel Leigh Sotheby, *The Typography of the Fifteenth Century, Being Specimens together with Their Watermarks* (1845), pl. Z, no. 92; and Karl Dziatzko, *Gutenbergs Früheste Druckerpraxis, Sammlung Bibliothekswissenschaftlichen Arbeiten*, Heft IV (1890), pp. 41–50 and pl. III.

4 Samuel Leigh Sotheby, *Principia Typographica to Which is Added an Attempt to Elucidate the Character of the Paper Marks of the Period* (1858), III, 102. Sotheby treats Caxton's watermarks at pp. 83–88 and pls. Qa–Qc.

5 On a suggestion of A. W. Pollard, Greg offered this hypothesis: "The middleman bought large stocks of paper from the manufacturer and sold comparatively small parcels of various sizes to printers. The inevitable result was that he was left with a number of oddments, remainders of various sizes, on his hands. These he simply stacked together and sold off cheap." (W. W. Greg, "On Certain False Dates in Shakespearian Quartos," *Library*, 2d ser., IX [1908], 395–396.) McKerrow concluded: ". . . it seems quite clear that many printers bought their paper in job-lots, and it is common to find a number of different watermarks in a book about the printing of which there appears to have been nothing abnormal." [Ronald B. McKerrow, *An Introduction to Bibliography* (1927), p. 101 n.] And in examining paper prices Marjorie Plant remarked: "It seems . . . as if the dealer was willing to sell off odd quires cheaply, and it may be for this reason that so many sixteenth-century books contain a number of different watermarks." (*The English Book Trade* [1939], p. 204.) But there was no reason to suppose that the varying prices applied to the same paper.

6 That a considerable part of the paper is Norman was shown by Edward Heawood in "Papers Used in England after 1600," *Library*, 4th ser., XI (1930–31), 263–299, 466–498, and "Further Notes on Paper Used in England after 1600," *ibid.*, 5th ser., II (1947–48), 119–149. That the *greater* part of it was Norman and Norman-Breton I shall attempt to show in some "Notes on Norman Papermakers" now in preparation.

7 H. Bourde de la Rogerie cites evidence from official surveys that in the eighteenth century practically all mills in Brittany had just one vat each. See *Les papeteries de la region de Morlaix, Contribution à l'histoire de la papeterie en France*, VIII (Grenoble, 1941), esp. 57–61.

8 Published archives of the eighteenth century mention Julien Sagory "marchand papetier à Morlaix," Yves Plassard "marchand papetier" apparently of the same place, and "Biard, marchand de papier et d'images à Brest." See *Inventaire-sommaire des archives départementales* . . . FINISTERE, B 4549, 4551, 4581. Nothing is known of such merchants in Normandy. In the seventeenth century there must have been English factors at Caen, Morlaix, and elsewhere. Nicolaï names some paper dealers and Dutch factors of the Angoumois. See Alexandre Nicolaï, *Moulins à papier du sud-ouest de la France* (Bordeaux, 1935), I, 49–50, 164, 170, 183.

9 Bourde de la Rogerie gives an account of many mills near Pleyber-Christ and Plourin, along the Jarlot and Queffleut just south of Morlaix [pp. 23 ff]. As late as the nineteenth century a writer in describing the valley of the Haute-Sée near Brouains and Sourdeval in Normandy speaks of "l'industrie qui a jeté mille usines à papier sur ses bords" (*La Normandie illustrée* [Nantes, 1852], II, pt. 6, 39). Nicolaï lists clusters of mills along the Charreau, Boëme, and other streams below Angoulême [I, 154], and points out that the clusters had been larger before the Revocation of the Edict of Nantes, 1685.

10 As in Raleigh's *History of the World* (1677) and Blome's *Cosmography and Geography* (1680–93). See Heawood, "Papers Used in England after 1600," p. 267; *Watermarks*, nos. 662, 663, 665a, 672, 678; and further examples can be cited.

11 The Oxford Pricelist of 1674 (among other evidence) shows that paper was imported from these three ports, of Normandy, Brittany, and the Angoumois respectively. See R. W. Chapman, "An Inventory of Paper, 1674," *Library*, 4th ser., VII (1926–27), 405–408.

12 See my "New Uses of Watermarks," esp. the tables, pp. 167, 172.

13 Hunter, pp. 435–440. This is part of a very informative chapter on "Present-day Papermaking by Hand in Europe."

14 The deckle is the frame that fits the two moulds of a pair and limits the edges of the sheet. See Hunter, figs. 297, 301. By metonymy the word also means the rough edges of the sheet itself. For definitions of terms see E. J. Labarre, *Dictionary and Encyclopaedia of Paper and Paper-making*, 2d ed. (Amsterdam, 1951).

15 The stay is the board to the left of the vatman on which he places the moulds for the coucher. Hunter, figs. 294, 305.

16 The post is the pile of felts and wet sheets. Ibid., fig. 303.

17 The bridge is the platform at the back of the vat along which the coucher returns the moulds to the vatman. Ibid., figs. 294, 298, 299, 305.

18 La Lande wrote: "Les operations . . . sont si promptes, qu'il se forme sept à huit feuilles par minute dans les grandeurs moyennes de papier, telle que *la Couronne*; en sorte qu'un Ouvrier peut faire huit rames dans la journée . . ." (Joseph Jérome Le Français de La Lande, *Art de faire le papier*, Institut de France, Académie des Sciences, *Descriptions des arts et métiers*. IV [Paris, 1761], 55.) Cf. Le Clert: "On estimait qu'un ouvrier, dans sa journée, pouvait faire de 7 à 8 rames de papier de 500 feuilles chacune, soit de 3500 à 4000 feuilles." (Louis Le Clert, *Le Papier* [Paris, 1926], I, 13.) Yet vatmen sometimes made even more. Schulte reckons that "Etwa 2500–3000 mal wird der Vorgang mit jeder Form täglich wiederholt . . ." (Alfred Schulte, "Papiermühlen- und Wasserzeichenforschung," *Gutenberg Jahrbuch* 1934 [Mainz, 1934], p. 22.)

19 Hunter, p. 185.

20 [A single cover suffices for the two moulds used in paper work; for . . . one of the two moulds is always uncovered at the moment when the other is dipped into the tub with its *cover*. [. . .] The vatman, removing the cover from on top of the first mould, immediately places it on the second mold, which one gives him to dip in its turn. . . . [T]he coucher lifts his mould, beginning on the good side; he gives it to the vatman right away before it has been dipped, and he finds on the *trapan* or plank of the tub a second sheet to dip that has formed while he was dipping the first one, and he takes it in passing before spreading the felt. Thus one sees, that because of two molds that are always in movement, there is no time wasted: while one mould is dipped, the

other sets; when the vatman passes a mould to the coucher, he [receives] another which is empty, on which he sets the cover that he has removed from the first one, and he dips it again.] La Lande, pp. 48, 53.

21 McKerrow, p. 100.

22 Le Clert, I, 16. La Lande says (p. 49): "Les formes & les couvertes se font dans toutes les Provinces ou il y a des papeteries; en Auvergne c'est le métier proper d'un grand nombre de gens qu'on appelle *Formaires*; il y en a sur-tout beaucoup à Ambert. . . ." For an instructive modern photograph see Georges Degaast, "Les vieux moulins à papier d'Auvergne," *Gutenberg Jahrbuch 1936* (Mainz, 1936), p. 11, fig. 3, "Le dernier 'formaire' d'Auvergne au travail."

23 Nicolaï mentions two such of the eighteenth century, both of La Couronne, below Angoulême: Pierre Laroche: "*Marchand papetier et faiseur de formes à papier*," at Moulin de l'Abbaye 1734, and Michel Gaillard, "fabricant de papier et de formes à papier," at Moulin de Beauvais 1747–49. [I, 166, 176, 183–84.]

24 Hunter, pp. 119–123.

25 Ibid., p. 264.

26 [Everyone knows the old way of making paper where one used two moulds or forms which the worker dipped one after the other in the pulp. The paper thus produced carried in equal parts the impression of each of the forms used, and that combination can be seen in each ream and *main* [quire] of this paper. These two varieties of watermark are sometimes so identical that they can only be distinguished from each other by the slightly different places they occupy on the sheet of paper. Sometimes the two varieties are divergent, see *cup, crown, Greek cross*; usually they are similar, offering among them only small differences of shape or size, see *pot, snake, head of beef*.] Charles M. Briquet, *Les Filigranes* (Paris, 1907), I, xix.

27 Sotheby, *Principia Typographica*, III, 15. "These *dots* are found frequently multiplied to a considerable extent in marks otherwise, to all appearance, of the same mould."

28 Briquet, I, xix; Schulte, pp. 22–25; Hunter, p. 266.

29 Cf. "New Uses of Watermarks as Bibliographical Evidence," pp. 155–158, 161–163. So diverse are the watermarks in the Cotes plays of 1640 that it was (luckily) not necessary to distinguish pairs of watermarks for the purposes of that article. See n. 15, p. 156: the Jesus and bird watermarks have now resolved themselves into pairs, and their association would seem to be that of companion vats.

30 Collation: A^2 $B\text{-}K^4$. Sir Walter Greg found in twelve copies seven formes variant, each in a different sheet. (*The Variants in the First Quarto of 'King Lear'*, ch. II.) Not counting second corrections in inner C or an altered catchword in inner K, and assuming equal numbers of uncorrected and corrected sheets (the totals being 42 and 45 among extant copies), that means 2^7 or 128 possible combinations within an edition of perhaps 1000 copies. But consider the watermarks. One paper with twin watermarks in 9½ sheets means 2^{10} or 1024 possible combinations, enough to allow a different watermark sequence for each copy. And now it may be pointed out that a pot can have its top or base in any one of the four leaves of a quarto gathering—and the pots so vary in position in examined copies of *Lear*. And even in an unwatermarked halfsheet the chain-pattern varies. Thus, in this simplest of cases, with only a third of the formes known to be variant and with only one pair of "identical" watermarks so far found, the possible combinations work out roughly as $2^7 \times 2^{10} \times 4^{10} = 2^{37} =$ 137,438,954,000. What chance is there for a duplicate Pide Bull among so many? What chance?

If such is the case with single-paper books, then what of books printed on a variety of papers, books with mixed watermarks! Clearly the possible combinations are astronomical, even if we assume only one or two pairs of marks within each edition-sheet. Let the reader perpend (if it does not terrify him) the case of the St. Jerome *Vitas Patrum* of Wynkyn de Worde with its fifty watermarks—whether this number includes pairs or not. He will need a ream of white paper for his problem, and a jar of paste.

Reading 44

Paul Needham

The Study of Paper from an Archival Point of View (1988)

A scholar and "archaeologist" of early printed books, Paul Needham (1943–) has unearthed an enormous amount of information by thoroughly examining their papers.[1] Attesting to the widespread use and accumulation of paper since the beginning of the fifteenth century, he points out that the study of paper as physical evidence will benefit scholars in the widest possible variety of historical disciplines. But, as he observes, "the world of scholarship seems to be broken up into more fiefs, principalities, and liberties than the Holy Roman Empire, and the flow of information between them is strongly discouraged by mental barriers."[2] This may be one more reason that "for the study of drawings and prints, the situation there is even more primitive." Needham wonders why watermarks are included at all if they are so poorly described and misidentified.

The status quo no longer holds: artists' papers have been carefully excavated by conservators, leading to valuable evidence about drawings and prints, including those of Rembrandt van Rijn (1606–1669),[3] Georges Seurat (1859–1891),[4] Edgar Degas (1834–1917),[5] J. M. V. Turner (1775–1851),[6] Winslow Homer (1836–1910),[7] John Singer Sargent (1856–1925),[8] John Marin (1870–1953),[9] and James Abbott McNeill Whistler (1834–1903),[10] among many others. Somewhat tempering Needham's criticism, Elizabeth Lunning (reading 45) makes the logical observation that there are disproportionately more sheets available for comparison in a few selected books than there are in many collections of prints and drawings.

[. . .]

I shall not attempt to prescribe in detail how, precisely, paper stocks are to be recorded and studied. But it may be useful to sketch out, very briefly, the rel-

---

From Paul Needham, "The Study of Paper from an Archival Point of View," *IPH Yearbook* 7 (1988): 124–26, 129–30, 132. Reproduced courtesy of the author.

evant categories of evidence. These categories define what, in principle, we need to know about paper stocks; and they are determined almost entirely by various aspects of the nature of paper moulds. As you will recognize, there is a close analogy with typographical study, where a knowledge of punches and matrices (and for later periods, of other technical processes) determines most of the categories of typographical evidence.

A. As just stated, we know that handmade paper was produced by pairs of moulds. Therefore; a paper stock is defined when, and only when, its twin moulds are defined. There is one complication, which I have referred to several times in print: sometimes several vats at the same mill seem to have produced paper of uniform quality and size. Their outputs were packed up and marketed together, and the reams of paper sold from these mills therefore contain paper made by multiple pairs of moulds. They are what I have called multigeminate paper stocks.[1]

B. Because one side of the paper sheets touched the wires of the moulds, and received their indentation, and the other side did not, two sides of the sheet can be distinguished: mould side and what in English is called felt side. Anyone who knows the uses of distinguishing hair side and flesh side of parchment will see at once why it is useful, and indeed often crucial, to distinguish the two sides of paper. Many apparently difficult questions of collation and book structure can be answered almost automatically, simply by giving intelligent attention to the pattern of alternation of the two sides of the sheet, and not just the pattern of recurrence of watermarks. I should add that it is usually easier to tell mould side from felt side of paper, than hair side from flesh side of parchment.

C. We must also pay attention to the two sides of the sheet if we are to perceive and record the watermarks properly. All watermarks should be reproduced and considered uniformly, whether as seen from the mould side, or the felt side; and the reader or user must know what convention is being followed. If not, confusion will invariably result. I shall mention later a strong argument for recording watermarks always from the mould side.[2]

D. Likewise, we must know which side of the sheet the watermark is in—I am speaking now of the commonest early tradition of watermark placement, whereby the wiremarks which made the watermarks were usually sewn in the center of one half of the mould. Various paper historians have used different conventions for expressing this—and not a few have failed to record the information, with inevitable ambiguity and confusion. The simplest terminology, I think, was Allan Stevenson's, which defined watermarks, when seen from the mould side, as mR (mark in the right half) or mL (mark in the left half). Matters become more complicated later in the sixteenth century, when countermarks and cornermarks commonly occur in sheets, but the principle of description remains the same: we need to know not only what the marks look like, but where they are on the sheet.

E. With regard to this, one other piece of information seems to me to be both simple and useful: a statement of which chain-space or spaces the watermark lies in. A very simple formula can supply this: for instance, one may use 'mL3' to indicate a watermark centered in the left half of its sheet, and in the third chainspace counted from the nearest edge, 'mL3|4' to indicate one sewn on the chain dividing the third from the fourth chain-space.

This suggests one other piece of information equally simple and useful: a count of the chain-lines of the sheet, i.e., how many chains made up its mould. Paper historians have sometimes elaborated related but more elaborate formulas: e.g., measures of the average distance between chain-lines; measures of the distances between each chain-line, all the way across the sheet; counts of the number of wire lines over a given distance, a measure of wire-line 'density.' These measures can all be useful, but they have a common problem. Paper moulds were often not perfectly rectilinear objects, with tight, even wires and chains. The measurements across one part of the mould can be significantly different from the measurement across another part of the mould. But it is obvious that a mould containing 11 chains cannot be identical to one containing 12 chains, even if the wiremark is the same in both cases. What we must have in such a case is the same wiremark, used first on one, then on another mould; and that means the same wiremark forming part of two separate paper stocks. I might add that this is just the kind of situation that comes up frequently in paper study: it is by no means purely theoretical. A record simply of the wiremark or watermark will not reveal this. From this circumstance emerges another principle of paper study which is worth keeping in mind: its basis is not, in fact, the study of watermarks—filigranologie, or Wasserzeichenforschung—it is the study of moulds, including their wiremarks.

F. Finally, I should mention what is perhaps the most neglected single aspect of paper study, at least in the fifteenth century: paper sizes.[3] I must confess that the importance of recording, and thinking in terms of, paper sizes has come upon me only slowly over the years, but I see now that it really is fundamental. It is also, like so much in paper study, very simple. In the first several centuries of European papermaking, there seem with few exceptions to have been only four sizes, whose approximate dimensions may be given as follows:

CHANCERY:	ca. 32 × 45 cm
MEDIAN:	ca. 35 × 52 cm
ROYAL:	ca. 43 × 62 cm
IMPERIAL:	ca. 49 × 74 cm

As for the study of drawings and prints, the situation there is even more primitive. A statement about watermarks is a standard feature in the description of drawings and prints in any respectable catalogue. But it seems to me to be essentially an empty, or formalistic tradition. These paper descriptions are invariably vague and imprecise; comparisons are made to such worthless albums as those of

Heawood[4] and Churchill.[5] No reliance can be placed on these paper descriptions, and I do not know of any art historian who succeeded in using paper evidence competently as a tool of scholarship. What I wonder is, why then do they bother to mention and reproduce watermarks at all? And yet it is not difficult to imagine the potential uses that could be made of paper evidence in these areas. Many artists' notebooks have been broken up over the centuries, and marketed leaf by leaf—often, indeed, the leaves themselves have been cut up into smaller vignettes. The paper of these drawings can be the primary clue to reconstructing such broken items, and to reestablishing their original structure and sequence. As for prints, of which numerous impressions were often made, even over a century or more, it must be obvious that a knowledge of paper is fundamental to the task if identifying and defining such distinct impressions.

[. . .]

[. . .] But I hope to have persuaded you of a few general points. First: the study of paper is a unified discipline, whose principles are grounded in a knowledge of how paper was manufactured and marketed. There is only one valid way to study it: not one way for the bibliographer, a second for the codicologist, a third for the student of prints and drawings, and so forth. Second: however complex individual cases may be, these principles are not arcane and difficult. Third: many valuable investigations into paper, touching on a very wide array of disciplines, are still waiting to be made. It can be guaranteed that those willing to learn the principles of paper study will make interesting and important discoveries.

Notes to Introduction

1 Claire Bustarret of the Institute of Texts and Modern Manuscripts in Paris proposes an archaeological approach for the codicological description of literary manuscripts: "L'archéologie du papier et la génétique textuelle, de Blaise Pascal à Marcel Duchamp," *IPH Congress* 11 (1996): 23–28.

2 Paul Needham, "The Study of Paper from an Archival Point of View," *IPH Yearbook* 7 (1988): 123.

3 Nancy Ash and Shelley Fletcher, *Watermarks in Rembrandt's Prints* (Washington, D.C.: National Gallery of Art, 1998).

4 Karl Buchberg, "Seurat: Materials and Techniques," in *Georges Seurat: The Drawings* (New York: Museum of Modern Art, 2007), 31–41.

5 Roy Perkinson, "Observations on the Dating of the Fourth State of Degas's *Edouard Manet, Bust-Length Portrait*," in *Looking at Paper: Evidence and Interpretation; Symposium Proceedings, Toronto, 1999*, ed. John Slavin et al. (Ottawa: Canadian Conservation Institute, 2001), 49–52; Roy Perkinson, "Degas's Printing Papers," in *Edgar Degas: The Painter as Printmaker*, ed. Sue Welsh Reed and Barbara Stern Shapiro (Boston: Museum of Fine Arts, 1984), 255–61.

6 Peter Bower, *Turner's Papers: A Study of the Manufacture, Selection and Use of His Drawing Papers, 1787–1820* (London: Tate Gallery, 1990); Peter Bower, *Turner's Later Papers: A Study of the Manufacture, Selection and Use of His Drawing Papers, 1820–1851* (London: Tate Gallery; New Castle, Del: Oak Knoll Press, 1999).

7 Roy Perkinson, "Observations on the Drawings of Winslow Homer," *Book and Paper Group Annual* 5 (1986): 1–8; Martha Tedeschi, Kristi Dahm, Judith Walsh, and Karen Huang, *Watercolors by Winslow Homer: The Color of Light* (Chicago: Art Institute of Chicago, 2008).

8 Annette Manick and Antoinette Owen, "Bringing Back Something Fine," in Erica Hirshler and Theresa Carbone, *John Singer Sargent: Watercolors* (Boston: Museum of Fine Arts Publications; New York: Brooklyn Museum of Art, 2013), 205–24.

9 Martha Tedeschi, Kristi Dahm, Ruth Fine, Charles Pietrazewski, and Christine Conniff-O'Shea, *John Marin's Watercolors: A Medium for Modernism* (Chicago: Art Institute of Chicago, 2011).

10 Margaret F. MacDonald, Grischka Petri, Meg Hausberg, and Joanna Meacock, *James McNeill Whistler: The Etchings, a Catalogue Raisonné* (Glasgow: University of Glasgow, 2011), online at http://etchings.arts.gla.ac.uk; Nesta R. Spink, Harriet K. Stratis, and Martha Tedeschi, *The Lithographs of James McNeill Whistler: A Catalogue Raisonné*, with David Kiehl, Katherine A. Lochnan, and Nicholas Smale and the assistance of Britt Salvesen (Chicago: Art Institute of Chicago, 1998); Harriet Stratis, "Whistler's Prints and Their Transfer Papers: A Technical Study," *Whistler Review* 2 (2003): 18–26; Martha Smith, "Hunting for Old Papers with James McNeill Whistler," *Book and Paper Group Annual* 16 (1997): 89–90.

Notes to Reading

1 'Four Strasburg Incunables incorrectly assigned to Anton Koberger, *British Library Journal*, 6 (1980): 130–43 at p. 131 and figs. 5–15.

2 This was Allan Stevenson's recommendation; but the watermark reproductions in his various writings do not always follow the convention. See also Paul Needham, 'Johann Gutenberg and the Catholicon Press,' *Papers of the Bibliographical Society of America*, 76 (1982): 395–456 at p. 453.

3 [The author has extensively researched historical paper since this publication. See Paul Needham, "Aldus Manutius's Paper Stocks: The Evidence of Two Uncut Books," *Princeton University Library Chronicle* 55, no. 2 (1994): 287–307; "Res papirea: Sizes and Formats of the Late Medieval Book," in *Rationalisierung der Buchherstellung im Mittelalter und in der frühen Neuzeit: Ergebnisse eines buchgeschichtlichen Seminars, Wolfenbüttel, 12.–14. November 1990*, ed. Peter Rück and Martin Boghardt (Marburg an der Lahn: Institut für Historische Hilfswissenschaften, 1994), 123–45; and "Concepts of Paper Study," in *Puzzles in Paper: Concepts in Historical Watermarks*, ed. D. W. Mosser et al. (New Castle, DE: Oak Knoll Press, 2000), 1–36.—Ed.]

4 [Edward Heawood (1863–1949), *Watermarks, Mainly of the 17th and 18th Centuries* (Hilversum: Paper Publications Society, 1950).—Ed.]

5 [W. A. Churchill (1865–1947), *Watermarks in Paper in Holland, England, France, etc. in the XVII and XVIII Centuries and Their Interconnection* (Amsterdam: M. Hertzberger, 1935).—Ed.]

Reading 45

Elizabeth Lunning

Characteristics of Italian Paper in the Seventeenth Century (1989)

The paper conservator Elizabeth Lunning (1948–2014) tried out the bibliographer's approach to describing paper and discovered one obvious advantage in the examination of northern Italian account books versus northern Italian drawings: there are simply more individual sheets for comparison in the former group than in the latter.[1] *Starting with the premise that watermarks can be false friends, Lunning transformed her cognitive experience of French and Italian papers into a holistic description of them using clear and evocative vocabulary. Perhaps forecasting* The Paper Sample Book,[2] *Lunning warns that when it comes to describing paper, "there is the risk that imagination will contribute more than observation"—or in the words of America's quintessential author and humorist, Mark Twain, "You cannot depend on your eyes when your imagination is out of focus."*[3]

[. . .] Watermarks, which were originally used by the papermaker to identify his product,[1] attract attention with their combination of charm and clarity, and in a similar way catalogues of them display a tantalizing order, suggesting that a paper can be dated, or at least classified, more easily than a work of art. Perhaps it is this initial optimism, the reasonable sense that a watermark, at the point its shape was cast in pulp, indicated a paper's place and moment of fabrication, that accounts for the disenchantment that often settles in after a brief encounter with these catalogues.

The history of paper can only be partially illuminated by watermarks, however. Their numbers and variety are so large that it is likely no inventory of them

---

From Elizabeth Lunning, "Characteristics of Italian Paper in the Seventeenth Century," in *Italian Sketches of the Renaissance and Baroque*, ed. Sue Walsh Reed and Richard W. Wallace (Boston: Museum of Fine Arts, 1989), xxii–xliii.

will ever be complete.[2] This may not be important to paper historians, who often make isolated studies of the appearance, evolution, and disappearance of a mark, but for art historians attempting to reach some conclusion about a sheet on the basis of a single mark, the deficiencies of watermark catalogues are bound to be frustrating. Under these circumstance, new collections of watermarks often seem random or erratic, so that the print scholar's task, to paraphrase Edo Loeber, feels like looking for a needle in an ever-expanding haystack.[3] Watermarks are most helpful to the art historian when they are recorded in large numbers, using an artist or a publisher as a focus. This comparative approach finesses the truly difficult task of trying to determine precisely where and when a sheet originated, and develops instead a good and dependable indication of the papers in an artist's repertoire.[4]

A catalogue such as this one, which represents some eighty artists and spans a period of nearly two centuries, thus presents a problem, in that any attempt to document the watermarks would surely serve primarily to enlarge the haystack. At the same time, the breadth of this study, in conjunction with the regional character of its approach, suggests a much broader question: What characterizes an Italian paper? Or, more cautiously: Why is it sometimes possible, in the absence of a watermark, to state that a paper is Italian? This kind of question, which is rarely posed explicitly by paper historians, may in many situations be more meaningful for the print scholar, not only because paper is distinguished by many features other than its watermark, but also because these features relate more directly to the appearance of a sheet than the watermark does.

The appearance of a sheet of paper in transmitted and reflected light is the result of characteristics imparted at every step in its manufacture—the retting and beating of the rags, and the formation, sizing, and pressing of the sheet—shaped by the structure of the mould and altered in different degrees and ways by time and, possibly, by conservation. In spite of the confusing variability of the material, certain sheets that possess a combination of traits, each of which is only moderately distinctive, and certain other sheets that show one or two very peculiar features, can be identified with some confidence as Italian. These characteristics merit some attempt at description because they are paradigmatic, that is, all papers show variations or modifications of them. Even in cases where the variation renders a sheet truly anonymous, an explicit reference to those qualities of which the viewer is usually only marginally or intuitively aware can prove interesting.

Whether the objective is to pinpoint the origin of a watermark or to classify a sheet more generically, any study of paper must take into account the ease with which the material was transported, and this often poses a serious problem for the researcher. A paper used in Holland, for example, might easily have been made in another country and exported several years prior to its use. However, in Italy during the period covered in this catalogue, the industry was so firmly and successfully established that there was little, if any, need to import paper. In his thoughtful and valuable study of paper and watermarks, *Les Filigranes*, Charles Briquet stated rather unequivocally that it was extremely unusual to find foreign paper in Italy.[5] This circumstance, in conjunction with the hypothesis that state and municipal archives, when located in or near a papermaking center, are composed largely of papers from that center,[6] makes it reasonable to further postulate that many Italian archives contain almost exclusively Italian papers. Therefore, it was in an effort to develop a paradigm for Italian paper which is not dependent on watermarks that this writer began to examine papers in the Archivio di Stato, Florence.

[. . .]

Figures 1, 2, and 5 illustrate several features that appeared separately or in a modified form in most of the sheets surveyed [note omitted]. Sheets that united these features were commonly found in the archives and these can be called paradigmatic, if not ideal Italian papers [note omitted]. Their distinctive characteristics can be described as follows:

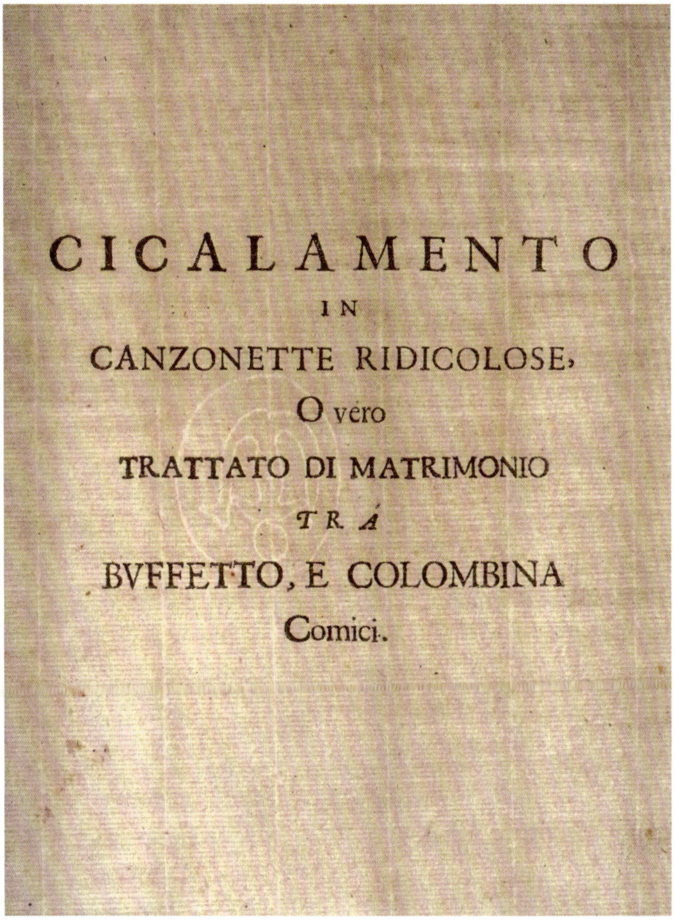

Figure 1
Transmitted light photograph of title page to *Cicalamento in Canzonette Ridicolose*, published in Florence in 1646 (enlarged detail, 160%). Typ 625.46.259, Houghton Library, Harvard University.

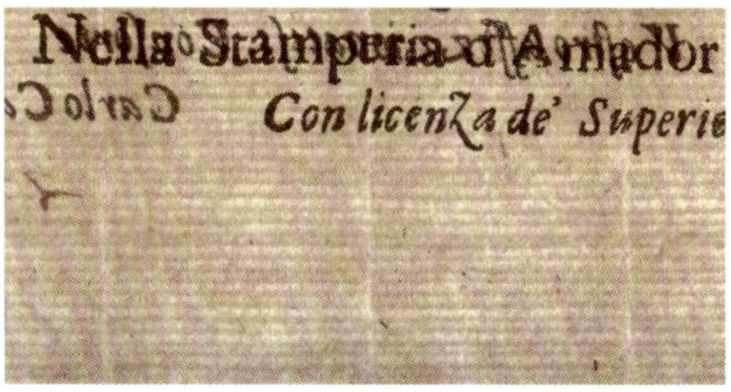

Figure 2
Transmitted light photograph of a page from *Cicalamento in Canzonette Ridicolose*, published in Florence in 1646 (enlarged detail, 160%). Typ 625.46.259, Houghton Library, Harvard University.

Figure 3 (at right)
Transmitted light photograph of title page to *Sibyllarum Duodecim Oracula*, published in Paris in 1586 (detail, actual size). Arthur and Charlotte Vershbow.

Figure 4 (below)
Enlarged detail of fig. 3.

Figure 5 (at bottom)
Raking light photograph of a sheet from Stefano della Bella's *Le Nozze degli Dei*, published in Florence in 1637 (enlarged). Arthur and Charlotte Vershbow. This photograph illustrates the wire side of a piece of Italian paper, that is, the side in contact with the mould during formation.

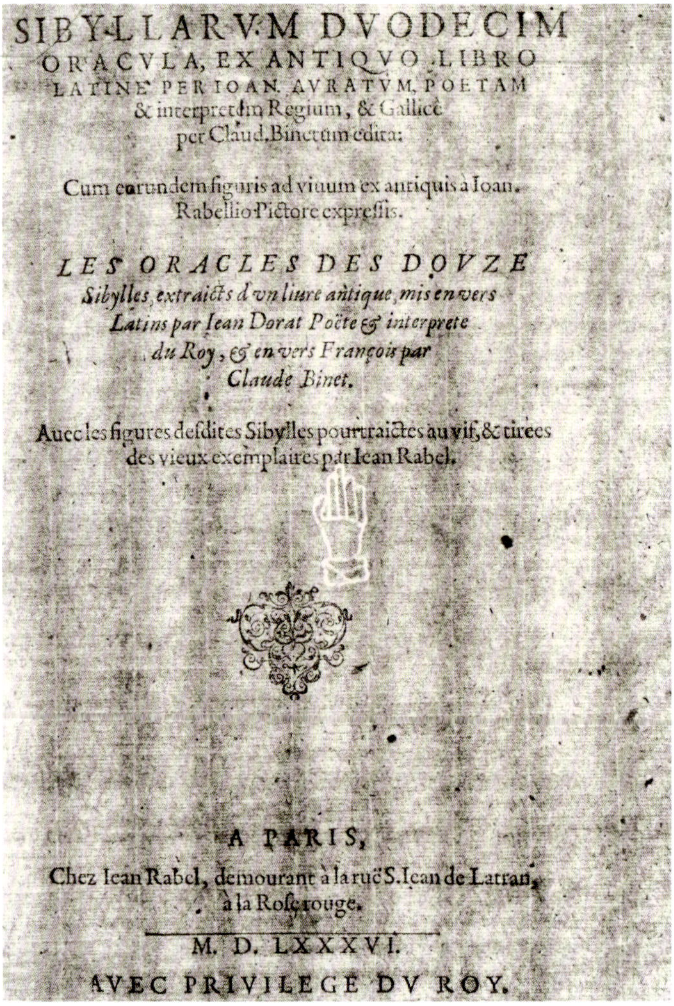
SIBYLLARVM DVODECIM
ORACVLA, EX ANTIQVO LIBRO
LATINE PER IOAN. AVRATVM, POETAM
& interpretem Regium, & Gallicè
per Claud. Binetum edita:

Cum earundem figuris ad viuum ex antiquis à Ioan.
Rabellio Pictore expressis.

LES ORACLES DES DOVZE
Sibylles, extraicts d'vn liure antique, mis en vers
Latins par Iean Dorat Poëte & interprete
du Roy, & en vers François par
Claude Binet.

Auec les figures desdites Sibylles pourtraictes au vif, & tirees
des vieux exemplaires par Iean Rabel.

A PARIS,
Chez Iean Rabel, demourant à la ruë S. Iean de Latran,
à la Rose rouge.

M. D. LXXXVI.
AVEC PRIVILEGE DV ROY.

Iean Rabel, demourant à la ruë S. Iean de Lat
à la Rose rouge.
M. D. LXXXVI.
AVEC PRIVILEGE DV ROY.

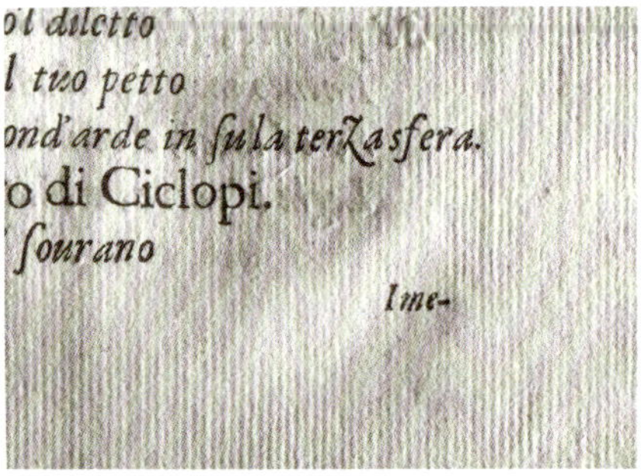

Most of the paper examined (eighty percent) showed chain lines at intervals of roughly three centimeters. It is worth noting that this moderately wide interval was maintained even in cases where a narrower one might be expected, for example, in fairly thin papers and in paper contained in very small books for which the original sheets may have been smaller.[7] By contrast, only four of the French papers showed a distance of three centimeters or more between the chain lines.

The shadows that sat adjacent to the chain lines (caused by an increased thickness of pulp) were extremely irregular and diffuse, and in places they disappeared entirely. Because the interval between the chains was not interrupted by uniform and continuous shadows, its width seemed to be even greater. In contrast, very regular and visible shadows often appeared in French paper, an example of which is shown in figure 3.

The watermark was anchored to a chain line. In the Florentine books and papers only one watermark was not centered on a chain, although it should be noted that a few showed countermarks that were located between two chains.[8] On the other hand, the French watermarks of this period were almost invariably placed between two chains.[9]

The chain line that bisected the watermark was often not accompanied by a shadow. The shadows on a sheet (the characteristic that defines "antique" laid paper) were caused by the suction of the ribs as the papermaker drew the mould out of the vat [. . .]. The chain wire (fig. 6) was almost invariably sewn to the ribs so that the shadows fell regularly on either side of the chains. This highly unusual characteristic, the absence of a shadow on the chain that bisected the watermark, appeared in a majority of the Florentine papers.[10] It was first noticed in the papers in the archives, where it was possible to examine large areas of blank paper, and it was then observed in the book papers, which often had large margins. In prints with narrow margins, however, it could be extremely difficult to see, especially when it occurred in conjunction with the faint and irregular shadows described

Figure 6
The face of a mould covering (detail). The fine sewing wires that attach the covering to the ribs beneath are visible spanning the chain wires at every fourth or fifth laid wire.

Figure 7
The ribs on the reverse of the mould in fig. 6, and the sewing wires that hold the covering in place (detail).

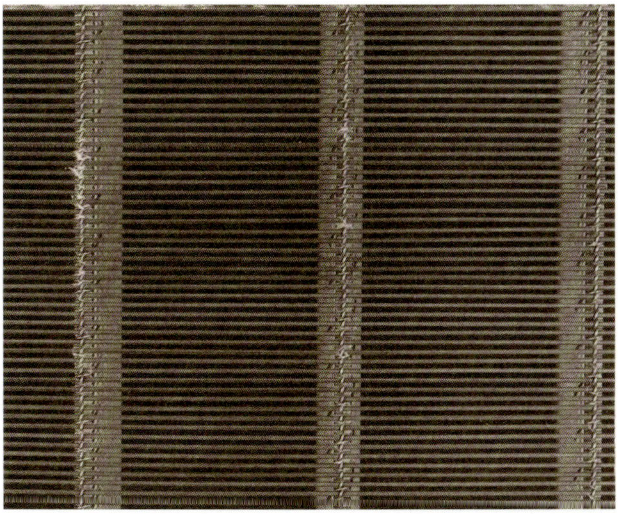

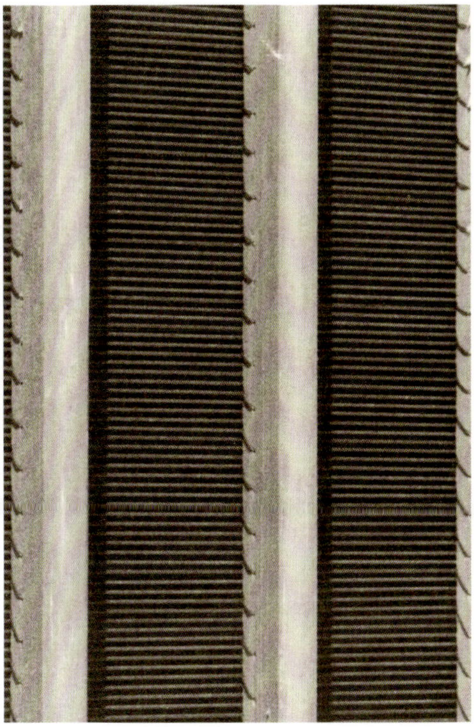

above. Nevertheless, in sheets where the watermark was obscured by printing ink, its presence and location could be inferred from this lack of a shadow.

The watermark was often attached to the mould covering between two ribs that were more widely separated than all the others on the mould.[11] It is tempting to conclude from this that the ribs were separated in order to facilitate the sewing on of the watermark, which was the last step in the fabrication of the mould covering and a step that would be repeated as the watermark, merely a piece of shaped wire, deteriorated in the course of use. A typical expanded interval of this kind was between 4.3 and 5.2 centimeters; when bisected by the shadowless chain it therefore created two fields smaller than the rest, that is, between 2.4 and 2.6 centimeters. Because watermark catalogues customarily illustrate only the chains immediately adjoining a mark, in illustrations of Italian marks the distance between these chains may not be representative of the average interval. Of the sixty-five Florentine papers on which a watermark appeared and of which measurements were taken of every interval, thirty-nine showed this irregularity; twenty showed a standard interval across the entire sheet; and six showed intervals so irregular that it was impossible to judge what effect the placement of the watermark may have had on the location of the ribs.[12]

The translucent lines left by the laid wires were often very clear and easy to see (figs. 1 and 2).[13] This trait appeared in many sheets and was independent of the fineness and the spacing of the wire. Thus, even when wires of a relatively narrow gauge had been placed close together, their pattern was often distinct and legible. Because it was obscured by any irregularity in the formation of the sheet this characteristic was best observed in large sheets, and because of its subtlety it was often most apparent at the moment a sheet was first glimpsed in transmitted light, and sometimes became harder to see on close examination.

The gauge of the laid wires in most moulds varied, but, in a sheet that showed this effect, which could be called "definition," by and large the wires were not just wide, but wider than the spaces between them, with the ratio of light to dark in the sheet sometimes exceeding two to one. Regardless of the width and independent of the ratio, however, the juncture between light and dark in the laid lines was crisp and neat, especially as compared with the more irregular, even discontinuous edges visible in the French paper (figs. 3 and 4). This definition, in addition to requiring a certain refinement of the pulp, was the result of what seems to have been an unusually high degree of contrast between the light areas of the paper and the shadows. The light areas, in other words, were extremely bright in transmitted light, suggesting that as the pulp sat on the mould it barely covered the laid wires; on the other hand, the shadows were often relatively dark and opaque. In addition, in certain sheets the boundary between light and shadow was so clearly delineated it seems possible that the laid wires on some moulds were rectangular in cross section rather than round; for with a series of round wires the thickness of the pulp would have diminished gradually as it settled around the circumference of the wires, whereas with a more angular wire the pulp would have risen sharply at each edge, creating a more acute impression of light and shade.[14]

[. . .]

Finally, Italian paper often showed significant texture in raking light, the reflection of the mould covering (fig. 5). This texture could even survive the pressure of intaglio printing. [. . .] At the same time these papers show very little felt patterning. A felt pattern [. . .] is the impression of the fibers (and sometimes the weave) of the felts that were interleaved between and pressed with the sheets following formation. When partially dry the sheets were pressed again without felts, and a third time following the application of sizing.[15] The unusual degree of gross texture caused by the mould covering, in conjunction with the near absence of fine texture caused by the felts, suggests both that the laid texture was obscuring some of the felt texture, and that the paper was not pressed very strenuously. In this context it is interesting to note that many of the archive papers appeared to have been glazed or burnished, a procedure that compressed the surface of the sheet and produced a slight sheen, rendering it more suitable for writing.[16]

In this discussion of Italian paper it is not coincidental that, with the exception of the fine texture left by the felts, the features described relate to the construction of the papermaker's mould, the shape of which was cast in the sheets formed on it. Characteristics that derived from the quality or the treatment of the pulp itself, such as color, transparency, and crispness or rattle, are both difficult to measure within the confines of a reading room, and infinitely more mutable, since they were the result of conditions at the mill that changed from month to month, and possibly from day to day. These would have included the composition of the rags, which varied according to the care taken to separate both rags of different qualities and rags of different weights. Grading by weight insured that a group of rags would all be reduced to pulp at the same time, for without such a precaution the stronger ones would still have contained knots or lumps while the finer ones would have been overly beaten. (In a more scientific sense, it seems there is little doubt that paper pulp was almost exclusively composed of linen and hemp until the eighteenth century.)[17] The degree to which the rags were retted and beaten, as well as the mix of water and fiber in the vat during the formation of each sheet, the temperature of the water, and the skill of the craftsmen—to name only a few variables—would also have affected the appearance of a sheet. Further, it is possible that some of these highly variable attributes had changed with time, depending on the conditions to which the sheets had been exposed and on the treatment they had received from conservators and restorers. Not surprisingly, then, this survey failed to reveal any clear pattern that would account for the presence or absence of these qualities.

[. . .]

Color was not included in this investigation because of the complexities involved in studying it. However, this characteristic deserves some general comment because of its effect on the viewer's perception of a print, and because of the attention accorded it by conservators and collectors, who are often uncertain about the degree to which the value, hue, and chroma of a paper are affected by age and conservation. Very broadly, the colors of the French and Italian sheets all fell within the same range and were distributed similarly within that range. More specifically, the following observations were made about the *Guardaroba Medicea*, whose papers were almost certainly untouched by either the devastating flood of the Arno in 1966, or the conservators who came afterward. Most of the sheets could be called cream or dark cream; some were darker still, nearly buff; and a few were bright off-white. The latter, a color that was very hard to attain prior to the discovery of chlorine in 1774, and one that was achieved only with the most meticulous and costly sorting and washing of rags, was an unexpected refinement among papers destined for office use. Of the darker papers, most showed a significant number of shives. When shives occurred in conjunction with the slight or severe reddish color associated with deterioration, it was difficult to conclude whether the darkening stemmed from the shives, from the sizing, or from some other negligent treatment of the pulp, of which the shives were merely symptomatic.[18] Many of the creamier sheets also contained shives. Of these, some were almost buttery in color and some warmer, but in transmitted light they all seemed to take on more color or to appear darker than would be anticipated by their color or brightness in reflected light. In a very few cases the opposite occurred: when held to the light a sheet of rather conventional color seemed to lose color and take on, instead, a brilliance—an occurrence for which there is no obvious explanation but which may relate to the transparency of the sheet relative to its thickness.

Trying to qualify the appearance of a paper or a class of papers is always a delicate business. Papermaking was, first of all, a craft, and for all the conventions of the craft, it also allowed for atypical practices. Further, it was not at all unusual for papermakers to move from town to town, or even from country to country, so that it is possible to anticipate the existence of papers showing traits associated with the countries of origin of both the paper and the papermaker.[19] Finally, the material itself is sometimes so foggy that, in attempting to find something worthy of description in it, there is the risk that imagination will contribute more than observation. Nevertheless, French and Italian papers of the seventeenth century seem to constitute at least one spectrum, at one end of which is a paradigmatic Italian paper. In reflected light this paper conveys immediately and conspicuously a sense of its ribby texture, and in transmitted light it clearly shows the horizontal stripes of its laid lines,[20] with a strong contrast between light and dark. At the other end of this spectrum is a French paper, which is smoother in reflected light and cloudier or more subdued in transmitted light. As a result, it is visually more delicate than the Italian sheet, and, in the absence of strong laid lines, it seems to be oriented vertically, along chain lines that are accentuated by clear shadows. As with any spectrum the number of points in between is infinite, and many papers will be difficult to locate within the array. In these cases the characteristics of the pulp—its composition, consistency, color, and transparency—are unlikely to reveal much about the origin of the sheet. However, they still contribute as much as, or more than, the mould does to the appearance of the sheet, and for this reason they continue to merit serious attention.

Notes to Introduction

1 Paper historians can argue that papers intended for writing purposes would have been glazed and sized more heavily than those intended for printing; however, this does not seem to have affected the application of Lunning's acute observations to the prints themselves. Lunning's quantifications are similar to those of David L. Vander Muelen, "The Identification of Paper without Watermarks: The Example of Pope's *Dunciad*," *Studies in Bibliography* 37 (1984): 73–81.

2 The germ of *The Paper Sample Book* began in 1983 in an attempt to describe the papers of Degas's prints. See Sue Welsh Reed and Barbara Stern Shapiro, "Degas's Printing Papers," in *Edgar Degas: The Painter as Printmaker* (Boston: Museum of Fine Arts, 1984), 255–61.

3 Mark Twain, *A Connecticut Yankee in King Arthur's Court* (New York: Harper and Brothers, 1989), 462.

Notes to Reading

1 Until the eighteenth century Western paper was made by drawing a mould through a vat containing a slurry of water and the pulp of retted and beaten rags. As the water drained through the mould, the fibers locked together, reflecting in their alignment the construction of the mould. The mould was a rectangular wooden frame whose face was spanned by a covering of "laid" wires that almost invariably ran parallel to the long side of the frame. The laid wires were held in place at intervals by a wire or pair of wires woven up and down between them and at right angles to them. These "chain" wires customarily coincided with and were sometimes attached to a series of wooden ribs located beneath the covering for support (figs. 8 and 9). Finally, the mould was often identified or decorated with a configuration of bent wire—a flower, a hand, or a monogram, for example—sewn onto the covering. This configuration is called a watermark when its shape, like that of all the wires on a mould, appears in a sheet of paper as a thin or transparent area in transmitted light. For more on both the history and the process of papermaking, see Hunter 1978.

2 A good but daunting indication of the number of surviving watermarks is given in Piccard 1961, a watermark catalogue comprising twenty-two volumes (to date), of which three alone document marks depicting "four-footed" animals.

3 Loeber 1982, p. 52.

4 A good example of a highly focused gathering of watermarks is appended to Robison 1986, pp. 215–239.

5 Briquet 1907, vol. 1, p. xxiv. Although the assumption that Italy did not import large quantities of paper is supported by the fact that it continued to be exported from Venice and Genoa even in the seventeenth century (when the industry was beginning to decline), small amounts may indeed have been imported. However, for this reason, and because a drawing requires only a single sheet for its execution, it is less reasonable to assume that drawing papers from the period were also largely Italian. It

also seems very likely that foreign paper was used more frequently in Venice, a city of numerous publishers and intense maritime activity.

6 Briquet 1907, vol. 1, p. xiv.

7 The outside dimensions of the Italian books examined ranged from 46 x 32 cm to 12.7 x 8.8 cm (closed); the French books, also closed, ranged from 43.5 x 28 cm to 15.6 x 11 cm.

8 A countermark was a subsidiary mark, often a pair of initials, usually located at the center of one half of a sheet. In these cases the main mark was located at the center of the other half.

9 With very few exceptions this distinction is also apparent in the French and Italian watermarks illustrated in John Krill's appendix to Russell 1975, pp. 314–339. Watermark appendices in many other catalogues of prints and drawings also confirm the overwhelming likelihood that an Italian mark of this period would have been bisected by a chain line. See, for example, Massari 1983, vol. 1, pp. 309–316, or Lewis and Boorsch 1985, pp. 215–224.

10 For two very interesting and provocative articles on the construction of moulds as it is reflected in paper, see Briquet 1955 and Gerardy 1970.

11 In an article about the watermarks of the English papermaker John Tate (d. 1507), Stevenson (1967; pp. 20, 25) describes a mould with this construction as Italianate, possibly Genoese. For anyone interested in the uses to which other disciplines put watermark research, this scrupulous article can stand as a model of the bibliographer's method.

12 This characteristic did not appear consistently in the Venetian and Neapolitan books.

13 In both French and Italian papers the appearance of the chain lines seemed totally unpredictable. they were sometimes very clear, sometimes nearly invisible; they were sometimes brighter than the laid lines, sometimes not; in some sheets the path of the chain lines interrupted that of the laid lines, but in others it did not; and no sheet gave the slightest true indication of how the chain wires had been attached to the ribs.

14 In trying to describe this effect, as it often becomes more difficult to see the more intensely a sheet is scrutinized, so it becomes more difficult to understand the more carefully it is described. In this writer's experience, readers will recognize exaggerated examples of this trait if it is described as looking like a bamboo window shade.

15 There are very few good accounts of the entire process of making paper by hand, and any two will inevitably disagree on many points. Three pressings seem to have been standard, but some mills were more exacting, turning and rotating the sheets after every pressing to further reduce the texture left by the felts.

16 Generally (and unexpectedly), apart from having been glazed, the archive papers were not especially different from the Italian book papers or from the papers seen in many prints, and books that included etchings were often composed of the same type of paper throughout. This writer is less familiar with Italian drawing papers but would be surprised if many were not smoother and finer than the majority of the sheets in this survey.

17 For interesting appendices about paper fibers, illustrated with photomicrographs, see Valls i Subirá 1978.

18 One color was observed in the archives only. It was dark, nearly buff, with absolutely no red or pink in it.

19 Italian papermakers went, among other places, to Spain, and a glance at the watermarks in Valls i Subirá 1978 indicates clearly that there was not only a great

deal of Italian paper in Spain, but that Spanish marks were very similar in style to Italian ones.

20 Because the European papermaker held the mould with one hand on each of the short sides, dipping the near (long) side first into the vat, the ribs on the mould almost invariably ran parallel to the short sides in order to facilitate the movement of the mould through the slurry of pulp and water. For this reason it is sometimes helpful to think of the chain lines on an uncut sheet of paper as running vertically, or parallel to the short sides, and the laid lines as running horizontally, or parallel to the long sides.

References

Briquet 1907

Briquet, Charles Moise. *Les filigranes: Dictionnaire historique des marques du papier dès leur apparition vers 1282 jusqu'en 1600*. 4 vols. Geneva, 1907.

Briquet 1955

Briquet, Charles Moise. "Notions pratiques sur le papier." *Briquet's "Opuscula": The Complete Works of Dr. C. M. Briquet without "Les Filigranes."* Hilversum, Holland, 1955 (Vol. 4 of *Momumenta chartae papyraceae historiam illustrantia*, E. J. Labarre, ed.), pp. 310–320.

Gerardy 1970

Gerardy, Theo. "Die Ursache der sog. Kannelierung des handgeschöpften Papiers." *Papiergeschichte* 20 (1970): 38–40.

Hunter 1978

Hunter, Dard. *Papermaking: The History and Technique of an Ancient Craft*. New York, 1978 (First edition New York, 1943).

Lewis and Boorsch 1985

Lewis, R. E., Michal Lewis, and Suzanne Boorsch. *The Engavings of Giorgio Ghisi*. Exhibition cat., The Metropolitan Museum of Art, New York, 1985.

Loeber 1982

Loeber, Edo G. *Paper Mould and Mouldmaker*. Amsterdam, 1982.

Massari 1983

Massari, Stefania. *Giulio Bonasone*, 2 vols. Rome, 1983.

Piccard 1961

Piccard, Gerhard. *Die Wasserzeichenkartei Piccard im Haupstaatsarchiv Stuttgart*. 22 vols. Stuttgart, 1961 (Repr.: 1987).

Robison 1986

Robison, Andrew. *Piranesi, Early Architectural Fantasies: A Catalogue Raisonné of the Etchings*. Washington, D.C., Chicago, and London, 1986.

Russell 1975

Russell, H. Diane, Jeffrey Blanchard, and John Krill. *Jacques Callot: Prints and Related Drawings*. Exhibition cat., National Gallery of Art, Washington, D.C., 1975.

Stevenson 1967

Stevenson, Allan. "Tudor Roses from John Tate." *Studies in Bibliography: Papers of the Bibliographical Society of the University of Virginia* 20 (1967): 15–34.

Valls i Subirá 1978

Valls i Subirá, Oriol. *The History of Paper in Spain*. Madrid, 1978.

Zonghi and Gasparinetti 1953

Zonghi, Aurelio, Augusto Zanghi, and A. F. Gasparinetti. *Zonghi's Watermarks*. Hilversum, Holland, 1953. (Vol. 3 of *Monumenta chartae papyraceae historiam illustrantia*, E. J. Labarre, ed.).

Reading 46

ELIZABETH LUNNING AND ROY PERKINSON

The Print Council of America Paper Sample Book: A Practical Guide to the Description of Paper (1996)

With practicality as their main objective, the paper conservators Elizabeth Lunning (1948–2014) and Roy Perkinson (1940–) devised a tool called The Paper Sample Book *to describe the color, texture, and thickness of the papers typically used in prints and drawings. It contains twenty-six vintage and new samples of papers of varying color, thickness, and texture mounted in a sturdy trifold reference chart. The success of* The Paper Sample Book *was achieved by channeling human tendencies and systematizing the knowledge and experience of curators and paper conservators. Several factors make the approach practical:*

- *The vocabulary used to designate color, thickness, and texture is restricted but is initially determined by comparison to familiar materials. Thus "thin (2)" corresponds to the thickness of a page from the* New York Times.
- *Material-specific terminology that is familiar to the stakeholders is modified according to numerical degrees of difference. Other than "cream,"*[1] *colors are called by their proper color names; texture and thickness are differentiated according to numerical degrees and by using simple adjectives such as* moderately *and* very.
- *The range of samples in all three categories has been preselected to correspond to typical structural and appearance properties of artists' papers. Thus the viewer is presented with a controlled "frame of reference within which distinctions can be made."*[2]

From ELIZABETH LUNNING and ROY PERKINSON, *The Print Council of America Paper Sample Book: A Practical Guide to the Description of Paper* (n.p.: Print Council of America, 1996), 7–11.

Lunning and Perkinson's methodology can be extended to produce customized paper sample books compiled specifically for the prints or drawings under consideration. For this purpose, conversion of their color samples into Universal Color Names, Munsell Notations, and reflectance data would facilitate reproducibility. A micrometer and scraps of newspapers, grocery bags, and telephone directories are all that is needed to imitate the thickness samples. Reproducing the samples of textures is more difficult; however, raking light images or reflectance transformation imaging[3] *of their surfaces might provide suitable surrogates.*

Echoing Labarre's (reading 39) belief that the human senses are often superior to instrumentation, the authors conclude, "In the end, however, the sample book is not a scientific tool, and we do not propose that it should replace one. There are already a daunting number of instruments for characterizing paper: colorimeters, glossmeters, folding endurance meters, tensile strength gauges, micrometers, and sizing testers, to name a few. But paper history is also replete with stories that illustrate the extraordinary sophistication of the senses: the vatman who judges the weight of the sheets he will produce by observing the color of the water in the vat; another vatman who, by tapping the side of the beater, knows when the pulp is beaten enough to make tracing paper; or the printer who holds a dampened sheet of paper against his cheek to determine the correct moment for printing an intaglio plate. And it is our opinion that the senses are the first tools, and very useful ones, with which to approach a work of art."[4]

As for the use of measuring instruments, Part IV commenced with mention of the mechanical or structural properties of paper, which include basis weight or grammage, thickness or caliper, formation (the uniformity of distribution of the fibers within the sheet), directionality or grain (the predominant orientation of fibers), two-sidedness (felt side vs. wire side), surface texture or finish, and porosity. For paper conservators, a thorough assessment of a paper's physical properties is a critical component of the pretreatment examination since they usually predict the paper's behavior during treatment. To this end, paper conservators and conservation scientists have selectively adopted the testing methods used in the fields of papermaking and printing to quantify these properties. Particularly useful are the TAPPI *(Technical Association of the Pulp and Paper Industry)* Standard Test Methods, *industry-wide procedures for evaluating, measuring, and describing pulp, paper, packaging, and the raw materials of papermaking.*[5] *The degree to which the data obtained from the* TAPPI Standard Test Methods, *such as pH*[6]*, or the results obtained from a folding endurance tester, can be used to predict a paper's aging behavior has been hotly debated, as will be seen in Part V.*

CATEGORIES OF DESCRIPTION

Color

It is perhaps the contrast between the darkness of an ink and the brightness of the paper that lies beneath it that has contributed most to the persistence of the term "white" to describe the paper, and for that matter, "black" to describe materials that are often quite clearly tinged with brown, blue, or green. For many purposes these descriptions are justifiable, even suitable, for traditionally a majority of prints have been *conceived* as images composed of black on white. But the fact remains that "white" is not a very enlightening term for an observer trying to understand or describe an artist's paper.

The late editor of *The New Yorker*, William Shawn, once defined the task of the reporter as the assembly, objectification, and communication of information that in a fragmentary form has little meaning but in the aggregate might, with luck, illuminate an issue. In this spirit the authors once amassed 105 paper samples illustrating color and attempted to arrange them in a single spectrum running from light to dark. Once organized, however, the samples resolved themselves into five subtle but distinct groups, each tinged with a different color: red, yellow, green, blue, and gray.

The spectrum arranged in this book is a condensed form of the "yellow" spectrum, meaning that even in the darker samples there is very little of the red/orange tint associated with deterioration, and even in the lighter samples there is very little of the coolness or grayness associated with certain nineteenth-century papers or with the effects of some conservation treatments. Although some of these colors, especially the darker ones, might resemble those suggestive of acute deterioration, this is by and large an arrangement of colors of paper in good condition, and it is the spectrum that one encounters most frequently in works of art on paper.

One way to look at the spectrum we have arrayed here is to say that, in terms of color, *Cream* (1) would represent the highest achievement of a sixteenth-century papermaker. A sheet of this color was produced only as a result of the most fastidious technique: the papermaker's rags were first carefully sorted according to both color and weight, then meticulously cleaned and put into piles to rot and soften slightly, and finally pounded with wooden stampers to produce a pulp.[1] Inattention at any stage except the last would produce a sheet of increased color. Cream (3), which was an office or ledger paper, illustrates clearly what happened when the rags were not assiduously cleaned or sorted to remove dark fibers. The sheet is by and large a conventional cream, but the inclusion of so much minute, dark, extraneous matter has darkened the overall appearance and probably accounts for its very slightly pink tinge.

The ambition to produce a whiter paper seems to have motivated many of the developments in papermaking and *Blued White, Bright White, and White* illustrate those developments. Curiously, this ambition does not appear to have been as strong in Asia, and it is no coincidence that one of our darker samples, *Beige* (1), is a Japanese paper, while *Cream* (2) is a French paper made to imitate Japanese paper (*simili japon*). It is a reflection of the nineteenth-century fashion for things Japanese that late in the century deluxe editions of prints were often issued either on Japanese paper or on *simili japon*. Toulouse-Lautrec's lithographs of Mlle Lender (of the 1890s) were printed on both papers.

In the mutability of its color, *Blued White* is one of the most curious papers in this array: when seen in isolation it looks cream; in other situations, it assumes a cooler and grayer cast. This ambiguity is perhaps the best indication that it is a "corrected white" paper, that is, a paper to which a small quantity of bluing has been added, in the same way that it might be added to laundry. For centuries, papermakers have used many blued dyes and pigments to produce paper with an unmistakably blue color, but there is some uncertainty about when these colorants were first added to pulps to enhance the whiteness of a sheet. In documents relating to a trial in 1765, however, the English papermaker James Whatman explicitly talks about the use of bluing, as an innovation, to neutralize the yellow cast of his paper. By contrast, *Light Blue*, which contains more blue pigment than needed merely to counteract yellowness, was deliberately produced as a blue paper. Paper, with this delicate, robin's egg color first appeared in the 1770s, and became very fashionable among Rococo printmakers and book publishers.

In this context, it is interesting that as early as the fifteenth century, papermakers realized that if considerable amounts of blue color were to be added to a pulp it was both unnecessary and even foolish to use their best raw materials. Unlike the *Light Blue* sample, which was made from refined white pulp, one will often find that early blue sheets are lumpy, irregularly formed, contain miscellaneous fibers and generally have the earmarks of a utilitarian product rather than a highly crafted one. For example, the blue papers used by Hendrik Gotzius for his prints, Giovanni Battista Tiepolo for his drawing, and Rosalba Carriera for her pastels were probably made principally for wrapping sugar, lace, or candles, rather than as an artist's materials. (Vendors of these items preferred blue paper because it made their products appear "whiter.")

Chlorine was discovered in 1774, and within fifteen years papermakers were putting it to use whitening light rags and lightening dark or soiled ones. The *White* sample in our spectrum could not have been achieved prior to the discovery of chlorine, and *Bright White* would have been unimaginable. Indeed, *Bright White* is the whitest paper we could find that does not contain fluorescent brighteners. Fluorescent brighteners, which are present in many contemporary bright white papers (and are commonly found in laundry detergents) are materials that absorb ultraviolet light, which is invisible to the human eye, and re-emit it at the blue end of the visible range, thereby serving the same function as a bluing agent. The presence of fluorescent brighteners in a sheet can easily be determined by examination with an ultra-violet light, which will cause the sheet to glow, or fluoresce, a brilliant bluish white color.

White, which is absolutely typical of the secondary (backing) papers used in *chine collé* in the nineteenth century, has a color that is slightly anomalous. It is difficult to see in a sample of this size, especially since its position here in the spectrum is fixed, but it is an anomaly the reader is likely to encounter, for example, in certain lithographs by Degas, and it is worth trying to describe. When one tries to locate this color within a lineup of samples, there is a tendency to place it high up in the column because of the near total absence of yellow or cream. Once situated there, however, it appears to be surprisingly dark, or gray, in comparison to a genuinely white sheet or even a very light cream. Furthermore, in certain lights or juxtaposed with certain colors, the sample, which earlier appeared grayish, takes on a warmth that suggests the presence of the smaller amount of pink. The description of very subtle or complex colors like this can tax one's vocabulary, and when readers encounter them they may be surprised by the aptness of color language, in which it is perfectly acceptable to speak of a cooler (bluer) red, a warmer (redder) green, and so on.

Beige (2) is a modern paper to which a powdered pigment was added in order to mimic the appearance of an aged paper. A paper containing a pigment often appears to be more opaque than sheets of comparable thickness that derive their color from the fiber alone. An example of the difference can be seen by comparing *Beige* (2) and *Cream* (3). In each case, the color of the first seems to come as a reflection off the surface of the sheet, while the color of the second, due only to its fibers, seems to emanate from the interior.

At the bottom of the sample array, Brown, is a modern, machine-made, wood-pulp paper. This paper is called "kraft," from the German for "strength" because the highly alkaline process, invented in 1884, produced paper that was notably stronger than that produced by other methods of the period. The brown color is a side effect of the process. We have included it here because it is not unusual to find a work of art executed on paper of approximately this color. In *early* brown papers, however, the color derives from the rags and not the chemistry of manufacture. These papers, lumpy and not quite as dark as the kraft paper, can be seen in the drawings of Guercino, for example. In the second half of the nineteenth century, a comparable color appears in papers used for drawings (some can be seen in the work of Eastman Johnson), and Whistler occasionally produced pastels on paper nearly this dark. Rauschenberg's lithograph *Kitty Hawk* (1974) is printed on two sheets of kraft paper laminated together in manufacture with a tarry substance to render the paper waterproof.

Notes to Introduction

1 The authors admit that *cream* can conjure up a broad range of colors but is still narrower in its connotation than *ivory*. Lunning and Perkinson, *The Print Council of America Paper Sample Book*, 2.
2 Ibid., I.
3 Reflectance transformation imaging (RTI) is a software-manipulated digital imaging technique that produces a virtual topographic map of any surface as it appears from many different lighting positions. http://CulturalHeritageImaging.org.
4 Lunning and Perkinson, *The Print Council of America Paper Sample Book*, 4.
5 TAPPI Standard Testing Methods are available from TAPPI, 15 Technology Parkway South, Norcross, GA 300024. www.tappi.org.
6 pH is a measurement of the relative activity of hydrogen ions in solution, which is taken as a measurement of acidity or alkalinity. See *Guidelines for pH Measurement in Conservation* (Ottawa: Canadian Conservation Institute, 2007).

Note to Reading

1 In the late seventeenth century, the Dutch invented the Hollander beater, a machine for reducing rags to pulp. Use of this device eventually became an indispensable part of papermaking, and brought many more changes in the physical and visual characteristics of paper than can be discussed here. Diminished strength, due to shortened fiber length, is usually cited as the most notable change, but by eliminating the necessity for fermenting the rags, with consequent darkening of the pulp, the Hollander also led to the production of white papers.

Reading 47

Judith Walsh

Paper Survey (1999)

Over a period of four years, the paper conservator Judith Walsh (1951–) examined more than one thousand sheets of paper used by Georgia O'Keeffe (1887–1986), constituting about 90 percent of the works included in the artist's catalogue raisonné. Walsh used the terminology and the parameters suggested in The Paper Sample Book *(reading 46) and also adopted Fontenelle's (reading 40) cognitive approach—feeling, smelling, and listening to O'Keeffe's papers. By synthesizing her observations and, more important, by assigning them descriptive names as the sorceress of Lucanus (reading 39) demands, Walsh came to instinctively "know" O'Keeffe's papers.*

From 1994 onward, I accompanied Barbara Buhler Lynes whenever possible to see Georgia O'Keeffe's works on paper, recording observations on the materials that constituted each work, paying attention not only to the paper and the media but, importantly, to the artist's manipulation of both. Over four years, information was gathered at close hand on more than a thousand sheets of paper used by O'Keeffe for her drawings, watercolors, pastels, and quick sketches.

We came to think of this project as comparable, in its scope and approach, to the efforts of natural historians, who gradually acquire an understanding of complex natural phenomena through the interpretation of simple, meticulously recorded, observations. In our case, each sheet of paper was measured, its thickness recorded using a hand-held micrometer, and its color and surface texture noted. The feel of the sheet to the hand was described; sheets were flexed to hear their "rattle," and we sometimes identified sizing or fixative by smell. Judgments

---

From Judith Walsh, "Paper Survey," in vol. 1 of *Georgia O'Keeffe, Catalogue Raisonné*, by Barbara Buhler Lynes (Washington, D.C., National Gallery of Art; Abiquiu, N.Mex.: The Georgia O'Keeffe Foundation), 25–26. Reprinted by permission of the National Gallery of Art, Washington, and the Georgia O'Keeffe Foundation.

had to be made about the manufacturer's projected use for the paper—for example, was it sold as drawing paper or as stationery?

O'Keeffe's career covered a long time-span, so I was prepared to find a broad range of papers used; I also expected that her use of them would be experimental, as was the case with many of her modernist contemporaries. Instead, however, we discovered that O'Keeffe consistently chose papers that were stunning for their simplicity: they tend to be almost invisible to the viewer, not competing with nor imparting texture to her painting and drawing. Her sheets are almost all of medium thickness, with slight to moderate texture, and pale cream in color. And the sheets she chose for each medium were designed for that medium—she did not put watercolors on charcoal paper, or pastels on watercolor paper.

O'Keeffe's sketch drawings fall into two categories: drawings on sheets from sketchbooks, and quick notations that she made constantly. The sketchbook sheets tend to conform to the narrow range of art papers she used for watercolor or charcoal. Her spontaneous sketches, in pencil or ballpoint pen, however, tend to be on bright white paper; she apparently made them on any scrap of paper she had to hand—notepads from hotels or airplanes, paper used for typing or correspondence, even rolls of paper designed for adding machines. She used blank paper from the back of her checkbook and, in one case, coarse brown wrapping paper from a magazine she received in the mail. These mostly brighter white papers are distinguished by their "found" quality; they were not specifically selected for a formal artistic expression.

Special attention was paid to the sheets used for formal works on paper. Their thickness was found to fall within a very narrow range: over the 35-year period between 1915 and 1950, for example, she consistently chose medium-weight papers for watercolors, with fewer than twenty sheets falling outside the range of .15 to .25 mm. (As a guide, very thin tissue paper may be as little as .05 mm, while thick illustration board can measure sixty times as much, or 3 mm.)

O'Keeffe was also consistent in choosing papers of relatively little texture. Twentieth-century technology makes papers of a wide range of textures available. Novelty papers that had the texture of cloth or wood grain or animal skin were readily available to O'Keeffe in Texas in 1916. They were understandably not of interest to her, but even among papers specifically made for artists, everything from a very rough watercolor paper to a smooth, polished card were produced. O'Keeffe chose sheets within a narrow range of the possible choices. She consistently chose a paper with the minimum texture practicable for the medium. In charcoal, she chose "Ingres" type drawing papers from various manufacturers. These are laid papers designed for use with pencil or charcoal; they show an even, moderate texture to give the friable media some hold or purchase on the surface of the sheet. In pastel, she experimented with many sheet types and finally settled on an evenly coated, gritty paper something like sandpaper, which was specifically designed for pastel work. In watercolor, O'Keeffe rejected the traditional handmade rag watercolor paper she had used in school. In fact, among her studio materials is a folder she inscribed: "Whatman paper—No good for painting." She sold the vintage watercolor papers it contained to an art supply shop in Santa Fe in the early 1970s. Instead of traditional watercolor paper, starting from 1915, her watercolors most often appear on a lesser quality sheet, an esparto fiber paper sold by the manufacturer as "watercolor cartridge." This paper she apparently selected for its innocuous surface texture and drab cream color.

Since the beginning of the nineteenth century, artists have had an increasingly wide range of choice in paper color. Almost any hue imaginable was available in the early years of the twentieth century: one has only to think of Paul Gauguin's prints on brilliant acid yellow paper, or of the range of colors found in Picasso's collages from 1913–14, to give context to O'Keeffe's choices in paper color. Except for a few singular examples in charcoal and pastel, she consistently chose neutral, cream-colored papers for her formal work. The overwhelming number of her charcoals and watercolors, and most of her pastels, are on pale cream paper. (A few sheets that seem to fall outside this rule are deceptive. For example, the 1917 series *Light Coming on the Plains* series [. . .] may appear to be on ocher paper, but it started out as cream. The newsprint O'Keeffe used for that series has darkened considerably since 1917.)

The neutral attributes of O'Keeffe's papers give a support for her ethereal images that is almost invisible. There is evidence that this was a highly conscious decision. Two watercolors from 1917—*Untitled (Colorado Landscape)* [. . .] and *Untitled (Long Lake, Colorado)* [. . .]—are, unusually, both on moderately textured, pinkish handmade paper with a mottled surface from a variegated fiber and a subtle, sturdy texture, with a handsome deckle edge. These sheets are conspicuous among O'Keeffe's papers of this date for their relatively exaggerated texture, color, and character. They come from a sketchbook inscribed to her by Alfred Stieglitz at 291[1] in June 1917, which he had apparently purchased in Munich some years earlier. She obviously treasured this sketchbook, keeping it (and another just like it) throughout her life. She even used it as a portfolio for other sketches but, in spite of the fact that it was an early token of affection from the man who encouraged and supported her at this early, critical moment, she never used more than a few sheets from it for her own work.

Observations on the way O'Keeffe prepared paper for use or storage were also collected during our research. Cut edges, torn edges, deckle edges, adhesive residue from a pad or drawing block, even the marks left by tack holes, paper clips, or creases, were all recorded. By comparing dimensions, some full sheets that had been quartered by the artist could eventually be reconstructed; sheets that had been bundled by rolling or clipping together were re-grouped. We found that one of O'Keeffe's persistent working habits was to use a group of sheets of the same type and approximately the same size when she worked on an image in series. When she used a block of paper, the sizes are, of course, quite regular. Often, however, we found that sheets differed by small fractions of an inch, with one or two irregularly torn or cut edges. It was clear that O'Keeffe took large sheets, folded them in halves or quarters, then slit or cut them along the folds to provide a short stack of papers of roughly equal size, very often about 9 x 12 inches. This pattern is found in her work from about 1915 to the 1960s. Techniques particular to different media—such as erasures, underdrawing, or corrections to the design—were also recorded.

As the number of works we examined grew, the database we amassed became a powerful tool for interpreting O'Keeffe's habits and preferences as an artist. With the author's dating of the works, we could arrange details chronologically, thereby giving evidence of the development of her aesthetic. We were able, for example, to compare a paper O'Keeffe used for a particular medium at a particular period with all the papers she ever used. In this way, we discovered that specific paper characteristics could be seen to fit perfectly with effects the artist was apparently seeking. When the sought-for effect appeared to change, so did the paper type. Oddities in her work could then be understood as a concession by the artist to a technique required by the particular image she was working on.

For example, O'Keeffe used newsprint only once for watercolor, for the 1917 series of three works, *Light Coming on the Plains*. The suffusing light of dawn that dispels the deep blue of night was achieved by adding ox-gall or another dispersal agent to the wet watercolor, causing one color to separate from the other. The slightly fuzzy cartridge paper that O'Keeffe routinely used during this period did not have a slick enough surface to allow the paint to move as needed to achieve this effect. New newsprint, the same color as her preferred cartridge paper, was a good substitute for it when she needed a slippery surface. In this case, as in others, the artist's choice of papers for formal works was seen to be deliberate and meaningful.

Our analysis, as well as giving critical information about the paper and media O'Keeffe used, became indispensable to the catalogue in another way. Inevitably, questions of authenticity were raised over some works and, with a view to assisting in settling such doubts, a decision was made to expand our technical research beyond those papers used by O'Keeffe for her formal work. We examined many hundreds of sketches held by The Georgia O'Keeffe Foundation, blank sheets found among the artist's studio materials, and even papers used for her correspondence. A private collector generously made available to us 35 cartons and several flat portfolios of O'Keeffe's studio materials, including many types of paper. Each sheet, brush, pastel stick, and tube of paint was catalogued. O'Keeffe's published letters offered information on which papers were of interest to her; those used by Stieglitz in his publications were considered, as were sheets used by artists within his circle. We looked at contemporary photographs taken of O'Keeffe at work and of her studio. Preserved tax records were consulted to review her sales receipts for art supplies. Notations she made on technique in books from her own library were transcribed, and queries about technique and other aesthetic concerns were culled from her correspondence with Caroline Keck,[2] her longtime conservator. Finally, a few artists who knew O'Keeffe offered reminiscences of her working methods.

At the same time, an effort was undertaken to identify any questionable sheets by name and manufacturer, which required direct comparison with identified papers of known vintage. Information on the paper trade was gathered from published sources, the corporate archives of paper manufacturers, and from the vintage Paper Sample Collection at the National Gallery of Art.[3] In several important cases, we were able to identify questionable sheets and ascertain their date of manufacture and availability in the U.S. (previously, it has been common practice to identify and date papers by watermarks alone).

It is our hope that this kind of technical connoisseurship—the broad study of art materials and studio practice as a foundation for the investigation of an artist's materials and method—will come to be recognized as an important tool for art historians both in the understanding of an artist's achievement and in the assessment of authenticity of works of art. Moreover, our experience from this cooperative connoisseurship has been to lead us to a deeper appreciation of the works of Georgia O'Keeffe. We came to have an intimate level of knowledge of her and of the art she produced. By increasing our understanding of the process by which she created her pictures, we have been able to discover how deliberately she pursued her unique vision and developed her personal aesthetic.

Notes to Reading

1 [Alfred Stieglitz (1864–1946), American photographer, promoter of photography, and, later, proponent of modern art, was instrumental in showing O'Keeffe's art at his gallery 291 in New York. His many photographs of O'Keeffe are well known. Stieglitz and O'Keefe married in 1924.—Ed.]

2 [Caroline Keck (1908–2007) was a well-known American conservator of paintings.—Ed.]

3 [Judith Walsh, Marian Peck Dirda, and Michael Skalka, *Paper Sample Collection Catalogue* (Washington, D.C.: National Gallery of Art, 1999).—Ed.]

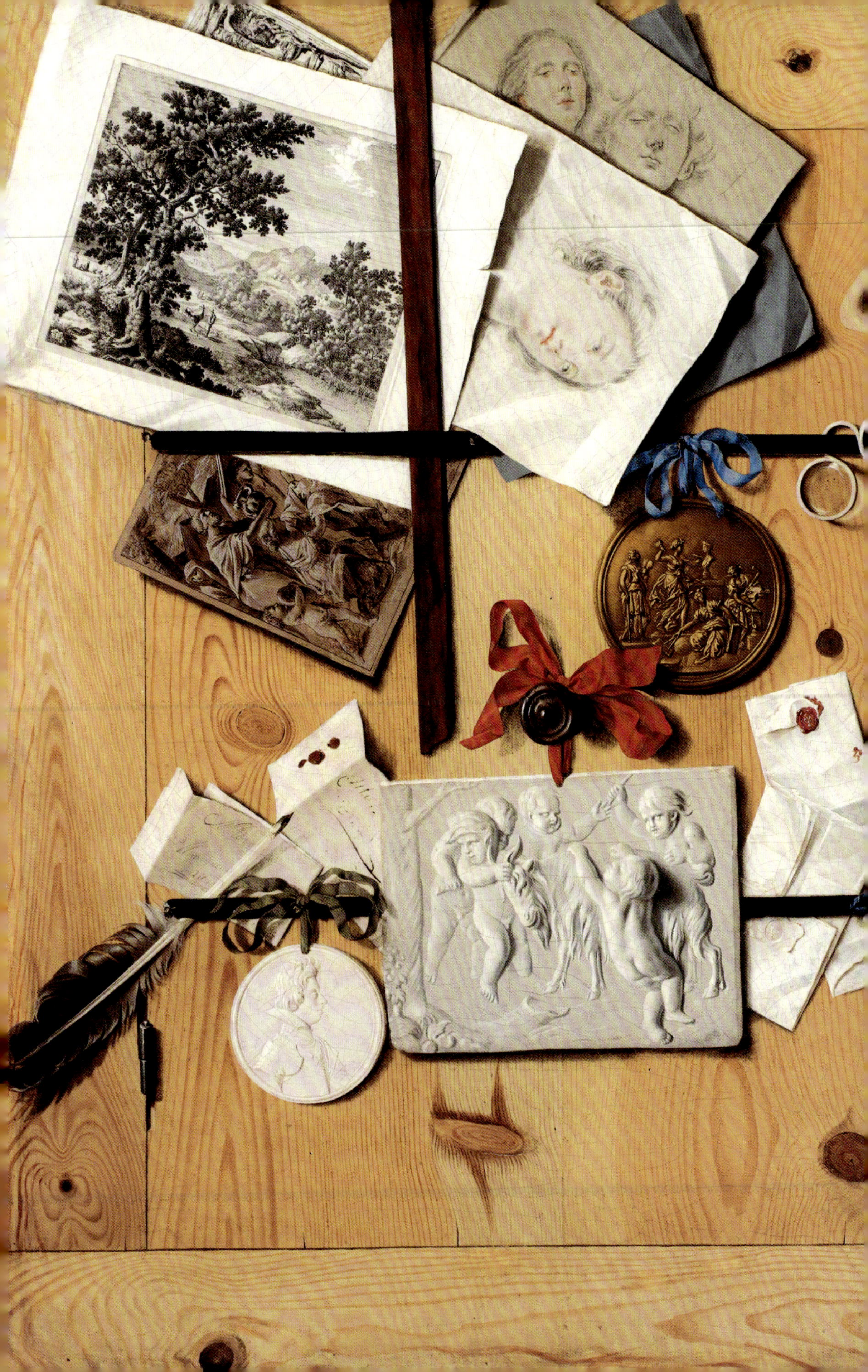

PART V

Deterioration and Change: Papers "Yellow'd with Their Age"

As demonstrated in Parts I through IV, drawn and printed marks and the paper on which they appear transmit meaning through their materiality and facture. By extension, deterioration of their substance, along with a change in their appearance, alters their meaning. Paper conservators carefully observe and systematically evaluate the telltale symptoms of deterioration and change that can result from accidents, age, human intervention, or even artist intent. In the words of the restorer Max Schweidler (1885–?) (reading 91), "The restorer should feel in his workshop—his clinic—like a medical chief of staff who recognizes with an experienced eye whether treatment or an operation is indicated."[1] From experience, conservators can visualize the work's original appearance and, using their other senses, virtually feel and hear what must have been its previous physical and chemical characteristics. For example, conservators know by touch and sound that a now-flaccid gelatin-sized paper once had a satiny surface and a crisp "rattle," or by sight that the blue trees on a hand-colored Currier and Ives lithograph were once a deciduous green, or that the cadaver-like cheeks of a pastel portrait began as a healthy rose. Smell can detect time spent in a dank basement, either by the lingering musty odor of mildew or by the pungent prophylactic fumes of thymol. Occasionally, conservators have no need for empirical evidence from which to extrapolate original condition: hypothesizes are confirmed by an untouched companion print fortuitously stored in an album or verified by the pristine perimeter of a watercolor protected by an overlapping window mat.

Sometimes the symptoms of deterioration can be used to verify an artwork's original materials: a blackened crust on white heightening is a sure indication of lead white, brownish halos surrounding greens can substantiate verdigris, a white crystalline growth or "bloom" supports the presence of a wax-containing medium such as Crayola® crayon or Paintstik®, and a paper darkened to a characteristic orange-tan signals the use of alum-rosin sizing.

Trompe l'Oeil Still Life (detail). Italian, 18th century. Oil on canvas, 31 7/8 × 46 1/4 in. The Frances Lehman Loeb Art Center, Vassar College, Poughkeepsie, New York. Gift of Henrietta Roig, mother of Josephine Roig Humphrey, class of 1945, 1990.14

Symptoms of deterioration can be used to diagnose root causes, which are often categorized as "internal" or "external." This classification of sources was proposed as a means of predicting paper permanence by the chemist Bertie Lee Browning (1902–1999) (reading 51). According to Browning, internal causes of paper deterioration result from the papermaking process itself and include poor-quality pulp, residues of processing chemicals, and unstable sizing agents, culprits presciently identified by John Murray (reading 48) almost two centuries ago. If this category is expanded to include the medium of prints and drawings, then internal causes of damage encompass what is popularly called "inherent vice," whereby the very materials used by the artist will inevitably deteriorate. Whether the result of the artist's intent, experimentation, or ignorance, unstable materials jeopardize the longevity of works on paper. This assumes that deterioration is not a desirable outcome. Some artists, especially those who do not regard their creations as precious commodities, deliberately create their works with the visual and physical consequences of disintegration in mind, such as Dieter Roth (1930–1998) or Joseph Beuys (1921–1985), who planned for the gradual decay of their food-based works.[2] Intentional or not, internally generated deterioration can be significantly retarded by preventive conservation, as introduced in Part III, whereby a protective environment for collections is created and maintained.

In the absence of such stewardship, external sources of deterioration proliferate. Environmental conditions such as high temperature and humidity levels, particulate and gaseous air pollution, and prolonged exposure to bright, full-spectrum light, or more properly termed, radiant energy, all act as external agents of change via the mechanisms of oxidation, acid hydrolysis, and photochemical reaction. Another kind of external damage specific to paper is a phenomenon known as acid migration, first described by William Barrow (reading 54), whereby paper deteriorates as a result of contact with acidic materials such as poor-quality wrapping paper, leather bindings, corrugated cardboard, and wood products. Other external sources of deterioration include insects and rodents, natural disasters, and thoughtless human actions.

Because sound treatment decisions depend on precise determination of causes, symptoms need to be carefully evaluated. A severely darkened paper can be originally made from groundwood pulp with alum-rosin sizing, the result of prolonged exposure to light, acid migration from contact with poor-quality storage or framing products, or previous bleaching with potassium permanganate; or the paper may be intentionally a chocolate-brown.[3] Each symptomatic source points to a different course of treatment or even to no treatment at all. Some signs of wear and tear that seem to call for repair—fingerprints, folds, and creases—may impart "authenticity," or, as described by Walter Benjamin (reading 34), "the essence of all that is transmissible from its beginning, ranging from its substantive duration to its testimony to the history which it has experienced."[4] Hence certain stains and ragged edges may not be candidates for treatment, for example, cultural artifacts salvaged during the Holocaust or Renaissance prints having didactic mechanical functions.[5] "Accidents" fraught with meaning often influence treatment decisions, as discussed in Part VII.

Part V, focusing on the deterioration of and changes in paper, begins by tracing the realization that changes in paper manufacture caused a marked decrease in its durability and permanence, two very different but crucial properties defined by Browning (reading 51). With the use of inferior rags and later the substitution of plant fibers; the increased degree of chemical and mechanical processing required; and the addition of deleterious substances such as alum-rosin sizing, certain coatings, and excessive amounts of clay fillers, papermaking embarked on a journey down the "road to Avernus," indeed a ruinous destination.[6]

The negative response from the public, represented here by Murray (reading 48) and Balzac (reading 49), accelerated with the growing problem. The demand for more permanent paper for newly created products, however, outpaced that for the preservation of existing paper. Libraries, in particular, were increasingly concerned with the usable life span of their holdings, especially newspapers both old and new. The New York Public Library predicted their total extinction with the erection of "bronze tablets to mark the whilom site of that pre-Cambrian fossil 'the newspaper room'" so beloved by Baker (reading 62).[7]

In the mid-twentieth century, in response to paper's downward slide, the papermaking industry conscripted a growing cadre of paper chemists, many of whom would go on to exert enormous, some might argue undue, influence on broad-based preservation programs in libraries and archives, as well as on the design of single-item treatments. One popular means of assessing the relative mechanical and optical properties of paper was artificial or accelerated aging followed by testing to measure mechanical strength. It was the dogmatic application of the findings of artificial aging to broad-based preservation programs and individual treatments that ultimately proved most controversial. As paper chemistry was incorporated into graduate-level conservation training programs, a new generation of paper conservators emerged having a better awareness of, and perhaps an overreliance on, paper testing data. This led to the temptation to conflate the goal of optimal paper permanence and the desire for sympathetic and appropriate conservation treatments.

Part V is not intended to be a chronology of industrial developments in papermaking or a primer in paper chemistry. Much like Part VI, it explores how paper conservation approaches have been influenced by ongoing scientific research and the relatively recent birth of the discipline of conservation science, as represented by Robert Feller (readings 63 and 64). While the integration of scientific knowledge into conservation practice has not been smooth or consistent, the critical importance of effective communication and proactive relationships across disciplines has been made crystal clear.[8]

Notes

"Papers yellowed with their age" from Shakespeare, *Sonnet 17* (1609).

1 Max Schweidler, *The Restoration of Engravings, Books, Drawings, and Other Works on Paper*, trans., ed., and appendix Roy Perkinson (Los Angeles: Getty Conservation Institute, 2006), 204.

2 According to Jan Schall, some artists will "allow nature to reclaim a work of art" to the extent that it may "disappear without a trace." Jan Schall, "Curating Ephemera: Responsibility and Reality," in *(Im)permanence: Cultures In/Out of Time*, ed. Judith Schacter and Stephen Brockmann (Pittsburgh: Center for the Arts in Society, 2008), 16. For a fuller discussion of artist intent and deliberate impermanence, see Miguel Angel Corzo, ed., *Mortality Immortality?* (Los Angeles: Getty Conservation Institute, 1999).

3 Winslow Homer used dark brown papers that were pigmented with anthracite coal and charcoal; see Roy Perkinson, "Observations on the Drawings of Winslow Homer," *Book and Paper Group Annual, American Institute for Conservation of Historic and Artistic Works* 5 (1986): 2.

4 Walter Benjamin, "The Work of Art in the Age of Mechanical Reproduction," in *Illuminations*, ed. Hannah Arendt, trans. Harry Zohn (New York: Schocken Books, 1969), 219.

5 Jane E. Klinger, "The Importance of the Provenance of Damage," in *Edinburgh Conference Papers 2006: Proceedings from the Fifth International Conference of the Institute of Paper Conservation and the First International Conference of the Institute of Conservation, Book and Paper Group, 26–29 July 2006*, ed. Jaques Shulla (London: Institute of Conservation, 2007), 51–58; and Kimberly Nichols, "Physical Qualities of Early Prints," in *Altered and Adorned: Using Renaissance Prints in Daily Life* (Chicago: Art Institute of Chicago; New Haven, Conn.: Yale University Press, 2011), 99.

6 In the words of Verner W. Clapp, "The Story of Permanent/Durable Book-Paper, 1115–1970," 8 (reading 1), and also, ironically, pounced on by Nicholson Baker (reading 62). Mentioned in Virgil's *Aeneid*, Avernus is a crater lake west of Naples, Italy, whose name, meaning "without birds," is taken from Greek. It was said that birds quickly perished if they flew over the lake. It is plausible that the birds died from toxic fumes emitted from the lake's volcanic source. Avernus was thought to be the entrance to hell but now generally refers to a cursed place.

7 H. M. Lydenberg, "Present Discontents with Newsprint Stock," in *Papers and Proceedings of the Forthieth [sic] Annual Meeting of the American Library Association Held at Saratoga Springs, N.Y., July 1–6, 1918* (Chicago: American Library Association, 1918), 213.

8 The uneasy alliance between science and conservation is explored in depth by Salvador Muñoz-Viñas, *Contemporary Theory of Conservation* (London: Butterworth Heinemann, 2005), 115–46.

Reading 48

John Murray

Practical Remarks on Modern Paper (1829)

A self-described philosopher of physics and chemistry, the outspoken John Murray (1786–1851) is often cited as the primary instigator of the public's outrage over short-lived paper. In a letter from 1823 he rails, "Allow me to call the attention of your readers to the present state of that wretched compound called Paper*. . . . Every person in the habit of writing on 'Bath wove post,' must have been sensible of what I complain.*[1] *Specimens there are, that being folded up, crack at the edges, and fall asunder; others, that being heated at the fire, disintegrate and tumble to pieces."*[2] *In spite of his histrionics, Murray provides a clear road map for an emerging corps of paper chemists who will follow his recommended avenues of research for well over a century: (1) the nature of the fibers used (Murray focuses on the cotton rags in use at the time); (2) the treatment to which they are subjected in order to be transformed into suitable half stuff; and (3) chemical residues of processing, sizing, and finishing substances.*[3] *Murray's methods of analysis—touch, taste, and smell—are second nature to the printer (reading 49) and purveyor of paper (reading 51).*

Chapter III

The various sources of decomposition and decay in modern paper, we have already detailed in the preceding pages, in describing the manipulation of paper-making, and have also given testimony in favour of hand-made paper, over that manufactured by machinery. In process of time the latter may certainly, in this inventive and ingenious age, receive improvements that may obviate all objections, and the vast quantity of work done by the machine, compared with that effected by manual labour, will raise its value in the estimation of the paper-makers, as one machine will throw off as much paper as fifteen men can make in a similar period. I should

From John Murray, *Practical Remarks on Modern Paper* (London: T. Cadell, 1829), 69–79.

be sorry that any opinion of mine should be construed into a sentiment hostile to the diffusion of mechanical aids: it is only by such ingenious applications of mental prowess that we can compete with the progressive demand which our literature claims from us. It is by the ingenuity of our mechanical arts, conjoined with our local advantages of insular situation, and our canals, and our coal-mines, that we can withstand our continental neighbours, and compete for the prize of superiority. We would, therefore, suffer mind to bring its full stores to the contest, and avail ourselves of its wonderful and varied resources. On the Continent of Europe, manual labour is of much cheaper obtainment than with us, and a substitute must therefore be found in British genius, lest we be worsted in the struggle.

The *second* source of evil[1] must be sought for in the mechanical consistency of the pulp. Let it never be forgotten, that a fibrous structure is essential to a lasting paper-material, and the fibre of linen rags is in every respect stouter and more durable than that of cotton, while the introduction of earthy substances will ill compensate for such a want; but should the cohesive and fibrous texture of the vegetable, whether linen or cotton, be reduced by any means whatever to a mere jelly, it is a matter of no consequence what be the source of the paper-pulp: our paper may as well be tablets of plaster or stucco, as any thing else; but damp will speedily cause decay, and, like quicklime, when water is poured upon it, it will crumble to atoms. The simple fact, that hair, or any fibrous substance mixed up with lime, imparts cohesive properties to it, affords a proof of the folly and ruinous consequences of wove paper, while, at the same time it imparts a useful hint for its improvement. We have made a great many experiments on the comparative cohesive strength of the parenchymatous fibre of a variety of trees and shrubs, &c., as those of the *mulberry*, *birch*, *willow*, *bramble*, *hemp*, *flax*, *nettle*, *Phormium tenax*, *&c.*, and are inclined to prefer *unbleached flax*, *bramble* and *nettle*, all of which would very much increase the strength and durability of paper. The New Zealand flax or *Phormium tenax*, is too rare for such a purpose.[2] The birch, willow, bramble, and nettle, will be found, if we mistake not, powerful and valuable auxiliaries to the paper manufacturer, as the supply of several of the others, might not be as available, or in sufficient quantity, or so economically subservient; but a *small proportion* of unbleached flax added to the half-stuff,[3] would most materially improve the cohesive strength of the paper tissue, provided the beating process were not carried to its maximum,—the great source of injury. These materials should also have animal fibrous substances, as shavings of pelt or skins, added, and such addenda would, we are certain, improve the texture and consistency of the paper produced from it.

A *third* source of evil consists in the adventitious mixture of earthy substances in the pulp: even snow must be cast into shade by the exterior whiteness of modern paper, and if the bleaching gas, or its association with lime, will not produce the effect, *gypsum* must needs be added to accomplish the end, as well as to impart weight and consistency. Display is the order of the day, for show all else must be sacrificed; and for the exhibition of an hour, permanence and durability must be abandoned. No earthy matter whatever should be suffered to enter into the composition of the pulp,—every species of terrene material is equally deleterious and mischievous; brittleness and want of cohesion and consistency, are their invariable accompaniments.

A *fourth* cause of the disintegration and decay of paper is to be found in the *alum*[4] which is used in the size. In the case of printing and machine papers, it is added at once to the pulp in the beating engine, but formerly, and occasionally even yet, a certain number of sheets are dipt by the workman into the sizing tub, containing diluted gelatine or glue, formed from shavings and cuttings of pelt or skins, to which a certain quantity of alum has been added. Now, as has been already premised, alum being a *bisulphate*, or, in popular language, containing a proportional excess of acid, the chloride of lime employed to bleach the half-stuff, or the already formed pulp, is decomposed by this means, and the free hydrochloric acid (muriatic acid) let loose to corrode the paper-tissue, or free chloride is evolved to form the same mischievous acid, and effect the same process of destruction. Now, chlorine, when *perfectly dry*, seems inert in its discharge of colour, or in the process of decomposition, but, by a hygrometric change in the atmosphere introducing humidity; damp imparts to it the power to destroy the surface it is in contact with. Besides, the portion of muriate of lime that would be formed on the contact of damp air, from the chloride of lime remaining united with the paper material, would always, from the highly *deliquescent* character of that salt, impart to such paper the susceptibility of, and tendency to, absorb moisture, and to contract damp. We should therefore recommend the substitution of the *acetate of alumina* for the alum, as is employed in the process of calico printing; and in the case of a neutral salt, such a decomposition as that we have adverted to would not occur, and one evil at any rate would be completely prevented. In several specimens of India paper,[5] the size used seems to be merely that of alum, which is proved in the efflorescent incrustation already referred to.

The *fifth* and chief source of the disintegration and ruin of modern paper, is the process of bleaching, in whatever stage it may be applied, or in whatever form. It may be used, whether as chloride of lime or gas. This destructive agent too generally clumsily, or incautiously, and unwarily employed: I have specimens of paper, where the sheets have evidently been "fumigated" after being completely fabricated; and the chlorine gas is in a state of such loose combination, that, as soon as it is slightly dated on a metallic plate, the gas is disengaged almost to suffocation, and evolves white clouds on the approach of a feather dipped into ammonia. The gas ought not to be employed otherwise than to bleach the rags after being shred by the cutters. Great care should be taken to have the chlorine much attenuated in its bleaching strength, by dilution with water; and the rags should be thoroughly washed afterwards. Indeed frequent affusions of water will be found necessary to steer completely clear of all danger. We can on no account whatever approve of *chloride of lime* as a bleaching power; but we see much less danger in employing *chloride of soda*, because, in such a case, were bisulphate of alumina to be used in the size, and a small portion of the chloride of soda remain in combination, a decomposition, by the excess of acid in the alum, of the chloride of soda, would form a sulphate of soda which, so far from constituting a salt of a deliquescent kind, would give rise to one of an *efflorescent* character, and repel and give off moisture rather than absorb it; and, if the chloride of soda were otherwise formed into a muriate, a pure muriate of soda would have no tendency to attract moisture; for, when rock-salt weeps in damp weather, and common salt becomes moist, they do so, in all probability, from being associated with *muriate of magnesia*, an ingredient in brine springs, and salt water.

Among the best specimens of modern paper which I have examined, are the patent bank-note paper of Messrs Roberts and Crawford, and of Messrs Wise and Brenchley.[6] As I have not had an opportunity of examining their specifications, I am of course unable to state their ingredients or mode of manipulation; but I have stated the chemical principles on which a durable and good paper can alone be manufactured; and it is hoped that the conditions will be duly weighed by the manufacturer. Our modern books and manuscripts are in a state of the most rapid decay. The greater part of them will not outlive half a century; many of them not half that period. An ichorous matter,[7] more rapid far than the tooth of Time,[8] is hurrying them onwards to the land of shadows, and the "tomb of all the Capulets."[9] We have had in possession letters that have become a *carte blanche* in the course of a few years; and others, already perfectly illegible, have been shown to us. A French gentleman submitted to me a letter, of the date of the 18th March 1815, written on *papier velin*,[10] which was in a state of complete integrity, though almost constantly exposed to friction in the pocket; and another on *English* paper, in a most ruinous condition, and crumbling away, though preserved with the greatest care, being one from a departed friend. The latter was distinctly aluminous to the taste. By chemical reagents I cannot detect alum and chloride of lime in the case of any French papers I have tested, otherwise than in such minor quantities that the causes of deterioration and decay are in a very diminished ratio, indeed, compared with paper of British manufacture. For instance, "Le Nouveau Testament," &c. edition stéréotype, publiée par les soins de M. Frederic Leo, &e. à Paris, 1816. I have singled out this date, by way or contrast with the Bibles issued by the British and Foreign Bible Society of that year. The former is the typography of M. Firman Didot, of Paris; the latter that of the Oxford University press. I may also, in farther proof of the superiority of French printing papers, refer to that noble work, "Dictionnaire de l'Académie Française,[11] 4te, 2 tomes, à Paris, 1814, cinquième edition." The paper is also that called laid. The vignette above, includes the words "à l'Immortalité"; and, though this expression is what may be called truly *French*, it would be mockery in us to bring our tomes into competition as to their respective duration. I do not see why we should be so far behind hand with our Gallic neighbours, or why we should either invite the scorn and execration of our posterity, or lose the record of all we account dear and valuable.

Notes to Introduction

1 "Bath wove post" was a popular English writing paper whose dimensions and weight were standardized for mailing.
2 *Gentleman's Magazine* 134 (July–December 1823): 21–22.
3 Lee E. Grove, "John Murray and Paper Deterioration," *Libri* 16, no. 3 (2009): 194–204.

Notes to Reading

1 ["the *second* source of evil": in chapter 2 Murray cites the first source of evil as the invention of the papermaking machine, which explains his apologetic defense of technological progress in chapter 3.—Ed.]
2 [Murray promoted the use of New Zealand flax in lieu of hemp and successfully cultivated it in the rugged climate of his seaside home in Stranraer, Scotland (*The Analyst* 5 [1836]: 169). A few years later, Murray published *An Account of the Phormium Tenax, or New Zealand Flax, Printed on Paper made from its leaves* (London: Renshaw, 1836). Its papermaking potential was deemed "promising" by the National Bureau of Standards in 1931 (Merle B. Shaw, George W. Bicking, and Martin J. O'Leary, "The Paper-Making Properties of *Phormium tenax* [New Zealand Flax]," *Bureau of Standards Journal of Research* 6, no. 3 [1931]: 411–20).—Ed.]
3 ["half-stuff" refers to a mass of processed fibers before it is mixed with a proportionate amount of water in the vat to make the "slurry." The "furnish" is the final combination of fibers and other additives, such as fillers, coatings, and sizing, which constitute the sheet of paper.—Ed.]
4 [The term *alum* refers to either of two chemical compounds used in papermaking: aluminum potassium sulfate or aluminum sulfate. The former is traditionally used to harden gelatin sizing; Murray is probably referring to this type of alum, since the more refined aluminum sulfate was just being introduced in the early nineteenth century. The latter was used to precipitate rosin as an internal sizing agent. See Irene Brückle, "Aspects of the Use of Alum in Historical Papermaking," in *The Institute of Paper Conservation: Conference Papers, Manchester 1992* (Leigh: Institute of Paper Conservation, 1992), 201–6.—Ed.]
5 [India paper, originally made in China and later imitated in England, refers to a thin, opaque paper typically used to produce high-density texts such as bibles or reference books.—Ed.]
6 [Roberts and Crawford are listed as papermakers in Edinburgh in the Scottish Book Trade Index (SBTI), www.nls.uk/catalogues/scottish-book-trade-index/reynolds-robertson; Wise and Brenchley were listed in a London Directory as "Papermakers and Makers of Bank Notes" (Ian Dye, "The Wise Family: Unlucky or Inept Papermakers?," *Quarterly*, no. 54 (April 2005): 1–11. Thanks are extended to Peter Bower for this reference.—Ed.]
7 ["an ichorous matter": ichor is the watery fluid that oozes from a festering sore; it originally referred to what was thought to be the lethal blood substitute in ancient Greek gods.—Ed.]

8 [Time is often personified as a sinister figure relentlessly gnawing away at the works of humans, surely an antithetical image for conservators.—Ed.]

9 [In Shakespeare's *Romeo and Juliet*, the Capulet family crypt is where the drugged Juliet slept, only to awaken to find that Romeo had killed himself, at which point she, too, committed suicide. Hence, the phrase implies that something precious is lost forever.—Ed.]

10 [*Papier vélin* is wove paper.—Ed.]

11 [The publication, as it appeared, is *Dictionnaire de l'Académie Françoise*, not "Française."—Ed.]

Reading 49

Honoré de Balzac

Lost Illusions (1837–1843)

Written as a serialized novel between 1837 and 1839, Lost Illusions, *by Honoré de Balzac (1799–1850), mirrors the seismic social shifts in post-Napoleonic France, as well as the upheaval in the papermaking industry. David Séchard is the son of a printer who, in order to pay off his unscrupulous father, is forced to perfect and sell his secret for transforming common weeds into paper. This reading opens with a discussion of the inferiority of cotton rags, especially as used in English paper, followed by a narrative of David's clandestine papermaking efforts.*

In the race to find substitute papermaking fibers, competition between Germany, England, and France was fierce. In addition to David's nettles and reeds, countless other substances were macerated. Joel Munsell (1808–1880), in his 1857 edition of A Chronology of Paper and Paper-Making, *lists over one hundred potential papermaking fibers, including such unlikely candidates as mummy cloth and cabbage stumps.*[1] *By Munsell's fourth edition, published in 1870, the list had grown to include frog spittle,*[2] *coconut husks, hibiscus, and the more promising esparto grass. Numerous products of experimentation survive in special collections today, among them Murray's 1836 book of paper made from New Zealand flax (reading 48). Matthias Koops (1752–1834) claimed to have produced paper made entirely from straw in 1800.*[3] *And the title paper of* A Treatise on Cobbett's Corn *(1828), by William Cobbett (1763–1835), does appear to have been made from corn husks and stalks.*[4] *While these items may be considered anomalies, exotic fibers from far-flung places were indiscriminately added to the pulp of papers throughout the nineteenth and twentieth centuries, and their dates of use have only been roughly outlined by the paper industry. For these reasons, some conservators are reluctant to undertake fiber*

From Honoré de Balzac, *Lost Illusions*, trans. Ellen Marriage, 1897 (London: J. M. Dent and Co.), 121–24, 194–95, 219, 248–49, 251–52, 256.

identification for dating and authentication purposes, depending instead on experts in forensics.[5] *Especially difficult is the quantitative assessment of a paper stock given the necessarily minute sample size available for analysis.*

'[A]lthough linen lasts so much longer than cotton, that it is in reality cheaper in the end, the poor would rather make the smaller outlay in the first instance, and, by virtue of the law of *Vae victis!*[1] pay enormously more before they have done. The middle classes do the same. So there is a scarcity of linen. In England, where four-fifths of the population use cotton to the exclusion of linen, they make nothing but cotton paper. The cotton paper is very soft and easily creased to begin with, and it has a further defect: it is so soluble that if you steep a book made of cotton paper in water for fifteen minutes, it turns to a pulp, while an old book left in water for a couple of hours is not spoilt. You could dry the old book, and the pages, though yellow and faded, would still be legible, the work would not be destroyed.

'There is a time coming when legislation will equilise our fortunes, and we shall all be poor together; we shall want our linen and our books to be cheap, just as people are beginning to prefer small pictures because they have not wall space enough for large ones. Well, the shirts and the books will not last, that is all; it is the same on all sides, solidity is dying out. So this problem is one of the first importance for literature, science, and politics.

[. . .]

'Lucien[2] told me that your father, with the intuition of a man of talent, had a glimmering of a notion of some way of replacing linen rags with an exceedingly common vegetable product, not previously manufactured, but taken direct from the soil, as the Chinese use vegetable fibre at first hand. I have classified the guesses made by those who came before me, and have begun to study the question. The bamboo is a kind of reed; naturally I began to think of the reeds that grow here in France.

'Labour is very cheap in China, where a workman earns three halfpence a day, and this cheapness of labour enables the Chinese to manipulate each sheet of paper separately. They take it out of the mould, and press it between heated tablets of white porcelain, that is the secret of the surface and consistence, the lightness and satin smoothness of the best paper in the world. Well, here in Europe the work must be done by machinery; machinery must take the place of cheap Chinese labour. If we could but succeed in making a cheap paper of as good a quality, the weight and thickness of printed books would be reduced by more than one-half. A set of Voltaire, printed on our woven paper and bound, weighs about two hundred and fifty pounds; it would only weigh fifty if we used Chinese paper. That surely would be a triumph, for the housing of many books has come to a difficulty; everything has grown smaller of late; this is not an age of giants; men have shrunk, everything about them shrinks, and house-room into the bargain. Great mansions and great suites of rooms will be abolished sooner or later in Paris, for no one will afford to live in the great houses built by our forefathers. What a disgrace for our age if none of its books should last! Dutch paper—that is, paper made from flax—will be quite unobtainable in ten years' time. Well, your brother told me of this idea of your father's, this plan for using vegetable fibre in paper-making, so you see that if I succeed, you have a right to—'

[. . .]

He drew several sheets of white paper from his pocket, brandished them in triumph, and laid them on his wife's lap.

'A ream of this paper, royal size, would cost five francs at the most,' he added, while Eve handled the specimens with almost childish surprise.

'Why, how did you make these sample bits?' she asked.

'With an old kitchen sieve of Marion's.'

'And are you not satisfied yet?' asked Eve.

'The problem does not lie in the manufacturing process; it is a question of the first cost of the pulp. Alas, child, I am only a late comer in a difficult path. As long ago as 1794 Mme. Masson tried to use printed paper a second time;[3] she succeeded, but what a price it cost! The Marquis of Salisbury tried to use straw as a material in 1800,[4] and the same idea occurred to Séguin in France in 1801.[5] Those sheets in your hand are made from the common rush, the *arundo phragmites*, but I shall try nettles and thistles; for if the material is to continue to be cheap, one must look for something that will grow in marshes and waste lands where nothing else can be grown. The whole secret lies in the preparation of the stems. At present my method is not quite simple enough. Still, in spite of this difficulty, I feel sure that I can give the French paper-trade the privilege of our literature; papermaking will be for France what coal and iron and coarse potter's clay are for England—a monopoly. I mean to be the Jacquart[6] of the trade.'

Harassed by money difficulties, beset with fears for his wife's health, stung to the quick by Lucien's disgrace, David had worked on at his problem. He had been trying to find a single process to replace the various operations of pounding and maceration to which all flax or cotton or rags, any vegetable fibre in fact, must be subjected; and as he went to Petit-Claud's office, he abstractedly chewed a bit of nettle stalk that had been steeping in water. On his way home, tolerably satisfied with his interview, he felt a little pellet sticking between his teeth. He laid it on his hand, flattened it out, and saw that the pulp was far superior to any previous result. The want of cohesion is the great drawback of all vegetable fibre; straw, for instance, yields a very brittle paper, which may almost be called metallic and resonant. These chances only befall bold inquirers into Nature's methods!

'Now,' said he to himself, 'I must contrive to do by machinery and some chemical agency the thing that I myself have done unconsciously.'

[. . .] David proposed to give his father indisputable proof of his discovery, while reserving his secret. He offered to give him an interest in the affair in return for money paid down; a sufficient sum to release him from his present difficulties, with or without a further amount of capital to be employed in developing the invention.

'And how are you going to prove to me that you can make good paper that costs nothing out of nothing, eh?' asked the ex-printer, giving his son a glance, vinous, it may be, but keen, inquisitive, and covetous; a look like a flash of lightning from a sodden cloud; for the old 'bear,' faithful to his traditions, never went to bed without a nightcap, consisting of a couple of bottles of excellent old wine, which he 'tippled down' of an evening, to use his own expression.

'Nothing simpler,' said David; 'I have none of the paper about me, for I came here to be out of Doublon's way; and having come so far, I thought I might as well come to you at Marsac as borrow of a money-lender. I have nothing on me but my clothes. Shut me up somewhere on the premises, so that nobody can come in or see me at work, and—'

'What? you will not let me see you at your work then?' asked the old man, with an ugly look at his son.

'You have given me to understand plainly, father, that in matters of business there is no question of father and son—'

'Ah! you distrust the father that gave you life!'

'No; the other father who took away the means of earning a livelihood.'

'Each for himself, you are right!' said the old man. 'Very good, I will put you in the cellar.'

'I will go down there with Kolb. You must let me have a large pot for my pulp,' said David; then he continued, without noticing the quick look his father gave him,— 'and you must find artichoke and asparagus stalks for me, and nettles, and the reeds that you cut by the stream side, and to-morrow morning I will come out of your cellar with some splendid paper.'

'If you can do that,' hiccoughed the 'bear,' 'I will let you have, perhaps—I will see, that is, if I can let you have—pshaw! twenty-five thousand francs. On condition, mind, that you make as much for me every year.'

'Put me to the proof, I am quite willing,' cried David. 'Kolb! take the horse and go to Mansle, quick, buy a large hair sieve for me of a cooper, and some glue of the grocer, and come back again as soon as you can.'

[. . .]

At four o'clock in the morning David came out of the distillery; he had been careful to leave no sign of his occupation behind him; but he brought out some thirty sheets of paper that left nothing to be desired in fineness, whiteness, toughness, and strength, all of them bearing by way of water-mark the impress of the uneven hairs of the sieve. The old man took up the samples and put his tongue to them, the lifelong habit of the pressman, who tests papers in this way. He felt it between his thumb and finger, crumpled and creased it, put it through all the trials by which a printer assays the quality of a sample submitted to him, and when it was found wanting in no respect, he still would not allow that he was beaten.

'We have yet to know how it takes an impression,' he said, to avoid praising his son.

'Fonny man!' exclaimed Kolb.

The old man was cool enough now; he cloaked his feigned hesitation with paternal dignity.

'I wish to tell you in fairness, father, that even now it seems to me that the paper costs more than it ought to do; I want to solve the problem of sizing it in the pulping trough. I have just that one improvement to make.'

'Oho! so you are trying to trick me!'

'Well, shall I tell you? I can size the pulp as it is, but so far I cannot do it evenly, and the surface is as rough as a burr!'

'Very good, size your pulp in the trough, and you shall have my money.'

[. . .]

'My adored Eve,—I am writing to you the first letter on my first sheet of paper made by the new process. I have solved the problem of sizing the pulp in the trough at last. A pound of pulp costs five sous, even supposing that the raw material is grown on good soil with special culture; three francs' worth of sized pulp will make a ream of paper at twelve pounds to the ream. I am quite sure that I can lessen the weight of books by one-half. The envelope, the letter, and samples enclosed are all manufactured in different ways. I kiss you; we shall have wealth now to add to our happiness, everything else we had before.'

Notes to Introduction

1 Joel Munsell, *A Chronology of Paper and Paper-Making* (Albany, N.Y.: J. Munsell, 1857).

2 Frog spittle is a foamy mass of wispy green algae that grows on still water.

3 Matthias Koops, *Historical account of the substances which have been used to describe events and to convey ideas, from the earliest date, to the invention of paper. Printed on the first useful paper manufactured soley [sic] from straw* (London: T. Burton, 1800; 2nd ed. London: Printed by Jaques & Co., 1801). Analysis of the four different papers in the second edition indicates that wood and cotton fibers were used in addition to straw (Walter Rantanen, "Report of Fiber Analyses to New York University," IPS FI02856-10, Integrated Paper Services, December 28, 2010). I am grateful to Morgan Adams for sharing the results of her analysis of Koops's Neckinger Mill papers.

4 William Cobbett, *A Treatise on Cobbett's Corn* (London: William Cobbett, 1828). I am grateful to Alan Balicki and Olga Marder for confirming that the title page is made from a virgin plant fiber presumed to be corn.

5 One well-known fiber analyst with ties to the paper conservation community is Walter Rantanen of Integrated Paper Services, 3211 East Capitol Drive, Appleton, WI 54911; http://ipstesting.com.

Notes to Reading

1 [*Vae victus!* or "Woe to the vanquished!" is taken from Livy's account of Gaul's conquest of Rome in 387 BCE. The Romans were forced to pay tribute in gold that had been weighed out in tampered scales; however, they were in no position to protest. Balzac's argument that better-quality paper is more economical in the long run foreshadows an argument also promulgated in the campaign for permanent paper.—ED.]

2 [Lucien is a childhood friend of David Séchard and also the brother of Eve, David's devoted wife. As an ambitious writer in Paris, Lucien has managed to incur substantial debt and disgrace himself.—ED.]

3 [No record of recycled papers in 1794 could be found.—ED.]

4 ["Marquis of Salisbury": No less a personality than Charles Dickens (1812–1870) describes the efforts of many, including the Marquis of Salisbury, to turn straw into paper, even going so far as to claim to be writing on it himself. Charles Dickens, "Splitting Straws," *Household Words*, no. 207 (March 11, 1854): 85–89.

5 ["Séguin straw": Marc Séguin (1786–1875) was a French engineer recognized for designing and constructing the first suspension bridge in Europe. From a prominent Annonay family, he was naturally involved in the region's extensive papermaking industry.—ED.]

6 [David is probably referring to Joseph Marie Jacquard (1752–1834), inventor of a programmable and mechanized loom that revolutionized textile manufacture in France.—ED.]

Reading 50

WOLFGANG SCHLIEDER

Friedrich Gottlob Keller, Inventor of the Wood-Pulping Machine: Contributions to His Biography from Letters (1977)

It was the prophetic observation of the entomologist René Antoine Ferchault de Réaumur (1683–1757) that some species of wasps use digested wood to produce their paperlike nests that inspired the young Friedrich Gottlob Keller (1816–1895) to design and patent in 1844 a grinding machine that produced papermaking fibers directly from logs.[1] *The technology for producing mechanically processed wood pulp, known simply as groundwood or mechanical woodpulp, transformed commercial papermaking. By the late 1800s inexpensive wood fibers, in the form of groundwood or further refined using chemical soda or sulfite processes, had replaced cotton rags for papermaking. The shorter wood fibers, combined with their integral noncellulosic ingredients, especially lignin, which is popularly described as the amorphous cement that binds and stiffens the bundles of cellulose fibers, led to drastically reduced durability and permanence. An internal sizing process developed in the early twentieth century, which used finely powdered pine pitch rosin precipitated with aluminum sulfate, created a virtual internal combustion reaction of acid hydrolysis. No doubt, this imagery is behind the prize-winning 1987 documentary* Slow Fires,[2] *part of a national campaign for the duplication and/or mass deacidification of acidic books or, conversely, "the most successful piece of library propaganda ever created."*[3]

Often demonized, wood pulp in and of itself does not necessarily produce a short-lived paper, just as the use of rags as a source of cellulose does not guarantee a permanent paper. For example, Barrow's (reading 54, table 1) rag paper of 1857 with a pH of 4.3 may not necessarily have been a victim of acid migration as he postulates. Advances in refining wood pulp to extract high-quality alpha-cellulose and the introduction of neutral or alkaline sizing agents, typically alkyl ketene dimer (AKD),

From WOLFGANG SCHLIEDER, *Der Erfinder des Holzschliffs Friedrich Gottlob Keller: Beiträge zu seinem Lebenbild aus Briefen* (Leipzig: VEB Fachbuchverlag, 1977), 110–15. Reproduced courtesy of the author.

that bond with cellulose to increase water resistance have resulted in improvements in both the lifetime of paper and the health of the environment. Conservation-grade papers and boards made from alpha-cellulose are less expensive alternatives to rag-based paper products, confusingly differentiated as "museum grade," for housing collections of prints and drawings.

F. G. Keller's story of the history of his invention and explanation of a drawing of the wood grinder drawing [explanation and drawing not included here]. Draft of an attachment to a letter to E. Grosse, Jena, dated November 18, 1890 (HH, B 6a).

As has often been mentioned, I got the idea of making paper from wood fibers from wood-fiber wasps' nests. But to break down the wood into the tiny fibers required in papermaking, I first thought of reducing mechanically shredded wood, wood shavings and such, by boiling it with the addition of soda, potash, etc., to the point that the substance that binds the fibers together dissolved and it became possible to make the fibers still finer by further treatment. Yet repeated experiments over the course of nearly two years, which I was able to undertake only in addition to my regular work, were unsuccessful. I was convinced that the effect of higher temperatures than the mere boiling point permits might make it possible to achieve this. However I had no access to such steam apparatus, so I had to abandon the idea. Needless to say, it was difficult for me to give up a hope I had nourished for years. As I was struggling with this, it occurred to me one day how, as a boy of eight or nine, I had once occupied myself by making necklace beads from cherry pits by grinding them on sandstone, then the next day made the observation that when the sandstone surface I had been using had dried, flakes lifted off that were pliable. This was not possible if it was only the cherry pit residue, which could only be powdery, and I was puzzled. I soon convinced myself, however, that I had not only sanded off the substance of the cherry pits but also fibers from the little piece of wood used to hold them, which lent pliability to the dried residue. As I mentioned, this only occurred to me again after some eighteen years. Having now a better understanding of it, I could not refrain, despite my other obligations, from immediately making an experiment on an ordinary stone used for sharpening tools. After roughly two hours I had ground off a small quantity of material, enough to make paper tests by means of an appropriate sieve. But before I got to it, by chance, since as a small quantity of the material mixed with water had dropped onto a cloth nearby that absorbed the water and I was able to peel off the wood fibers from the surface of the cloth with a knife, a first small piece of wood paper had formed, which I then treated with further pressing and drying. I enclose it for your obliging inspection. My delight in the fact that I now believed I had finally found the way to solving the problem after years of experiments to no avail was very great, and not even Columbus, after long wanderings, can have rejoiced more greatly when he discovered the first indications that let him suspect that the longed-for land could not be very far away.

Encouraged by this success, I first made a small apparatus consisting of a small grindstone roughly 20 centimeters in diameter and 15 centimeters thick, that could be affixed to a lathe. I placed this in a box of the appropriate size filled with water. With this apparatus, I—or rather for the most part my blessed wife (who always faithfully and tirelessly stood beside me)—ground off a larger quantity of material, during which time I made myself a paper form the size of a quarto sheet as well as a kind of wedge press and other necessary implements. With it I produced roughly eighty sheets of the above-mentioned size, among them colored ones and others mixed with fibers from tree bark. In terms of quality, these experiments were satisfactory. Also it seemed hopeful, almost with certainty, that wood substance produced in this way would be less expensive to obtain than papermaking substance from rags. With these results, I first sought the opinions of experts, also contacts who might help me exploit my discovery or help to fund it. As for the latter, I was told that the practicality of the process could only be roughly appraised after seeing the results of larger experiments. That I should therefore attempt to make a larger quantity, several reams of ordinary size, and for this purpose might use the existing equipment of a paper mill. But this could only be done to the extent that I could have the finished abraded material formed into paper, as the production of the material could not be accomplished there owing to the absence of a larger grinding apparatus. Even if I had had such an apparatus, I would not have done it, so as not to reveal the process owing to its novelty. I therefore decided to follow the advice I had been given only in part, namely to produce the material in my own home and then take the finished material to a paper mill for forming. For this purpose I made myself a wooden stand with a water chest of the size visible in the enclosed sketch. On it I mounted, submerged into the water of the trough, a grinding stone roughly 60 centimeters in diameter and 25 centimeters thick. Above this stone I attached a hollow box with a side opening into which the wood could be placed crosswise, above which, to press it against the grinding surface of the stone, there was a weighted lever. To the ends of the grinding stone's axle I affixed two cranks with which to turn it.

This highly primitive apparatus was now ready for grinding. However, I immediately discovered that it took a lot of strength to set it in motion. Even with a piece of wood of only very modest thickness, my wife and I were unable to turn the stone. I was therefore required to reduce the surface of the wood to be ground little by little until we were able to overcome the resistance. When grinding this very narrow wood surface so many splinters and fringelike fibers were produced that barely half of the material was usable, and to separate this from the rest with the help of a sieve was very difficult and time-consuming. In this manner it would thus have taken a great deal of time before I collected enough material to be able to have a ream of paper made from it. So I had the idea of introducing the wood in the form of cylinders of the appropriate length, and in such a way that wood, also rotating, but slower, be placed in contact with the rotating stone. This meant that only a narrow wood surface was presented for grinding, which was naturally much easier to manage. Also, it had the great advantage that the formation of splinters was completely eliminated. In this way I was able to obtain a larger quantity of material, though with a great deal of work, from which I obtained nearly five reams of paper. Otherwise, following the first but better method, with larger, immobile wood surfaces, I would have had to use some sort of motor for power, and would also have subjected myself to the danger of being exposed. [. . .]

Notes to Introduction

1 Keller's original wood grinding machine is preserved in the Gellert Museum in Hainichan, Germany, according to Peter Burger. I am grateful to Irene Brückle for bringing this publication to my attention.

2 *Slow Fires: On the Preservation of the Human Record* (American Film Foundation, 1987).

3 Nicholson Baker, *Double Fold: Libraries and the Assault on Paper* (New York: Random House, 2001), 184.

Reading 51

B. L. Browning

The Nature of Paper (1970)

Bertie Lee Browning (1902–1999) was an organic chemist at the Institute of Paper Chemistry (IPC) in Appleton, Wisconsin.[1] *Like the other scientists represented in Part V, Browning was a popular practitioner of knowledge transfer through his numerous alliances with paper and book conservators.*[2] *He served as an instructor in the groundbreaking seminar held at the IPC, "Scientific Approach to the Preservation of Paper Artifacts," October 18–22, 1971, and his book,* The Analysis of Paper, *remains a standard reference for paper conservators.*[3] *Probably because Browning's research is so readily accessible, his classification system for internal and external causes of paper deterioration and his explanation of their mechanisms via oxidation, acid hydrolysis, and photochemical deterioration still serve as the scientific underpinnings for current collection care policies in museums and libraries.*[4]

In this reading, Browning explains the difference between durability and permanence, a critical distinction for conservators who distrust apparently irrelevant measurements of durability such as folding endurance and tear resistance.[5] *For example, it can be rightly asked if extending the useful life of paper, as determined by fold endurance after accelerated aging, by over 300 percent is a meaningful treatment outcome if the item is unlikely to ever be folded.*

STABILITY OF PAPER

Durability and Permanence

Paper is by its nature a perishable material. It can be burned, it decays, it can be attacked by insects and rodents, and it is degraded or destroyed by chemical

---

From B. L. Browning, "The Nature of Paper," in *Deterioration and Preservation of Library Materials: The Thirty-Fourth Annual Conference of the Graduate Library School, August 4–6, 1969*, ed. Howard W. Winger and Richard Daniel Smith (Chicago: University of Chicago Press, 1970), 30–36.

reaction with many materials. Nevertheless, in the absence of seriously deteriorative influences, paper endures for many years or even for many centuries. Any paper product must last long enough to serve its intended function. Many uses of paper do not demand any great degree of permanence; once the use requirement has been satisfied, the paper has no further utility. On the other hand, in papers used for record keeping, legal documents, library books of permanent value, art works, and similar applications, it is vital that the paper retain its original character without alteration over periods of many decades or centuries.

[. . .]

Paper is complex in composition and structure. The profound effects of source and nature of fibers, of the response of fibers during processing and stock preparation, of variables during the complicated processes of sheet forming, pressing, and drying on the paper machine, of the addition of sizing, filling, coating, and other additives, of the introduction of impurities from raw materials or in processing—all of these contribute to the fabrication of a product which cannot be described or evaluated simply. The sensitivity of measurable physical properties of paper to variations in moisture content, as well as to formation and relaxation of internal stresses which may be induced by mechanical treatment and by temperature or humidity cycling, contributes to difficulties in interpretation of test results. More research will surely be necessary to elucidate the behavior and properties of paper under various conditions of storage and use and to evaluate the relative contributions of factors which influence permanence.

Papermakers prefer to distinguish between the concepts of durability and permanence. Durability is the property of resisting deterioration by use, that is, the handling to which paper may be subjected, the hazards incident to heavy use of books in libraries, or the ability of sandpaper to hold together until the user finishes his work. Permanence, on the other hand, refers to the degree to which a paper retains its original qualities during storage [1]. In a sense, permanence may be defined as the retention of durability, whether the latter be small or great. A paper may be durable but not permanent that is, durability may be lost rapidly because the paper contains an excessive amount of acids, or it may be permanent but not durable.

The quality of durability is built into a paper when it is made, that is, by choice of fibers, conditions of manufacture, and the inclusion of additives which may enhance resistance to damage during use. Permanence is determined by the composition of the paper and by external conditions to which it is subjected during storage or use.

Factors Influencing Permanence

The factors which influence permanence may be divided into: (*a*) internal factors, which are established primarily by the composition of the paper; and (*b*) external factors, which are determined by conditions during storage and use, for example, temperature, relative humidity of the atmosphere (which controls the hygroscopic moisture content of the sheet), exposure to light, atmospheric contaminants (sulfur dioxide, nitrogen oxides, ozone), and the existence of biological hazards (decay by bacteria or fungi and attack by insects or rodents).

The control of internal factors is in the province of the manufacturer, who must choose the raw materials and processes which are requisite for production of paper having the necessary stability for intended uses. Control of external factors is the responsibility of the user. Methods for control of atmospheric contamination and biological deterioration are obvious and well known, even if burdensome. Hence, for any selected paper, the additional important conditions for maintaining stability available to the user are control of temperature and relative humidity.

Permanence of the fibers is governed largely by the resistance of cellulose to deterioration. Pure cellulose is a relatively stable organic compound; when stored under optimal conditions it can be preserved for centuries or millennia without severe deterioration. Nevertheless, it is subject to alteration by several types of chemical attack, of which the most important from the viewpoint of permanence are oxidation, hydrolysis, and photochemical reactions.

Cellulose is oxidized by atmospheric oxygen, although under normal conditions the rate of oxidation reactions is slow. Oxidation leads to formation of carbonyl (reducing) groups and of carboxyl (acidic) groups at one or more locations in the glucose units which comprise the cellulose chain molecule. Oxidation also results in formation of peroxides [2], which contribute to further oxidation and side reactions. Ozone in the atmosphere, even in minute concentrations, is a very active deteriorative reagent. The formation of oxidation products may be accompanied by immediate scission of the polymeric chain molecule, or it may result in a sensitivity to further deterioration by a variety of influences. The poor stability of "oxycellulose" has been abundantly established [3].

The hydrolysis of cellulose by acid attack leads to decrease in length of the polymeric chain molecule through random scission of the chains. The strength of the fiber is decreased; deterioration may proceed so far that the fiber becomes very weak and brittle and finally can be reduced to a fine powder. Hydrolysis may be caused by the presence of acids introduced into the paper during manufacture, for example, from the alum used in sizing or from some wet-strength resins. Oxidized cellulose or the acidic hemicelluloses associated with the cellulose in wood and wood pulps contain carboxyl (acid) groups which are a part of the polysaccharide chain molecules. Wood contains acetyl groups; hydrolysis of these groups in wood or in pulps from which they have not been removed introduces additional acidity. External sources of acids arise from the atmosphere of urban or industrial areas that may contain sulfur dioxide (which is readily oxidized to sulfuric acid) and oxides of nitrogen (which form nitric acid); these atmospheric sources contribute to acid deterioration.

The extent to which hydrolytic reactions proceed depends upon the amount (or concentration) of acidic substances originally present in the paper and upon the acids introduced later through oxidation of cellulose and hemicelluloses or through absorption of acidic gases arising from atmospheric contamination. The amount of acidic materials in paper introduced subsequent to manufacture may be large in comparison with that originally present. The contribution of such acids to deterioration is minimized if the paper contains buffering materials (a "buffer solution" is one which resists change in pH when acids or alkalies are added). Many papers exhibit a considerable buffering capacity; a high buffering capacity can contribute much to stability of paper during aging by partial neutralization of acidic substances formed or introduced.

Deterioration of cellulose by the chemical reactions of oxidation or hydrolysis proceeds more rapidly with increase in temperature. There is a rule-of-thumb (and useful approximation) among chemists that the rate of a reaction is doubled for each increase in temperature of 10° C. Many studies on aging of paper (for example [4]) have confirmed the profound effect of temperature and suggest the importance of this factor in the preservation of paper.

Hydrolysis and oxidation generally proceed more rapidly in the presence of moisture. Paper is hygroscopic and has a moisture content of 5–7 percent in equilibrium with relative humidities in the range of about 50 percent. The effect of moisture content during natural aging has not been well established, although it has been shown that aging reactions at temperatures in the range of 80°–120° C are much more rapid in the presence of moisture [4,5]. Deterioration is promoted by the presence of many metallic salts, particularly those of iron and copper which are unavoidably introduced in small but measurable amounts during manufacturing processes.

Photochemical degradation, that is, the degradation produced by light, is not important in many uses of paper where prolonged exposure to strong light is not to be expected. Cellulose is degraded photochemically, but lignin is especially subject to deterioration from this source. Papers containing groundwood are observed to have poor stability upon exposure to light, particularly to strong ultraviolet sources.

Prediction of Permanence

The prediction of permanence is necessary in evaluation of papers for which this property is important. The manufacturer needs to determine the probable permanence of his product; the user wishes to know whether the paper meets his requirements [4]. Any practical predictive test must depend on some accelerated aging test. If accelerated aging is to have any predictive validity, it is necessary to demonstrate that there is a meaningful and reproducible relationship between such aging and aging under "normal" conditions.

An accelerated aging procedure based on heating the paper in an oven at circa 100°C was developed at the Swedish Government Testing Institute and at the U.S. Bureau of Standards in the period 1925–35 [6,7]. Rasch and Scribner [8] considered the heating test (seventy-two hours at 100° C) as a valid means for estimating the relative aging quality of papers. Later investigators concluded that accelerated aging under these conditions is empirically equivalent to approximately twenty-five years (a range of about eighteen to forty-eight) aging at normal temperatures.

The oven aging test can be valid only if the reactions leading to deterioration are essentially identical at circa 100° C and at "normal" temperatures. The profound effect of moisture content on rate of aging [4,5] suggests the necessity of taking this factor into account. In any case, the reactions in question can be only those whose rates are dependent on temperature; changes associated with photochemical effects, access to atmospheric contaminants, and other extraneous factors are not evaluated in oven aging.

It is necessary to establish that deteriorative reactions proceed in a regular and predictable way as a function of temperature. The relation of reaction rate to temperature has been determined for many systems; this is the scope of chemical kinetics. In the case of chemical reactions which can be studied in a fundamental way, it has been shown that a reaction rate is related to the temperature at which the rate is measured. The rate constant, k, at one temperature is described by an appropriate relationship which yields a linear plot of chemical composition versus time of reaction.

The rate constant, k, is then used to describe the temperature dependency of reaction rate according to the exponential form of the Arrhenius equation:

$$k = Ae^{-E/RT}$$

where k = rate constant; A = a frequency factor; e = the base of the natural system of logarithms (2.71828 . . .); E = activation energy in kilocalories per mole; R = the gas constant; T = absolute temperature in degrees Kelvin (degrees centigrade + 273). The Arrhenius equation is usually used in its logarithmic form:

$$\log k = \log A - E/2.303RT.$$

For any specific reaction, log A and R are considered constant, and E is the activation energy which is constant if only one reaction occurs within the temperature range employed. Hence, a plot of logarithm of k versus $1/T$ will yield a straight line having a slope which is the activation energy E (fig. 1).

The changes of a paper property (for example, folding endurance or development of color) with time are not determined by any single chemical reaction whose

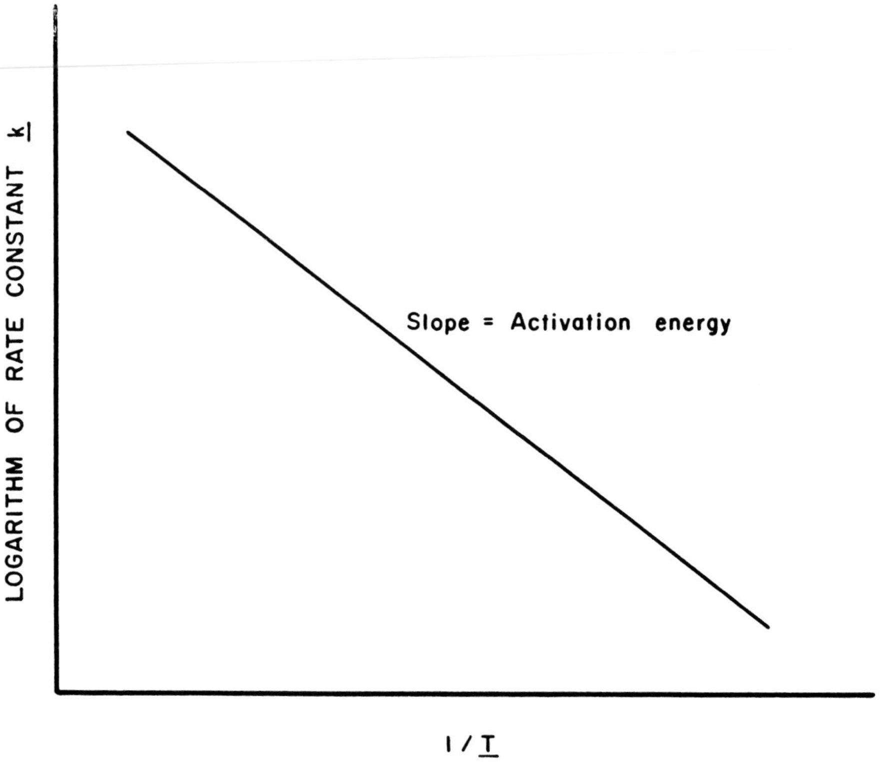

Figure 1
Relationship of the logarithm of the rate constant and the reciprocal of absolute temperature.

effects can be isolated and controlled. Hence, applications of conventional kinetic treatment of aging data on paper are largely empirical. In any case, the results of kinetic treatment of data can be undertaken with confidence only if both the rate plots and the Arrhenius plots are linear.

The rate plots for loss of folding endurance are linear or nearly so for many papers if the reaction is treated as first order; that is, the logarithm of the property is plotted versus time of aging. However, many papers give nonlinear plots, and it is evident that not all papers exhibit the same type of behavior [4]. Other physical properties (for example, zero-span tensile strength, tearing energy) have also been plotted according to first-order kinetics; it appears that this is the most suitable general relationship for these tests, but again many rate plots are nonlinear [4]. The development of colored substances during aging is shown by a plot of specific absorption coefficient versus time of aging; these plots are also linear for some papers and nonlinear for others.

If the Arrhenius plots (log k versus $1/T$) are linear, it becomes possible to extrapolate the relationship obtained upon aging at elevated temperatures and to determine the rate at a specified lower temperature, for example, 20° C or 70° F. One can then predict the value of the measured property after any specified time

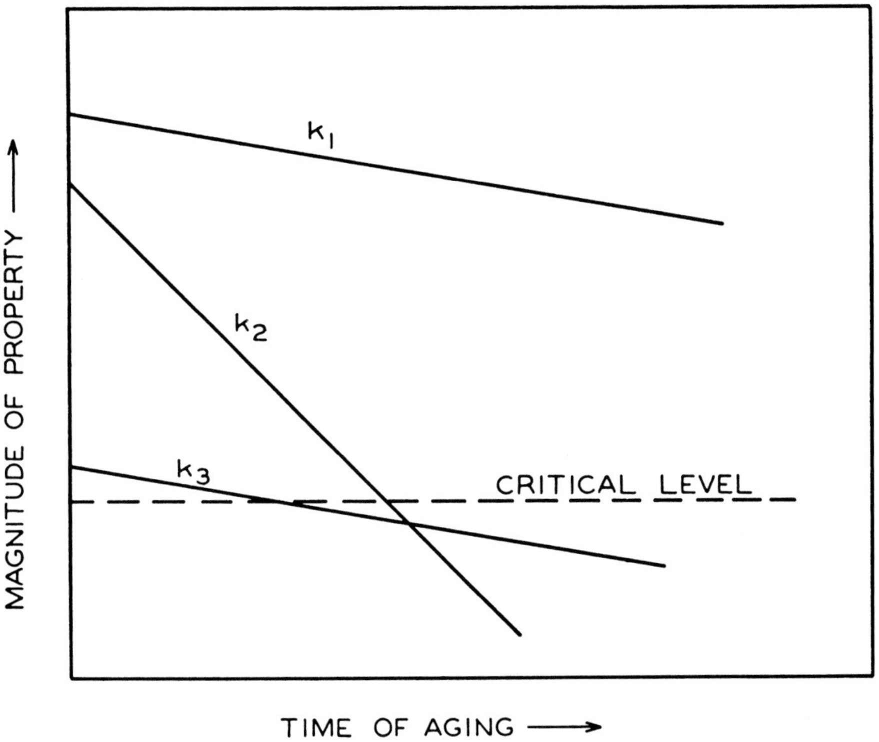

Figure 2
Idealized values for rate constants of three papers as determined by accelerated aging.

of aging or determine, as in the case of folding endurance, the time for this to be reduced to one-half or other fraction of its original value. It has been shown that the Arrhenius plots for aging of paper are in fact approximately linear, but deviations are large enough that accurate predictions of changes taking place at lower temperatures can be established in only a very approximate manner. Nevertheless, the effect of temperature on aging rates is profound, and the relationship found, even if only an approximation, can be of great service in comparing stability of paper and in estimating rates of deterioration as a function of temperature during storage or aging [4]. Interpretation of aging data in this way assumes that all factors (including moisture content of the paper) other than temperature are held constant.

The relative stability of papers can be expressed by comparing the rates of change (rate constants k) in physical or other properties with time of aging. However, a strength property of a paper at any given time depends upon the initial value as well as upon the rate of change. For example, if two papers change at identical rates, indicated by rate constants k_1 and k_3 in figure 2 [9], the one having the higher initial strength will maintain a higher strength throughout its life. Two papers may both have high initial strength values (designated by rates k_1 and k_2 in fig. 2), but the one having the higher rate of strength loss (k_2) will deteriorate more rapidly than the other (k_1). A high initial strength is desirable but alone is not sufficient; a paper (indicated by rate k_3) may have lower initial strength than another (indicated by rate k_2) but maintain a usable strength for a longer time. If it is possible to establish a critical level below which the paper loses its essential utility, the information furnished by the rate constants and by this critical level may serve for comparative evaluations of useful life.

Notes to Introduction

1 In 1989 the IPC, along with the Dard Hunter Paper Museum, relocated to Georgia and became the Institute of Paper Science and Technology. The museum was renamed the Robert C. Williams Paper Museum in 1996.

2 Dr. Browning was a mentor of Richard D. Smith, president of Wei T'o Associates, Inc. I am grateful to him for sharing his reminiscences of Dr. Browning.

3 B. L. Browning, *The Analysis of Paper* (New York: Marcel Dekker, 1969).

4 Sally Robbia points out that the knowledge that paper deteriorates from a combination of internal chemical processes and external environmental factors was published as early as 1911 by Hans Wrede ("Atmospheric Effects on Paper," *Paper* 2, no. 6 [January 25, 1911]: 9–11). See Sally Robbia, "William James Barrow: A Biographical Study of His Formative Years and His Role in the History of Library and Archives Conservation from 1931 to 1941" (PhD diss., Columbia University, 1999), 82.

5 The distinction between durability and permanence remains murky when applied to conservation. See Ellen McCready, "Definitions of Permanence and Durability," *Alkaline Paper Advocate* 2, no. 4 (October 1989): 42–44.

Notes to Reading

1 American Paper and Pulp Association. *The Dictionary of Paper*. 3d ed. New York: American Paper and Pulp Association, 1965.

2 Marracinni, L. M., and Kleinert, T. N. "Aging and Color Reversion of Bleached Pulps. I. Peroxide Formation during Aging." *Svensk Papperstidning* 65 (February 28, 1962): 126–31.

3 Ott, E.; Spurlin, H. M.; and Grafflin, M. W., eds. *Cellulose and Cellulose Derivatives*. 2d ed., pt. 1, chap. 3. New York: Interscience Publications, 1954.

4 Browning, B. L., and Wink, W. A. "Studies on the Permanence and Durability of Paper. I. The Prediction of Paper Permanence." *TAPPI* 51 (April 1968): 156–63.

5 Richter, G. A., and Wells, F. L. "Influence of Moisture in Accelerated Aging of Cellulose." *TAPPI* 37 (August 1956): 603–8.

6 Hall, G. "Permanence of Paper." *Paper Trade Journal* 82 (April 8, 1926): 52–58.

7 Rasch, R. H. "A Study of Purified Wood Fibers as a Papermaking Material." *Bureau of Standards Journal of Research* 3 (September 1929): 469–506.

8 Rasch, R. H., and Scribner, B. W. "Comparison of Natural Aging of Paper with Accelerated Aging by Heating." *Bureau of Standards Journal of Research* 11 (December 1933): 727–32.

9 Browning, B. L. *Analysis of Paper*. New York: Marcel Dekker, Inc., 1969.

Reading 52

S. KOHLER AND G. HALL

Acidity in Paper (1925)

How to accurately measure and properly interpret the pH of paper is a contentious issue for conservators.[1] *Because of the destructive nature of TAPPI (Technical Association of the Pulp and Paper Industry) hot and cold extraction techniques, paper conservators are usually limited to surface measurements using a permeable membrane electrode attached to a pH meter or nonbleeding indicator strips, both of which are pressed against a drop of deionized water on the surface of a print or drawing—a procedure not without the risk of staining and tideline formation.*[2] *The values obtained from surface pH measurements of paper may be misleading since an alkaline surface may belie an acidic core or vice versa. Superseding the debate on the best way to measure and interpret pH, however, is the pragmatic decision on what course of action to take in the face of an acid reading of below 7 or 6.6, the pH of pure cellulose. One frequently recommended option is chemical intervention via aqueous or nonaqueous deacidification, whereby an alkaline reserve is precipitated within the structure of the paper. Despite a long-standing scientific justification for deacidification, however, paper conservators have not wholeheartedly embraced the various alkalization processes that have been developed and widely advertised. Several theories can be advanced for this reluctance. First, many items do not come under the jurisdiction of conservation until darkening and embrittlement are manifest. At this point little can be done to reverse the damage: to be effective, deacidification must occur before acid hydrolysis produces such symptoms. Second, conservators are wary of the unpredictable visual and physical changes brought about by deacidification. And third, many gelatin-sized rag papers were probably acidic at the time of manufacture but remain strong and unblemished centuries later.*[3]

From S. KOHLER and G. HALL, "Acidity in Paper," *Paper Industry* 7, no. 7 (October 1925): 1059.

Because of these concerns, many conservators opt for a more "conservative" approach to treating acidic paper: washing the item to reduce acidity, encapsulating it against buffered tissue, or lining it with calcium-containing uda-gami. Many scientists argue that such partial treatment approaches are, in fact, not safer, since the root cause of deterioration is neither eliminated nor effectively counteracted. According to the chemist Margaret Hey, "a 'sandwich filling' of acidity will remain in the body of paper to continue destroying it."[4] *While theoretically the outcome desired by the chemist—optimal permanence for the paper—is the same as that of the conservator, deacidification, the most effective "cure" for acidic prints and drawings, is not always justified.*

The success of early intervention can be seen in the carefully designed and integrated mass deacidification programs currently under way at large research libraries and archives.[5] *Options for dealing with already acidic materials include large-scale duplication and, more recently, paper-splitting processes, with special collections receiving case-by-case consideration.*

During an investigation with regard to the durability of paper, recently completed at the Swedish Government Testing Institute in Stockholm, it became apparent that the acidity of a paper has influence upon its durability.

In general, the presence of acid reacting substances in a paper is due to one or more of the following facts:

(a)—The fiber material itself reacts acid on account of acid reacting substances imparted during the manufacturing process.

(b)—Acid reacting salts of alumina, usually sulphate of alumina, are added to the paper with rosin sizing.

(c)—The paper is surface sized with acid reacting glue.

(d)—The paper is surface sized with glue to which sulphate of alumina has been added as a preservative.

(a) and (c) have ordinarily but a slight influence on the total acidity; in rosin sized papers, acidity is due mainly to (b), and in surface sized papers mainly to (d). For papers both rosin and surface sized, the acidity will consequently be a combination of (b) and (d).

The quantitative determination of acidity of paper is generally made by the extraction of the acid substances with water and titration of the extract obtained with alkali, using a suitable indicator.

All methods suggested up to the present refer to testing of the paper cut into small pieces. However, taking into consideration the difficulty in extracting by water such a hardly pressed and sized fiber sheet as paper, this way of testing seems to cause incomplete and unreliable dissolving of the acid substances present in the paper. Furthermore, no attention has been paid to the various origins of the acidity; therefore, the results obtained from papers with different kind of sizing are not directly comparable. Herzberg[1] has adopted the following method, suggested by Vandevelde[2] for the determination of the "total" acidity.

Ten grams of paper, cut into small pieces, are extracted with 100 cc. distilled water 24 hours at room temperature with repeated agitation. The extract is drawn off, the paper stock washed with 25 cc. cold distilled water and the total volume titrated with n/10 $Ba(OH)_2$, using phenolphthalein as indicator. The Acidity Number, or the "Acidity Coefficient," is defined as the number of cc. n/10 $Ba(OH)_2$ required per 100 grams of paper.

A thorough investigation of this method showed that the acidity numbers of "double" sized papers, i.e., papers both rosin and surface sized, determined according to this method, almost entirely referred to the acidity of the surface size, while the internal acid reacting matters remained mainly intact. The above mentioned investigation with regard to the durability of paper, at the same time proved that the "external" acidity, due to the surface size, is almost harmless as regards the permanence of the paper, while the "internal" acidity may affect the durability seriously and requires rigid control.

Notes to Introduction

1 Technically pH is a measurement of the relative activity of hydrogen ions in solution. Thus, for surface measurements only, the electrons of readily water-soluble compounds in paper are being detected. See M. Strlič, J. Kolar, D. Kočar, T. Drnovšek, V.S. Šelih, R. Susič, and B. Pihlar, "What Is the pH of Alkaline Paper?," *e-Preservation Science* 1 (2004): 35–47; www.e-PreservationScience.org.

2 TAPPI T509 *Hydrogen Ion Concentration (pH) of Paper Extracts—Cold Extraction Method*; TAPPI T435 *Hydrogen Ion Concentration (pH) of Paper Extracts—Hot Extraction Method*; TAPPI T529 *Surface pH Measurement of Paper*. Indicator pens containing pH-sensitive solutions are also available but are generally not used directly on works on paper. J. Miller and E. McCrady, "Comparison of pH Pens on the Market," *Alkaline Paper Advocate* 3, no. 5 (1990): 53–55.

3 Timothy Barrett and Cynthea Mosier, "The Role of Gelatin in Paper Permanence," *Journal of the American Institute for Conservation* 34 (1995): 173–86; Catherine H. Stephens, Timothy Barrett, Paul M. Whitmore, Jennifer A. Wade, Joy Mazurek, and Michael Schilling, "Composition and Condition of Naturally Aged Papers," *Journal of the American Institute for Conservation* 47 (2008): 201–15.

4 Margaret Hey, "The Washing and Aqueous Deacidification of Paper," *Paper Conservator* 4 (1979): 66.

5 A description of the Library of Congress deacidification program and selection criteria is available at www.loc.gov/preservation/about/deacid/index.html.

Notes to Reading

Kohler, in charge Paper Section, Government Testing Institute, Stockholm, Sweden. Hall, Chemical Engineer, with the American Writing Paper Company, Holyoke, Mass., formerly with the Government Testing Institute, Stockholm, Sweden.

1 Herzberg, *Papierprüfung*, 5th ed., Berlin 1921, p. 172. [Wilhelm Herzberg (1861–1945), a German paper chemist, is most famous for developing a multiuse stain for the identification of specific paper fibers and the methods used to refine them. Herzberg was also among the first to propose that heat could be used to accelerate paper aging: "Einfluss höherer Wärmegrade auf die Festigkeitseigenschaften von Papier," *Papier-Zeitung* 24, no. 61 (1899): 2343.—Ed.]

2 *Revue des Bibliothèques et Archives de Belgique*, 1906, part 4, p. 77. [The article cited is "Notes de chimie sur les papiers et les encres," by the Flemish chemist A. J. J. Vandevelde, who identified acidity as a source of deterioration in paper and ink.—Ed.]

Reading 53

Arthur David Baynes-Cope

Superstition and Science in Conservation (1982)

Arthur David Baynes-Cope (1928–2002), a master of British understatement—"A well-baked telephone directory is much easier to tear in half than an unbaked one"—was principal scientific officer at the British Museum from 1960 to 1984. Calling for conservators to "trust fully" in the knowledge of scientists, the well-intentioned Baynes-Cope famously advocated the use of soluble nylon (N-methoxymethyl nylon) for document conservation,[1] *a substance that has since been shown to be unsuitable for conservation work of any kind.*[2] *This is by no means the first instance in which a widely used material or process has been discarded after widespread acceptance. Indeed, the conservator-scientist relationship has been tempered over time, as can happen when alluring cure-alls are prematurely or indiscriminately endorsed, a practice beginning as early as the 1899 Zapon conference in Dresden, when it was recommended that deteriorating documents be impregnated with a cellulose nitrate solution.*[3]

[. . .] Perhaps it is appropriate that I be talking today in Stockholm on a matter that has for some time past been a worry to me, the use of unwise methods and materials and the continuation of unsound ideas in the conservation of library and archival material. Two years ago, I was reading an account of somebody's practical work on the conservation of a work of art on paper. The man concerned described the successive treatments and included one which was, in my view, not certainly safe and possibl[y] of doubtful utility, either in the short or the long term. What struck me was not that he was ignorant of the chemical reactions he was using, for

From Arthur David Baynes-Cope, "Superstition and Science in Conservation," *Tidens Tand* 1 (1982): 153–54, 169–70. Courtesy of the School of Conservation, Denmark.

that is common enough, but the real reason behind the step he took. I thought that he was being blindly superstitious and not only was he being blindly superstitious, but those who were publishing the work had let this pass without comment. I had then an able, objective and cynical assistant and an account of his work will come later in this lecture, but I knew that his work alone would not suffice, for I had for a long time realised that what we lacked was not the power to discover the truth but a pulpit from which it could be proclaimed.

When I was offered the opportunity to lecture here in what, I might add, is officially a holiday, I felt that I would attempt to preach rather than to teach. That is, I would not merely tell you what I knew to be wrong and what to be correct, but I would tell you what I believed to be right and what I believed to be wrong, and as far as in me lay, to say why some things were right and others wrong.

In order that I should the better understand what I am talking about, I consulted the Oxford English Dictionary. Of the definitions given, perhaps the second is the most helpful to us today. "Superstition: — A tenet, scruple, habit etc., founded on fear or ignorance". One of the problems that I have encountered is that of an able craftsman who was talking scientific rubbish on the subject of acidity in paper. We were alone together, so I shut my laboratory door and said "No-one can overhear us. Be honest, do you really know what 'pH' means?" He looked very lost for a moment and said, "No, I do not, but other people talked about it and I was afraid to look ignorant." So those of us whose job it is, or has been, to establish sound learning and sound practice must accept all that definition and accept that fear alone will lead good men into untenable situations. They can be led out again only if they can trust fully in our knowledge and in our compassion.

[. . .]

There are quite a number of dubious beliefs in conservation. Why do document conservators persist in the belief that soluble nylon N-methoxy methyl nylon stabilizes soluble inks? I impregnate filter paper with soluble nylon when I separate the components of a soluble ink such as Quink Permanent Black by paper chromatography. Soluble nylon is an excellent, the best, consolidant for loosely attached pigment inks, such as we have on ostraka but one of these days, some repairer will drop a clanger because the alcohol in which the soluble nylon is dissolved will itself dissolve the ink. This has happened with a water colour. A man consolidated a fragile water colour with soluble nylon, not realising that a green tree in the picture was painted with a mixture of Prussian Blue and Gamboge. Gamboge is soluble in alcohol and the man was left with a blue tree.

[. . .]

[. . .] I hope that you have realised that conservation is not yet an exact science and that it will not be an exact science as long as we accept without question all the stories that we are told. Blue trees do well enough in fairy tales, but not on water colours.

Notes to Introduction

1 A. D. Baynes-Cope, "Einige Verwendungmöglichkeiten von löslichen Nylonarten in der Dokumentenkonservierung," in *Dauerhaftigkeit von Papier: Vorträge des 4. Internationalen Graphischen Restauratorentages, Bückeburg, 1979*, ed. H. Bansa, G. Brannahl, C. Kiettewesch, and O. Wächter (Frankfurt am Main: Klostermann, 1980), 151–57.

2 Catherine Sease, "The Case against Using Soluble Nylon in Conservation Work," *Studies in Conservation* 26 (1981): 102–10; S. B. Hanna and N. J. Lee, "The Consequences of Previous Adhesives and Consolidants Used for Stone Conservation at the British Museum," in *Early Advances in Conservation*, British Museum Occasional Paper 65, ed. Vincent Daniels (London: British Museum, 1988), 89–102; E. DeWitte, "Soluble Nylon as a Consolidation Agent for Stone," *Studies in Conservation* 20 (1975): 30–34.

3 An account of the 1899 conference appeared in *Zentralblatt für Bibliothekwesen* 16 (1899): 555–57. For the uses of Zapon, see Ernst Schill, *Anleitung zur Erhaltung und Ausbesserung von Handschriften durch Zapon-Imprägnierung* (Dresden: Verlag Apollo, 1899).

Reading 54

W. J. Barrow

Migration of Impurities in Paper (1953)

The subject of historical revisionism, William J. Barrow (1904–1967) is no longer regarded as a pioneer paper chemist and innovative conservator but is seen as an "essentially self-taught tinker [sic]" who developed the processes of cellulose acetate lamination and deacidification in a well-intended campaign to preserve brittle and acidic paper documents.[1] *Despite this humbling reassessment, Barrow's description of the phenomenon of acid migration had a significant and positive impact on the establishment of preventive conservation policies for the storage of paper-based materials. Conversely, it can be argued that a misplaced fear of acid migration has resulted in the overzealous destruction of irreplaceable decorative drawing mounts and albums of prints whose arrangement and sequencing reflect contemporary modes of patronage and collecting (see reading 81).*[2] *Because Barrow's research appeared widely in periodicals read by librarians and archivists, like that of Browning (reading 51) and Feller (reading 63), the term* acid migration *is almost as familiar as the overused* acid-free. *Now it is known that physical contact with an acidic material is not required for acid migration since the hydrophilic characteristics of paper make it absorb volatile acids especially in higher relative humidity. In one experiment a neutral paper decreased in pH to the same extent whether it was in physical contact with acidic newsprint or just exposed in a closed environment with volatiles from newsprint, an especially worrisome finding given that the relative humidity in packed and closed document storage boxes increases as external humidity decreases.*[3]

Like the case of Baynes-Cope (reading 53), that of Barrow can serve as a cautionary tale about the allure of simple solutions to complex problems. In the 1930s the National Bureau of Standards began to develop a method of providing auxiliary support for weakened documents by lamination between thin sheets of transparent cellulose acetate film as an alternative to silking, a process that required considerable

From W. J. Barrow, "Migration of Impurities in Paper," *Archivum* 3 (1953): 105–8.

manual skill and time.[4] *Barrow streamlined and improved the process by designing a special laminating press and, based on his research on the internal causes of deterioration, added a preliminary deacidification step and the insertion of high-quality tissue to increase tear strength and decrease gloss. Endorsed by leading figures in the manuscript restoration field, including Verner Clapp (reading 1) and eventually even the venerable H. J. Plenderleith (1898–1997) of the British Museum,*[5] *the "Barrow Method" of restoring deteriorated documents was in full force by the mid-1950s and continued well into the 1970s. The Barrow Method was fast and easy, an observation echoed, albeit in a pejorative way, by Sir Hilary Jenkinson (1882–1961), the father of archive administration in England: "The process is popular with Scientists because it is new and uses modern materials, is economical of money and labour, and is worked largely by machine. The archivist should, to my mind, reply to these allurements that the materials have not yet stood the one test which he can accept—that of time—and that no one has suggested what plan of action will be feasible supposing that it does after all go wrong. It is not unknown for one generation of Scientists to prove a theory to their satisfaction and the next to disprove it."*[6] *Only slightly less chilly a few years later, he allowed, "We are very grateful to our Colleagues in America (who have not, as we have, the excuse of fragmentary Documents and Seals of medieval date clamouring for treatment) for making on a large scale what we regard as a most interesting experiment by which we hope presently to profit."*[7]

Jenkinson's distrust was well founded. Despite attempts to standardize the process, there remained considerable variation in protocols from institution to institution. In addition to inconsistent selections and procedures—many documents in good condition were laminated and documents in poor condition were not deacidified beforehand—conservation problems arose from the heat and pressure of the presses and the chemical breakdown of the cellulose acetate, resulting in the release of acetic acid, the "vinegar syndrome," which affected not only the document itself but also its neighbors via Barrow's very own phenomenon of acid migration. But the temptation to seize a tantalizing cure felled even Jenkinson's protégé Roger Ellis (1910–1998), who effused that lamination was "the most spectacular development in document repair"[8] *but added, "Consult the chemist; sometimes a dangerous guide but always a valuable ally."*[9]

When put into historical context, Barrow's choice of cellulose acetate can be seen as an example of the natural tendency to opt for a material that is simply better than its predecessor, in this case, cellulose nitrate, even though it may have been used for an entirely different operation. As the conservator-chemist Velson Horie (1950–) points out, when a substance is shown to be significantly better than its predecessors, it can easily become the default treatment without an examination of the properties and interactions important for this new application.[10]

The result of placing rotten apples in a basket containing good apples has been known for centuries. Many archivists and librarians have become aware recently that a similar principle applies to papers. Frequently, documents stored in contact with paper made of ground wood fibers (newsprint) show signs of discoloration. This is also true of other types of low-grade papers as well as those containing acids and other injurious compounds. These low-grade papers or other materials causing this discoloration are usually file folders, interleaving sheets, end sheets, book markers, cover boards, etc.

The aesthetic value of a document naturally suffers when discoloration takes place. Discoloration, too, makes it obvious that certain compounds do migrate from one paper to another when stored in contact. This migration takes place gradually over a period of years. There is a question as to whether the physical strength of these discolored areas has been adversely affected. The principal object of this study was to determine this factor. However, it was difficult to find blank sheets suitable for testing which have been subjected to this type of migration. This accounts for the limited number of tests in some of the following experiments.

A booklet of 1857 offered a good example for testing as it was composed of rag paper with a low-grade paper cover. The badly yellowed cover was made of 20% straw and 80% hemp and jute. The appearance of this booklet indicated that it had been stored under normal conditions. However, the cover had stained not only the sheet in contact with it, but also had slightly discolored the second sheet. The folding endurance and acidity tests of these two types of papers may be found in Table 1.

These test data indicate that the cover contained certain compounds that migrated to the adjoining sheets and caused physical damage. This migration and damage decreased in the second sheet. The fourth sheet, which contained writing, was unavailable for testing and therefore the third sheet could not be comparatively evaluated for decrease in physical strength.

Since the pH of the cover was only 5.3,[1] the compounds that caused deterioration could not be considered particularly acid. Judging from the appearance of the cover, it was originally intended for wrapping packages. In the raw state, the fibers of which it was made ordinarily have lignins, waxes, resins, etc., which were seldom removed before being manufactured into this type of paper. It is generally accepted that this non-cellulose material breaks down into other compounds over a period of years, and causes injury to the physical strength of the fibers. The

Table 1

	No. of Folds Before Breaking	pH
Cover		5.3
First sheet next to cover	16	4.3
Second sheet from cover	116	4.6
Third sheet from cover	328	4.6

Table 2

	No. of Folds Before Breaking	pH
First sheet next to end sheet	0	4.7
Second sheet from end sheet	9	4.9
Third sheet from end sheet	10	4.8

Table 3

	pH
Ground wood sample used in interleaving	4.5
Sample 1 (1790)	
portion not in contact	4.8
portion in contact	4.6
Sample 2 (1789)	
portion not in contact	4.8
portion in contact	4.5

migration of these compounds from the cover to the adjoining sheets is the most probable cause of the deterioration found in this case.

Ground wood, frequently called newsprint, is another type of low-grade paper which causes discoloration and deterioration to adjacent papers. The specimens in the next test were blank sheets in a record book of circa 1760 which had been in contact with an end sheet composed of ground wood. It was estimated that this volume had been rebound about thirty-five years ago, and the end sheets were inserted at that time. The effect of the ground wood sheet on the record papers may be found in the test data of Table 2.

Again the test data indicated, as in Table 1, that the physical strength of the record paper was adversely affected by remaining in contact with a low-grade paper of almost no chemical purification. These data also show that the migrating compounds were somewhat acid in nature.

There was also some increase in acidity found in two other papers of 1789 and 1790, which had been interleaved with ground wood sheets about twenty years prior to the test. Unfortunately, there were insufficient blank sheets for a fold test, but the results of the test for acidity may be found in Table 3.

These specimens were also badly discolored in the areas in contact with the ground wood paper. I have always found very brittle ground wood sheets to be quite acid, and, in one sample, the pH was as low as 3.2. Other workers have made similar findings, and several of them attributed this acid to a product resulting from a chemical change in the lignin of the ground wood.

Other acids also seem to have a tendency to migrate. Sulfuric acid, present in many iron gall inks, migrates to adjoining sheets to give a brown writing in reverse.

This is frequently found in manuscripts, and is sometimes called "mirror writing". In sufficient quantities, this acid is very injurious to paper and produces a brown discoloration over a period of time. Occasionally the strength of the acid is sufficient not only to form reversed brown writing on the back of a document, but it also continues to migrate through the paper to the second sheet forming brown writing not reversed. Some people have looked upon this brown writing as a transfer of the ink. However, in my article, "Black Writing Inks of the Colonial Period"[2], evidence was given to support the fact that this comes as a result of sulfuric acid gradually migrating. Mitchell and Hepworth[3] and Waters[4] have also shown that the sulfates and chlorides in writings do migrate.

Some of the workers in this field think that this migration of sulfuric acid stops after a few weeks or months. An interesting example which appears somewhat contradictory to this conclusion may be found in a Princess Anne County, Virginia, U.S.A., manuscript record book, *Deeds, Wills, etc.*, 1735–1740, now on deposit in the Virginia State Library. This book was rebound circa 1800 with an end sheet having a watermark of that date. This end sheet now has brown writing in reverse from the original black writing made in 1735.

Chlorine and alum are among the injurious residual compounds often found in low-grade papers as the result of the procedures used in papermaking. The work of both Mitchell[3] and Waters[4] has shown that chlorides in ink do migrate from the pen strokes in manuscripts. For a number of years it has been accepted that chlorine left in paper is harmful to the fibers.[5] If samples were available for testing alum migration, one might expect results similar to those of other injurious compounds heretofore described.

Other interesting examples of migration, which may or may not be injurious to paper, are often found among manuscripts. Recently, among the records of 1794, U.S. Circuit Court, Virginia District, one came to light which consisted of a charred manuscript mounted on one half of a sheet of blank paper. The other half of the blank sheet was folded over the charred document. This mounting most likely took place around 1794 in order that this document could be presented as evidence in court. During storage, various compounds of the charred document migrated to the unmounted portion of the blank sheet causing a dark discoloration. This migration occurred in all areas with the exception of those which contained writing. The iron ink blocked the migration, and thus white writing in reverse was produced on a dark background. Much of the original charred document has crumbled away, but the blank sheet in contact with the charred document has preserved most of the original text in reversed writing.

The data assembled in this study indicate that many compounds migrate from one paper to another when remaining in contact over a period of time. The injurious compounds are usually either residual compounds left in the paper from the paper making processes, or non-cellulose compounds found with many fibers in the raw state. As a rule, they damage the document both aesthetically and physically. To eliminate any future damage to documents of value, it is recommended, therefore, that laboratory tests be made on such materials as file folders, end sheets in books, interleaving sheets, sheets used for mounting, or any other similar materials stored in contact with documents of value.

These materials should contain no ground wood fibers, and the acidity on the pH scale should be not less than pH 5.0. An aging test of 72 hours at 100°C. followed by a loss of not more than 25% folding endurance is also desirable, and would indicate adequate longevity of the material being used. These suggested tests were the results of a conference with the late Mr. B. W. Scribner[6] and the late Mr. Charles G. Weber, Paper Section, National Bureau of Standards. Of course, the tests should conform to accepted standard testing methods.

The possibility of using book papers containing a small amount of calcium carbonate was also discussed during this conference. The calcium carbonate in many instances would neutralize the migrating acidic compounds. These papers would have value for lining boxes, interleaving sheets, etc. Of course, high-grade, all-rag papers of a high degree of permanency would be excellent to use in a number of cases, but are quite costly when used in large quantities.

The Library of Congress, the Virginia State Library, and others have used file folders conforming to the above specifications for several years. All of these institutions report satisfactory results. Additional information regarding size, type, and other characteristics is provided in an article by W. J. Van Schreeven,[7] State Archivist of Virginia.

It is of interest to note that file folders conforming to the above tests have, at times, been purchased at approximately the same cost as some of those made of ground wood. The latter have a useful expectancy of from fifteen to twenty years, while those containing fibers of higher purification should last several times longer, and with no expected injury to the document.

In view of the results obtained in this study, it should be mentioned that undesirable results may be expected when newspaper clippings, photostats (containing residual hypo), cellulose nitrate film, etc., are filed in contact with documents of value. This is also true of the corrosive action of paper clips, pins (made of ferrous metals), rubber bands, and other materials less commonly used.

Notes to Introduction

1 Sally Roggia, "William James Barrow: A Biographical Study of His Formative Years and His Role in the History of Library and Archives Conservation from 1931 to 1941" (PhD diss., Columbia University, 1999), 1. The significance of Barrow's lifelong champion, Verner Clapp, has likewise been reassessed by William J. Crowe: "Verner W. Clapp as Opinion Leader and Change Agent in the Preservation of Library Materials" (PhD diss., Indiana University, 1986).

2 Marjorie B. Cohn, *A Noble Collection: The Spencer Albums of Old Master Prints* (Cambridge, Mass.: Fogg Art Museum, Harvard University Art Museums, 1992), 11.

3 John Slavin and Jim Hanlan, "An Investigation of Some Environmental Factors Affecting Migration-Induced Degradation in Paper," *Restaurator* 13, no. 2 (1992): 91. See also Anika Knop, Gerhard Banik, Udo Schade, and Irene Brückle, "Paper and Board in Closed Boxes: Alteration of Water Sorption Capacity during Cyclic Temperature Changes," *Restaurator* 28, no. 3 (2007): 218–24.

4 Silking is a process whereby a deteriorated document is lined using a fine silk gauze applied with a starch-based paste. Claire S. Marwick, "An Historical Study of Paper Document Restoration Methods" (PhD diss., American University, 1964), 67. Silking, as detailed in William Berwick's article titled "The Repairing and Binding of Archives," published in the 1916 *Annual Report of the American Historical Association.*

5 H. J. Plenderleith was the author of an early book on paper conservation, *The Conservation of Prints, Drawings and Manuscripts* (Oxford: Oxford University Press, 1937), which appeared only one year before the first edition of Max Schweidler's *Die Instandsetzung von Kupferstichen, Zeichnungen, Büchern, usw.* (see reading 91), as well as a comprehensive book on art conservation, *The Conservation of Antiquities and Works of Art* (London: Oxford University Press, 1956), reissued with A. E. A. Werner in 1971.

6 "The English Archivist: A New Profession," inaugural lecture for a new course in archive administration delivered at the University College, London, 14 October 1947 (H. K. Lewis Ltd.), 253–54.

7 "The Principles and Practice of Archive Repair Work in England," *Archivum* 2 (1953): 302.

8 Roger Ellis, "Latest Information for the Private Owner and Smaller Repository," *Archives* 6, no. 29 (1963): 3.

9 Roger Ellis, "The Principles of Archive Repair," paper read at the opening by Sir Hilary Jenkinson, deputy keeper of public records, of the first course of instruction in record and document repair to be held in the London School of Printing and Graphic Arts. Published London: The School, 1951, 8.

10 Velson Horie, "Keynote Presentation: Does What We Want Exist?," *Proceedings of Symposium 2011—Adhesives and Consolidants for Conservation* (Ottawa: Canadian Conservation Institute, 2011), 2, www.cci-icc.gc.ca.

Notes to Reading

1 [Not specified, it is assumed that the pH measurement was obtained via cold extraction.—Ed.]

2 Barrow (William J.), Black Writing Inks of the Colonial Period, *The American Archivist*, XI (October 1948), 291–307.

3 Mitchell (Charles A.), and Hepworth (Thomas C.), *Inks, Their Composition and Manufacture* . . . Fourth edition, London, 1937.

4 Waters (Campbell E.), *Inks*, National Bureau of Standards, *Circular* C, 426, August 7, 1940.

5 Kantrowitz, (Morris S.), Spencer (Ernest W.) and Simmons (Robert H.), *Permanence and Durability of Paper, an Annotated Bibliography of the Technical Literature from 1885 A.D. to 1939 A.D.*, United States Government Printing Office, Division of Tests and Technical Control, *Technical Bulletin*, No. 22, 1940.

6 [Bourdon Walter Scribner (1885–1952) observed that the "validity of this test as a measure of the relative resistance of paper to deterioration from internal causes is a debatable matter" but that the results correlated with "exposures to

sunlight, exposures at more normal temperature and a 4-year period of natural aging." A. E. Kimberly and B. W. Scribner, *Summary Report of Bureau of Standards Research on Preservation of Records*, Bureau of Standards Miscellaneous Publication No. 144 (Washington, D.C.: U.S. Department of Commerce, 1934), 21.—Ed.]

7 Van Schreeven (William J.), The Filing Arrangement of the Archives Division, Virginia State Library, *The American Archivist*, XI (July 1948), 248–251.

Reading 55

PHILIP LUNER

Paper Permanence (1969)

The wood and paper chemist Philip Luner (1925–) was among the first to question the validity of applying the Arrhenius equation to extrapolate data derived from the accelerated aging of paper, pointing out that other deterioration mechanisms likely occur at elevated temperatures. Concerns also focused on the aging methodology, particularly the lack of moisture, as well as the papers being tested, typically Whatman #1 filter paper, which barely resembles any of the broad range of papers used in prints and drawings.[1] *Even more complications arise when evaluating the aging of adhesive/paper systems, as degradation products from the adhesive can affect the degradation mechanisms of the paper.*[2]

Salvador Muñoz Viñas (1963–) cites the proposition that paper abides by the Arrhenius laws as an example of a mistaken hypothesis. Echoing Jenkinson, he concludes, "Unless a material has passed the test of real-time aging, it is not possible for scientists to guarantee that it will age in a certain way."[3] *It is the clear-thinking Robert Feller (readings 63 and 64) who elucidated the three purposes of aging tests, not necessarily just on paper, and the application of their results to conservation: (1) "to establish in a conveniently short time the relative ranking of materials, or physical combinations of materials, with respect to their chemical stability or physical durability"; (2) "to estimate or 'predict' potential long-term serviceability of material systems under expected conditions of use"; (3) "to elucidate the chemical reactions involved (the 'mechanism' of the degradation) and the physical consequences thereof. An important facet of this effort is the disclosure of the overall pattern of deterioration, that is, whether the processes accelerate in time, whether there is an induction period, or whether a number of distinct stages are observable before failure occurs. The ultimate objective of this third area of investigation is the development of tech-*

From PHILIP LUNER, "Paper Permanence," *TAPPI* 52, no. 5 (May 1969): 803–5. Reproduced courtesy of TAPPI.

niques that can monitor the extent of degradation and methods by which the useful lifetime of the materials can be extended."[4]

Prediction of Paper Permanence

A number of environmental conditions affecting the loss in paper permanence have already been mentioned, along with some of the difficulties associated with the use of the folding endurance test. While conditions for aging and testing are still not standardized, it is, nevertheless, important to discuss the basis for predicting the performance of paper on aging.

An accelerated aging test increases the intensity of deterioration so that the deterioration of paper corresponding to normal conditions for extended times is obtained in a relatively short time. For paper permanence studies, elevated temperatures with or without moisture have been used to accelerate the intensity or deterioration. The Arrhenius equation,[1]

$$k = Ae^{-E/RT}$$

or in logarithmic form

$$\log k = \log A - \frac{E}{2.303RT}$$

where

k = specific rate constant
A = frequency factor (usually treated as a constant)
E = activation energy, kcal/mole
R = gas constant
T = absolute temperature, °K

has then been used to describe the temperature dependence of the reaction rate. For paper, in most instances, the rate constant refers to a mechanical property of paper. For this equation to be used in predicting the rate of aging, a number of conditions have to be fulfilled. The rate plot must be linear and the activation energy independent of temperature, so that a plot of log k vs. $1/T$ yields straight lines. Once E is obtained, the rate constant k can be evaluated at any other temperature. Fairly extensive data on activation energies [1–3] for paper strongly suggest that an E value for *each* paper is necessary in order to predict its permanence with any accuracy.

There are several difficulties inherent in the use of the Arrhenius equation which are not readily apparent. For papers, plots of log k vs. $1/T$ have been found to be fairly linear. From these results it has been inferred that the mechanism of permanence loss does not change throughout the temperature range employed. However, it should be recognized that the mode of failure of paper may change on heating. This may easily be recognized by examining the ruptured areas microscopically. All papers, when extensively aged, fail by fiber failure during folding, while in unaged or slightly aged papers, bond failure may also occur. Indeed, the nonlinear relationship often encountered between the logarithm of the fold value and aging time may reflect this difference in failure mode.

In view of the existence of different modes of failure, it would not be surprising if the activation energy should vary with temperature. However, in most instances this is not the case. This is most likely caused by the fact that other reactions occur simultaneously. Cross-linking and dehydration may be two such reactions. The resulting activation energy is then the average of several simultaneous competing reactions involving chemical as well as physical processes, and predictions based on activation energies for papers are uncertain. Another severe shortcoming of the Arrhenius relationship for predicting paper permanence is the accuracy necessary in determining *k* and, hence, *E*. The large inherent variability in the folding endurance test, as well as the frequent necessity for empirically linearizing the rate data, makes it difficult to obtain activation energies with the precision required to calculate rate constants which can be used for predicting permanence with any great degree of reliability. Table I gives one such example. Considerably more information is needed on the chemical and physical reactions contributing to

Table I

Estimation of Permanence of Papers at 20°C (4)

Sample	By extrapolation of Arrhenius plot	Range: 95% Confidence limits	Range: 50% Confidence limits
	Estimated Time to Reach One-Half Original Folding Endurance, Years		
A-2	714	196–2606	523–975
C-2	2159	79–58, 910	975–4791
F-2	1529	97–24, 173	788–2972
H	452	83–2486	301–681
	Estimated Time to Reach a Specific Absorption Coefficient of 60 cm²/g, Years		
A-2	835	357–1951	681–1024
C-2	709	171–2937	502–998
F-2	179	113–285	160–201
H	314	125–788	151–391

loss in paper permanence over a temperature range before the Arrhenius relationship can be used with confidence to predict paper permanence.

Notes to Introduction

1 Vincent Daniels notes that scientists acknowledge this discrepancy but prefer to work with standard testing materials in order to obtain repeatable results. Vincent Daniels, "The Changing Face of Paper Conservation Science," *Paper Conservator* 25 (2001): 67.

2 I am grateful to Norbert S. Baer, author of numerous aging studies of adhesive/paper systems, for his assistance. See N. S. Baer and N. Indictor, "Use of the Arrhenius Equation in Multicomponent Systems," in *Preservation of Paper and Textiles of Historic and Artistic Value*, Advances in Chemistry 164, ed. J. C. Williams (Washington, D.C.: American Chemical Society, 1977), 336–51.

3 Salvador Muñoz Viñas, *Contemporary Theory of Conservation* (London: Butterworth Heinemann, 2005), 184.

4 Robert L. Feller, *Accelerated Aging: Photochemical and Thermal Aspects* (Los Angeles: Getty Conservation Institute, 1994), xvi.

Literature Cited

1 Browning, B. L., and Wink, W. A. *Tappi* 51 (4): 156 (1968).

2 Gray, G. G., *Tappi* 52 (2): 325 (1969).

3 Millett, M. A., Western, L. J., and Booth, J. J., *Tappi* 50 (11): 74A (1967).

Note to Reading

1 [Proposed by Svante Arrhenius (1859–1927) in 1889, the formula correlates the speed of deterioration with increases in temperature. The rate has been grossly generalized to double with every 10° Celsius increase.—Ed.]

Reading 56

American National Standard for Information Sciences

Permanence of Paper for Printed Library Materials Z39.48-1984 (1984)

ANSI Z39.48–1984, directed to publishers and promoting the use of permanent paper for printed materials,[1] illustrates one long-awaited government response to the steady stream of complaints about paper deterioration heard throughout the nineteenth and twentieth centuries. It came at a time of extensive research in the field of paper chemistry that coincided with the first generation of paper conservators educated in graduate-level art conservation training programs in North America. Given the priorities of science faculties and national funding organizations, the paper science curriculum in these programs emphasized improving the longevity of paper assessed according to the common industry standards pH, folding endurance, and tensile strength. Having spent countless hours listening to the rhythmic clicking of MIT Fold Endurance testers and inserting strips of Whatman chromatography paper into the jaws of Instron paper testers, it is no wonder that some budding conservators placed disproportionate faith in data, especially pH, as determinants for treatment and benchmarks of success. It required experienced conservators and dedicated science instructors to synthesize the noble goals of the paper chemist and devise sympathetic treatments to meet the equally noble goals of the paper conservator. For example, a paper chemist's recommendation for the most efficient removal of discoloration from paper using an ethanol pretreatment followed by very hot water may differ drastically from a conservator's strategy.[2]

The objective of this standard is to establish criteria for permanence of uncoated paper; that is, paper that should last at least several hundred years without significant deterioration under normal library use and storage conditions. Actual observation and laboratory test results suggest that paper meeting the requirements for pH,

From American National Standard for Information Sciences, *Permanence of Paper for Printed Library Materials,* Z39.48-1984 (New York: American National Standards Institute, 1984), foreword, 8.

alkaline reserve, and freedom from groundwood described in this standard should survive hundreds of years as have papers made in the 14th and 15th centuries. Librarians and other persons concerned with the preservation of library materials are all too familiar with the speed at which acidic paper embrittles. This embrittlement has made probable the loss of the original hard copy format of much of the published record from the 19th and 20th centuries. By defining permanent paper and encouraging its use, it is hoped that the selection of such paper by publishers will be simplified, and that comparable future preservation problems can be prevented.

[. . .]

Accepted parameters exist for the production of paper used for printing based on end use. It is assumed that characteristics such as opacity, thickness, and others are specifically required by the publishers, but since they do not necessarily affect the permanence of the paper, they are not included here. Tests of artificial aging are also excluded because of unresolved technological problems. It is, however, widely recognized that acid content is a primary indicator of paper permanence; measurement of pH has, therefore, been included.

There are a number of factors beyond the scope of this standard that affect the shelf life of books. The environmental conditions under which materials are stored are highly significant determinants. [. . .] Another factor is the method by which books are bound. The use of permanent materials for binding and the choice of proper binding styles contribute greatly to a book's shelf life. So too does the use of printing inks that do not contain acidic or highly alkaline components.

Libraries and their patrons will be the primary beneficiaries from the use of permanent paper in printed library materials, but other book purchasers and readers of printed materials also will benefit from the longer-lasting product. Permanent paper should be used for all publishing of an important nature, especially as such paper becomes more widely available and competitively priced.

Publishers should indicate the use of permanent paper in the publication itself as set forth in this standard in Section 5, "Compliance."[1] This indication, which should also appear in advertising and reviewing media, will both promote the use of permanent paper and provide librarians with a basis for considering the physical quality of materials when planning for future preservation needs.

4. Minimum Requirements

Uncoated permanent paper shall meet all of the following requirements:

(1) *pH*. Minimum pH of 7.5 in accordance with the cold extraction method described in TAPPI T509 om-83.

(2) *Folding Endurance*.[2] Minimum cross-direction folding endurance of 30 double folds at 1-kg tension (25 replicates in accordance with TAPPI T511 su-69[3]). This requirement applies to 30- to 60-lb paper.[4]

(3) *Tear Resistance.*[5] All results listed in the following table are based on tear resistance tests using Elmendorf, 10 replicates, 8-ply tears in accordance with ANSI/TAPPI T414 om-82.

Weight of Paper (pounds)	Minimum Average Machine-Direction Tear Resistance (grams)
60[6]	50
50	40
40	32
30	24

(4) *Alkaline Reserve.* Minimum alkaline reserve equivalent to 2% calcium carbonate based on oven dry weight of the paper (see TAPPI UM531 for qualitative testing; or ASTM D 3290-81, Subsection 11.4, for quantitative testing).

(5) *Paper Stock.* The paper shall include no groundwood or unbleached pulp (see ANSI/TAPPI T401 os-74).

Notes to Introduction

1 The standard was updated nationally in 1992 and internationally in 1994; American National Standards Institute (October 26, 1992). "Permanence of paper for publications and documents in libraries and archives," NISO Press, http://webstore.ansi.org/RecordDetail.aspx?sku=ANSI%2FNISO+Z39.48-1992+(R2002); "Information and documentation—Paper for documents—Requirements for permanence," International Standard ISO 9706:1994, International Organization for Standardization, Geneva.

2 Compare the recommendations of the paper chemist Margaret Hey, "The Washing and Aqueous Deacidification of Paper," *Paper Conservator* 4 (1979): 10–23, with those of the papermaker Timothy Barrett, "The Role of Gelatin in Paper Permanence," *JAIC* 34 (1995): 173–86.

Notes to Reading

1 ["Compliance": This allows for the use of the mathematical symbol for infinity ∞ within a circle. Items so marked are excluded from current mass deacidification programs in libraries and archives.—Ed.]

2 Although folding levels and tear resistance are not directly related to paper permanence, minimum values for each are suggested to ensure reasonable durability in the finished paper.

3 It should be recognized that the results obtained with TAPPI T511 su-69 have a considerable degree of inherent variability.

4 All basis weights of paper referred to in this standard are figured on 500 sheets measuring 25 inches by 38 inches. It should be noted that papers of lighter weight than 30 pounds can also be manufactured to meet the minimum requirements for permanence.

5 [See note 3.—Ed.]

6 [See note 4.—Ed.]

Reading 57

ANTOINETTE DWAN

Paper Complexity and the Interpretation of Conservation Research (1987)

A paper conservator who undertook advanced coursework in paper physics, testing, and fiber microscopy, Antoinette Dwan (1952–) was among the first practicing conservators to wrestle with the dilemma of how to analyze and apply the results of paper research. She stresses that each sheet of paper is unique, as is its response to the testing of its chemical and physical properties. Her remarks concerning the repeatability of fold endurance results add a note of caution regarding their relevance to the development of treatment modalities. In response, it can be argued that trends in changes brought about by artificial aging can be meaningful but only if correctly interpreted.

Introduction

The testing of materials and the interpretation of research are fundamental issues in paper conservation. This paper presents the properties of paper as a material and shows how paper variables relate to paper testing and the interpretation of research.

Although paper degradation and its mechanisms and products are not fully understood, it is generally recognized that the chemical reactions of hydrolysis, oxidation, and cross-linking are involved. Physical properties associated with degradation are loss of strength, embrittlement, and loss of bonding within the paper matrix, among others. There are several options for testing either the chemical or physical properties of paper. Chemical testing methods provide information on the molecular level, while physical tests describe the total overall structure.

From ANTOINETTE DWAN, "Paper Complexity and the Interpretation of Conservation Research," *Journal of the American Institute for Conservation* 26, no. 1 (1987): 1–2, 10, 12–15. Reprinted with the permission of the American Institute for Conservation of Historic and Artistic Works, 1156 15th Street NW, Suite 320, Washington, DC 20005, info@conservation-us.org, www.conservation-us.org.

Paper conservators are presented with explanations of chemical degradation processes, and should question why they are often measured using physical rather than chemical methods.

Paper is a composite material: physical tests provide information about the integrated paper network rather than isolated contributing components. This is important because in addition to the chemical composition of the sheet, paper properties are dependent on the network matrix, the bonding within that matrix, and the fibers' properties responsible for the bonding. These categories are interrelated and within each, variables exist that affect the paper structure as a whole.

The composite nature of paper limits the usefulness of testing isolated components, either chemical or physical. It is difficult to relate a single component back to the overall structure. The usefulness of chemical testing may be limited by lack of correlation to the overall physical structure of paper. Likewise, physical tests cannot provide information on a chemical reaction, or predict permanence. For example, loss of strength (a physical measurement) in a sheet of paper cannot be interpreted as resulting from the removal of cations (a chemical reaction) by wash water. The results of a particular test, chemical or physical, may be interpreted in a number of ways, and making the proper interpretations and conclusions depends upon the total research design and expertise of the researcher.

Paper scientists use the combination of tests that fit their project design and that provide the necessary data to answer their specific inquiry. It is important to realize that standards established by TAPPI (Technical Association of the Pulp and Paper Industry) or ASTM (American Standards and Testing Methods) are guidelines developed to provide consistency and accountability within the paper industry, and were not developed specifically for research. Although the procedures discussed in the standards provide useful information, the methods themselves are not definitive. The data require interpretation within the framework of probable reactions and the test methods used for the evaluation.

[. . .]

Interpretation of Test Data

The previous discussions are intended to show the composite structure of paper, the many variables affecting the testing of that structure, the sensitivity of the variables to their prior history and testing conditions, and the possible effects of this history on test data. The following discussion addresses the interpretation of test results with emphasis on the fold endurance test, which is commonly used in testing for conservation research.

Valid paper conservation research must be based on the relationship of the inquiry to the research design. The sample material, sample population size, conditioning of the samples, testing methodology, and appropriate instrumentation for testing must be carefully chosen. The data must be analyzed and interpreted with appropriate consideration given to possible ambiguity caused by the samples' manufacturing techniques and history, and the testing methodology.

Test results that demonstrate either an increase or decrease in strength properties, as measured by fold endurance for example, could be interpreted in several ways:

> If hydrolysis occurs, one could record a low fold endurance; if cross-linking occurs, one could record a high fold endurance; if both processes are occurring the results are unpredictable.
>
> If a paper has a large number of aldehyde groups, data may show an increase in strength, but this could be due to cross-linking.
>
> In a paper with high internal bonding that is aged, it may cross-link more than another sheet with less initial bonding affecting the fold endurance values differently.[1]

An increase in strength, as measured by fold endurance, does not necessarily mean improvement in paper properties. For example, an increase in strength measurement could be attributed to cross-linking caused by an increased production of aldehyde groups. Test results do not provide answers, they show changes. Researchers should address other possible interpretations of data and provide an explanation for their interpretation either through theory or additional supporting tests. For example, cross-linking may be identified by changes in wet tensile as a percent of dry tensile test results, or by changes in the degree of polymerization.

Understanding the choice of instrumentation used in a project is important to overall evaluation of the results. Each test method is useful for obtaining a particular type of information. For example, fold endurance testing has been used because of its ability to detect changes earlier than other physical methods. This information can be useful in conjunction with data from other physical methods, or as a general indication of trends useful in designing a research project. However, by itself, fold endurance is not a reliable research tool.[2]

As early as 1939, the Institute of Paper Chemistry made a comparison of three fold endurance instruments and concluded that in all cases the percent deviation was very high.[3]

Casey discusses the wide variation in individual test results which "has caused some people to consider the fold endurance test useless."[4] He discusses the number of tests needed to obtain significant data and the magnitude of differences in data necessary for significant data and the magnitude of differences in data necessary for significant results:

> It is important to keep the variability of the folding-endurance test in mind when evaluating test results. An average of 5 to 10 measurements should be used and a difference between two samples of *only 10 to 20 percent is seldom*

> *significant.* The precision of any test method can be increased by using more tests to obtain the average, but for one paper sample with a folding endurance of 1000, it was calculated that over 6000 tests would have to be made to produce an average reproducible within 10 folds, at the 3 δ assurance level (99.7% of the time).[5]

The sample population of many research projects reported in the conservation literature is often as low as ten, and differences in data less than 20%. The interpretation of these projects should be questioned.

In 1956 Hoffman found that there is a greater deviation in high fold than in low fold values.[6]

In 1964 and 1966 Kahlson and Lindholm investigated the cause of wide test variation in fold endurance data. They determined that motor warm up and mechanical frictional heat caused a local rise in temperature at the point of fold. The resulting drop in local relative humidity at the sample test site greatly reduced fold endurance. They discovered that using a fan to maintain constant air circulation around the specimen could reduce variability.[7] Current TAPPI standards require a fan as part of the instrumental control. Even with a fan, variability is still high. TAPPI standards state that the degree of precision for "repeatability" (same lab, same instrument, same operator, same sample) is 20% and for "reproducibility" (different lab, different operator, same instrument, same sample) is 30% at the 95% probability level for the MIT instrument. The standard further states:

> This method is very susceptible to small errors in the adjustment and calibration of the instrument and in the relative humidity of the test room. *Limits as much as two times those shown [20% and 30%] may be expected if the instructions are not followed meticulously.*[8]

Fold endurance tests are very sensitive to moisture. In their summary of all tests of physical properties, Crook and Bennett state in reference to fold endurance:

> Changes of relative humidity, when the temperature is held constant, influence paper properties to different degrees. Of the properties studied, folding endurance was the most affected . . . The outstanding fact about folding endurance is its extreme dependence upon relative humidity. The maximum *rate of increase of folding endurance was greater than 10% for each 1% R.H. increase of humidity.* The influence of temperature change, at a constant relative humidity, also varies from paper to paper. The most marked effect is on folding endurance. . . .[9]

Controversy exists over the type and amount of loading during the fold test and the magnitude of discrepancy caused by the weight. The MIT tester allows for variable weights, the Schopper is set for one load setting. Luner discusses the problem of using a standard load on sheets of various basis weights and tensile strength.[10] Cardwell et al. also discusses the effect of equally loading differing papers during aging tests.[11] Casey states that using the same 1 kg weight on all samples "is unfair to papers of low tensile strength; it is not a true test of flexibility is such cases."[12] Kohler recommends not using a standard load, but one that is a percentage of the initial tensile strength of the paper.[13] Crane comments on the use of standard loads on samples before and after aging tests:

> Due to the strong dependence of folding values on load, one questions the value of comparing the very low values encountered after aging to the unaged sample under the same load.[14]

In many paper conservation research projects, the same weights are used on the control samples as on the tested samples. The use of the same weights may produce unreliable test results.

Fold endurance results are influenced by formation and furnish variables. Different values are obtained depending on how the paper was manufactured. Fold endurance rises in the early stages of beating, but drops with further beating and wet pressing.[15] It increases with basis weight up to a point, but decreases with higher weights.[16] Crane found that fiber orientation, as a result of fiber composition and drying tension, influences fold values.[17] He also reported that the presence of fines affects the relationship of fiber length to fold endurancc.[18] Given this information, it is questionable that two different papers can be compared accurately using this test.

In the conclusion of his study of fold endurance testing, Crane comments:

> Fold endurance is related to certain network properties. It does not correlate as clearly as do other tests such as tear and tensile with specific characteristics of the web. The relationship between folding endurance and viscoelastic properties is complex. Again, viscoelastic behavior is not as easily related to folding endurance as other strength properties.
>
> While there is little argument that folding endurance is important in paper and papermaking, the fold test will likely remain an unpopular one. It is generally felt that fold will roughly parallel changes in the more easily measured strength values. Consequently, it seems unlikely that fold endurance will see any surge in popularity in describing paper strength.[19]

The following general statements can be made regarding the interpretations of data and evaluation of research based on fold endurance values:

The test method was designed for checking continuity of production within the paper industry.

By itself it is not a research tool, although it may be useful in initial planning of a project.

Careful preconditioning of samples and careful control of temperature and humidity during testing is necessary for reliable data.
Data may be questionable if the same load is used before and after artificial aging.
A large sample population is necessary due to large variability in test results.

The same type of investigation that has been used in the discussion on fold endurance tests can be applied to other instruments and research designs in order to better evaluate paper research projects and interpret the data.

Conclusion

Paper research is difficult to conduct, interpret, and judge due to the composite structure of paper and the variability due to each sheet's prior history. Each sheet of paper is unique in its physical and chemical properties and its response to testing those properties.

There are many ways of obtaining information in conservation, from single observations to shared information among colleagues, and from mock-ups stuck in windows to full research projects. All of these approaches are valid and can give significant information within the context of their specific limits. Much of our information is of necessity based on limited observations, and always will be because of the nature of conservation and the many unknowns that can never be repeated or simulated for a major research endeavor. However, many of our current treatment practices such as water bathing, enzyme use, deacidification, and bleaching are based upon published research which requires careful review. The topics discussed in this paper should assist in critically reviewing results of paper research projects and the provided interpretations. It is important that research projects related to paper conservation have the involvement of people with expertise in material sciences research. This background is necessary for proper research design, choice of test methods and instrumentation, and for insightful interpretation of data.

Notes to Reading

1 Wilson, W., personal communication, May 1983.
2 Ibid.
3 T. Crane, "The Folding Endurance of Paper," Seminar on paper properties and their measurement, Institute of Paper Chemistry, 1982.
4 J. P. Casey, editor, *Paper Testing and Converting*, vol. IV of *Pulp and Paper* (New York: Interscience Publishers, Third edition, 1981), p. 1806.
5 Ibid., p. 1807. (Emphasis added.)
6 Ibid., p. 1806.
7 Crane, "The Folding Endurance of Paper," p. 3.
8 *TAPPI* standard 511 su 69. (Emphasis added.)

9 D. M. Crook and W. E. Bennett, *The Effect of Humidity and Temperature on the Physical Properties of Paper* (The British Paper and Board Industry Research Association, 1965), pp. 1.7–1.8. (Emphasis added.)

10 P. Luner, "Paper Permanence," *TAPPI*, vol. 52 (May 1969), p. 801.

11 R. Cardwell, L. Lyon, and P. Luner, "Evaluation of a Fold Endurance Procedure for Paper Aging Studies," *TAPPI*, vol. 55 (February 1972), p. 229.

12 Casey, *Pulp and Paper*, p. 1804.

13 Ibid.

14 Crane, "Folding Endurance," p. 10.

15 Casey, *Pulp and Paper*, p. 1339.

16 Ibid., p. 1805.

17 Crane, "Folding Endurance," p. 5.

18 Ibid.

19 Ibid., p. 6.

Reading 58

VINCENT DANIELS

The Discolouration of Paper on Ageing (1988)

Recipient of the prestigious Plowden Medal in 2007,[1] Vincent Daniels (1948–) regularly engaged with practitioners of paper conservation throughout his career at the British Museum and the Victoria and Albert Museum. Known for his approachability and wide-ranging expertise, Daniels, like Robert Feller, was able to explain complex scientific concepts to a broad audience. Other influential chemists with an early following among paper conservators include Helen Burgess (1951–1999), Margaret Hey, Françoise Flieder, and Ludovico Santucci (1927–2003). In this reading, Daniels discusses a number of possible causes—from gravy to iron particles—of foxing in paper, an unpredictable phenomenon observed in both old and new papers that still puzzles caretakers of collections and bedevils papermakers and publishers.

Introduction

A certain amount of yellowing/browning of paper is considered to be a natural consequence of ageing. It is only when discolouration is irregular or extensive that it becomes a problem on works of art or other artefacts. This article attempts to classify the majority of stains encountered in the author's experience and give general mechanisms for the formation of the stains, concentrating on phenomena not previously described in the conservation literature. In many cases yellowing is an early stage of what could end up as browning and the term discolouration will be used to cover these descriptions.

From VINCENT DANIELS, "The Discolouration of Paper on Ageing," *Paper Conservator* 12 (1988): 93, 99–100.

The Nature of the Discolouration

Paper is a composite material made of many substances including cellulose, hemicelluloses, lignin, rosin, alum, glue, starch, dyes, pigments, fillers and the ageing products of all these materials. Yellow/brown degradation products may be obtained from several of these materials either separately or in combination with one another. It is possible to consider the origin of discolouration without being specific as to the matrix material. Any substance is coloured if it absorbs light in the visible region of the spectrum. Structures which absorb light in organic molecules (chromophores) are composed of alternating single and double bonds (conjugated), but atoms such as nitrogen, oxygen and sulphur can also aid in colour production. As degradation proceeds, more double bonds can be produced. Small conjugated systems absorb light in the ultraviolet, but as they get bigger the absorption moves into the visible region and violet light is absorbed, the material in question appearing yellow. Some undegraded materials already contain chromophores which absorb ultraviolet light and need only a small amount of extra degradation to produce discolouration. Degradation produces carbonyl groups (C=O) and carbon-carbon double bonds (C=C).

Hon[1,2] has reviewed the types of chromophore responsible for yellowness in paper, and in his list he includes furans, quinines, quinomethides and stilbenes. Tannin-type compounds can form highly coloured complexes with metals present, such as iron. Of course, metals such as iron can corrode and produce coloured compounds without any reaction with the paper.

The mechanism of degradation of paper cellulose has been studied by Lewin[3], who attributes the main source of yellowness to condensed furan derivatives. Papers which contain many aldehyde groups tend to discolour easily, and these groups result from most of the degradation mechanisms, that is acid hydrolysis, oxidation, alkaline hydrolysis, autohydrolysis and pyrolytic chain scission.

[. . .]

Foxing

One of the first types of local discolouration which springs to mind is foxing. The apparently trivial description, 'brown spots on paper', is probably the best description of foxing, as, in practice, the spots initially described as foxing in fact seem to be caused by a wide range of phenomena including gravy stains, iron particles and fungi. With the wide range of discolouration mechanisms available, it is surprising that there are some people who maintain that foxing is caused by one mechanism only, either metal particles, usually iron, or micro-organisms. Using the above definition of foxing, both mechanisms must be allowed, either acting separately or together.

Baynes-Cope[4] describes foxing as often having two fluorescent zones, one, at the centre, orange-brown and the other, at the edge, pale blue; he also presents evidence to support a biological mechanism for foxing. Nol[5] has isolated fungi from several spots on postage stamps and has further isolated and regrown the fungi reproducing the effects of foxing. In the samples, hyphae are found over all the paper but colouration is not associated with high concentrations of hyphae. The brown stain was phenolic in character.

Iron-produced stains generally fluoresce differently, and under ultraviolet light, the iron-impregnated areas appear darker than the surrounding paper. On the basis of the appearance under ultraviolet light, it may be possible to define foxing to exclude one of the two mechanisms above, but if iron-stained spots were to produce degradation products which provide food for micro-organisms, then the micro-organisms are more likely to grow on the spots and there would be many cases where evidence for both mechanisms exists.

Tang[6] has shown that a definite relationship exists between brown stains on paper and the local presence of iron or copper in the paper. Some transition metals, including iron and copper, are catalysts for the oxidative degradation of many polymers, including cellulose.[7] Oxidation is accelerated by high humidities and results in the discolouration and weakening of paper.

Iron particles occur in some types of paper because they are worn away from steel parts of the papermaking apparatus and find their way into the paper pulp. Iron can corrode in paper to produce corrosion products which are brown. At the same time, iron or copper catalyse oxidation, and this also causes brown stains.

Iron particles are quite common in paper, but many may not have corroded. They are easily detectable by low voltage radiography. Using X-rays generated at 10kV potential they are visible, on a radiography plate, as white specks on a black background. Pure iron should not corrode below about 70% RH, but in the presence of ions such as chloride, storage needs to be at 40% RH or lower to avoid corrosion. Low RH storage would inhibit oxidative degradation and acid hydrolysis too and is the best precaution against foxing and discolouration generally.

Note to Introduction

1 The Plowden Medal, named for the conservator Anna Plowden, CBE, is awarded by the Royal Warrrant Holders Association.

Notes to Reading

1 Hon, D. N-S., 'Yellowing of Modern Papers', *Preservation of Paper and Textiles of Historic and Artistic Value II*, Advances in Chemistry Series 193, Washington: American Chemical Society, pp. 119–41.

2 Hon, D. N-S., 'On Possible Chromophoric Structures in Woods and Pulps—A Survey of the Present State of Knowledge', *Polymer Plastic Technology*, 12(2), 1979, pp. 159–79.

3 Lewin, M., 'The Yellowing of Cotton Cellulose Pt. II and III,' *Textile Restoration Journal*, 1965, pp. 935–42 and 979–86.
4 Baynes-Cope, A.D., 'Some Observations on Foxing at the British Museum Research Laboratory', *International Biodeterioration Bulletin*, 12(1), 1976, pp. 31–3.
5 Nol, L., Henis, Y. and Kenneth, R.G., 'Biological Factors of Foxing in Postage Stamp Paper', *International Biodeterioration Bulletin*, 19(1), 1983, pp. 19–25.
6 Tang, L.C., 'Determination of Iron and Copper in 18th and 19th Century Books by Flameless Atomic Absorption Spectroscopy', *Journal of the American Institute of Conservation*, 17, 1978, pp. 19–32.
7 Williams, J.C., Fowler, C.S., Lyon, M.S. and Merril, T.L., 'Metallic Catalysts in the Oxidative Degradation of Paper', *Preservation of Paper and Textiles of Historic and Artistic Value*, Advances in Chemistry Series 164, Washington: American Chemical Society, 1977, pp. 37–61.

Reading 59

PAUL M. WHITMORE

The Mechanisms of Chemical Deterioration of Paper (1994)

Paul Whitmore (1955–), director of the recently relocated and renamed Art Conservation Research Center, within the Center for Conservation and Preservation at Yale University, discusses the complexity of the physical deterioration mechanisms of paper and the difficulty of predicting the aging of already existing papers and those of the future. Whitmore would later apply his expertise in organic chemistry to the extrapolation of photochemical changes in colorants.[1]

Besides addressing questions about the cellulose scission chemistries themselves, the influence of other processes on those scission chemistries must be more thoroughly explored. Some of these interactions may be direct, such as the effect of cellulose oxidation on subsequent hydrolytic scission. Other processes may have a less direct influence on the scission chemistries. For example, chemistries that generate acidity will affect the rate of hydrolysis by adding the catalyst for that scission chemistry. Other chemistries may act to reduce the amount of paper acidity, by forming volatile compounds that may evaporate before affecting the cellulose scissions. Transport processes may be important in the effect of acidity that may be introduced from external sources, such as from the degradation of a size or from exposure to air pollutants. None of these processes are as well understood as the basic cellulose chemistries, yet they will have an impact on such fundamental issues as: the effect of lignin on cellulose degradation; the chemistries and rates

From PAUL M. WHITMORE, "The Mechanisms of Chemical Deterioration of Paper," 38–39. Reprinted, with permission, from *Proceedings, ASTM Institute for Standards Research: Workshop on the Effects of Aging on Printing and Writing Papers: July 6–8, 1994, Philadelphia, PA.*

of acid development in paper; the efficacy and adequacy of "alkaline reserves" in paper; and the ability of accelerated aging tests to duplicate the appropriate combination of these processes as they occur in natural aging.

Conclusion

This last unresolved issue—understanding both natural and accelerated aging—is the key to the ultimate success of our efforts to predict the future of our cultural heritage recorded on a paper medium. This will affect not only the most immediate decisions about the type of paper to manufacture for future books and manuscripts, but also decisions about how best to treat these materials and the collection of documents that come to us from the past. Whether we choose to make these forecasts based on accelerated aging tests of papers that the world has never seen before—which includes paper that has been "treated" in an effort to prolong its life—or based on studies of historic papers in an effort to exploit whatever made those papers robust enough to survive the rigors of time, the task remains the same: discover the means to monitor the chemical changes, and use these methods to achieve a thorough, detailed understanding of paper degradation chemistries.

Note to Introduction

1 Paul Whitmore, "The Scientific Examination of Works of Art on Paper," in *Scientific Examination of Art: Modern Techniques in Conservation and Analysis* (Washington, D.C.: National Academies Press, 2003), 27–39.

Reading 60

HELMUT BANSA

Accelerated Ageing of Paper: Some Ideas on Its Practical Benefit (2002)

Helmut Bansa (1936–), director of the Institute for Restoration of Books and Manuscripts, Bavarian State Library, and editor of the frequently cited Restaurator, *attempts to bridge the gap between the application of paper aging data to conservation treatments and the production of permanent paper, two very different operations having separate goals and metrics for assessing successful outcomes.*

Introduction

Accelerated ageing tests are widely used in conservation research, either to predict how long a paper would last or to what extent a certain treatment would prolong the period during which an object, or its material, will keep the qualities needed to be used for its intended purpose. In the context of a library and archive the "object" is a book, an archival file or a piece of art, the "material" is mainly paper.

Accelerated ageing is to expose the object in question to a higher level of energy, mostly heat, but also to light or other radiation and aggressive pollutants. It is presumed that the higher energy promotes the processes that together interact to produce symptoms recognized as "ageing", that is harming a material's useful qualities. Moreover it is presumed that these processes take place in the same relationship to each other whether at a high or low energy level. It is also presumed that the reaction, (i.e. the ageing), of a material is proportional to the level of energy used to age it, that is the higher the energy level, the more rapid the decay, and this in direct proportion.

It is well known that these last two assumptions are not true.

[. . .]

From HELMUT BANSA, "Accelerated Ageing of Paper: Some Ideas on Its Practical Benefit," *Restaurator* 23 (2002): 106, 111, 115. Used by permission of DeGruyter.

How to Interpret Ageing Tests

For practical use the following statements, derived from the above reported observations and deliberations, may be pertinent. They are not totally new[1]:

- The method of accelerated ageing does not matter a great deal. It is much more important to make the correct use of the results.
- Accelerated ageing tests never allow absolute, they only allow relative conclusions.
- Accelerated ageing can only give some information on whether one certain quality, once certain conservation treatment done in order to improve the actual quality, influences the ageing behaviour for the better or worse.
- The relative change, the observation that a certain quality brought into the paper by a certain conservation treatment changes the ageing behaviour of the object for the better or worse, must be true not only for one pair of objects that are exactly the same except for the very quality in question, but also for more samples, and these different samples should be as different from each other as possible.

Transferring these statements to paper production, in order to define stability classes, could be done by

- defining two standard papers,
  - one of undoubtedly high permanence: pure rag of high DP,[2] slightly alkaline production, a certain (quite small) amount of (precipitated) $CaCO_3$;
  - the other of undoubtedly low permanence: high lignin (and hemicelluloses) containing fibres, acid production, high amount of inert filler (clay)
- describing by fixed numbers for the reduction rate of apt chemical, optical and mechanical qualities occurring during accelerated ageing according to a certain, arbitrarily, but reasonably fixed method, the place of the paper in question between the two standard papers.

The author does not maintain this to be a practical method. But he insists on it being a scientifically sound one, possibly the only scientifically sound one among those based directly on accelerated ageing.

Notes to Reading

1 Bansa, H., & H. H. Hofer: *Die Aussagekraft einer künstlichen Alterung von Papier für Prognosen über seine zukünflige Benutzbarkeit*. Restaurator 6 (1984): 21–60. —English version: *Artifical aging as a predictor of paper's future useful life*. Abbey Newsletter Monograph Suppl. I (1989).

2 [DP = degree of polymerization, an average measurement of the length of cellulose and thus popularly interpreted as an indication of strength.—Ed.]

Reading 61

HENK J. PORCK

Rate of Paper Degradation: The Predictive Value of Artificial Aging Tests (2000)

Henk J. Porck (1953–), conservation scientist and curator of the historical paper collection of the National Library of the Netherlands, here summarizes the various artificial aging methods beginning at the end of the nineteenth century and underscores the lack of agreement on the interpretation of the data produced.[1] *He concludes with the ominous observation that there exists among scientists a "deadlock characterized by uncertainty and doubt" (p. 9). No doubt, this situation provided the opening, as well as plenty of ammunition, for Nicholson Baker's attack the following year on the library and archive preservation community (reading 62).*

DISCUSSION

Evaluation

'Artificial aging' is a test in general use and the term a common one in conservation research. This seems to suggest (i) that there exists a normal, natural aging process, (ii) that this process can be artificially accelerated without changing in its nature, and (iii) that by using artificial aging tests meaningful information on this natural aging process can be obtained. The present study shows that all of these suggestions can be seriously questioned.

Insufficient knowledge on natural aging

In the first place there is no generally valid description of the normal, natural aging process of paper. There are so many paper-dependent internal factors, partly as yet

From HENK J. PORCK, *Rate of Paper Degradation: The Predictive Value of Artificial Aging Tests* (Amsterdam: European Commission on Preservation and Access, 2000), 25–28. Reproduced courtesy of the author and the Royal Netherlands Academy of Arts and Sciences (KNAW).

unknown, as well as external factors varying with time and place that influence the stability of paper that it seems impossible to predict unambiguously the natural aging of a particular sheet of paper (Roberson 1981; Wilson 1983; Fellers 1989; Feller 1994a,b; Gurnagul 1994; Norris 1994). Still more complex is the situation in a stack of paper, like a book or a file, let alone a complete library or archive collection. A reliable judgement on the nature and rate of natural aging can actually be made only afterwards, i.e. deduced in retrospect. Moreover this is only possible if detailed information is available on the original material condition of the object and on the storage conditions to which the object has been exposed in the course of time. [. . .]

Simulation by stimulation?

In the second place it is probably only wishful thinking that natural aging of paper can be simply accelerated by exposure to artificial, laboratory conditions, and that in this way certain forms of damage, familiar from the conservation practice, can be reliably simulated. The complex character of the natural aging process, which is partly still obscure, can never be completely achieved experimentally. Even if natural aging could be simplified to the effects of the paper degradation reactions as known today, it is highly doubtful that artificial stimulation of these reactions, by means of certain adjustments of environmental parameters in a climate-chamber, will result in an identical condition of the paper in the short term as would result from the natural aging process in the long term. Such an assumption is as yet insufficiently verified by scientific research. The current uncertainty in this regard is clearly illustrated by the multitude of standard and 'home' methods for artificial aging mentioned above. Artificial aging is performed at different temperatures, different humidities, sometimes in the presence of air pollutants or other 'stimuli' and sometimes not, and for different periods of time. Even the standard methods for artificial aging, developed in order to counteract this large variation, do not resolve this: the international ISO-standards, for example, developed in the eighties, leave the possibility open to age artificially at different temperatures and relative humidities.

Relevance of artificial aging

All this implies that quite a few objections can be put forward against the idea that artificial aging presents relevant data for the determination of the nature and rate of the natural aging of paper. The studies into the reliability, i.e. predictive value of artificial aging, as described above, have not yet resulted in a conclusion that is generally endorsed, nor into the use of a generally accepted method. The findings of the few investigations performed in this respect have not yet sufficiently proven in practice that the theoretical basis for the application of artificial aging is indeed valid. Nor have these studies resulted in adjustments of current methods. The limited scale of the research into the comparison of artificial and natural paper aging is remarkable. However complex this kind of investigation may be, it is one of the few possibilities to attune the procedure of artificial aging to the natural aging process. In the field of leather conservation, for example, this approach appears to have been rather successful (Larsen 1994).

Problem of reliability

The small number of research projects that have tried to verify the predictive value of artificial aging analysis strongly contrasts with the widespread use of this analysis in practice. Even so, this does not mean there is no concern about the issue. The problem of the reliability of artificial aging has been raised frequently in the specialist literature (Arney 1980; Bansa 1984, 1992, 1998a; Mokken 1987; Erhardt 1988; Ströfer-Hua 1990; Wypych 1990; Brown 1991; Feller 1994a; Shahani 1994; Havermans 1997b). The present insecurity about the value of artificial aging can for instance be illustrated from the fact that the international standard ISO-9706 for permanent paper, in which a prediction of the durability and permanence would be extremely desirable, does not include an artificial aging test (Berkhout 1994).

In view of the different objectives with which artificial aging is applied, the Arrhenius-test, aimed at a quantitative statement on permanence, i.e. rate of paper decay, will probably be the most debatable. In the case of other applications, like the determination of the qualitative effect of a conservation treatment, the investigation of a certain degradation reaction or the simulation of specific kinds of damage, the strict prerequisite for the Arrhenius-test, namely that all natural deterioration reactions must be accelerated by one and the same factor, does not need to be fulfilled. In these cases it will be more likely that artificial aging may produce trustworthy indications. Still, here too reservations should be made because on the basis of the present state of knowledge, it cannot be excluded that unrealistic effects will occur as a consequence of undesirable reactions, i.e. reactions absent in the natural aging process.

Deadlock

With regard to artificial aging there seems to have developed a kind of deadlock in the conservation science community, and sometimes even an obvious reluctance to exchange experience and to strive for a joint approach to tackle the problem. This reluctance, which perhaps also caused the limited response to the discussion planned for the present study, can be explained to some extent. A critical discussion of such crucial and historically deeply rooted issues may lead to changes in opinions that can have far-reaching consequences. It is possible that decisions taken in the past, either in conservation policy or conservation practice, based on the results of artificial aging tests, are undermined in such a discussion. On the other hand the current deadlock is a cause for serious regret, as it obstructs a solution of the many questions on artificial aging which are still left unanswered. In the meantime the natural aging of paper proceeds without mercy.

Conclusions

In summary, four conclusions can be drawn on the basis of this evaluation of the current state of affairs with respect to the theory and practice of artificial aging of paper:

The rate of paper deterioration and other quantitative aspects of the natural aging of paper, such as durability and permanence, cannot be reliably predicted by means of the present artificial aging tests.

The application of artificial aging can be relevant and meaningful with respect to:

a. the determination of the qualitative effect of a certain paper conservation treatment;
b. fundamental research into paper degradation reactions;
c. the simulation of specific types of paper damage.

The amount of research into the predictive value of artificial aging of paper is not sufficient and the results are not adequately used for the improvement of artificial aging methods.

Reluctance in discussion and maintenance of a status quo characterize the conservation science community with regard to the problem of artificial aging. As a consequence the solution of this problem is seriously hindered.

Recommendations

The present study is intended as an overview and evaluation of the current state-of-the-art rather than an elaboration of solutions. However, several recommendations can be made which may contribute to the improvement of the current situation of uncertainty and doubt.

The exchange of knowledge and experience in the field of artificial aging analysis of paper among conservation scientists and policy makers needs to be improved. [. . .]

Research to verify the hypothesis that artificial aging produces meaningful data to predict the effects of natural aging of paper should be continued and subsequently be 'translated' in improved, i.e. more reliable artificial aging tests.

Comparative investigation of natural and artificial aging of paper can be performed in parallel with long-term studies into natural paper deterioration, by means of artificial aging treatments and subsequent analysis of part of the initial test material. The findings can be compared in a later stage, by a next generation of conservation scientists, with those of the naturally aged material.

It is recommended that statements on the rate of the natural aging of a certain kind of paper are not only limited to predictions for the future by means of artificial aging. In some cases, a retrospective approach is possible, using historical sources that present detailed information on the quality of a specific type of paper in the past.

In the application of the current artificial aging tests for the determination of the relative effects of a certain paper conservation treatment, for the investigation of a specific degradation mechanism, or for the simulation of a certain form of paper decay, a 'mild' artificial aging method is to be preferred because of the inevitable risk of undesirable degradation reactions, that is to say reactions that do not occur under natural conditions. Artificial aging conditions should therefore in principle be kept as closely as possible to normal, natural conditions.

As false expectations may be evoked by the term 'artificial aging' it should be considered:

a. to limit the use of the term 'aging' to the genuine process of natural aging;
b. to use another term for the so-called artificial aging test, like 'artificially induced degradation (AID-) test', which reflects the way in which these tests are commonly used in practice: choosing and adjusting the environmental conditions in such a way that a measurable deterioration of the material takes place in a relatively short time.

Note to Introduction

1 The historical paper collection was started in 1972 with the purchase of the collection of the noted paper historian Henk Voorn.

References

Arney, J.S., Jacobs, A.J. (1980). Newsprint Deterioration. The Influence of Temperature on the Relative Contribution of Oxygen-independent and Oxygen-dependent Processes in the Total Rate. *Tappi* 62 (1), p. 75–77.

Bansa, H., Hofer, H.H. (1984). Die Aussagekraft einer künstlichen Alterung von Papier für Prognosen über seine zukünftige Benutzbarkeit. *Restaurator* 6, p. 21–60.

Bansa, H. (1992). Accelerated Aging Tests in Conservation Research: Some Ideas for a Future Method. *Restaurator* 13, p. 114–137.

Bansa, H. (1998a). Aqueous Deacidification—with Calcium or with Magnesium? *Restaurator* 19, p. 1–40.

Berkhout, S.A. (1994). Beoordeling en vergelijking van drie duurzaamheidnormen voor papier. TNO (TNO-report PK 94.7009), Delft.

Brown, R.P. (1991). Survey of Status of Test Methods for Accelerated Durability Testing. *Polymer Testing* 10, p. 3–30.

Erhardt, D. [Smithsonian Center for Materials Research and Education, Suitland, MD] (1998): personal communication.

Feller, R.L. (1994a). Accelerated Aging: Photochemical and Thermal Aspects. The Getty Conservation Institute (Research in Conservation 4), Marina del Rey.

Feller, R.L. (1994b). Aspects of Chemical Research in Conservation: The Deterioration Process. *Journal of the American Institute for Conservation* 33, p. 91–99.

Fellers, C., Iversen, T., Linström, T., Nilsson, T., Rigdahl, M. (1989). Aging/Degradation of Paper. A Literature Survey. FoU-projekter för papperskonservering (Report 1E), Stockholm.

Gurnagul, N., Howard, R.C., Zou, X., Uesaka, T., Page, D.H. (1994). The Mechanical Permanence of Paper: A Literature Review. In: *Proceedings of the Workshop on the Effects of Aging on Printing and Writing Papers*. ASTM Institute for Standards Research, Philadelphia, p. 235–257.

Havermans, J. (1997b). Durability of Information—From Stone to Polymer. In (Fellos-Jensen, G., Springborg, Pl., eds): *Care and Conservation of Manuscripts 3. Proceedings of the Third International Seminar on the Care and Conservation of Manuscripts Held at the University of Copenhagen 14th-15th October 1996*. The Royal Library, Copenhagen, p. 99–106.

Mokken, M., Steemers, T., Lingbeek, N. (1987). *Versnelde papierveroudering*. Een literatuurstudie in het kader van de HBO-opleiding Papier-en Boek-restauratie, Amsterdam; Vakbibliotheek conservering (L 689), Konink-lijke Bibliotheek, Den Haag.

Norris, T.O. (1994). Overview of Paper Deterioration. In: *Proceedings of the Workshop on the Effects of Aging on Printing and Writing Papers*. ASTM Institute for Standards Research, Philadelphia, p. 14–21.

Roberson, D.D. (1981). Permanence/Durability and Preservation at the Barrow Laboratory. In (Williams, J.C. ed): *Preservation of Paper and Textiles of Historic and Artistic Interest. II*. American Chemical Society, Washington, DC, 45–55.

Shahani, C.J. (1994). Accelerated Aging of Paper: Can It Really Foretell the Permanence of Paper? In: *Proceedings of the Workshop on the Effects of Aging on Printing and Writing Papers*. ASTM Institute for Standards Research, Philadelphia, p. 120–139.

Ströfer-Hua, E. (1990). Experimental Measurement: Interpreting, Extrapolation and Prediction by Accelerated Aging. *Restaurator* 11 (4), p. 254–266.

Wilson, W.K., Parks, E.J. (1983). Historical Survey of Research at the National Bureau of Standards on Materials for Archival Records. *Restaurator* 5, p. 191–241.

Wypych, J. (1990). *Weathering Handbook*. Chemtec Publishing, Toronto, p. 197–221.

Reading 62

Nicholson Baker

Double Fold: Libraries and the Assault on Paper (2001)

Double Fold *by the novelist and investigative journalist Nicholson Baker (1957–) rocked the library and archive preservation world on its publication in 2001.*[1] *Seizing upon the dissension in the scientific ranks concerning the validity of artificial aging and the means of quantifying degradation, such as the now-infamous "double fold," whereby a paper is folded back and forth in a ninety-degree angle to measure brittleness, and the inconsistent implementation of large scale duplication and deacidification programs, Nicholson single-handedly forced preservation administrators to reassess their policies. While conservators, but not hapless preservation administrators, managed to duck the shotgun swath of Nicholson's rage, they, too, were forced to examine the degree to which they had come to rely on numerical or irrelevant outcomes in designing treatment and assessing results.*[2]

On behalf of the preservation community, Vandals in the Stacks? A Response to Nicholson Baker's Assault on Libraries *was published the following year by the archivist Richard Cox.*[3] *Unfortunately, the book did not receive the coverage it deserved, no doubt due to Cox's more measured tone and methodical writing style, not to mention the book's drastically higher price tag.*[4] *Of more importance than Nicholson's many provocative and highly debatable pronouncements, however, was the lesson absorbed and promoted by Cox: "The chief value of Baker's* Double Fold *may be that it serves as a warning for individuals like me that we cannot take for granted that we or our discipline will be understood or appreciated by external observers. . . .*

From Nicholson Baker, *Double Fold: Libraries and the Assault on Paper* (New York: Vintage Books, 2001), 6–9.

Perhaps it will take a rant from outside our community to wake us up about what and how we need to communicate."[5]

[. . .]

Vilified though it may be, ground-wood pulp is one of the great inventions of the late nineteenth century: it gave us cheap paper, and cheap paper transformed the news. "All that it is necessary for a man to do on going into a paper-mill is to take off his shirt, hand it to the devil who officiates at one extremity, and have it come out 'Robinson Crusoe' at the other," wrote the founder of the New York *Sun* in 1837.[1] But there were never enough shirts, and in 1854 rag shortages lifted the price of newsprint to alarming heights. The arrival of the brothers Pagenstecher,[2] who in the eighteen-sixties imported a German machine that shredded logs to pulp by jamming their ends against a circular, water-cooled grinding stone, brought prices way down—from twelve cents a pound in 1870, to seven cents a pound in 1880, to less than two cents a pound in 1900.[3] The drop gave Pulitzer and Hearst the plentiful page space to sell big ads,[4] and allowed their creations to flower into the gaudy painted ladies they had become by the first decade of the twentieth century.

There's no question that wood pulps are in general weaker than rag pulps; and old newsprint, especially, tears easily, and it can become exceedingly fragile if it is stored, say, on the cement floor of a library basement, near heating pipes, for a few decades. But the degree of fragility varies from title to title and run to run, and many fragile things (old quilts, old clocks, astrolabes, dried botanical specimens, Egyptian glass, daguerreotypes, early computers) are deemed worth preserving despite, or even because of, their fragility. The most delicate volume I've come across (a month of the Detroit *Evening News* from 1892), though the pages were mostly detached, and though it shed flurries of marginal flakes when I moved it around, could nonetheless be page-turned and read with a modicum of care—there was an interesting article, with two accompanying etchings,[5] about a city shelter for "homeless wanderers."[6] (Sinners slept on wooden bunks without bedding, while the newly converted got cots with mattresses, and a reading room.)

Old newsprint is very acidic—and so? Our agitation over the acid in paper is not rational. Just because a given page has a low pH (a pH of 7 is neutral, below that is acidic) doesn't mean that it can't be read. There are five-hundred-year-old book papers that remain strong and flexible despite pH levels under five, a fact which has led one conservation scientist to conclude that "the acidity of the paper alone is not necessarily indicative of the state of permanence of paper."[7] It is difficult, in fact, to get a meaningful measure of how alkaline or acidic a paper actually is, since chemicals on the surface behave differently than those held within; the standard scientific tests (which often rely on a blender) don't discriminate. It's true that, all things being equal, pH-neutral paper seems to keep its properties longer than paper that is made with acid-containing or acid-forming additives; scientists have been making this observation, on and off, for more than eighty years.[8] But saying that one substance is stronger than another is not the same as saying that the weaker substance is on the verge of self-destruction. A stainless-steel chair may be more durable than a wooden one, but the wooden one isn't necessarily going to collapse the next time you take a seat.

Can't scientists foretell with a fair degree of certainty how long a newspaper collection of a given age will last? No, they can't; there has never been a long-term study that attempted to plot an actual loss-of-strength curve for samples of naturally aging newsprint, or indeed for samples of any paper.[9] Years ago, William K. Wilson, a paper scientist, began such a study at the National Bureau of Standards.[10] For three decades he recorded the degradative changes undergone by a set of commercial book papers; then somebody decided to clean out the green filing cabinet in which the papers were stored—end of experiment. "That raised my blood pressure a little," Wilson told me.

In the absence of real long-term data, predictions have relied on methodologically shaky "artificial aging" (or "accelerated aging") experiments, in which you bake a paper sample in a laboratory oven for a week or two and then belabor it with standardized tests. With your test results in hand, you can, by applying a bit of chemist's legerdemain called the Arrhenius equation, come up with what appears to be a reasonable estimate of the number of years the sample will last at shirtsleeve temperatures. But the results of these sorts of divinatory calculations, invoked with head-shaking gravity by library administrators, have been uniformly wrong, and they are now viewed with skepticism by many paper scientists.[11] The authors of the *ASTM Standards*, for example, write that the use of the Arrhenius equation to predict the life expectancy of paper is "an interesting academic exercise, but the uncertainty of extrapolation is too great for this approach to be taken very seriously."[12] William Wilson points out that you can't predict how long an egg will last in the carton by putting it in boiling water for five minutes. Paper has a complex and as yet ill-charted chemistry, with many different molecular and mechanical processes under way concurrently; one Swedish researcher wrote that it is a "naive hope" to think that we can estimate "the life length of books by means of accelerated aging tests and [the] Arrhenius approach."[13]

In a way, however, all surviving newspaper collections, in and out of libraries, are taking part in an immense self-guided experiment in natural aging—an experiment that confutes the doctrine of newsprint's imminent disintegration. Peter Waters, former head of the conservation lab at the Library of Congress, told me that he sees no reason why old ground-wood pulp paper can't hold its textual freight for "a hell of a long time" if it is properly stored. He notes that most of the cellulose-sundering chemical reactions that can happen to a book or newspaper volume seem to take place in the first decade or so of its life; fifty years of handling paper (Waters is a master bookbinder) have taught him that the rate at which paper loses strength decreases significantly over time—the curve of observed decay levels out. There is a very good chance, then, that a volume of the New York *World* that is doing okay at age ninety will be in pretty much the same shape when it is a hundred and eighty, assuming someone is willing to take decent care of it.

Notes to Introduction

1 The hardbound version of *Double Fold* (2001) is on permanent paper; the first paperback Vintage Books edition (2002) is not.
2 Klaus Hendricks (1937–1996) of the Canadian Conservation Institute also proposes more pragmatic means of assessing paper permanence: "Permanence of Paper in Light of Six Centuries of Papermaking in Europe," *Environnement et conservation de l'écrit, de l'image et du son: Actes des Deuxièmes Journées Internationales d'Études de l'ARSAG* (Paris: ARSAG, 1994), 131–37.
3 Richard J. Cox, *Vandals in the Stacks? A Response to Nicholson Baker's Assault on Libraries* (Westport, Conn.: Greenwood Press, 2002); printed on permanent paper complying with Z39.48-1984.
4 *Double Fold* is available in paperback ($14.25) and in Kindle ($13.99); *Vandals* is available in hardcover ($102.95) and as a digital download for only $5.95.
5 Cox, *Vandals in the Stacks?*, 196, 194.

Notes to Reading

1 "All that it is necessary": New York *Sun*, June 1, 1837, quoted in Frank M. O'Brien, *The Story of the Sun* (New York: George H. Doran, 1918), p. 124.
2 [Rudolf Pagenstecher (1838–1903) and Albrecht Pagenstecher (1839–1926) were the first to import Keller's wood-pulping machine into the United States. However, the logs are inserted lengthwise, not from their ends as Baker states.—Ed.]
3 brought prices way down: Alfred McClung Lee, *The Daily Newspaper in America* (New York: Macmillan, 1937), pp. 742–45.
4 [Joseph Pulitzer (1847–1911) published the *New York World* in competition with the *New York Journal*, published by William Randolph Hearst (1863–1951).—Ed.]
5 [The illustrations are certainly wood engravings, not etchings.—Ed.]
6 interesting article: Detroit *Evening News*, January 10, 1892.
7 "the acidity of paper alone": Klaus B. Hendriks, "Permanence of Paper in Light of Six Centuries of Papermaking in Europe," in *Environnement et conservation de l'écrit, de l'image et du son: Actes des Deuxièmes Journées Internationales d'Études de l'ARSAG* (Paris: Association pour la Recherche Scientifique sur les Arts Graphiques [ARSAG], 1994), p. 136. See also Otto Wächter, "Paper Strengthening: Mass Conservation of Unbound and Bound Newspapers," *Restaurator* 8 (1987).
8 scientists have been making this observation: Sally Roggia cites *Paper Makers Monthly Journal* (June 1920), which summarizes a work by Aribert and Bouvier: the "most frequent and most harmful chemicals remaining in the paper are free acids and free chlorine"; it is "important to avoid acidity in the alum." Sally Roggia, "William James Barrow: A Biographical Study of His Formative Years and His Role in the History of Library and Archives Conservation from 1931 to 1941," Ph.D. diss., Columbia University, 1999.
9 [In 2000, ten libraries entered into a 100-year natural-aging program whereby fifteen samples of custom-made papers will be stored and tested by four research laboratories at regular intervals. www.loc.gov/preservation/scientists/projects/100-yr_nat_aging.html.—Ed.]

10 Wilson: William K. Wilson and E. J. Parks, "Comparison of Accelerated Aging of Book Papers in 1937 with 36 Years Natural Aging," *Restaurator* 4 (1980): "Unfortunately the papers were accidentally discarded several years ago." Elsewhere, Wilson and Parks write, "Very few organizations can maintain a project [such as testing naturally aging paper] that spans 25–50 years, especially when there is no pot of real gold at the end of the trail." "Historical Survey of Research at the National Bureau of Standards on Materials for Archival Records," *Restaurator* 5 (1983). In 1998 the American Society for Testing and Materials announced a one-hundred-year natural-aging experiment; books containing fifteen different kinds of test paper have gone into ten libraries in North America. The object is to develop an accelerated-aging test that better correlates with natural aging.

11 "now viewed with skepticism": In *Artificial Aging as a Predictor of Paper's Future Useful Life*, an *Abbey Newsletter* Monograph Supplement, Helmut Bansa and Hans H. Hofer find that "there may be at best an accidental agreement between the results of artificial aging at high temperatures and natural aging" (Provo, Utah: Abbey Newsletter, 1989). See also Wilson and Parks, "Comparison of Accelerated Aging," in which the data suggest that "either the number of samples is less than adequate to provide a valid statistical population or the accelerated aging method used in 1937 does not fully simulate natural aging, or both." Later (p. 47) Wilson writes: "Don't try to predict permanence in years" (47). E. Ströfer-Hua, after an experiment that demonstrated the flaws of oven aging, concluded, "History can only happen; it cannot be simulated in advance." E. Ströfer-Hua. "Experimental Measurement: Interpreting Extrapolation and Prediction by Accelerated Aging," *Restaurator* 11 (1990).

12 "an interesting academic exercise": American Society for Testing and Materials, "Standard Test Method for Determination of Effect of Moist Heat (50% Relative Humidity and 90°C) on Properties of Paper and Board," *Annual Book of ASTM Standards, 1998*, vol. 15.09, D 4714, Appendix (Conshohocken, Pa.: ASTM, 1998).

13 "naive hope": Tom Lindstrom, "Discussion Contribution: Slow Fires—Its Paper Chemistry, Physics, and Biology," in *Paper Preservation: Current Issues and Recent Developments*, ed. Philip Luner (Atlanta: Tappi, 1988). Glen G. Gray writes, "Although several accelerated-aging procedures or chemical specifications have been proposed, carefully controlled experiments and many years of natural aging would be required to verify predictions." Glen G. Gray, "Determination and Significance of Activation Energy in Permanence Tests," in *Preservation of Paper and Textiles of Historic and Artistic Value*, ed. John C. Williams (Washington, D.C.: American Chemical Society, 1977). Wilson and Parks write, "An accelerated aging test does not tell what a paper will be like after 25 or 50 years of storage. It only provides information concerning the ranking of different samples with respect to storage properties." William K. Wilson and E. J. Parks, "An Analysis of the Aging of Paper," *Restaurator* 3 (1979): 56.

Reading 63

ROBERT L. FELLER

The Deteriorating Effect of Light on Museum Objects: Principles of Photochemistry, the Effect on Varnishes and Paint Vehicles and on Paper (1964)

This reading is by one of conservation's most influential scientists, Robert L. Feller (1919–), founder of the Research Center on the Materials of the Artist and former conservator at Carnegie Mellon University.[1] *Feller's many publications have helped conservators understand picture varnishes, color, light damage, and polymer and paper deterioration. The scientific principles he established and his evaluations of treatments, such as the light bleaching of paper, and tools, such as the ISO Blue Wool fading standards, are of profound importance to the development of preventive conservation. While mention has been made throughout Part V of occasional instances of the misunderstanding and misapplication of scientific research, Feller put the word "'conservative' into conservation science, demanding cautious adoption and careful scrutiny as new materials or treatments are put into practice."*[2] *Representing Browning's (reading 51) third mechanism of photochemical deterioration, Feller's introduction of the concept of cumulative light damage to the museum community via the popular professional journal* Museum News *was instrumental in communicating the importance of establishing stringent lighting exposure policies.*[3] *We can also hear echoes of Russell and Abney (reading 67) in this reading.*

Feller's successor, Paul Whitmore (reading 59), continued the investigation of photochemical deterioration and the practical application of scientific research. One exciting development is Whitmore's micro-fading tester, which simultaneously causes fading by aiming an approximately seven-million-lux beam of light at a 0.4-millimeter-diameter sample spot and measures its fading rate relative to the ISO Blue Wool fading standards as introduced to conservators by Feller.[4] *The instrument,*

From ROBERT L. FELLER, "The Deteriorating Effect of Light on Museum Objects: Principles of Photochemistry, the Effect on Varnishes and Paint Vehicles and on Paper," *Museum News Technical Supplement*, no. 3 (1964): vi–vii.

constructed of off-the-shelf components, consists of a 75-watt xenon arc lamp that is filtered to remove the ultraviolet and infrared wavelengths to produce light similar to that in museum galleries. Since most light-sensitive colorants fade in the visible range of light, the intense beam of light works quickly, often detecting fading rates ranging from Blue Wool Standard no. 1 through no. 3 within ten to fifteen minutes. Not only can the micro-fading tester be used to assess the lightfastness of known artists' materials, but it can also be used to predict the effect of cumulative light exposure for works of art having unknown colored media. For example, a selection of watercolors by Georgia O'Keeffe (1887–1986) has been tested by the Getty Conservation Institute and the relative rates of fading of each color have been used to make informed decisions regarding the pictures' future exhibition.[5] The recurring dilemma for paper conservators and curators remains, however: given the steep slopes of fading rates of most organic colorants, whether to give preference to the display of an already-faded or a still-vibrant watercolor. The corollary quandary is whether the unexposed edges of a watercolor should be exposed in order to inform the public of the piece's original vibrancy or kept conserved in order to preserve such evidence.

Effect of Light on Paper

Influence of Wavelength

Cellulose does not absorb visible radiation and for this reason, one might at first expect that it would not be affected by wavelengths greater than 400 millimicrons. Yet, Launer and Wilson[1] have reported on the effect of radiation from 330 to as high as 440 millimicrons, and both Richter[2] and Kohler[3] have reported on the effect of radiation between 400 and 460 millimicrons. Rather than the cellulose itself, it is perhaps glue, rosin, and other constituents of ordinary paper that are able to absorb the violet and blue radiation and to sensitize the paper to deterioration. Be that as it may, the activity of light falls off rapidly at the higher wavelengths. In the deterioration of low grade paper, for example, the National Bureau of Standards' "factors of probable relative damage" were found to be only one-tenth as great at 480 millimicrons as at 400.[4]

Exposure Necessary to Cause Deterioration

Serious deterioration of paper is produced in 50 to 100 hours of exposure to full sunlight, an exposure of about 400,000 to 1 million footcandle hours, or 3600 to 9000 Langleys.[2] It is difficult, however, to judge the significance of this observation in terms of exposure in a museum. This is because the temperature of the samples is increased during exposure to direct sunlight and the deterioration of paper is markedly influenced by temperature as well as by humidity. The influ-

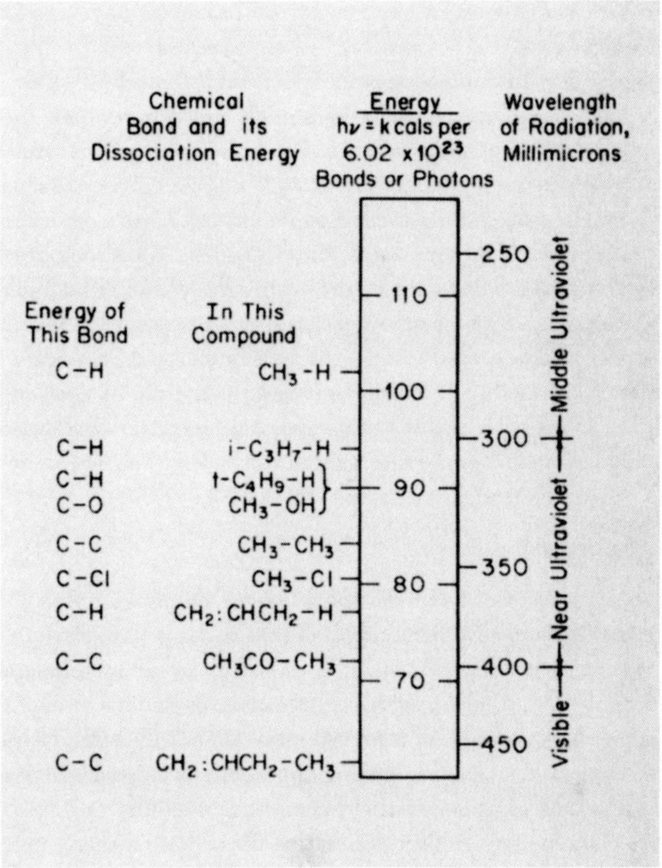

Figure 1
A chart relating the energy of radiation of different wavelengths to the energy of various chemical bonds

ence of temperature is of such importance that the heating of paper is used as an accelerated-aging test to predict its long-term durability.

Control of the temperature of the sample has frequently been neglected in accelerated-exposure tests. This has led to many experimental results that cannot be readily interpreted in terms of room conditions. With this problem in mind, Launer and Wilson built special equipment to expose paper to the light of a carbon arc in such a way that the temperature of the samples was kept at 30°C. From their results, a rough estimate may be made that rag papers can lose 25%, and soda-sulfite papers 15 to 50%, of their resistance-to-folding after an exposure at 30°C (86°F) and 58% R.H. to about 1,700,000 footcandle hours of 6500°K daylight.[5]

A finding of importance to the museum curator, verified by these authors and others, is the fact that, once exposed to light, paper may deteriorate at an accelerated rate when it is removed from the light and stored.[1,6] In other words, exposure to radiation is able to sensitize paper so that it exhibits an increased rate of deterioration upon subsequent storage.

Intrinsic and Extrinsic Factors That Influence the Deterioration

Papers of high quality are much more stable than those of low.[7] Launer and Wilson found that papers decreased in stability in the order: new-rag, refined sulfite, old-rag, soda-sulfite, and newsprint. Paper containing lignin was particularly unstable to light.

The presence of dyes in paper also increased the extent of deterioration in certain cases. The sensitivity of new-rag papers was found to be markedly affected by the presence of acid in the rosin size; old-rag and soda-sulfite papers were less affected. Richter reported that sized papers deteriorated faster than water-leaf papers when exposed to sunlight, but Launer and Wilson found that rosin did not affect the stability of any paper as long as the acidity was low.

In the absence of water vapor and oxygen, there is reduced scission of the cellulose molecules by radiation in the near ultraviolet.[1]

Yellowing and Bleaching

The processes of both yellowing and bleaching go on at the same time. Launer and Wilson clearly demonstrated that paper yellowed when heated during exposure to light, but was bleached when kept at a low temperature. The net result depended on whether the changes caused by light or by heat predominated. Lignified papers discolored, however, even when no heating of the paper occurred. In groundwood papers, bleaching apparently is caused by wavelengths longer than 385 millimicrons, yellowing by shorter wavelengths.[8]

Launer and Wilson found that the yellowing that had been induced by heat, even to the extent of scorching, could be bleached by exposure to light. This could again be nullified by further heating of the paper. They cautioned, however, against any hasty suggestion that light should be used to bleach discolored documents, since it also causes deterioration of the cellulose.

The complex reactions that take place during discoloration—the "color reversion" or the so-called "brightness reversion" of paper—have been reviewed by several authors[9,10,11] and have recently been further elucidated by MacMillan.[12]

Deterioration of Size

Light can cause deterioration of the size. Violet, blue, and yellow wavelengths have proven to be the most active; red, the least.[13,14] Upon exposure to sunlight, rosin- and starch-sized papers have been found to be less sensitive to the change in writing qualities than glue-sized.[2]

Summary

Whereas the best grade of paper is perhaps reasonably stable to light, the presence of lignin, rosin, and other materials can hasten yellowing and embrittlement. The shorter wavelengths of light are more destructive than the longer. Once irradiated, paper is more sensitive to deterioration upon subsequent storage. The rate of deterioration is sensitive to temperature, so much so that heating samples in an oven is a standard test to "accelerate the age" of paper.

Notes to Introduction

1 The center relocated in 2013 to Yale University as the Art Conservation Research Center.

2 From remarks made by Paul Whitmore (reading 59) on the presentation of the 2011 Robert L. Feller Lifetime Achievement Award to Robert L. Feller. www.conservation-us.org.

3 Also introduced in a Technical Supplement to *Museum News* was the Universal Color Language (see reading 38), which demonstrated how colors could be described using accurate, convenient, and easily understood terminology. It is unfortunate that this promising conduit to facilitate interdisciplinary communication was never widely adopted.

4 Paul M. Whitmore, Xun Pan, and Catherine Bailie, "Predicting the Fading of Objects: Identification of Fugitive Colorants through Direct Nondestructive Lightfastness Measurements," *Journal of the American Institute for Conservation* 38 (1999): 395–409; P. M. Whitmore, C. Bailie, and S. A. Connors, "Micro-Fading Tests to Predict the Result of Exhibition: Progress and Prospects," in *Tradition and Innovation: Advances in Conservation*, ed. A. Roy and P. Smith (Melbourne: International Institute for Conservation, 2001), 200–205; P. M. Whitmore, "Pursuing the Fugitive: Direct Measurement of Light Sensitivity with Micro-Fading Tests," in *The Broad Spectrum: The Art and Science of Conserving Colored Media on Paper*, ed. H. K. Stratis and B. Salvesen (London: Archetype, 2002), 241–43.

5 H. R. Morris and P. M. Whitmore, "'Virtual Fading' of Art Objects: Simulating the Future Fading of Artifacts by Visualizing Micro-Fading Tests Results," *Journal of the American Institute for Conservation* 46 (2007): 215–28.

Notes to Reading

1 Launer, H. F., and Wilson, W. K. *J. Res. Nat. Bur. Std.* 30: 55; 1943. *J. Am. Chem. Soc.* 71: 958; 1949.

2 Richter, G. A. *Ind. Eng. Chem.* 27: 177; 432; 1935.

3 Kohler, S. *Tek Tid. Uppl. C, Kemi* 65: 49; 1935. *Chem. Abstracts* 29: 6379; 1935.

4 National Bureau of Standards. *Preservation of the Declaration of Independence and the Constitution of the United States.* NBS Circular 505. Washington, D.C.: U.S. Government Printing Office. July 7, 1951.

5 This estimation is based on the following reasoning: the authors say that 80% of the energy of the arc light fell in the region between 3300 and 4400Å. Approximately 25.8% of the energy of 6500°K daylight falls in this range. If the intensity on the

paper was 0.4 watts/cm^2, then 0.32 watts/cm^2 lay between 3300 and 4400Å. For 6500°K daylight to provide this much energy, the total energy of the daylight would be 1.24 watts/cm^2. Moon, in *The Scientific Basis of Illuminating Engineering*, gives a figure of 85 lumens/watt for a 6500°K radiator. Therefore, it may be estimated that 1.24 watts/cm^2 equals 104 lumens/cm^2 or 97,500 lumens/sq. ft. intensity on the paper. Hence, about 1,660,000 footcandle hours of 6500°K daylight would be represented by 17 hours of arc light.

6 Stillings, R. A., and Van Nostrand, R. J. *J. Am. Chem. Soc.* 66: 753; 1944.

7 The superior quality of the paper is in a large measure indicated by the content of alpha cellulose.

8 Nolan, P., Van der Akker, J. A., and Wink, W. A. *Tech. Assoc. Papers* 28; 62; 1945. *Paper Trade J.* 121: 33; No. 11, 1945.

9 Luner, P., and Vershney, M. C. "The Color Reversion of Oxycellulose." Syracuse, N.Y.: Empire State Paper Research Institute, State College of Forestry. (Presented at meeting of Division of Cellulose, Wood, and Fibre Chemistry, American Chemical Society, Los Angeles, April 1–4, 1963).

10 Ott, E., and Spurlin, H. *High Polymers*, Vol. V., *Cellulose*, Part I, pp. 168–174. New York: Interscience Publishers, Inc. 1954.

11 Spinner, I. H. *Tappi* 45: 495; 1962.

12 MacMillan, W. R. *Colour Reversion in Oxidized Cellulose*. Thesis, Master of Applied Science. Canada: University of Toronto, 1958.

13 Herzberg, W. *Mittheilungen aus den Koenigtechnischen Versuchanstallen*. Vols. 5 and 6, 1896.

14 Zhereboff, L. P. *Bumazhnaia Promyshlennost* 1: 25; 1922. *Paper* 31: 143; No. 26, 1923. *Paper Trade J.* 76: 50; No. 11, 1923. *Chem. Abstracts* 17: 1716; 1923.

Reading 64

ROBERT L. FELLER

Concerning the Place of Science in the Scheme of Things (1989)

The birth of the discipline of conservation science is relatively recent, and its positive impact is exemplified by the work of the organic chemist Robert L. Feller. Paper conservators have benefited enormously from Feller's research, as epitomized by the previous reading. While not addressed specifically to paper conservators, this reading describes the broader role that science should play in conservation. According to Barbara Appelbaum (1947–), who has written extensively about treatment methodology, science contributes to a conservator's decision-making process in three ways. It provides information about materials and their properties, affords various analytical and imaging techniques, and presents a systematic method of inquiry that can be universally applied.[1] Thus science provides not only information but also the tools for its interpretation and, ultimately, its more effective communication to conservators and nonconservators alike.

[. . .]

I have been asked to make some remarks regarding the place of science. In fulfilling this obligation, I shall feel more comfortable being descriptive rather than prescriptive. Rather than suggest the way things ought to be, I shall attempt instead to describe the way things are and to point out certain deficiencies that remain.

Let us recognize at the very beginning that "science is here to stay". The pursuit of science and technology represents a strong subculture in our society to say the least. Both the artist and the art historian have been, and will continue for some time to be, markedly influenced by the practice and philosophy of the sci-

From ROBERT L. FELLER, "Concerning the Place of Science in the Scheme of Things," in *Training in Conservation, a Symposium on the Occasion of the Dedication of the Stephen Chan House, October 1, 1983, Institute of Fine Arts, New York University*, ed. Norbert S. Baer (New York: Institute of Fine Arts, 1989), 18–21, 23–24. Courtesy of Institute of Fine Arts, New York University.

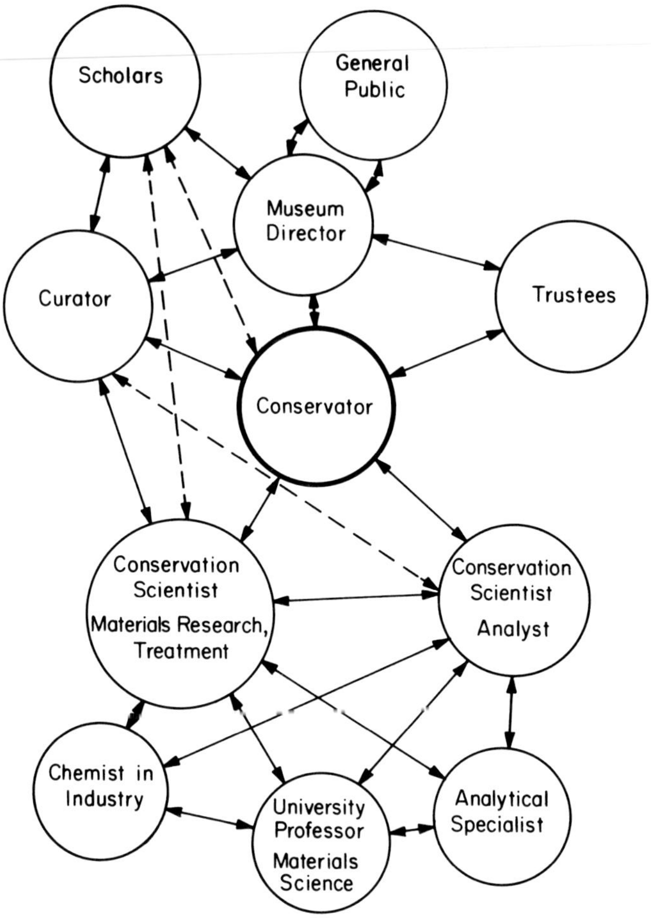

Figure 1
The conservator and those interacted with.

entist and engineer. Given this circumstance, the humanist—as well as the technologist—must make the best of it. We must try to utilize the tools of science and technology to advantage, to assist ourselves toward the goals that society considers worthwhile. The operative word is assist.

The diagram in Figure 1 was devised to emphasize the fact that this individual whom we wish to educate, the conservator, must be capable of understanding, communicating, and interacting with persons having a remarkably broad range of interests, concerns, and abilities. We are asking a lot of this young professional. It is not necessary to belabor the lessons that might be drawn from this doodle, but apropos of my obligations today, let me point out that the scientist is not at the center of the scheme. Instead, he finds himself slightly off to one side in a supporting role. In addition, there are several types of such persons indicated, some more directly connected to the conservator's needs than others. That is one of the points

I should like to make: science is not monolithic; there are many ways of practicing the profession. Moreover, some practitioners are in a better position to be of direct assistance to the conservator than others. The scientist is but one member of a team; science, but one of an array of disciplines and skills that must be called upon.

How did science get into the field of conservation? It was not one thousand years ago that we set forth upon this voyage but practically within the memory of our own lifetime. The idea was born in the minds of a number of colleagues that the care of artistic and historic works need not continue slavishly to employ methods and materials that come into the conservator's hands primarily through tradition. Better adhesives, coatings, packaging materials, fungicides, and insecticides exist today than were available to our forefathers. There are better analytical tools at hand with which to determine the composition of craftsman's mediums and the causes of their disintegration. Should not these tools be used? If society is ever to change its way of doing things, one must try. Those who set forth upon uncharted waters, must try. The vision held by such men as F.I.G. Rawlins,[1] Edward Forbes,[2] George L. Stout,[3] and others of their generation was simply that, where a technical understanding of the materials of construction could help to preserve the physical evidences of the past, it was their, as it is our, duty to bring these skills and this knowledge to the task.

That's why science and technology were introduced: in the hope that useful benefits might come about. Note the use of the earlier word, assist; now, hope. There is no guarantee that doing things the scientific way—whatever that may mean to you—will necessarily make things better. In considering *the* role of science, then, let us retain some of the sense of humility that our forebears had in these matters, emphasizing once again that this discipline has been introduced into conservation in the same spirit that it had been introduced into agriculture, forestry, animal husbandry, and medicine: in the hope—perhaps faith, today—that an objective analysis of the way the physical world seems to work will help us do a better job and reach our objectives most efficiently.

[. . .]

What are some of the specific things that we hope science will do for the field of conservation? First of all, it can render a technical service; the tools of the chemist-analyst can be used to analyze or identify the paints, metals, and other materials of the artist and artisan. It is in the role as an analyst that the chemist's and physicist's contributions to the field of art and archeology are perhaps most familiar. Considering that the field is represented by three functions—examination, preservation, and restoration—it is fair to say that science has principally benefited the first of these through the introduction of such apparatus and techniques as the polarizing and the scanning electron microscopes, radiography, infrared sensing devices, and x-ray fluorescence analysis.

Summed up in Table I are a number of roles in which the scientist may serve. Individuals may be called upon for specific technical expertise in corrosion,

Table I
Some Roles of the Scientist

Analyst
Technical Specialist
Translator-Interpreter of Technical Matters
Teacher
Product and Process Tester-Evaluator
Research and Development Investigator

mineralogy, entomology, the chemistry of paper and leather, the metallurgy of copper alloys. One may also find the chemist and physicist in the role of a translator-interpreter of technical matters; the conservator and curator frequently have need for assistance in interpreting literature that describes the results of chemical analyses or the possible uses and potential hazards of new materials. There is, of course, the role as teacher, imparting something of his methods, special understanding, and way of looking at phenomena. The scientist is also employed to test, compare, and evaluate products and processes. Last, but not least, there is the role as an investigator in research and development. For the reasons of insufficient time, money, and manpower, the field of conservation has yet to take full advantage of the scientist in this latter capacity as we shall surely see in the coming generation. If we agree that science was brought into this field not only to help in the examination of historic and artistic works but also to assist in doing a better job of preservation and restoration, we must learn to be patient. The latter objectives are challenging and the assignment formidable. Industrially- or academically-trained colleagues, whom we invite from time to time to assist in this effort, are not always cognizant of just how challenging the parameters of the task really are. The conservation scientist is not interested in introducing new methods of bleaching, new methods of lining, new methods of varnishing. Not at all. Rather, he is interested in introducing new and improved methods of bleaching, lining, and varnishing (a) *that will not lead to disastrously harmful results during the next generation and* (b) *that will not result in more difficulties in analysis, preservation, and restoration than are presently encountered in the use of traditional materials*. These are the constraints that make the attainment of acceptable solutions elusive.

Note to Introduction

1 Barbara Appelbaum, *Conservation Treatment Methodology* (London: Butterworth Heinemann, 2007), 42.

Notes to Reading

1 [Francis Ian Gregory Rawlins (1895–1969) was the first scientific adviser and deputy keeper of the National Gallery (London) Laboratory, founded in 1934.—Ed.]

2 [Edward Waldo Forbes (1873–1969), art historian and first director of the Fogg Art Museum of Harvard University, is best known for his hands-on "egg and plaster" course on the materials and techniques of Italian painting, a precursor to courses now offered by many graduate programs in art conservation.—Ed.]

3 [George Leslie Stout (1897–1978) was the first conservator at the Fogg Art Museum and later director of the Isabella Stewart Gardner Museum, Boston. After World War II he also served as a "monuments man," rescuing looted artworks.—Ed.]

A CENTURY
PUBLISHED
CHARLES ROGERS
F.R.S. & S.A.L.
VOL. II.

PART VI

Deterioration and Change: Media "Even in Their Partial Ruin Marvelous"

That works of art deteriorate in substance and change in appearance over time is generally acknowledged. Thanks to published information that describes these aging properties, art historians are able to more accurately interpret artistic trends with fewer misconceptions about prevailing workshop practices or color palettes. For example, by applying data obtained from the use of intense microbeams of light, it is now possible to extrapolate an artwork's original appearance using predictable fading rates.[1] Conservators evaluate the visual manifestations of change when distinguishing between artist intent, evidence of past use, natural aging, or accident. More precise identification of internally and externally generated causes of deterioration has allowed them to devise strategies to prevent or at least slow down degradation.

Part VI attempts to put into historical perspective the emerging awareness of and response to two media-related conservation issues: the sensitivity to light of watercolors and the chemical instability of iron-gall ink. Viewed as case studies, the readings follow a predictable pattern: the "problem" is initially noticed by someone with a special interest in the matter, for example, a lawyer who needs to enforce an indecipherable deed (reading 71) or a speculative collector whose watercolors fetch less at auction because of rumors of their fading tendencies, as apparently occurred at Christies London in 1886.[2] Invariably, the outcry, often occurring in a public forum, is followed by a knee-jerk reaction from another vested party, typically the manufacturer. Blame is then assigned. Historically, for artists' materials in general and watercolors specifically, blame was placed squarely on the shoulders of the artist, neatly sidestepping accountability on the part of the outraged patron or the profit-driven manufacturer (reading 68). Likewise, in the case of iron-gall ink, Walter S. Landis (reading 74) and an anonymous attorney (reading 71) suggest that inks should be ranked according to their aging properties. Whether artist or writer, the watchwords are *caveat emptor*.

Charles Rogers (1711–1784), *A Collection of Prints in Imitation of Drawings. To Which Are Annexed Lives of Their Authors with Explanatory and Critical Notes*, vol. 2, Frontispiece/Title page, "Time with great fury thrusts his destroying scythe into the midst of a heap of drawings…." London, Printed by J. Nichols, sold by J. Boydell, 1778. The Pierpont Morgan Library, New York.

While the conservation problems pertaining to watercolors and iron-gall ink are significant, it is the aftermath of the controversies about their permanence that has had the greatest impact on stakeholders: collectors, manufacturers, artists, and of course conservators. Whether investigations are conducted using the time-honored experimental method of Russell and Abney (reading 67), the instrumental analysis approach of current iron-gall ink research, or the DIY testing proposed by Sidebotham (reading 66), Ostwald (reading 69), and Davids (reading 70), the responsibility of how to eliminate or retard deterioration has become more equitably distributed. Conservators and collection care professionals have assumed prominent roles in the long-term preservation of works on paper.

Notes

"even in their partial ruin marvelous": John Ruskin, annotator, *Notes by Mr. Ruskin on His Drawings by the Late J. M. W. Turner, R.A. . . .* (London: Fine Arts Society, March 1878), 9.

1 Paul M. Whitmore, Xun Pan, and Catherine Bailie, "Predicting the Fading of Objects: Identification of Fugitive Colorants through Direct Nondestructive Lightfastness Measurements," *Journal of the American Institute for Conservation* 38 (1999): 395–409. See reading 62; also see Leo Jansen, ed., *Van Gogh's Studio Practice* (Amsterdam: Mercatorfonds, 2013).

2 Norman Brommelle, "The Russell and Abney Report on the Action of Light on Water Colours," *Studies in Conservation* 9, no. 4 (1964): 141.

Reading 65

Marjorie B. Cohn

Wash and Gouache: A Study of the Development of the Materials of Watercolor (1977)

This reading places in historical context the growing realization in the nineteenth century that watercolors were susceptible to fading when displayed alongside oil paintings in sun-filled galleries. It also serves as an example of the pattern of complaint (a collector feels betrayed by the impermanence of his or her prized possession), response (typically caveat emptor from the manufacturer), and eventually scientific investigation. Similar controversies continued throughout the twentieth century, also with positive trickle-down effects. Just one example is the extreme light sensitivity of digital prints made with brilliant transparent dye-based inks applied by ink-jet printers invented in 1976. In this instance, research conducted by Henry Wilhelm (1943–) resulted in improved ink formulations and the concurrent development of strict guidelines for the display of color digital prints.[1]

Wash and Gouache *is an example of the rewards of focused attention on the historical evolution and use of a specific material or technique, a research approach spearheaded by Marjorie Cohn (1939–).*[2] *That she is represented by four readings in this volume is a testament to her substantial contribution to conservation and art history.*

The earliest concern for the safety of watercolors, however, the concern that would prove most fruitful in the development of standards for the safe exhibition of many forms of art, arose from an awareness that light has a destructive effect on the materials of watercolor. As early as the 1820s warnings were given that the

---

From Marjorie B. Cohn, *Wash and Gouache: A Study of the Development of the Materials of Watercolor* (Cambridge, Mass.: Center for Conservation and Technical Studies, Fogg Art Museum, and Foundation of the American Institute of Conservation, 1977), 61–64. Reproduced by permission of the Harvard Art Museums.

growing fashion for watercolor exhibitions would damage paintings already on view and would require new standards for the materials of the art:

> the tints used for glazing are almost entirely composed of fugitive pigments . . . without the aid of which, in spite of all that is said to the contrary,[1] those brilliant and richly harmonious combinations cannot be produced. Hence, landscape painters must be content to study for the portfolio[2] . . . unless . . . they would forego the temptation of present applause for a more lasting though less brilliant fame, by confining themselves to . . . pigments less liable to change.[3]

Samples of pigments cited as fugitive in 1824 have been preserved unfaded in manuals such as those by Perrot (1834) and Phillips (1839) and are displayed in this exhibition. Gamboge, Indigo, Brown-pink, and above all Carmine bear witness to the fragile glories of the traditional watercolor palette.

Early warnings generally went unheeded in the rush to exhibit watercolors as freely as if they were as light-resistant as oil paintings.[4] The public was encouraged to value watercolors as highly as oils in every sense, so that by the 1880s, when the permanence of watercolor paintings became such a controversial question that a government commission was appointed to resolve the dispute, "the effect . . . of the scare was very distinctly felt, both by artists and dealers, in the increased difficulty with which they were able to dispose of water-colour drawings."[5]

The controversy did not arise overnight, however, nor had there been a sixty-year lapse in attempts to counter the dangers of fugitive pigments. Palliative suggestions were advanced throughout the nineteenth century, in particular the recommendation that light-sensitive colors be heavily gummed to retard fading.[6] In the main, however, the emphasis lay on the development of light-fast pigments, although in 1820 Francis Nicholson[7] expressed a strong and apparently well-founded distrust of newly introduced colors from most artists' suppliers:

> instead of discovering new and fine colours, of which we already have more than we want, the attention of the colour manufacturer [should be] directed to the means of making those we have as true and permanent as possible . . . discoveries having been less applied to the improvement of the substances . . . than to the means of making them by cheaper . . . processes, and . . . to the adulteration of such of them as are capable of affording . . . an increase in profit.[8]

His suspicions were both confirmed and countered by the greatest colorman of his time, George Field. Field was a color theorist, chemist, engineer, and manufacturer so obsessed with chromatic character that he saw in pigments a metaphor of sexual identity[9] and in the three primary colors an image of the Holy Trinity. Red, blue, and yellow united in white light were "symbols of Himself . . . AFTER THE PATTERN OF HIS OWN PERFECTION."[10] Field was also a very practical pioneer in the development of pigments and manufacturing apparatus;[11] and in his monumental

Chromatography of 1835 he published the definitive results of his longtime concern for the permanency of artists' colors.[12]

Field conducted various tests, which in general condemned the natural organic pigments traditionally used in watercolor washing; his verdict on Brown-pink: "Upon the whole, it is more beautiful than eligible [as a permanent pigment]."[13] He was also perhaps the first to complain in print about the survival of archaic pigment names: "Pinks are sufficiently absurd names of yellow colours."[14]

Field was particularly concerned for the relative permanence of pigments, recognizing that the absolute alteration of hues was not so significant as their alteration in relation to each other. This was not a new idea among artists; Hogarth commented upon the problem in 1753 in his widely read *The Analysis of Beauty*,[15] and Nicholson's discussion of the relative fading of the components of neutral tint has already been quoted. Field, however, devised and reported upon practical experiments to demonstrate the problem,[16] and he systematized on a scientific basis the widely recognized incompatibility of many pigments with each other: "colors, like characters, suffer contamination and disrepute from bad association."[17]

To conclude his *Chromatography* Field published lists of pigments, which were widely reproduced or referred to in nineteenth-century manuals.[18] His tables itemized permanent pigments, pigments that suffered from the action of various agents, and pigments that formed incompatible mixtures. In Field's eyes these tables justified his prefatory assertion that:

> pigments, and fine ones too, so abound, that nearly as much experience is requisite to a judicious selection . . . as was formerly required for their acquisition or production . . . respectable colourmen . . . purvey, regardless of necessary expense, the choicest and most perfect materials . . . so that the odium of employing bad articles attaches to the artist, if he resort to vicious sources or employ his means improperly.[19]

Thus the battle lines of responsibility for permanent painting were drawn; all-out war would erupt in fifty years.

In the interim, the proliferation of colors continued. Field had noted that "some sort of criterion of the durability and changes in colour in pigments [was] that time and fire produce similar effects. . . ."[20] In general it became the accepted rule that substances formed by fire, that is, by oxidization, would be relatively permanent to the effects of "sunshine; whilst those which are organic in their origin are peculiarly liable to its influence."[21] All manner of exotic earths and rare metals were investigated, so that by 1869 the editor of a new edition of *Chromatography* would complain that

> Strange preparations have been offered as pigments. . . . Some . . . do not retain their colours on drying; others [require] processes too nice, complicated, or

> expensive. . . . There are many colours . . . which exist only on paper. We have too often found the imaginations of chemical writers far more vivid than the colours they describe.[22]

Many of the new compounds reached the market, however, to the initial consternation of a public unaccustomed to the intensity of pure mineral pigments in watercolors. Turner in particular was criticized for his brilliant color combinations, especially when eyes accustomed to the reproduction of his paintings in black-and-white engravings first saw the gaudy harmonies of the watercolors he prepared as models for the engraver [. . .].[23] Turner's paintings inspired the judgement in the 1820s that "the powerful attraction of colours . . . is the vice of our modern school."[24]

Rapid development of mineral pigments in the first half of the nineteenth century was matched by continuing investigations of the chemistry of organic coloring substances. Some of the traditional organic pigments were refined and synthesized,[25] and entirely new dyes were created from coal tars, byproducts of the distillation of bituminous coal into coke, illuminating gas, and other substances required by England's burgeoning industrial and urban centers. Unfortunately, all of these so-called aniline[26] dyes, with the exception of Alizarin,[27] were shortly discovered to be fugitive in light. Their great beauty and low cost encouraged their continued use by unscrupulous colormen, however, so that manuals recommended that artists test their colors for the presence of analine dyes as adulterants.[28]

The notorious impermanence of organic aniline colors retarded the development of other synthetic organic pigments for decades,[29] but today new organic compounds are among the most permanent watercolor paints. For the most part, however, for the convenience and, perhaps, the peace of mind of artists, colormen have christened synthetic organics with the names of the traditional substances that they replaced.

By the 1880s the proliferation not only of actual colored substances but also of names for any given substance was such that F. W. Devoe[30] listed 156 names for oil colors, including seventeen that included "madder" and thirty-seven that were "ancient traditional names that according to authoritative texts were either no longer in use as the actual materials or available at that time."[31] Manual authors attempted to simplify the problem of pigment selection for the bewildered watercolorist by recommending "a working 'daily-bread' colour-box" complemented by "little 'private' or 'freak' colour-boxes,"[32] such as one reserved specifically for sky-painting that contained no less than twenty-two separate named colors. Some of these were virtual chemical duplicates (e.g., Purple madder and Alizarin purple madder).[33]

The simmering controversy broke out afresh in the 1860s. In 1868 the permanency of watercolors required justification to a new market.

> The practice of water-color . . . and the public exhibition and sales of water-color painting, are comparatively new in America; and there are a good many people who . . . ask anxiously, "Are the colors permanent?" . . . a piece of paper, though protected by a glass and a wooden back [!][34] seems to them, somehow, to possess less durability than canvas. . . . it is our purpose to show that all such prejudices are unfounded. . . .[35]

In 1869 Winsor & Newton's[36] color chemist rebutted critics by admitting that the palette "wants weeding; not only of the bad new colours, but of the bad old colours. This, however, must be a work of time, and depend, not on the colorman—for where there is a demand there will be a supply—but upon the artists themselves."[37] In 1878 many manuals were harshly criticized as "too evidently published in the interest of the trade; while many others recommend villainous pigments. . . ."[38] In 1880 in a speech before the Society of Arts, William Holman Hunt recalled his and Millais's absolute faith as students in 1848 in George Field.[39]

> That ours was but a blind reliance was proved after George Field's death, when some of the vermilions supplied in imitation of his blackened after a short time. . . .[40]

> It transpired that the producers of colours were no longer . . . superintending all their preparations personally; [they] had been supplanted by the proprietors of large factories, where each production goes through numerous irresponsible hands.[41]

> Academies give no such attention to material matters, and accordingly the student . . . cultivates the practice of art in ignorance of the nature of the materials he uses. . . . To the modern painter the only difference between one paint and its fellow is that one is a bright and another is a dull colour. . . .[42]

Finally in the Spring of 1886 the storm broke in the pages of *The Times* to an astonished and outraged public. Affection for the English old masters of watercolor and investments in their paintings were threatened by the disclosure that the colors were liable to fade during prolonged exhibition. In 1878 Ruskin[43] had admitted in print that Turner watercolors of the 1830s had deteriorated badly,[44] but in a letter of April 1886 he beguiled the nervous collectors reading their *Times*:

> Properly taken care of—as a well-educated man takes care also of his books and furniture—a water colour drawing is safe for centuries; out

> of direct sunlight it will show no failing on your room-wall till you need it no more. . . . is it for your heir that you buy your horses or lay out your garden?[45]

The British government, which had "continuously exhibited in full daylight for twenty or thirty years past"[46] its collection of master watercolors in the South Kensington Museum[47] was forced to answer "yes" to the sense of Ruskin's question. The Department of Art and Science commissioned Dr. Walter J. Russell, a chemist, and Capt. William de W. Abney,[48] a painter, to investigate the permanence of watercolors. In 1888 they published their *Report . . . on the Action of Light on Water Colours*, which was quickly acclaimed by all the leading authorities on the medium.[49] Although refined and supplemented over the years,[50] the validity of its methods and findings has been repeatedly reaffirmed.[51]

Notes to Introduction

1 www.wilhelm-research.com.

2 Other examples by Cohn include her research on Ingres's use of graphite ("Technical Appendix," in *Ingres Centennial Exhibition, 1867–1967: Drawings, Watercolors and Oil Sketches from American Collections, Fogg Art Museum, Harvard University, February 12–April 19, 1967* [Cambridge, Mass.: Fogg Art Museum, Harvard University, 1967], 241–49) and the brown inks favored by father and son Tiepolo (see reading 21).

Notes to Reading

1 In the same year, 1814, the Old Water-colour Society proclaimed, "when this art first began to develop new and extensive powers, the prejudices which probably originated in a contempt of its ancient feebleness, degenerated into a species of hostility . . . an opinion was industriously spread abroad, that these qualities were evanescent and the material on which these works were executed, so frail) and perishable, that the talents of the artists were rendered useless by the ephemeral nature of his productions . . . but no philosophical reasons ever were, or ever could be, adduced against the possibility of producing by means of water-colours, pictures equal in . . . permanency . . . to those executed in oil" (Roget, I, p. 417).

2 [Prior to the growing interior design trend of the late eighteenth century for the display of framed prints and drawings in domestic interiors, works on paper were generally stored in albums and portfolios and thus protected from excessive light.—Ed.]

3 Roget, I, 442, quoting an anonymous critic of 1824.

4 Oil paintings have often been executed in the same light-sensitive pigments as watercolors, but their colors are somewhat protected by yellowing films of binders and varnishes. Also, the use of the most fugitive organic pigments was discontinued in oil painting long before watercolorists, who sought the greatest possible transparency, would relinquish such pigments as Carmine and Indigo.

5 Taylor, p. 2.

6 Field, *Rudiments*, p. 129.

7 [Francis Nicholson (1753–1844) was an English artist and founding member of the Society of Painters in Water-Colours.—Ed.]

8 Nicholson, p. 48.

9 George Field, *Chromatics*; or, *An Essay on the Analogy and Harmorry of Colours* (London, 1817), pp. 56–57: ". . . the many requisites of a perfect pigment [are] never wholly united in the same substance: hence it is generally necessary to employ two pigments of the same colours if you would produce the fullest effect. . . . delicacy and depth in the beauty of colours are at variance in all pigments. . . . they are the male and female in beauty."

10 Ibid., p. 57.

11 In 1816 Field was awarded a gold medal for his "apparatus for preparing coloured lakes" (R. B. Beckett, ed., *John Constable's Correspondence*, 6 vols. [London: H.M.S.O. and Suffolk Records Society, 1966], IV, 171).

12 Colormen such as Ackermann and Winsor and artists such as Constable were subscribers to *Chromatography*. Field also sent a copy to Turner, "who did not think that colours were reducible to scientific rules and pronounced the book to be fallacious" (ibid., p. 171).

13 Field, *Chromatography*, p. 142.

14 Ibid., p. 84. [George Field (1777?–1854), like Munsell (reading 38), expressed his frustration with meaningless names of colors, specifically "pink." According to *Color: The Universal Language and Dictionary of Names*, English Pink is equivalent to Stil de Grain Yellow, which is akin to sap green.—Ed.]

15 William Hogarth, *The Analysis of Beauty* (London, 1753), p. 119, n. I. [William Hogarth (1697–1764) was an English painter and printmaker. Cohn is referring to the following quote from Hogarth's *The Analysis of Beauty*: "Therefore, how is it possible that such different materials, ever variously changing (visible after a certain time) should accidentally coincide with the artist's intention, and bring about the greater harmony of the piece, when it is manifestly contrary to their nature, for do we not see in most collections that much time disunites, untunes, blackens, and by degrees destroys even the best preserved pictures" (119).—Ed.]

16 Field reported an experiment that determined the relative discoloration on exposure of natural and artificial Ultramarine, Cobalt blue, and Prussian blue. Interestingly, he did not include Indigo, perhaps because he already felt certain of its impermanence (Field, *Chromatography*, pp. 110, 112–113).

17 Field, *Rudiments*, p. 41.

18 See, for example, Barnard, p. 10, and Cassagoe, p. 67. One nineteenth-century manual even recommended that the artist compensate for anticipated fading: "The upper lip being almost always in shadow is both darker, and less intense, and less bright in colour, than the lower lip; but as these reds will lose somewhat by time, they may be painted a little brighter in colour than they are in nature, to allow for this loss" (Day, p. 33).

19 Field, *Chromatography*, p. xi.

20 Ibid., p. 44.

21 Taylor, p. 10.

22 Thomas W. Salter, ed., *Field's Chromatography*, p. 115.

23 Finberg, *Life of Turner*, p. 290.

24 Ibid., p. 279.

25 The coloring matter of Indigo was isolated in 1826; the coloring matter of Madder was synthesized in 1868.

26 Aniline was named after the Romance languages' (ultimately the Sanskrit) word for Indigo, the first organic pigment to yield its coloring matter to chemical analysis.

27 [Alizarin is a natural red dye derived from madder; it was chemically synthesized in 1868 from coal tar.—Ed.]
28 Hiler, p. 95. Tests for many different kinds of adulterants were as common in nineteenth-century pigment handbooks as instructions for pigment preparation were in eighteenth-century manuals.
29 Henry W. Levison, *Artists' Pigments: Lightfastness Tests and Ratings*, p. 15. Levison presents a summary of the history and current status of permanent pigments; he has been a pioneer in the development of modem standards.
30 [F. W. Devoe and Company was an American manufacturer and purveyor of artists' materials from the late eighteenth to the early nineteenth century.—Ed.]
31 Ibid., p. 16.
32 Brown, p. 90.
33 Ibid., p. 73.
34 [The bracketed exclamation mark was inserted by Cohn. Wooden backs, often cedar shakes, cause significant damage to paper because of their acidity.—Ed.]
35 Albert Fitch Bellows, *Water-Color Painting*, pp. 8–9.
36 [Founded in London in 1832, Winsor & Newton remains a leading manufacturer of artists' materials.—Ed.]
37 Salter, p. 44.
38 Rollin, p. 3.
39 [William Holman Hunt (1827–1910) and Sir John Everett Millais (1829–1896) were English Pre-Raphaelite painters.—Ed.]
40 Hunt, II, 455.
41 Ibid., p. 454.
42 Ibid., pp. 452–453.
43 [John Ruskin (1819–1900), the English art critic, artist, and social commentator, bequeathed his collection of Turner watercolors to the nation with recommendations to prevent their fading from overexposure: "Each drawing with its own golden case and closing doors—with guardians in every room to see that these were always closed when no one was looking at *that* picture" (Joan Evans, *John Ruskin* [New York: Haskell House, 1952], 177). The frame for Turner's *The Falls of Terni* still exists and is fitted with a pull-down roller blind (Ann Sumner, *Ruskin and the English Watercolour from Turner to the Pre-Raphaelites* [Manchester: Whitworth Art Gallery, University of Manchester, 1989], 81).—Ed.]
44 Ruskin found the faded Turners (all from his own collection) "even in their partial ruin marvelous" and thought that their "colours . . . changed so harmoniously that I find in their faintness more to discover through mystery than to surrender as lost" (John Ruskin, annotator, *Notes by Mr. Ruskin on His Drawings by the Late J. M. W. Turner, R.A. . . .* (London: Fine Arts Society, March 1878), pp. 9; 29, no. 26.
45 Taylor, n.p. (prefatory page).
46 Sir William Armstrong, "The Permanency of Water Colours," *The Art Journal*, 1886, p. 253.
47 [The South Kensington Museum was officially opened by Queen Victoria in 1857 and was renamed the Victoria and Albert Museum on May 17, 1899, with the laying of the foundation stone for its expansion along Cromwell Road in London.—Ed.]
48 [On Russell and Abney, see reading 67.—Ed.]
49 See, for example, Roget, II, p. 128, and Redgrave, p. 254.
50 See, for example, *Burlington Fine Arts Club, Reports I. & II. Of a Sub-Committee Appointed to Test Certain Methods Devised for the Preservation of Drawings in Water*

Colour (London, 1895); J. A. MacInryre and H. Buckley, "The Fading of Water-Colour Pictures," *The Burlington Magazine* LVII (1930), 31–37; and Garry Thomson, "A New Look at Colour Rendering, Level of Illumination, and Protection from Ultraviolet Radiation in Museum Lighting," *Studies in Conservation* VI (1961), 49–70.

51 A summary of Russell and Abney's report, more accessible than the original publication, is available in Norman S. Bromelle, "The Russell and Abney Report on the Action of Light on Water Colours."

References

Barnard, George. *The Theory and Practice of Landscape Painting in Watercolours*. London, 1858.

Bellows, Albert Fitch. *Water-Color Painting: Some Facts and Authorities in Relation to Its Durability*. New York, 1868. Listed in the New York Public Library Union Catalogue under New York City. American Watercolor Society.

Bromelle, Norman S. "The Russell and Abney Report on the Action of Light on Water Colours." *Studies in Conservation* IX (1964), 140–152.

Brown, J. Hullah. *Water-colour Guidance for the Student, the Amateur, and the Occasional Colourist*. London: A. & C. Black, 1931.

Burlington Fine Arts Club. *Reports I & II of a Sub-Committee Appointed to Test Certain Methods Devised for the Preservation of Drawings in Water Colour*. London, 1895.

Cassagne, Armand Théophile. *Traité d'Aquarelle*. Paris, 1875. English and French watercolor paper samples.

Day, Clarence William. *The Art of Miniature Painting*. London, 1853. Winsor & Newton Co. catalogue.

Field, George. *Chromatography; or, a Treatise on Colours and Pigments, and of Their Powers in Painting, etc*. London, 1835. Wash samples.

Field, George. *Rudiments of the Painter's Art; or, a Grammar of Colouring . . .* London, 1850.

Finberg, A.J. *Life of J. M. W. Turner, R.A.* Oxford: Clarendon, 1939.

Hiler, Hilaire. *Notes on the Technique of Painting*. London: Faber & Faber, 1934.

Hunt, William Holman. *Pre-Raphaelites and the Pre-Rraphaelite Brotherhood*, 2 vols. New York: MacMillan, 1904.

Levison. Henry W. *Artists' Pigments: Lightfastness Tests and Ratings*. Hallandale, Fla.: Colorlab, 1976.

Nicholson, Francis. *The Practice of Drawing and Painting from Nature in Water Colours*. London, 1820.

Redgrave, Gilbert Richard. *A History of Watercolour Painting in England*. London, 1892.

Roget, John Lewis. *A History of the "Old Watercolour" Society*. 2. vols. London, 1891.

Rollin, Horace J. *Studio, Field, and Gallery*. New York, 1878.

Ruskin, John. *The Elements of Drawing*. London, 1857.

Salter, Thomas W., ed. *Field's Chromatography; or, Treatise on Colours and Pigments as Used by Artists*. London, n.d. [c. 1869].

Taylor, J. Scott. *A Descriptive Handbook of Modern Water Colours*. 2nd ed. London, 1887. Winsor & Newton Co. wash samples.

Reading 66

JOSEPH SIDEBOTHAM

Letters to the Manchester Literary and Philosophical Society (1877, 1881)

Up to the mid-nineteenth century, an artist's palette consisted of colors derived from minerals, plants, insects, and a few animal by-products. While a great deal of effort went into refining products intended for artistic purposes, color chemists were restricted to the technology and substances at hand. Thus some colors were difficult if not impossible to formulate or were prohibitively expensive. But in 1856, an eighteen-year-old chemistry student named William Perkin discovered a way of making never-before-seen colors easily and cheaply.[1] *Working with coal tar, a viscous, black liquid produced from the distillation of bituminous coal, Perkin discovered aniline purple while attempting to synthesize quinine for the treatment of malaria. Instead of quinine, he produced a dark oily sludge that, with further experimentation, dyed silk an unusual shade of purple. The color, called mauveine or mauve, took the fashion houses of London and Paris by storm. Unlocking the chemistry of mauve quickly led to other new colors—crimsons, blues, and greens—and earned their inventor a fortune, but its importance extended far beyond fashion. Within a decade, Perkin's dyes were being used as artists' colorants.*

Almost as quickly as the new aniline colors multiplied, disquiet arose over how swiftly they faded. Amateur photographer and natural scientist Joseph Sidebotham (1824–1885) voiced his concerns in a letter to the Manchester Literary and Philosophical Society. He observed that the popular transparent colors used to tint photographs were being taken up by watercolorists. Invoking the higher authority of science to sound the alarm, Sidebotham found his efforts at reform frustrated. Three years later, he again reported with great agitation that the colors were being used with poor results. As proof he submitted a light-exposure test sheet demonstrating that even

From JOSEPH SIDEBOTHAM, "Letter to the Manchester Literary and Philosophical Society," *Quarterly Journal of Science*, no. 54 (April 1877): 287; "Letter to the Manchester Literary and Philosophical Society," *Chemical News* 44, no. 1153 (December 30, 1881): 313–14.

gum-coated colors faded within a two-week period, contradicting the prevailing belief (see reading 65). Sidebotham proposed that artists boycott dye-based colors, thus forcing their discontinuation. The manufacturer Winsor & Newton on the other hand supported the consumer's freedom of choice (reading 68).

[. . .]

In a letter written from Mentone[1] to the Manchester Literary and Philosophical Society,[2] Mr. Joseph Sidebotham, F.R.A.S., says:—"My attention has been for some time directed to the growing use of the aniline colours for tinting photographs. Now I find they are being *extensively* used in paintings and water-colour drawings, and the colours regularly sold for that purpose. Anyone who knows the speedy alteration by light of nearly all of these colours will protest against their use, and a statement of this with the authority of some of our chemists would probably have the effect of causing them to be discontinued by all artists who care to think that their works should last more than a single year."

* * * *

Ordinary Meeting, November 29, 1881. R. Angus Smith Ph.D., F.R.S., &C, in the Chair.

The following letter from Mr. Joseph Sidebotham, F.R.A.S., was read:—

"Some few years ago, when I was at Mentone, I wrote to you concerning the aniline colours, so much used on the Continent in water-colour drawings. Since that time Holman Hunt[3] and others have called attention to the fading of these and some other colours prepared for artists and I hoped the practice of making and using these colours had ceased: however, I find this is not the case.

A friend of mine was on his way to the south of France, and I asked him to see if these colours were still sold and used, and he tells me they are extensively, and sent me cakes of them. He also sent the enclosed sheet of the colours on drawing-paper, half of it having been exposed to the light a fortnight, the other half covered up. He thought that by putting gum over the colours they might be made more permanent, and you will see he has put a band of gum across them: the colours exposed have faded in a great degree, some of them almost disappearing. The band of gum has retarded the fading, but the colours are even there much lighter and all the brightness gone. It is most desirable that artists should entirely give up the use of all these colours, and then the makers would cease to supply them. When we see the sad effects of a fortnight's light upon them, what can we expect to see in drawings hung on the walls of a room for a few years?"

Note to Introduction

1 For an account of Perkin's discovery, see Simon Garfield, *Mauve: How One Man Invented a Color That Changed the World* (New York: Norton, 2001).

Notes to Reading

1 [Sidebotham was staying in the Hotel de Menton in Menton, France, where he also reported on a species of singing mice.—Ed.]
2 [Founded in 1781, the Manchester Literary and Philosophical Society is one of the oldest learned societies in England.—Ed.]
3 [William Holman Hunt (1827–1910), English Pre-Raphaelite painter.—Ed.]

Reading 67

W. J. Russell and W. de W. Abney

Report to the Science and Art Department of the Committee of Council on Education on the Action of Light on Water Colours (1888)

So divisive became the controversy over the fading tendencies of watercolors that Sir William de Wiveleslie Abney (1843–1920), a British civil servant and photographic scientist, and the chemist Walter James Russell (1830–1909) were commissioned to conduct an investigation in which the skylights of the South Kensington Museum, today the Victoria and Albert Museum, would substitute for an instrumental light source. Their Report . . . on the Action of Light on Water Colours *remains one of the seminal documents in preventive conservation today.*[1] *In it the authors established the tripod of variables upon which today's exhibition lighting standards are balanced: (1) the intensity of the visible light, (2) the duration of exposure, and (3) the proportionate wavelengths emitted by the light source. As for the first two factors, Russell and Abney countered the then-prevailing notion that diffused daylight, as would occur in a domestic interior, is harmless (reading 65), although there would be some quibbling about how much light should serve as valid experimental lighting conditions, as will be seen in reading 68. They confirmed that heat and moisture increase the rates of fading. The* Report *forecast the potential for anoxic enclosures to prevent fading and, related to the third variable listed above, the use of slightly yellow glass or film to filter out ultraviolet radiation.*

Russell and Abney described three artificial light sources to which watercolors were exposed at the time: gaslight, the electric arc lamp, and the incandescent lamp. Just as viewers today respond strongly to the quality of light emitted by compact fluorescent lamps, light-emitting diodes, and "natural" light, the Holy Grail for celebrity architects, Gordon Wigan in 1883 recounted that contemporary museumgoers also had definite preferences:

From W. J. Russell and W. de W. Abney, *Report to the Science and Art Department of the Committee of Council on Education on the Action of Light on Water Colours* (London: Her Majesty's Stationery Office, 1888), 12, 14, 17, 20, 46–47.

> *In considering how best to light our picture galleries by electrical means, we may well begin by considering which system gives the kind of light best suited to our purpose. I unhesitatingly say that the arc system is really the best from this point of view. I know that to most people—I have myself been subject to the illusion—the light from an arc lamp is hard, cold, and ghastly; objects look strange, women look like corpses, their dresses look hard, cold, and ill-arranged as to color. In truth the light from an arc lamp is not a ghastly blue, but a pure white; it cannot be distinguished from the light reflected from a white cloud in the north, the light of all others most liked by painters. It in no way alters the colours from the shade and tone which they have by day. . . . The reason for the prejudice against this beautiful light is not far to seek. We have all become accustomed to a very yellow form of artificial light, so that our brain automatically applies a correction to what we see whenever we know that we are looking at objects illuminated by artificial light; and this correction is still made with the arc light, even though it is no longer required.*[2] *After a few evenings spent in a building lighted by it the illusion disappears, and we are able to enjoy the delight of daylight by night. But until the public at large have been trained to like, instead of hate, the arc light, it would be extremely unwise to introduce it into a picture gallery.*

Wigan continued:

> *It would be well, if possible, to conceal the [incandescent] lamps themselves from view. The reason is, that as yet they are curiosities to most people, who delight in staring at them, and thus fatigue their eyes so much that if afterwards they looked at pictures they would probably see them covered with black shifting images of the carbon filaments used in the lights.*[3]

[. . .]

Estimation of chemical action in a body.

X. In estimating the chemical action effected on a body by radiation there are thus two factors which have to be taken into account, viz., the intensity of the radiation acting, and the time during which it acts. To obtain the same amount of action in two cases the product of these two amounts must be the same. Thus if a certain tint be exposed to an intensity of radiation which we will call 100 and bleaches it in, say, one hour, then if a similar tint be exposed to an intensity 1, it will require 100 hours' exposure to it to effect the same bleaching. This has been fully proved by experiment. There is an idea abroad that if the light be very feeble a bleachable colour, no matter what length of exposure be given, will not fade. This, however, is not the case. The same proportion of the total energy absorbed by the body which, with an intense radiation, effects chemical decomposition, is expended with a feeble radiation in doing the same kind of work. [. . .]

[. . .]

Light to which water colours are exposed.

XIII. As to the light to which pigments in water-colour drawings are ordinarily exposed in a room, a few remarks must be made. There is no doubt that pictures are, as a rule, carefully protected from direct sunlight, but it is nevertheless true that the greater portion of the light they receive is reflected sunlight. On a bright day clouds reflect sunlight, and on a dull day the principal part of the diffused light is also sunlight which is reflected according to the laws of geometrical optics from particle to particle, a certain percentage eventually reaching the earth through the clouds. There is of course also a fair proportion of the light due to the sky, and this light is bluer than reflected or diffused and weakened sunlight. In cases where the windows of a gallery are in the vertical walls and have an uninterrupted view of the horizon, the blue light reflected is comparatively small, the light near the horizon being distinctly more like sunlight than is that nearer the zenith. In galleries lighted like those at South Kensington by skylights the light to which pictures are subjected is on the whole bluer. The artificial lights to which water colours are exposed are gaslight, the arc, and incandescence electric lights, and as we shall see presently the first and last are very deficient in blue rays.

[. . .]

Results from one source of light can be applied to another source of light.

XVI. Since, then, all sources of light emit the same rays, but of different intensities, which can be measured, it follows that if we know which rays are chemically active, and the amount of work which, when of a certain intensity, they perform, we can, from the work done by the light from one source, deduce the work that would be done by another. The most perfect manner of noting the action of light would be to expose for a given time the pigments to the action of the spectrum formed by an unvarying source of light, and to measure the amount of chemical action (fading of the colour in most cases) which had taken place in every part of the spectrum. When the relative intensities of the different parts of the spectra from other sources of light compared with this standard spectrum were known, then the length of time during which it would be necessary to expose the colour to any one of them to produce that same total effect could be calculated. Unfortunately for our experiments, even in full sunlight, which is the most powerful light we can work with, months are often required to effect a visible chemical action on some of the pigments; it was therefore useless to take a narrow slice of light, say 1/10 inch in width, and 1/2 inch in height, and form a spectrum with it on a surface of coloured paper four inches in width and one-half inch in height, and wait to see when the bleaching took place.

Use of coloured glasses in the experiments.

XVII. To avoid this impracticable method resort was had to the use of coloured glasses to ascertain the part of the spectrum which was most active in producing the fading action. The glasses used were red, green, and blue. It may be well here to say something regarding what is meant by the terms red, green, blue, &c. as applied to glasses. It is a very popular idea that light coming through coloured glass is really white light, which is in some way transmuted into red, green, blue, &c., as the case may be. There is no transmutation. The effect of color is merely produced by the abstraction of certain rays, or of a proportion of them, from the white light by the material of which the glass itself is composed. [. . .]

[. . .]

Joint effect of heat and light.

XIX. It might very properly be objected that although it has been shown that the dark rays do not affect chemical decomposition, it has not been proved that the heating effect they have on a pigment might not aid the rapidity with which the decomposition takes place. Direct experiments were undertaken with this object in view. The backs of papers coloured with pigments which we have proved to be fugitive were placed in contact with a tin containing boiling water and exposed to light, together with similar papers merely resting against wood. Some few of those colours which are affected by heat without light in an atmosphere *saturated with moisture* [. . .] did fade with very slightly greater rapidity where exposed as above, but with the majority the rate of change was, if anything, slower. Further experiment has also shown that if the dark rays be cut off from sunlight by proper means the rate of fading in colours freely exposed is not diminished. In our experiments in the open tubes which will be described presently, the temperature was only a very few degrees higher than the temperature of outside atmosphere, and therefore the experiment made by heating the pigmented paper by contact with a vessel at the temperature of boiling water was an extreme example of the effect of heat. A reference to the results of experiments shows that damp is often a factor in the rapidity of fading, and as heat tends to lessen the moisture present in the paper and pigment, it might be expected that in the majority of cases fading would result more slowly when the pigment was heated whether by radiation or by heat applied as above.

Deductions made from the [fact] that chemical changes take place in the blue rays.

XX. That fading should principally take place in the blue rays was to be expected, from experiments that have been conducted with other objects, and is of great practical importance. We have already stated that it is only those parts of the spectrum which are absorbed by a colour that can do work on it. Of all colours, the reds, yellows, and greens absorb principally in the blue part of the spectrum and the blues much less. Hence we may expect that the former pigments would fade more rapidly than the latter if they are fugitive. And what is more important, it locates the action of the spectrum to the region which is least luminous, and which varies enormously in the different kinds of light to which pigments are exposed. Of the different kinds of daylight, viz., sunlight and skylight, to which water-colours are exposed, sunlight is the safest when reduced to equal intensity, since it contains a far less proportion of blue light than does skylight.

Conclusions.

XVII. [. . .]

Mineral colours are far more stable than vegetable colours, and amongst those colours which have remained unaltered, or have only very slightly changed after an exposure to light of extreme severity, a good gamut is available to the water colour artist.

The presence of moisture and oxygen are in most cases essential for a change to be effected, even in the vegetable colours. The exclusion of moisture and of oxygen, particularly when the latter is in its active condition, as experiments to be described in our next report show, would give a much longer life, even to these, than they enjoy when freely exposed to the atmosphere of a room. It may be said that every pigment is permanent when exposed to light "in vacuo," and this indicates the direction in which experiments should be made for the preservation of water colour drawings.

The effect of light on a mixture of colours which have no direct chemical action on one another is that the unstable colour disappears and leaves the stable colour unaltered appreciably.

Our experiments also show that the rays which produce by far the greatest change in a pigment are the blue and violet components of white light, and that these, for equal illumination, predominate in light from the sky, whilst they are less in sunlight and in diffused cloud light, and are present in comparatively small proportion in the artificial lights usually employed in lighting a room or gallery.

The experiments have also shown that about a century of exposure would have to [be] given to water colour drawings in galleries lighted as are those at South Kensington before any very marked deterioration would be visible in them, if painted with any but the more fugitive colours; and that when the illumination is of the same quality as that of gaslight, or of the electric glow light rendered normally incandescent, and of the same intensity as that employed in those galleries, an exposure to be reckoned by thousands of years would have to be given to produce the same results. We have here not taken into account the action, if any, which might arise from the products of combustion where gaslight is the illuminant, and which our experiments so far have shown to have but a trifling effect, nor of any modification of hue which might be due to change in the whiteness of the paper on which the paintings were made, but simply to the change in the colours themselves. When it is determined what is the minimum illumination in which a water colour drawing can be well seen, the length of time during which the drawing will retain almost its pristine freshness can be easily calculated.

Since it is the blue light which causes the fading, it might be thought that the glazing of skylights with a glass of a slightly yellow tint should be adopted. It must be recollected, however, that in ordinary diffused sunlight this would entail an alteration in the brilliancy of the blues of a picture, and a change in their tone. It is well known to artists that a picture painted or illuminated by blue skylight looks colder than one illuminated by diffused sunlight, whether the diffusion be caused by white blinds to the gallery or by cloud. [. . .] When the illumination is from the blue sky a yellow glaze or blind might be useful in imparting the warmer tone of diffused sunlight, and at the same time it would reduce in intensity the destructive component of the light.

Notes to Introduction

1 Prior to Google Books, only a summary version of the *Report* was easily available: Norman Brommelle, "The Russell and Abney Report on the Action of Light on Water Colours," *Studies in Conservation* 9, no. 4 (1964): 140–45.

2 The color rendering index (CRI) of compact fluorescent lamps (CFL) and light emitting diodes (LED) is measured relative to still-familiar incandescent bulbs.

3 Gordon Wigan, "Electric Lighting for Picture Galleries," *Telegraphic Journal and Electrical Review* 13, no. 293 (1883): 2–3.

Reading 68

Winsor & Newton

Statement of Policy 1892 (1997)

The public's consternation about impermanent colors put relentless pressure on Winsor & Newton (W&N), the leading English manufacturer and purveyor of artists' materials, as suggested by two references to the overwhelming volume of private correspondence that the company hoped would finally cease with the publication of their Statement of Policy.[1] *Not surprisingly, W&N supports the artist's freedom to choose pigments known to be fugitive, neatly transferring the moral dilemma onto his or her shoulders. The collector remains blameless, since it is assumed that "no person who values a painting ever dreams of exposing it to the direct blaze of sunlight," a sentiment earlier expressed by John Ruskin (see reading 65). By disputing Russell and Abney's experimental methodology, W&N managed to distance themselves from its damaging findings.*[2] *Their contention that it is not valid to test materials using extreme conditions persists, as seen in Part V.*

Two criticisms are often levelled by thoughtless people at the heads of Artists' Colourmen, and of these we have decided, as one of the leading English firms, to take some practical notice. It is alleged—

1. That Artists' Colourmen are in the habit of selling Colours which are not permanent; and
2. That they keep Artists in ignorance of the Chemical Composition of the colours they sell.

The answer to the first of these two criticisms seems obvious enough. Manufacturng businesses exist, as a rule, not for the enforcement of moral laws on their customers, but for the satisfaction of the demands which those customers make;

From Winsor & Newton, "Statement of Policy 1892," in *Notes on the Composition and Permanence of Artists' Colours* (Middlesex, England: Winsor & Newton, 1997), 2–3.

and while, for instance, we continue to be asked for Carmine and Geranium Lake,[1] so long shall we continue to supply them. The Artist is, in our opinion, the sole judge of his right to employ such pigments, and we, who use our best efforts to supply him with all he requires, have no intention of excluding them from our list of manufactures. We do not assert that such colours are durable; all we do assert is that they are as durable as they can be made.

We have not hitherto deemed it incumbent on us to publish any information relating to the permanence of pigments. Instruction of this subject has always been attainable in Field's Chromatography,[2] published by us—a work which, in its various editions, has for more than half a century maintained its reputation as the standard authority on pigments. Lectures on the Chemistry of Painting have also long formed at the Royal Academy (as they should form everywhere) a part of the training of every art student. Indeed, we do not think it has ever been seriously suggested that it is the function of Artists' Colourmen to give the information to which we refer.

We propose, however, in the following pages, to classify colours in the order of their permanence, because answering questions on this topic has involved us in a considerable amount of correspondence; and because we shall thus, once and for all, dissipate the notion that we affirm all our colours to be durable.

The important question now presents itself: What experimental tests shall we take as a basis for the classification which we propose to issue? It may seem rather a naive idea for a manufacturer to publish his own tests of his own colours, but everybody who has had to deal with the subject knows that is really the only thing at present to be done. For, even supposing the tests hitherto conducted by scientific men to be, in respect of their applicability to the object we have in view, beyond criticism, there is (and can be for obvious reasons) no published guarantee that the colours upon which they experimented were identical in chemical structure or mode of preparation with those which any particular house is in the habit of selling.

We believe, however, it is a matter of opinion whether there are at present any investigations before the public which, with regard to their direct bearing on ordinary painting, and exclusive of scientific value in the abstract, can be considered satisfactory. The reason is that all experiments hitherto published have been based, practically without exception, on the method of exposure to direct sunshine. The value of these sunshine tests when used to prove anything more than the pros of permanence is a matter of the gravest doubt.

Adverse criticisms of the reasoning which, in this particular, constituted the only weak point in the otherwise admirable report of Dr Russell and Captain Abney have been endorsed by no less an authority on Light than Prof G.G. Stokes, the President of the Royal Society.[3] Prof Church, of the Royal Academy, has also thrown his vote in the opposing scale.[4] So that there seems to be now a more or less general consensus of opinion that pigments can be satisfactorily tested only by exposure to diffused daylight; and that sunshine exposures, tbough useful in their way as conferring special hall-marks of permanence, should be put out of court for being used as arguments against the durability of a colour. No person who values a painting ever dreams of exposing it to the direct blaze of sunlight, and, with the proviso already specified, no experimenter should therefore carry out his investigations under conditions other than those which obtain in the ordinary life history of a properly kept picture.

At the North London Colour Works careful testing of the durability of our colours has always gone hand in hand with colour manufacture, and it is by this method of daylight exposure that the pigments have always been tried. A well-lighted room with a northerly aspect was many years ago set apart for this purpose by the founders of the business, and from the results of experiments made in this manner we have arranged the classified lists.

The reply to Criticism No. 2—that Artists' Colourmen keep Artists in ignorance of the composition of their Colours—is no less obvious than that to the first. It is impossible in these days of analytical chemistry for the secrets of colour-making to consist to any great extent in the chemical composition of the colours. We, for our part, now attach little importance, as far as the success of business is concerned, to any concealment of the chemical composition of our colours and mediums; although we necessarily endeavour, as far as possible, to maintain the secrecy of those methods which long experience has taught us (as long experience alone can teach) are best adapted for the manufacture of the above-named commodities. We have, in fact, invariably been willing to divulge the composition of our pigments and vehicles to bona fide artists; but have, at the same time, been averse from taking any step which might result in rendering indirect assistance to enterprising competitors, enabling them thereby to reap the benefit of knowledge which we have ourselves often acquired only by much expenditure of time and capital.

Artists will no doubt see how difficult it has hitherto been for us to give them information which we wished to safeguard effectively from other sources of inquiry, and they will no doubt therefore also see the common sense of the conclusion to which our experience has forced us—the conclusion that we cannot treat artists in one way and the world at large in another; all must be treated alike. We have, therefore, decided to make a second new departure, and to obviate further misunderstanding as well as to lessen our private correspondence, by publishing a statement of the nature of our colours and vehicles.

Our position, broadly stated, is therefore this: That we make no concealment of the chemical structure of our pigments and mediums, but reserve to ourselves only those details which have reference to the manner in which these pigments and mediums are made. In plain words we reserve to ourselves no more than is necessary for any business conducted by private enterprise in these present times. From artists themselves we wish we could have no concealment at all; but only in a more ideal age can such things as this be possible. Nevertheless, with the Artists' Colourman in such a position as we have here endeavoured to define, the painter has little cause for regret. He is in possession, if not of all the interesting, yet of all the essential, information relating to the manufacture of those materials which, manifold in their variety and application, pass from his palette to constitute in his picture the physical basis of his craft. He knows, in fact, everything that may help him in compassing stability, as far as may be, for the complex fabric of his work; and there are no more earnest wishes than ours that success may attend him in his efforts to endow the fleeting aspects of Nature and the ephemeral creations of Imagination with the "living hues of Art".[5]

Notes to Introduction

1 The Winsor & Newton Archive of 19th Century Artists' Materials is now available as a searchable database hosted by the University of Cambridge, Hamilton-Kerr Institute: www.hki.fitzmuseum.cam.ac.uk/archives/wn/.

2 Most manufacturers of artists' materials today voluntarily disclose the relative permanence and composition of their colors. However, their standards of permanence often vary.

Notes to Reading

1 [Carmine is manufactured by precipitating a natural red dye derived from the *Dactylopius coccus* insect onto an inert particulate, usually alumina. Carmine topped Russell and Abney's list of the most light-sensitive colorants. Geranium Lake is precipitated from the red dye eosine, first synthesized in 1871.—Ed.]

2 [See reading 65.—Ed.]

3 [Sir George Gabriel Stokes (1819–1903) was president of the Royal Society, a fellowship of eminent scientists, when the Russell and Abney *Report* was published but had retired by the time of the Winsor & Newton Statement of Policy. As a public figure and authority on the physics of light, his opinion carried considerable weight.—Ed.]

4 [Arthur Herbert Church was professor of chemistry at the Royal Academy. Successive editions of his book, *The Chemistry of Paints and Paintings*, appeared in 1890, 1901, and 1915 (London: Seeley, Service and Co.).—Ed.]

5 ["living hues of Art": This quote is taken from the poem *Locksley Hall Sixty Years Later*, by Alfred Lord Tennyson, published in 1886. One wonders if the authors are aware that the poem expresses Tennyson's disillusionment with industrialization.—Ed.]

Reading 69

W. Ostwald

Letters to a Painter on the Theory and Practice of Painting (1904)

The Nobel Prize–winning chemist Wilhelm Ostwald (1853–1932) was not the first to use science to elucidate the properties of artists' materials, but unlike other authors he directed his attention to the rapidly growing cohort of enthusiastic self-taught artists.[1] *Guiding amateur artists in the selection of proper materials was a growing concern. Along with an increase of prepackaged traditional media and an influx of new ones of uncertain longevity came a widening distrust between the manufacturer and the consumer.*

When Letters *was first published in Germany in 1904, synthetic dyes were already incorporated into virtually all colored media used to create works of art: watercolors (see reading 65), oil paints, gouache, wax crayons, colored pencils, writing and drawing inks, pastels, and even the paper itself as "paper stainers" were mentioned as early consumers of the new aniline dyes. Notoriously light-sensitive "basic" dyes had a particular affinity for the acidic groundwood introduced into papermaking in the nineteenth century (see reading 73), while equally fugitive "fat" dyes were compatible with oil- and wax-based media. In Letter IV, presented below, Ostwald instructs his readers on how to test pastels for adulteration with synthetic dyes. His solvent test is not accurate, but it and the primitive accelerated-aging test using sunlight neatly foreshadow the ongoing efforts of the American Society for Testing and Materials (ASTM) working group D01.57 on W739, "Standard Specification for Artists' Pastels," as well as the disagreement surrounding the validity of artificial-aging testing, as explored in Part V.*

From W. Ostwald, *Letters to a Painter on the Theory and Practice of Painting*, trans. H. W. Morse (Boston: Ginn & Company, 1907), 35–36, 43.

My dear Sir:

You say that you have succeeded without any difficulty in making pastels according to my directions, but that you do not know what pigments to use in order to be sure of really durable results.

I will try to give you an answer which shall be as short and as definite as possible; but this will not end the matter, for the adulteration of pure pigments, by adding to them things which make them cheaper or more brilliant, is a very widespread evil.[1] There is, of course, no protection against unreliable colors, when one orders pigments which are stable and gets adulterated ones instead.

There is fortunately a fairly easy way of recognizing many adulterations, especially those "beautifiers" which consist of coal-tar colors.[2] Most of them are soluble in water or alcohol, while the pigments we want are not. Place some of the pigment which is to be tested for coal-tar derivatives (assuming that you have the raw pigment in powdered form) on several layers of white blotting or filter paper, piling the color up in a little heap with a hollow top, and drop water into this hollow until it runs through the pigment and into the paper. Then examine the paper, and you will easily recognize a soluble pigment if one is present, for it will be visible not only on the under side of the top sheet but also in the sheets below this. Make the same test with alcohol, and a third with a mixture of alcohol and a little aqua ammonia. If the liquid comes through uncolored in all these tests, you will be fairly safe in concluding that no coal-tar derivatives are present.[3] This conclusion is not an absolutely safe one, since many coal-tar colors, added in the form of insoluble "lakes,"[4] do not betray themselves under these conditions; here the chemical expert must decide.

[. . .]

Finally, let me give you the general method of testing pigments to be used in pastel, with respect to their stability under the action of light.[5] Cover a piece of pastel paper smoothly with the color in question, using a medium shade. Fix the color and expose it to the light so that half of it is protected. If you possess a photographic printing frame, place the colored strip in it, covering half of it with black paper or a folded piece of pasteboard, the paper lying half in and half out of this covering. If the test is made in summer in bright sunlight, an exposure of a few days will be sufficient to show whether the exposed part has changed its color compared with the part which is protected from the action of the light.

In making such a test you must avoid being deceived by a possible yellowing of the paper, for if this is made of poor material it may show very rapid changes as a result of the action of strong sunlight.[6]

Note to Introduction

1 I am grateful to Cynthia Connelly and Rebecca Ruderman for sharing their research on Wilhelm Ostwald.

Notes to Reading

1 [Adulteration of artists' materials was a widespread problem. See Leslie Carlyle, "Authenticity and Adulteration: What Materials Were 19th-Century Artists Really Using?," *The Conservator* 17 (1993): 56–60.—Ed.]

2 [Coal-tar colors are derived from distillates of coal. William Perkin synthesized the first coal-tar color, mauveine, in 1856.—Ed.]

3 [Ostwald's testing is unreliable, since the solubility of coal-tar colors varies. Many are not soluble in water; alcohol is the more broad-based diagnostic solvent. In addition, Ostwald's alcohol/ammonia solution would not detect several coal-tar colors. More effective would be an acid solution, which would dissolve basic aniline dyes. A. J. E. Brown, "The Use of Basic Dyes in Lithographic Inks and Their Influence on Conservation Treatments," in *Modern Works—Modern Problems?*, ed. Alison Richmond (n.p.: Institute for Paper Conservation, 1994), 6.—Ed.]

4 [A lake is defined as a pigment made by precipitating or fixing a dye on an inert colorless particle, typically aluminum oxide (alumina).—Ed.]

5 [Ostwald's testing method, using only a few days of exposure to sunlight, is less extreme than other methods but will indicate impermanence for some particularly fugitive pigments. Tests having similar methodologies are still performed by artists and conservators. See Mark D. Gottsegen, *The Painter's Handbook* (New York: Watson-Guptill, 2006), 140–53; see also Marion Boddy Evans, "How to Do a Simple Lightfastness Test for a Paint Color," http://painting.about.com/od/colourtheory/ht/lightfastness.htm.—Ed.]

6 [See reading 20 and Part V.—Ed.]

Reading 70

Thaddeus Davids

The History of Ink including Its Etymology, Chemistry, and Bibliography (1860)

An irrepressible American entrepreneur, Thaddeus Davids (1810–1894) made and lost a fortune in the cutthroat world of ink manufacture. Appended to his remarkable History of Ink, *in itself a stunning example of mid-nineteenth-century typographic art (fig. 1), are two press releases touting the results of an aging test conducted by James R. Chilton, a celebrated chemist whose analyses ranged from copper deposits, mattress stuffing made from processed moss, and lager beer to tinctures of sarsaparilla.*[1] *Davids seized upon a report published in* Scientific American *on April 12, 1856, about an aging test carried out by Dr. Chilton indicating that two of his inks had withstood months of direct weathering, a testing methodology Davids extolled as "ingenious."*[2] *A lithographic facsimile (fig. 2) of the aging test was promptly commissioned and widely distributed, even appearing on the obverse of a tradecard issued on the occasion of the 1876 Centennial celebration in Philadelphia. Dr. Chilton's favorable results were a critical component of an aggressive advertising campaign for Thaddeus Davids and Co. for the next fifty years.*[3]

Davids's claims reflect the public's growing concern about the permanence of writing ink, a ubiquitous medium for both writers and artists. Like those of watercolors, the conservation problems of writing ink—its tendency to fade or become corrosive—followed the same pattern of awareness, response, and, ultimately, an accumulation of invaluable scientific data and empirical observations to guide conservators in designing treatments and devising preventive conservation strategies.

From Thaddeus Davids, *The History of Ink including Its Etymology, Chemistry, and Bibliography* (New York: Thaddeus Davids and Co., 1860), 1–5.

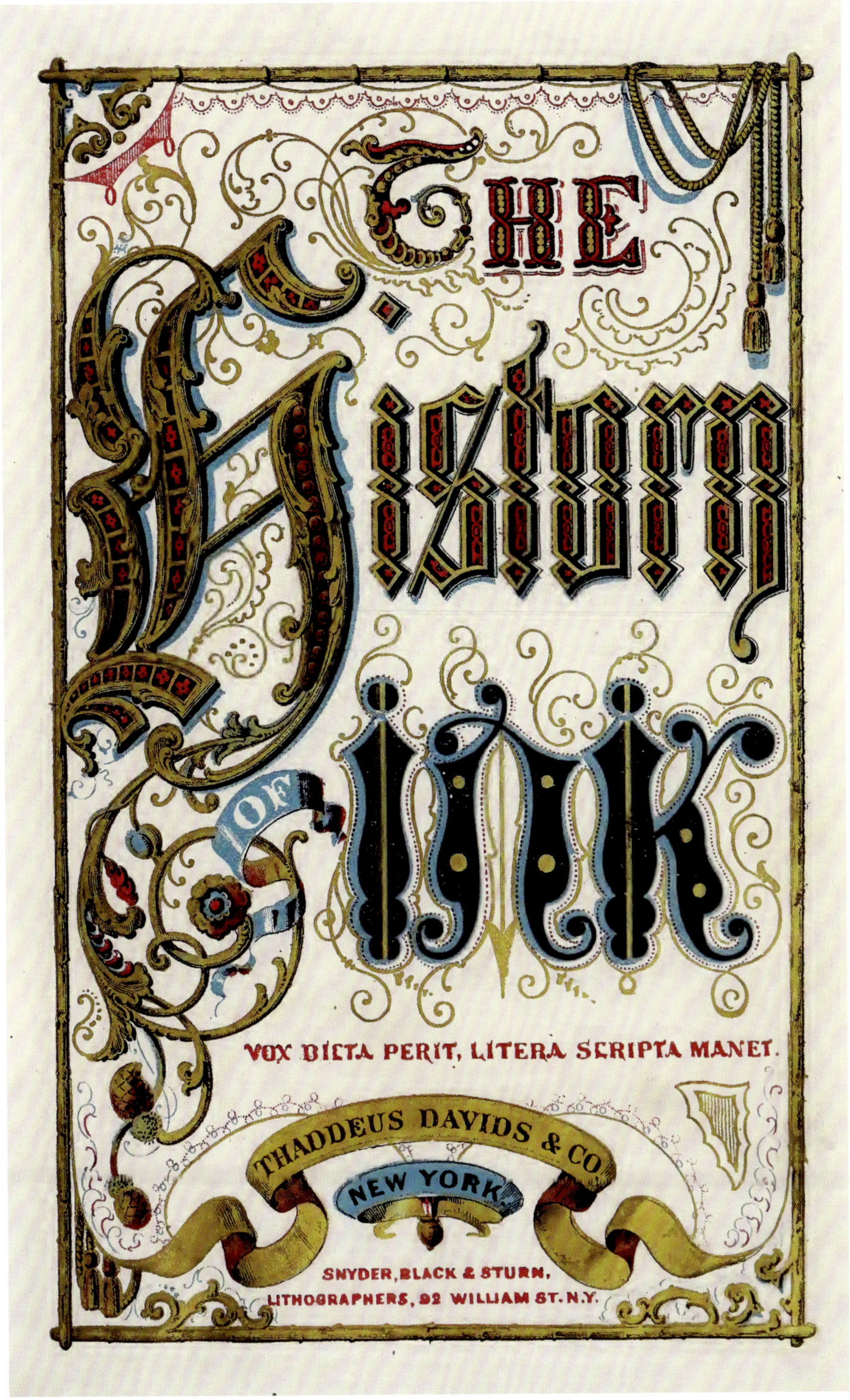

Figure 1
Title page from Thaddeus Davids, *The History of Ink*, 1860.

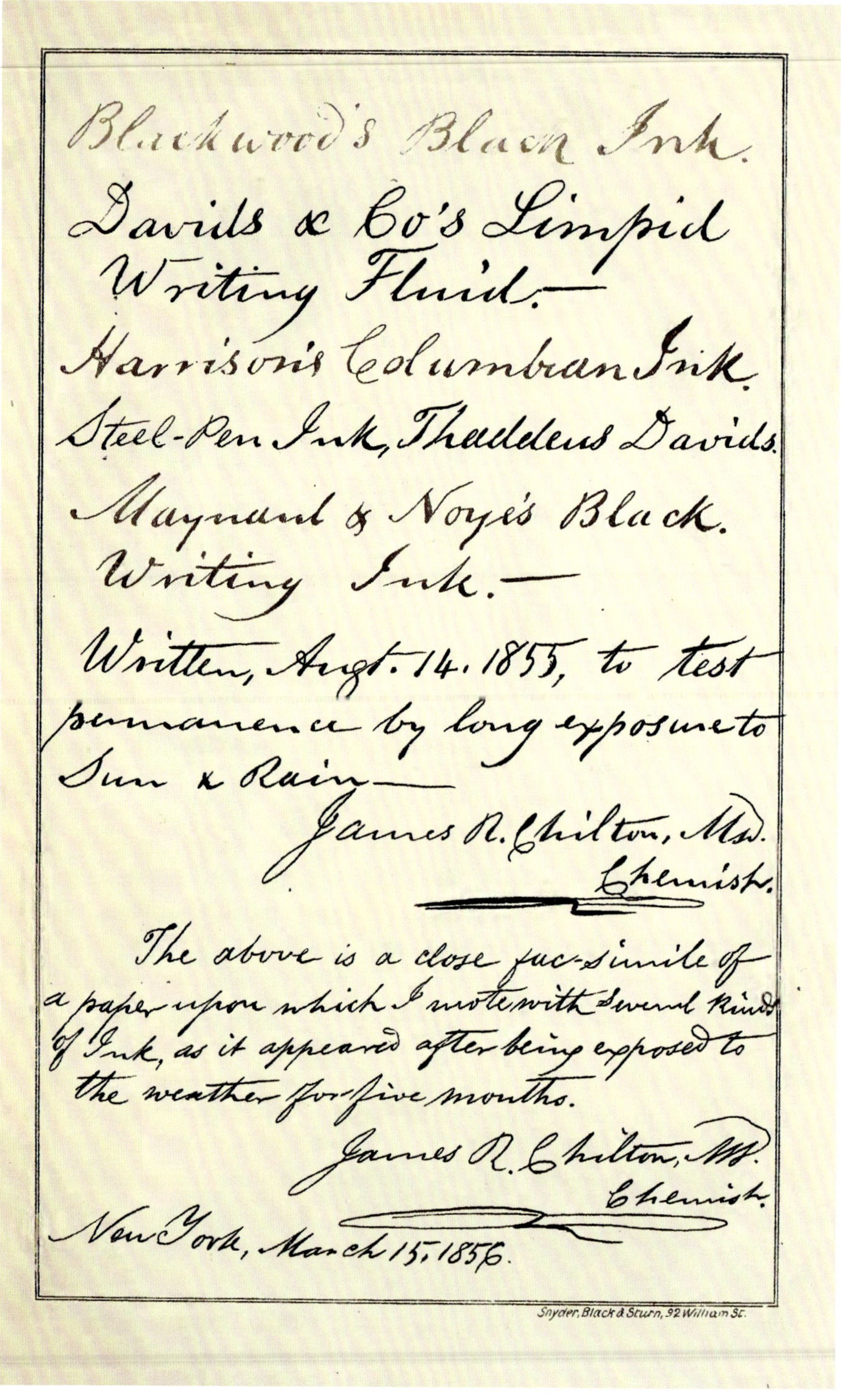

Blackwood's Black Ink.

Davids & Co's Limpid Writing Fluid.—

Harrison's Columbian Ink.

Steel-Pen Ink, Thaddeus Davids.

Maynard & Noyes's Black. Writing Ink.—

Written, Augt. 14. 1855, to test permanence by long exposure to Sun & Rain—

James R. Chilton, M.D.
Chemist.

The above is a close fac-simile of a paper upon which I wrote with several kinds of Ink, as it appeared after being exposed to the weather for five months.

James R. Chilton, M.D.
Chemist.

New York, March 15. 1856.

Snyder, Black & Sturn, 92 William St.

Figure 2
Facsimile of aging test, from Thaddeus Davids, *The History of Ink*, 1860.

Notices of the Press.

From the N.Y. Daily Tribune, April 13, 1856.
WRITING INKS.

Among the many discoveries of the age developing improvements in almost every branch of the arts, little has been done by chemical science to improve and perfect this "Art preservative of all arts."[1] An Ink perfectly black, limpid, and indelible, has yet to be discovered.

Nutgalls[2] appear to be the only really valuable ingredient in modern inks. . . . the value and permanence of which depend upon the proper quantity of gallic acid extracted from the galls being used in its preparation; but nutgalls are expensive, and ink-makers are apt to increase their profits by substituting logwood, which gives a large amount of coloring matter at a small cost.

A compound of bichromate of potash[3] and the extract of logwood, discovered by the German chemists a few years since, makes a limpid ink, black when first used, at a cost of less than three cents a gallon; but it is a vile dye, totally valueless as an ink, fading out on an exposure to the light, and becoming illegible in a few years, however securely kept. Still, its cheapness, ease of manufacture, and apparent good qualities when first made, caused it to be extensively manufactured and sold, particularly in the Western States. The recipe for its production has been sold by traveling vendors in every considerable town in the Northern States, as a very great secret, which each particular buyer would make his fortune by getting exclusive possession of. . . . at any price, from one to fifty dollars. A cute Yankee, near Boston, was victimized to the tune of $250 for the secret, which he calculated would make his fortune, by making ink for nothing.

The ink used by the ancients was made with powdered charcoal, mixed with mucilage and water, and is indestructible by time; but this ink is impracticable in this age, when barrels are used where a penful sufficed.

The value of permanent ink cannot be questioned; for state, town, and county records, and the like, it is of vital importance that the ink used should be legible in succeeding generations. Too little attention is paid in this matter by our registers, surrogates, town clerks, and court officers. A writer in the April number of The New England Historical and Genealogical Magazine, speaking of the town records of Bolton, Conn., and making suggestions for their preservation, etc., says: "Moreover the ink is fast fading out." This matter of permanent Ink is attracting much attention in Europe, particularly in the government offices, records, courts, etc.

Professor Sharp,[4] of Edinburgh, in a paper recently read before the Royal Society of Arts, gives the result of a series of elaborate experiments, designed to test the permanence of writing ink by exposure to the light and to the action of the elements.

Dr. Chilton, of New York, has submitted a number of the principal inks of this country to a similar trial, with a view to test their relative indelible qualities. The result of his experiments is before us in a finely-engraved fac-simile[5] of the original paper, upon which the Doctor wrote with four different inks, and then exposed the writing over five months to the action of the weather upon the roof of his laboratory.

This test appears to be conclusive, so far as it goes, as showing what will be the action of time upon manuscripts in various stages of exposure. The matter is suggestive, and of importance to all who would preserve legible records for future generations.

From the N.Y. Journal of Commerce, April 15, 1856.
EXPERIMENT ON WRITING INKS.

A very ingenious experiment to test the durable qualities of ink has recently been made by Dr. Chilton, the Chemist. He exposed a manuscript written with four different inks of the principal makers,[6] of this and other countries, to the constant action of the weather upon the roof of his laboratory. After an exposure of over five months, the paper shows the different kinds of writing in various shades of color. The English sample, Blackwood's, well known and popular from the neat and convenient way that it is prepared for this market, was quite indistinct.

The American samples, David's, Harrison's, and Maynard's, are better. The first appears to retain its original shade very nearly; the last two are paler. This test shows conclusively the durability of ink; and while, for many purposes, schools and the like, an ink that will stand undefaced for a year or so, is all that is necessary, yet there is hardly a bottle of ink sold, some of which may not be used in the signature as execution of papers that may be important to be legible fifty or one hundred years hence.

For State and County offices, and Probate records, it is of vital importance that the records should be legible centuries hence. We believe that some of the early manuscripts of New England are brighter than some town and church records of this century.

In Europe, at the present time, great care is taken by the different governments in the preparation of permanent ink. . . . some of them even compounding their own, according to the most approved and expensive formulas.

Manuscripts of the eleventh and twelfth centuries, now in the state paper offices of Great Britain, are apparently as bright as when first written; while those of the last two hundred years are more or less illegible, and some of them entirely obliterated. The engraved facsimile of Dr. Chilton's test, with his certificate of accuracy appended, is a credit to the art of lithographing.[7] So perfect is the imitation in shade and color of the original, that it will puzzle good judges to tell which is written and which engraved.

This method of testing the durability of ink is new and conclusive.

Notes to Introduction

1 On the elaborate title page (fig. 1) appear the words *VOX DICTA PERDIT, LITERA SCRIPTA MANET* (the spoken word perishes, the written word endures (see Part I).

2 A bit of sleuthing suggests that the test sheet (fig. 2) copied and widely distributed by Davids was not the actual one described in *Scientific American*. That test was dated May 9, 1855–August 1, 1855, for a total exposure of three months. Dr. Chilton attested that Davids's facsimile dated August 14, 1855–March 15, 1856, for a total exposure of five months, was "close." "Writing Inks," *Scientific American* 11, no. 31 (April 12, 1856): 244.

3 Dr. Chilton's tests were cited, albeit inaccurately, as late as 1907. *Tribune Almanac and Political Register* 19, no. 1 (January 1907): 389–91.

Notes to Reading

1 ["Art preservative of all arts": Davids is translating *Ars artium omnium conservatrix*, a once widely known phrase that appeared on the facade of the Haarlem residence, now destroyed, of Laurens Janszoon Koster, who is credited with the invention of movable type in 1440, twelve years before Johannes Gutenberg (1398?–1468). Davids's allusion is unclear.—ED.]

2 [Nutgalls are the spherical growths that form on various trees and shrubs in response to the sting of certain wasps in the process of depositing fertilized eggs. Gallotannic acids, contained in the galls, are extracted by mixing with water, boiling, or fermentation. A higher proportion of gallic acid has been found to produce the blackest inks when combined with a metallic salt, typically iron sulfate. See Cyntia Karnes, *How to Make Ink—Ingredients*. http://ink-corrosion.org/make-ink/ingredients.—ED.]

3 [Bichromate of potash or potassium bichromate darkens in the presence of tannins. Logwood is derived from the campeachey tree and turns blue in alkaline solutions and red in acidic solutions. It is extremely sensitive to light. See Cyntia Karnes, *How to Make Ink—Ingredients*, http://ink-corrosion.org/make-ink/ingredients.—ED.]

4 [Davids is probably referring to Dr. James Stark, not Sharp, who from 1842 to 1855 made and tested 229 different inks. See David N. Carvalho, *Forty Centuries of Ink; or, A Chronological Narrative Concerning Ink and Its Backgrounds* (New York: Banks Law Publishing Co., 1904), 101.—ED.]

5 [The "finely-engraved fac-simile" is a lithograph.—ED.]

6 [In fact, Dr. Chilton tested five different writing inks, not four.—ED.]

7 [It is unclear if Davids had Dr. Chilton's test results copied by a local lithographer, Snyder & Black of 92 William Street, located just a few doors down from his establishment on 127–129 William Street in New York or if the facsimile was commissioned by a less partial client.—ED.]

Reading 71

J.R.

Ink for Lawyers (1884)

J.R.'s frustration as an attorney echoes that felt by collectors of watercolors, but in this case it is the legal implications of disappearing inks that is of primary concern. J.R. calls for a ranking of permanence to be published, a request repeated by other beleaguered consumers (see reading 74).

This reading illustrates the confusion over the meaning of permanent *as the term is used to qualify writing inks. As Davids recounts in* The History of Ink, *"The most important peculiarity of the modern writing-ink, as contrasted with the ancient, naturally suggested the two names which it bore in the Latin and Greek of the middle ages, or (to speak more definitely,) the time of its invention and first employment. It was a Tinta, a DYE, or STAIN, which tinged and tinctured the material on which it was placed, entering among its fibres as coloring fluids do into cloth in the ordinary process of manufacture. It penetrated the substance of the paper (as caustics or powerful chemical solvents and corrosives act on the organic fibre): it bit in, or burned in,— and was therefore well named ENCASTON and Incaustum."*[1]

A writing ink, as differentiated from a simple suspension of super-fine particles of carbon black plus a binder, is indelible and cannot be removed by rubbing or washing. Having no guarantee of lightfastness, this is the meaning that prevails today. Hence, the marks left by solvent-based marker pens labeled "permanent" cannot be removed by laundering but are notoriously sensitive to light, to the extent that Robert Feller (readings 63 and 64) suggested that sunlight would be a more effective and labor-saving graffiti remover than scrubbing.[2] *Likewise, today's fluorescent "highlighter" inks, characterized by their transparency and added optical brighteners, are also advertised as permanent, an insidious claim for those who equate the*

J.R., "Ink for Lawyers," *The Law Times: The Journal and Record of the Law and the Lawyers* 78 (December 13, 1884): 123.

highlighting of texts with defacement.[3] *The demand for permanent marking inks to accentuate notable passages in books, however, has existed for over a century: in 1906 Samuel Bagster and Sons advertised their eight different colors of Bible-marking inks as noncorrosive and not liable to change color over time.*[4] *Currently on the market, Memory Markers are also touted as archival, surely straining that adjective's definition.*[5]

INK FOR LAWYERS—There is a question most important to solicitors, but to which I think sufficient attention is not given, viz., the use of ink for engrossing deeds, of assured permanence. Some deeds have lately come before me of recent date which have faded to a light brown. I am told that Field's[1] non-corrosive ink, and I presume Mordan's[2] and Stephens',[3] are durable; but they have not yet had the test of time. It would be conferring a boon on the profession if some person, who is competent, but not biased in favour of any particular sort, would insert a paper in the LAW TIMES setting forth the composition and relative merits of different inks, which would enable solicitors to choose one that would be durable. It seems very strange that, with all our chemical science of the present day, we cannot get an ink equal to that which was used by the copyists in the monasteries before the invention of printing.

Notes to Introduction

1 Thaddeus Davids, *The History of Ink including Its Etymology, Chemistry, and Bibliography* (New York: Thaddeus Davids and Co. 1860), 15. See reading 70.
2 R. L. Feller, "Felt-Tipped Markers and the Need for Standards of Lightfastness for Artists' Colorants," *Bulletin of the IIC-AG* 8, no. 1 (1967): 25.
3 Sir Hilary Jenkinson (1882–1961), father of archive administration and management, has this to say about those who defile the pages of books: "The habit of writing in the margin is so ingrained in some persons that nothing short of a greased or blackened margin will retrain them." "Physical Defence of Archives," *A Manual of Archive Administration*, new and rev. ed. (London: Percy Lund, Humphries and Co., 1937), n. 3, 59.
4 Bible Marking Ink (Samuel Bagster and Sons, Ltd., London, 1906), in George Allen's *List of Publications* (London: Ruskin House, 1906), 26.
5 Memory Marker is manufactured by American Crafts.

Notes to Reading

1 [George Field (1777?–1854); see reading 65. Field's Chemical Non-Corrosive Ink was advertised as a "true gall ink, most admirable for Legal and other important Documents; Writing turns a splendid and permanent black." *The Newfoundlander*, June 5, 1883, 4.—ED.]
2 [Silversmith Sampson Mordan (1790–1843) and his sons manufactured a variety of mechanical pencils and pens, as well as alkaline inks. Their writing accessories are highly collectible.—ED.]

3 [Dr. Henry Stephens (1796–1864) invented Stephens Ink in 1832, an indelible "blue-black writing fluid," which he described as a "carbonaceous black writing fluid, which will accomplish the so long-desired and apparently hopeless task of rendering the manuscript as durable and as indelible as the printed record." His son, Inky, aggressively marketed the ink, which was officially adopted for use by the British government. http://bbc.co.uk/radio4/history/making_history.—Ed.]

Reading 72

John Murray

Practical Remarks on Modern Paper (1829)

It should come as no surprise that the provocateur John Murray (1785?–1851) was quick to demand improvements in ink manufacture as well as papermaking (see reading 48). He observed that the very property that confers permanence to iron-gall ink, its ability to burn into the paper, also leads to its eventual demise. But Murray's concern was with an immediate consequence of using indelible ink: it corroded the steel knives used to sharpen quill nibs—and had the potential to poison mischievous schoolboys. The caustic effect of writing inks was a nuisance that pervaded everyday life before the development of alkaline dye–based inks in ballpoint pens and impervious Teflon® roller-ball nibs. Not many of us today would link the visual metaphor of tooth decay with spattered ink, as did the poet-dentist Solymon Brown in 1866:

Whene'er along the ivory disks are seen
The filthy footsteps of the dark gangrene,
When caries comes, with stealthy pace to throw
Corrosive ink-spots on that bank of snow,
Brook no delay, ye trembling, suffering fair,
But fly for refuge to the Dentist's care.[1]

Writing-inks often contain many noxious ingredients. The recommended addendum of *corrosive sublimate*[1] cannot be too severely censured, as children at school frequently suck their pens; and, in our boyish days, we can well remember to have seen some of our compeers who made no scruple of drinking the ink from the stand. *Sulphate of copper*[2] is another adjunct, and evidence of our attainments in the scale of improvement; and this is the cause why the *pen-knife* is so promptly

From John Murray, *Practical Remarks on Modern Paper* (London: T. Cadell, 1829), 97–98.

blunted in mending the pen—the sulphuric acid of the sulphate of copper, in virtue of its superior affinity, leaving its previous combination with the copper for the steel. We should venture to recommend the *carbonate of iron*, instead of its sulphate; as, in such a case, there would be no combining acid sufficiently powerful to evolve the chlorine, or gnaw the paper tissue. Carbonaceous matter, in the form of Indian ink, must always be the prerequisite of a durable and permanent ink.

Note to Introduction

1 Solymon Brown, *Transactions of the Connecticut State Dental Association 1864–5* (Hartford: Wiley, Waterman and Eaton, 1866), 45.

Notes to Reading

1 [A few grains of corrosive sublimate—mercuric chloride—were added to writing ink to prevent mold growth. Whole cloves could also be dropped into the bottle, a preferable fungicide since even minute amounts of mercuric chloride are highly poisonous.—Ed.]
2 [Sulfate of copper, also known as blue copperas, was often used in addition to iron sulfate; however, it does not increase the production of color. See Cyntia Karnes, *How to Make Ink—Ingredients*, http://ink-corrosion.org/make-ink/ingredients.—Ed.]

Reading 73

SIGMUND LEHNER

Ink Manufacture including Writing, Copying, Lithographic, Marking, Stamping and Laundry Inks (1902)

First published in German in 1885, Sigmund Lehner's Ink Manufacture *quickly became the standard reference in the ink industry. In it the author attributes the fading and corrosiveness of inks to two technological "advancements": the introduction of machine-made wood pulp papers and steel nibs. Recalling Abbot Johann Tritheim's (1462–1516) campaign advocating more durable parchment over paper (see reading 1), Lehner decries the destructive effects of chlorine and lime residues in wood pulp papers, which, in his opinion, contribute to ink fading, a theory not mentioned elsewhere and apparently never pursued.[1] Lehner describes the chemical action of caustic inks on steel nib pens, even ones coated with more noble metals (see reading 74) and notes the superior longevity of goose and raven quills. The prejudice for more resilient quills persisted due to a "certain unyielding rigidity in steel pens."[2] Nonetheless, steel nibs had gained precedence by the mid-nineteenth century, with only their names suggesting avian origins: today's finest dipping pen is still called a crow quill.*

[. . .]

Old deeds and other parchments show us that even in the early Middle Ages the art of ink making had been brought to a high degree of perfection. Letters 800 years old, indeed, are less faded in many cases than those of a much more recent date. When we see a MS. which has become almost illegible in fifty years or less, we are inclined to think that the knowledge of ink manufacture has gone backwards instead of forwards. We must not forget, however, that the durability of an ink depends not merely on itself, but also on the substance it is written on. Many of the substances now used in preparing paper, such as lime and chlorine, have, even when present in very small quantities only, a destructive effect upon an ink which

---

From SIGMUND LEHNER, *Ink Manufacture including Writing, Copying, Lithographic, Marking, Stamping and Laundry Inks* (London: Scott, Greenwood, and Co., 1902), 2–4.

if used on pure parchment would last as long as the monastic productions of the Middle Ages have done. All our bleached papers contain chlorine, which sooner or later destroys the paper. It is thus useless to write documents intended for long preservation with indelible ink[1] if paper is used to write them on. Our modern books, printed without exception upon paper that has been bleached with chlorine, will disappear in a few centuries by the disintegration of the paper. This defect attaches particularly to a sort of paper much used now on account of its cheapness, that made from wood pulp. This in a few years turns brown, and becomes so brittle that it breaks when folded. Printing ink, on the other hand, which depends for its colour on carbon, is practically imperishable. People are now becoming alive to the disadvantages of such paper, and are printing important books on paper which is free from every trace of free chlorine.[2] The durability of unbleached vegetable fibre is well exemplified by the linen wrappings of the Egyptian mummies. Woven four thousand years ago, it is now scarcely brown, and is still fairly strong. If the ancient Egyptians had written on such stuff with good ink the writing would be plain to this day.

The manufacture of ink has been carried on empirically until very lately. The first to apply scientific knowledge to the manufacture was Lewis,[3] at the end of the eighteenth century. Since his time Berzelius[4] and Bottcher[5] have done a large amount of work at the subject.

Changes in the methods of manufacture of ink have been accompanied by two other movements which have had great influence on it. One of these is the substitution of smooth machine-made paper for the grey or brown handmade paper.[6] This glazed paper[7] requires for its manufacture a very powerful bleaching of the raw material, with the result that it invariably contains chlorine or lime. It has thus become important to find inks that will resist the slow action of these substances. This may appear a very simple matter, but is in reality a difficult one. The second movement was the substitution of the steel pen for the quill. While in former times the goose-quill only was used for writing and the raven-quill for drawing, the steel nib is now almost universally employed for both. The substance of a quill is extremely resistant to chemical action. Steel, on the other hand, is readily corroded by very many things. Even very dilute acid in the ink will soon make a nib unusable. Before inks were invented which would not act on iron, many attempts were made to devise a protective coating for the nib, by covering it with copper, silver, or even gold. This coating, however, soon gets worn off the tip of the nib, even in using the smoothest paper.

We have called the manufacture of ink an art. Many may smile at the designation, but we shall adhere to our nomenclature. It is not easy to make an ink fulfilling every requirement that may be demanded of it, especially as we are even now to a large extent in the dark as to the exact chemistry of ink manufacture, although distinguished chemists have not thought it unworthy of them to work at the subject.

Notes to Introduction

1 Other paper properties that have been considered as contributing factors are the amount of alkaline reserve in the paper and the degree and type of its sizing. Birgit Reissland, "Iron-Gall Ink Corrosion: Progress in Visible Degradation," in *Contributions to Conservation: Research in Conservation at the Netherlands Institute for Cultural Heritage (ICN)*, ed. Jaap A. Mosk and Norman H. Tennent (Amsterdam: Earthscan, 2002), 113–18.
2 "The Pen—Its Construction and Use," *Scientific American* 15 (1866): 145.

Notes to Reading

1 [Lehner is using "indelible ink" to specify iron-gall ink.—Ed.]
2 [Almost a century before the publication of the *American National Standard for Information Sciences, Permanence of Paper for Printed Library Materials; Z39.48-1884* (see reading 56), Lehner endorses the precautionary practice of printing "important" books on better-quality paper.—Ed.]
3 [Lehner is referring to William Lewis (1708–1781), author of *Commercium Philosophico-Technicum; or, The Philosophical Commerce of Arts: Trades and Manufactures* (London: printed by H. Baldwin for the author and sold by R. Willock, 1763/1765) or, more specifically, Lewis's chapter 6 on black colors, which was translated into German under the title *Historie der Farben: Erst Abteilung, von den schwarzen Farbe*, trans. J. H. Ziegler (Zurich: Heidigger, 1766).—Ed.]
4 [Lehner must be referring to Jöns Jakob Berzelius (1779–1848), the famous Swedish chemist who observed that the element vanadium forms a dark compound when combined with gallotannic acid.—Ed.]
5 ["Bottcher" refers to R. Böttger, inventor of a black aniline ink.—Ed.]
6 [Second-quality handmade rag paper is traditionally known as "grey" or "brown" and would have had better aging properties than machine-made wood pulp paper.—Ed.]
7 [A "glazed" machine-made paper has a smooth, shiny surface produced by running it through two high-pressure-heated metal rollers or calenders. One wonders how this writing surface compares to the gelatin tub-sized paper extolled by Clapperton (reading 41).—Ed.]

Reading 74

Walter S. Landis

The Electrolytic Corrosion of Electroplated Objects (1911)

The frustration of Walter Savage Landis (1881–1944), an American professor of metallurgy, is palpable as he describes how his costly and "improved" gold-plated nibs corrode more quickly than his pedestrian steel nibs. His apprehension was no doubt aggravated by the rapidly changing technology of inks and papermaking (see reading 73), which had an effect not only on how the action of writing in pen and ink felt but also on how it looked.

The subtle shift in appearance of aged inks was noted by a literary critic writing in 1903. In deconstructing the popular letters of Hester Lynch Thrale (1741–1821), a prolific British diarist and friend of Samuel Johnson, an "essentially modern" postscript is singled out because it is overly informal in tone by prevailing eighteenth-century customs. But not only did Mrs. Thrale's postscript sound too modern; it struck the critic as visually anachronistic as well, since it was "embalmed in that old brown ink, like a fly in amber, a phrase we might expect to belong exclusively to steel pens dipped in the modern non-corrosive ink, or the blue-black [ink], whose advertisements sputter over large and small railway stations at the present day."[1] Perhaps she has in mind Stephens's Blue-Black Writing Fluid (see reading 71), whose broadside advertisements were ubiquitous in early-twentieth-century England. Conservators, too, are highly attuned to the historically appropriate appearance of inks. This presents a dilemma when determining the extent of intervention for active iron-gall ink corrosion.[2] A "conservative" approach may not be chemically optimal in ensuring the longevity of the drawing but can preserve its expected historical appearance. This leads one to wonder if the ultimate goal of long-term preservation is compromised by a partial treatment.

From Walter S. Landis, "The Electrolytic Corrosion of Electroplated Objects," *Transactions of the American Electrochemical Society* 19 (1911): 59–62.

For several years past the author has been using a certain well-known brand of steel writing pen with great satisfaction. The demand made upon the pens is not severe, they being used only for signatures, corrections and such occasional writing,[1] though being moist with ink for several hours at a time. Under such light and easy service the life of the pens was always at least a week, and frequently several weeks elapsed before they reached that scratchy stage demanding replacement.

This past winter, on attempting to purchase a fresh supply of this old, reliable brand, it was found that the local stationer did not have the plain steel ones in stock, but could furnish the same style and brand gold-plated. The cost of this plated pen was about one-half more than the plain steel ones. Thinking that the manufacturer had improved the life of the pen by this plating, a box of the new style was purchased. In use under practically the same service that the plain steel ones had been subjected to, the plated variety lasted rarely more than two days. The corrosion of the points and between the nibs was excessive, one pen frequently not writing more than five hundred words before requiring replacement. The life of the pens being so very short, I determined to make an investigation of the cause. The results of this investigation are here produced.

Microscopic and chemical tests showed that the pens were probably first copper-plated and then flashed with gold. The total thickness of the yellow metal coating was 0.006 mm., and the pens carried approximately 0.25 milligram of gold each. The plating was excellently done and was absolutely continuous as far as could be ascertained by chemical test. The pens could be thrown into sulphuric acid and left there for some time without loss of weight. I therefore came to the conclusion that the corrosion was electrolytic in its nature, and formulated the following hypothesis to account for it: The friction of the pen point on paper rapidly wears off the plating on the point, thus exposing the steel. An electrolytic couple, *Steel—Ink—Gold*, is set up and corrosion of the steel takes place with great rapidity.

Some idea of the electrochemical nature of this couple may be gained from the following: A pen was stripped of its plating by an 18-hour immersion in strong potassium cyanide solution. This stripped pen and a plated pen were then connected in series with a high resistance voltmeter and the two pens dipped into various kinds of ink, contained in a small cell about the size of a thimble. A potentiometer was not available at the time the experiments were made, so that I had to content myself with measurements made on closed circuits of 300 to 500 ohms' resistance.

When the two pens were dipped into "Barnes' National Jet Black Ink," the voltmeter showed a reading of 0.045 volt at once, dropping in three minutes to 0.020 volt, and in five minutes to 0.015 volt, at which point it remained constant for some time. When taken out and redipped, the whole series repeated itself. Vibration of the pens, while immersed, gave a continuous reading of about 0.045 volt.

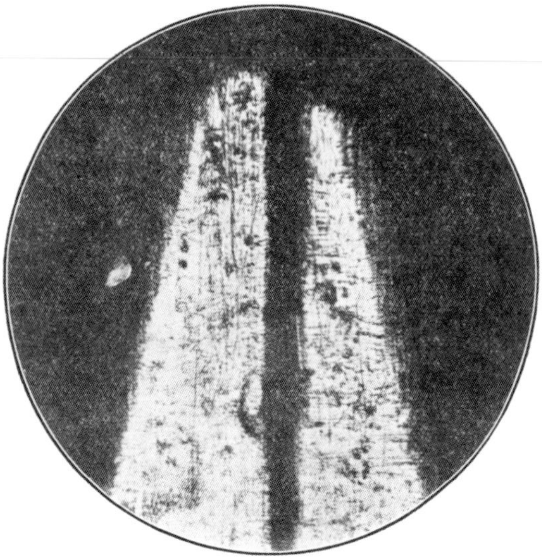

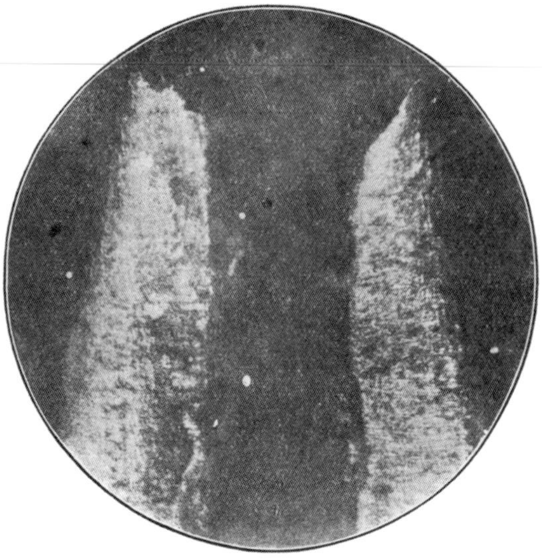

Figure 1
Points of new gold-plated pen.

Figure 2
Points of corroded gold-plated pen.

The resistivity of the ink was approximately 112 ohms per cm. cube. Index of corrosion, 0.00040.

In "Stafford's Universal Writing Fluid," the series was 0.08, 0.02 and 0.008 volt; vibrating the pens maintained a constant electromotive force of 0.022 volt. The resistivity of this ink was 68 ohms per cm. cube. Index of corrosion, 0.00033. In a rather old and faded brand of this ink, the reading under vibration of the pens was 0.016 volt. No measure of the resistivity of this faded ink was made.

In "Waterman's Ideal Ink," the series was 0.088 and 0.040 volt. A steady electromotive force of 0.120 volt could be maintained by vibrating the pens in the ink. The resistivity of this ink was 68 ohms per cm. cube. With a freshly stripped pen against a fresh gold-plated one, a similar result of 0.12 volt was measured. Index of corrosion, 0.00176.

These experiments seem to corroborate my original hypothesis as to the cause of the excessive corrosion of the pens. When in use, the continued dipping of them into ink and the vibration occasioned as they rub over the paper maintains a short-circuited couple with the maximum electromotive force noted, and causes consequent corrosion of the steel point as the anode of this couple.

The "index of corrosion" as given above for the various writing fluids is the product of the maximum electromotive force found for the vibrated pens multiplied by the conductivity of the various inks. This index assists in choosing the least corrosive ink in case the plated pens have to be used.

While the corrosion of the plated pens is not a matter of great commercial importance, yet there is to be gained from the experience a certain wholesome lesson on the application of electroplating to the protection of less noble metals:

Do not plate with a nobler metal a material that is likely to be subjected to wear on one spot and which spot is to be moistened with liquids. For, as soon as the plating is worn through on that spot, accelerated corrosion will take place, and in the end the resulting corrosion will be far worse than without the plating.

Electrometallurgical Laboratory.
Lehigh University,
March 1, 1911.

Notes to Introduction

1 Miriam A. Ellis, "Some Unedited Letters of Mrs. Thrale," *Fortnightly Review*, no. 440 (August 1, 1903): 271–72.

2 Cindy Connelly Ryan, Elmer Eusman, Holly Krueger, Linda Stiber Morenus, and Heather Wanser, "Optimizing Ink Corrosion Treatment Protocols at the Library of Congress," in *Edinburgh Conference Papers 2006: Proceedings from the Fifth International Conference of the Institute of Paper Conservation and the First International Conference of the Institute of Conservation, Book and Paper Group, 26–29 July 2006*, ed. Shulla Jaques and Barbara Venables (Leigh: Institute of Conservation, 2007), 195–202.

Note to Reading

1 [This statement would seem to indicate that Landis was in the habit of using a typewriter. By 1911, mechanical typewriters with standard QWERTY keyboards were widely available.—Ed.]

Reading 75

Otto Posse

The Conservation of Manuscripts, According to the International Conference for the Conservation and Repair of Old Manuscripts at St. Gallen (1899)

In 1898 Franz Ehrle (1845–1934), Vatican librarian, organized the International Conference for the Conservation and Repair of Old Manuscripts at St. Gallen, Switzerland. The event and this proclamation in particular are considered by many conservators to be the call to arms to address the widespread and intractable problem of iron-gall ink corrosion.[1] *Despite widespread political upheaval in Europe, research on the chemistry of iron-gall ink corrosion continued steadily throughout the twentieth century. It accelerated in 1993 with the establishment of the Ink Corrosion Project, or InkCor, at the Netherlands Institute for Cultural Heritage with partners in Slovenia, France, and Germany.*[2] *InkCor was catalyzed by the recognition that drawings by Guercino (from 1618?–1622) disproportionately suffered ink-induced corrosion and that there existed no explanation for the cause of the problem or consensus on how to treat it.*[3] *It is worth noting that InkCor brought together fine art conservators and library/archive conservators, thereby swelling the ranks of stakeholders and encouraging broad-based cooperation. A symposium on ink corrosion hosted by the Netherlands Institute for Cultural Heritage in 1997 represented the first major conference dedicated to the subject since St. Gallen.*[4] *In 2000 the University of Northumbria hosted a second comprehensive conference on iron-gall ink, with subsequent meetings organized by a European network.*[5] *While numerous symposia and a dedicated website on iron-gall ink have clarified the mechanism of corrosion, indecision persists regarding treatment.*[6]

Two avenues of treatment—stabilization and chemical intervention—are followed. The first, which includes impregnation with Zapon (a cellulose nitrate solu-

From Otto Posse, "Handschriften-Konservierung: Nach den Verhandlungen der St. Gallener Internationalen Konferenz zur Erhaltung und Ausbesserung alter Handschriften von 1898," *Restaurator*, Suppl. 1 (1969): 7. Used by permission of DeGruyter.

tion), silking, cellulose acetate lamination—these three procedures are no longer recommended—and encapsulation, paper splitting, lining, and mending, simply provides additional physical support for papers weakened by corrosion. The second treatment, which includes deacidification, chelation, or reduction of iron (II) ions, arrests the actual deterioration mechanism. It is clear that no quick fix exists for the wide variety of drawings and documents executed in iron-gall ink. What is remarkable is the far-reaching influence of notable personalities like Father Ehrle and Dr. Russell and Captain Abney on the establishment of broad preservation policies in collecting institutions today.

While it has long been known that, like manuscripts treated with reagents,[1] other valuable manuscripts also appear to be threatened in their very existence by damaging after-effects of the ink itself, so it must be recognized as a providential act that the papal Curia has issued an exhortation, in the hope that it will have universal resonance, that we all come together like good Samaritans in the effort to preserve the earliest transmissions of our ancient literary classics.

Notes to Introduction

I am grateful to Jennifer Badger for sharing her research into the history of the manufacture, deterioration mechanisms, and treatment approaches of iron-gall ink.

1 In fact, what was reprinted in *Restaurator* was the text of a talk delivered by Otto Posse to a conference of German archivists in Dresden on September 18, 1899, the so-called Zapon conference. D. G. Vaisey, "E.W.B. Nicholson and the St. Gall Conference, 1898," *Bodleian Library Record* 9, no. 2 (March 1974): 101. A detailed description of the St. Gallen meeting, including participants and resolutions, was published the following year by Franz Ehrle, "Die Internationale Konferenz in St. Gallen am 30. September und 1. Oktober 1898 zur Beratung über die Erhaltung und Ausbesserung alter Handschriften," *Zentralblatt für Bibliothekswesen* 16 (1899): 27–51.

2 Johan G. Neeval and Birgit Reissland, "The Ink Corrosion Project at NICH: A Review," *Abbey Newsletter* 21, no. 5–6 (1997): 88.

3 R. Van Gulik and N. E. Kersten-Pampiglione, "A Closer Look at Iron Gall Ink Burn," *Restaurator*, no. 15 (1994): 173–87.

4 *Iron-Gall Ink Corrosion: Proceedings, European Workshop on Iron-Gall Corrosion, June 16–17, 1997* (Rotterdam: Museums Boijmans van Beuningen, 1997).

5 *Iron Gall Ink Meeting Postprints, September 4th & 5th 2000*, ed. A. Jean Brown (Newcastle: University of Northumbria, 2001); Transitional Metals in Paper (MIP), http://miponline.org.

6 http://ink-corrosion.org.

Note to Reading

1 ["like manuscripts treated with reagents": In an attempt to restore faded iron-gall inks or to reveal palimpsests, effaced writing found below later texts, it was common

practice to "fume" manuscripts using the foul-smelling vapors of ammonium or hydrogen sulfide, which reacts to form dark compounds with the ferric salt residues. Alternatively, a series of chemicals in liquid form, potassium ferrocyanide or silver nitrate among others, were applied to react with residual iron salts or chlorides. See Edwin O'Neill, "Restoration of Obliterated Ink Writing," *Journal of Criminal Law and Criminology (1931–1951)* 27, no. 4 (November–December 1936): 575. In the nineteenth century, many early manuscripts on parchment were sacrificed in this manner as part of the international race to find the earliest surviving biblical texts.—Ed.]

PART VII

Treatment: Limits and Limitations

Part VII explores the ways in which paper conservation treatments are designed, with emphasis on the decision-making process, especially the determination of appropriate treatment outcomes, the risks and benefits of various treatment approaches, alternatives to physically or chemically "aggressive" treatments, and the pros and cons of no intervention at all. The limits and limitations of treatment will usually emerge during these deliberations and influence what paper conservators can do versus what they choose to do.

According to the *Guidelines for Practice* of the American Institute for Conservation of Historic and Artistic Works (AIC), to which professional paper conservators subscribe, "The conservation professional should only recommend or undertake treatment that is judged suitable to the preservation of the aesthetic, conceptual, and physical characteristics of the cultural property. When nonintervention best serves to promote the preservation of the cultural property, it may be appropriate to recommend that no treatment be performed."[1]

The identification of the aesthetic, conceptual, and physical characteristics—or values—of a work on paper that must be respected and preserved is a crucial step in designing a suitable treatment. The task is made much more difficult, however, by the fact that the status of prints and drawings has shifted over time and will continue to do so. Although many categories of works of art on paper no longer function as originally intended, many are now appreciated for their aesthetic and evidentiary value. One example is the increased regard (and financial value) accorded Italian Renaissance cartoons, unassuming preparatory drawings that are often tattered and unfinished but technically revelatory (reading 79)—hence the trend toward minimal intervention that favors signs of former function and handling. In the words of the art historian Ernst van de Wetering (1938–), "Minimalism in conservation and restoration can be defended in various ways. The most significant argument is surely the need to preserve the many-layered documentary

Edwaert Colyer or Collier (c.1640–c.1702), *Trompe l'Oeil Portrait of a Lady*. Oil on canvas. Ateneum Art Museum, Finnish National Gallery, Helsinki, Finland / The Bridgeman Art Library.

evidence that every historic object bears. No one can foretell what sort of questions will be asked of the object in connection with future attempts at interpretation; in examining (art) historical objects one all too often discovers that essential traces have been made unintelligible or destroyed through treatment because their meaning went unrecognized."[2]

If treatment is indicated, the degree to which it may or certainly will alter the work's visual record is highly subjective. In reading 78 we see Bonnardot's mid-nineteenth-century "hands-off" approach to the restoration of a devotional statuette worn smooth by "pious lips" or a wax- and tear-spattered Book of Hours, which presages the "idea of patina" so eloquently expressed by the Belgian humanist Paul Philippot (1925–), who disputes the simplistic goal of restoring an artwork's original appearance in favor of maintaining its "present state of original materials" or its "'second historicity,' the span of time it has crossed to appear before us"[3] For Philippot, "'patina' is precisely the 'normal' effect that time has had on materials. This is not a physical or chemical, but a *critical* concept. Patina is nothing more than the combination of these 'normal' alterations as they affect the appearance of the work without disfiguring it—precisely because it is the result of 'normal' alterations."[4] This approach is echoed and further elucidated by Barbara Appelbaum (1947–), who goes so far as to recommend against an assessment of a work's condition at all since the process inevitably privileges its original appearance. Instead she favors a determination of the "state that best embodies an object's values."[5] This would seem to have been the objective of the treatment designed for Leonardo da Vinci's *Virgin and Child with Saint Anne and Saint John the Baptist* in the National Gallery of London after it suffered a shotgun blast in 1987 (reading 79).

The scale used for assessing the "fine art" value of a print or drawing is a sliding one, since the number of candidates eligible to be included in museum collections has increased exponentially, often the result of technological developments. Among these are hand-drawn architectural drawings after the introduction of computer-aided design in in the late twentieth century. The paper conservator Elizabeth Lunning (1948–2014) writes, "Sometimes we change our opinion about something as basic as what constitutes a work of art. I see this a lot with works on paper—prints, drawings, and photographs—and the shifting tastes are often reflected in how a work is stored. It might be something as simple as an illustrated letter from an artist that moves from the files to drawing storage after his death. Or it might be something like albums of Piranesi prints which get moved from libraries, where they sat as sort of travel books, to print rooms" (see readings 45 and 46).[6] Is the degree to which visual alteration as an intentional outcome of treatment in such newly elevated but now functionless items therefore less acceptable? The difficulty of assessing the intrinsic value of archival materials as separate from works of art on paper is especially thorny and is considered in reading 77.

The minimalist approach to treatment, which aims to avoid the alteration of meaningful visual attributes, can also be extended to include chemical ones. For example, some ramifications of routine paper conservation treatments can modify the paper's cellulosic structure on the molecular level, as occurs in repeated washings, or change a medium's "archeometric fingerprint," as is the case in the phytate treatment of iron-gall inks.[7] Chemical changes also challenge the long-held principle of reversibility, an outdated but tenacious precept of many paper conservators.

While philosophically the minimalist, or noninterventionist, approach is widely endorsed by the conservation profession, as Carlo James observes, in practice "systematic nonintervention is not the answer, whether intervention is reversible or not."[8] And Burns (reading 82) echoes that even master drawings "must sometimes be subjected to a full range of interventive conservation treatments."[9] First of all, it must be recognized that in most instances a print or drawing is brought to a conservator to be visually improved. Stated more bluntly, "The life of any work of art depends on how it looks."[10] For example, a print disfigured by brown corrugated cardboard striations or scattered foxing is typically a candidate for conservation because of its visual defects. While the goal of washing said print can be honestly claimed to be the removal of the acidic by-products of degradation, it is disingenuous to claim that its subsequent light bleaching has any such beneficial purpose. In extreme cases of cosmetic procedures, the goal of improved appearance is justified as a last-ditch preservation effort without which the work will be relegated to the dustbin. Another possible consequence of nonintervention is that without regular repetition the highly refined hand skills required for some complex interventive treatments may become rusty from lack of practice.[11] Finally, as the paper conservator Jane McAusland (reading 89) points out, some consumers of conservation services will demand different outcomes, requiring specific skill sets that are acquired and refined only through repetition.

Notes

1 *Guidelines for Practice*, American Institute for Conservation of Historic and Artistic Works, Treatment: 21: Suitability. 1994. www.conservation-us.org.

2 Ernst van de Wetering, "The Surface of Objects and Museum Style," originally presented at the Third International Seminar for Restorers, Veszprém, Hungary, 1981; published under the title "Die Oberfläche der Dinge und der museale Stil," *Maltechnik Restauro* 2 (1982), in *Historical and Philosophical Issues in the Conservation of Cultural Heritage*, ed. Nicholas Stanley Price, M. Kirby Talley Jr., and Alessandra Melucco Vaccaro (Los Angeles: Getty Conservation Institute, 1996), 421.

3 Paul Philippot, "La notion de patine et la nettoyage des peintures," *Bulletin Institut Royal du Patrimoine Artistique* 9 (1966). Published as "The Idea of Patina and the Cleaning of Paintings," in Price, Talley, and Melucco Vaccaro, *Historical and Philosophical Issues in the Conservation of Cultural Heritage*, 373.

4 Ibid. Original emphasis.

5 Barbara Appelbaum, *Conservation Treatment Methodology* (Oxford: Elsevier, 2007), 173.

6 Elizabeth Lunning, "Restoration: Storage and Art over Time; A Conversation with a Conservator," in *Deep Storage: Collecting, Storing and Archiving in Art*, ed. Ingrid Schaffner and Mattias Winzen (New York: Prestel, 1998), 245.

7 Irene Brückle, "Aqueous Treatment in Context," in Gerhard Banik and Irene Brückle, eds., *Paper and Water: A Guide for Conservators* (Oxford: Elsevier, 2011), 430.

8 C. James, C. Corrigan, M. C. Enshaian, et al., *Old Master Prints and Drawings: A Guide to Preservation and Conservation*, trans. and ed. Marjorie Cohn (Amsterdam: Amsterdam University Press, 1997), 211.

9 Thea Burns, "Preserving Master Drawings: A Cultural Perspective," *Paper Conservator* 25 (2001): 108.

10 Judith Walsh, "Special Conservation Problems for Collectors—The Cleaning of Drawings," *Drawing* 8, no. 1 (1986): 5.

11 Joanna Kosek, "Washing Paper in Conservation," in Banik and Brückle, *Paper and Water*, 315.

Reading 76

Arthur David Baynes-Cope

Ethics in Paper Conservation (1982)

The British chemist Arthur Baynes-Cope (1928–2002) sought to adapt the code of ethics for archive conservation, as written by Roger Ellis (1910–1998) and Sir Hilary Jenkinson (1882–1961), to the conservation of works of art on paper. While an interesting example of an early attempt to formulate guidelines for professional paper conservators, Baynes-Cope's value-laden distinction between functional but insensate documents and nonfunctional drawings that are "comparable only to a song composed for the joy of singing" was quickly outmoded. Indeed, the nature and extent of a procedure appropriate for archival materials as opposed to that for works of art on paper have become increasingly blurred since the days of the Ellis-Jenkinson Rule 4: "The process of repair should not diminish in any way the aesthetic appearance of non-archival material." That aesthetic uniqueness is also recognized by Keiko Keyes (reading 85).

While Baynes-Cope was not the first to recognize the wisdom of "passive," or preventive, conservation, his sardonic comment that "display is the cruellest and most selfish form of storage" seems particularly relevant given the current relentless demand for artworks by popular artists or schools. His view that preventive conservation should be a shared responsibility is taken up in Part VIII.

[. . .] I went to the Shortened Oxford English Dictionary and looked up the word Ethics and I found, to my relief, that it could be defined quite simply: "The science of human duty in its widest extent". I said that I read this with relief, for it seemed to me to approve my own outlook, a rather old-fashioned one, in that I tend to think in terms of a duty to be done, of privileges rather than rights.

From Arthur David Baynes-Cope, "Ethics in Paper Conservation," *Tidens Tand* 1 (1982): 144–46. Courtesy of the School of Conservation, Denmark.

In the field of paper conservation, we consider a range of materials which cover a number of purposes. At one end of the range is the austere archival document. A document which was drawn up in the course of an administrative or executive transaction of which itself formed a part, and which is preserved by those engaging in the transaction, and their successors, without limit of time, as evidence in law. At the other end of the range, a few strokes of a master's pen, comparable only to a song composed for the joy of singing. There are rules, clear and concise, for the way in which archives may receive conservation. They were drawn up by Roger Ellis and Sir Hilary Jenkinson,[1] each great archivists and I have added to them because as a chemist, a Chartered Chemist, a Fellow of the Royal Institute of Chemistry, my Institute lays on all of us that we, in our use of chemistry, have a duty to our fellowmen, and since we now more consciously use chemistry in the conservation of paper, it is necessary to make conservators realise that they have a duty which extends beyond that which is custumar[ily] owed to the document itself alone.

The Ellis Jenkinson Rules are:

1. No process of repair shall be allowed to remove, diminish or obscure in any way the documents' value as evidence.
2. No process of repair shall be used which would in any way damage or weaken the materials of which the document is made.
3. The process of repair should not interfere in any way with any subsequent treatment that the document may need.
4. The process of repair should not diminish in any way the aesthetic appearance of non-archival material.
5. No process of repair should be irreversible.
6. No biocide may be used which will, when the document is stored, used or handled under normal conditions injure in any way, temporarily or permanently the persons engaged in work with that document.

These rules are perhaps known well only to those who carry out active conservation on archives, but they deserve a wider appreciation from other conservators.

Now I have used the term Active conservation, and I am going to bring in some more definitions. Active Conservation is that branch of conservation which deals with the actual treatment of the objects themselves, its counter part being Passive Conservation, which is concerned with the provision of a safe environment for the object, which itself is not being treated. Thus Passive Conservation deals with storage and display. I can add two more definitions, of my own devising: Conservation is the science of fulfilling the real needs of objects. Display is the cruellest and most selfish form of storage, because it is done for [one's] own pleasure rather than for the needs of the object.

So now we are bringing more people into the group who have a duty towards the object, the archivist or curator who controls the custody of the object, the architect who designs the repository, the designer who creates the exhibition, perhaps the Council Committee who selects and employs them all, even the body which wishes to borrow material for a great exhibition. Each has a duty to the object and to those of the others with whom he must collaborate, and it must be duty unselfishly performed.

Note to Reading

1 [Roger Ellis (1910–1998), a disciple of Sir Hilary Jenkinson, was keenly interested in emerging techniques of document repair and taught the subject at the London School of Printing. His personal area of expertise was sigillography, or the study of seals attached to documents.

Sir Hilary Jenkinson (1882–1961) is considered the father of modern archive administration and management. He published one of the earliest articles on the conservation of seals, "Some Notes on the Preservation, Moulding and Casting of Seals," *Antiquaries Journal* 4, no. 4 (October 1924): 388–403.—Ed.]

Reading 77

IRENE BRÜCKLE

The Practice of Looking in Paper Conservation (2001)

Irene Brückle (1960–) is a professor of paper conservation at the Stuttgart State Academy of Art and Design and has written extensively about conservation treatments, technical connoisseurship, and ethical decision making. In this reading she wrestles with how to make the distinction between a work of art on paper and "non-art" artifacts. This crucial distinction can determine not only conservation parameters for an item but even its survival.[1]

Art and Archives

The terms 'art' and 'archives' have long been used to classify different types of flat paper artefact; and these words reflect, and to some extent determine, our viewing habits. Except in the case of certain avant-garde artworks, a print or drawing is usually a self-contained entity that can be stored, framed, and exhibited, and thus presents a well-defined, often intimate area of viewing.

Archives are usually measured by the cubic volume or linear feet of their holding, where only items of singular importance may receive special treatment. Archives present a diverse and layered viewing experience, because individual documents are accessed through stacks of boxed or bound materials which continuously open the view to additional documents. Conservators have developed different treatment approaches for archives and art materials based on their inherent distinction. Paper conservator Keiko Keyes wrote that discolouration is more distracting when it is associated with a work of art than with an archival document.[1]

From IRENE BRÜCKLE, "The Practice of Looking in Paper Conservation," *Paper Conservator* 25 (2001): 113–14.

> This interplay of medium and paper produces certain visual effects which are usually irrelevant to the archivist, but which are essential to the appreciation of works of art.

Artworks, she implied, best unfold their meaning when they are undisturbed by certain symptoms of ageing, while non-art artefacts contain documentary information which does not require the specialized visual perception crucial to the appreciation of art. This distinction rests on a divide between artworks and archival documents which was principally a construct of nineteenth-century thought concerned with the schematically driven classification of the various aesthetic experiences under the aegis of the burgeoning sciences. Our society has inherited the association of art with aesthetic enjoyment, and archives with historical information. This divided viewing habit has definite practical application. It is exemplified when conservators compare a representative specimen from either category: a watercolour landscape painting of the kind viewers might expect to find in a museum and a newspaper clipping readers might encounter in an archive (for example, Figs. 1, 2).

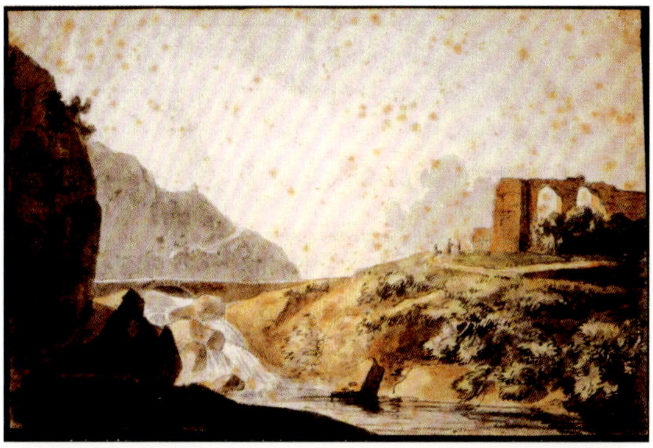

Figure 1
Attributed to Louis Gabriel Moreau (1740–1806), *Landscape*. Watercolour over traces of graphite. Reproduced courtesy of the author.

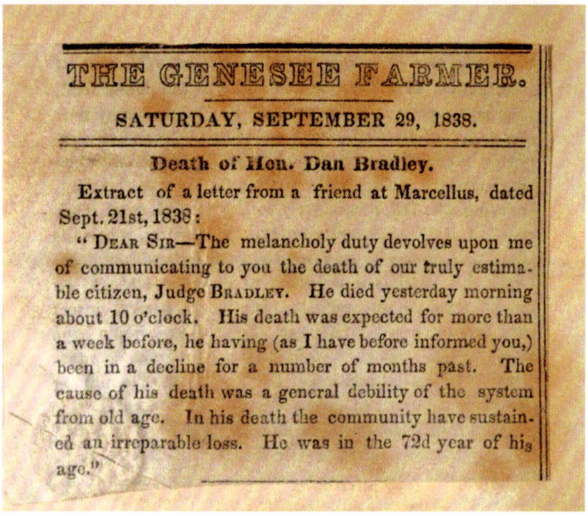

THE GENESEE FARMER.

SATURDAY, SEPTEMBER 29, 1838.

Death of Hon. Dan Bradley.

Extract of a letter from a friend at Marcellus, dated Sept. 21st, 1838:

"DEAR SIR—The melancholy duty devolves upon me of communicating to you the death of our truly estimable citizen, Judge BRADLEY. He died yesterday morning about 10 o'clock. His death was expected for more than a week before, he having (as I have before informed you,) been in a decline for a number of months past. The cause of his death was a general debility of the system from old age. In his death the community have sustained an irreparable loss. He was in the 72d year of his age."

Figure 2
Anonymous author, *Death of Hon. Dan Bradley*. Newspaper clipping dated 29 September 1838, pasted on the inside cover of a ledger book. Reproduced courtesy of the author.

Both are disfigured by foxing stains. While viewers will probably find the stains scattered across the light blue sky above the landscape distracting and will desire their removal, the stains on the newspaper clipping will unlikely bother readers eager to extract from it information about Judge Bradley's death, providing that the original newspaper is preserved well enough to be consulted, or a readable copy is made available.

Note to Introduction

1 The subjective denial of "intrinsic" value of archival materials can result in their disposal. National Archives and Records Administration, "Intrinsic Value in Archival Material," Staff Information Paper 21, 1982, Washington, D.C., 1999. www.archives.gov/research/alic/reference/archives-resources/archival-material-intrinsic-value.html. The words *litera scripta manet* appear on the seal of the National Archives and Records Administration.

Note to Reading

1 K. Keyes. 'The unique qualities of paper as an artifact in conservation treatment', *The Paper Conservator* 3 (1978): 5–8.

Reading 78

MITCHELL S. BUCK

Book Repair and Restoration (1918)

The words of the American poet Mitchell S. Buck (1887–1959) are scarcely heard in his translation and annotation of selections from the 1858 second edition of Essai sur l'art de restaurer les estampes et les livres *by Alphonse Bonnardot (1808–1884), one of very few treatises on washing, bleaching, and repairing prints and drawings available at the time.*[1] *As the anecdotes in this reading suggest, Bonnardot's prescient notion of the value of patina and evidence of past use is remarkably consistent with modern conservation theories.*

The determination of what constitutes acceptable alteration as opposed to unacceptable defacement, however, is subjective and depends on the vicissitudes of an object's past life and, it would seem, gradually emerges with the passage of time. For example, political parallels can be drawn between Bonnardot's letter bloodied from raging mobs in the sixteenth century and documents damaged during the devastation of the Holocaust in the twentieth—both items bearing witness to horrific moments in history. Conversely, the aesthetic value (or, one can argue, its fame and financial value) of Leonardo da Vinci's drawing Virgin and Child with Saint Anne and Saint John the Baptist *far outweighed the historical evidence of the shotgun blast it suffered in 1986 (reading 79). Finally, "it is precisely their wormholes, tears, stains and other damage" that aid in the understanding of many Renaissance prints that were frequently used in ways that "variously altered, misused, adorned and adored them."*[2] *On the one hand, Bonnardot justifies restoration for acts of carelessness; on the other, he seems comfortable with exempting efforts to remove Courier's spilled ink, surely an act of sheer carelessness but modulated when viewed through the blurry lens of history.*

From MITCHELL S. BUCK, *Book Repair and Restoration: A Manual of Practical Suggestions for Bibliophiles* (Philadelphia: Nicholas L. Brown, 1918), 25–28.

[. . .]

Before discussing the means of attacking stains which may blemish a book or a precious print, I am going to say that, in certain cases, it might be very desirable to allow them to remain. If I possessed, for example, a missive addressed to Charles IX during the night of Saint Bartholomew,[1] and stained with bloody fingerprints, I would take great care not to disturb these marks which, supposing their authenticity established, would increase tenfold the value of the autograph. If the custodian of the Laurentian Library at Florence should efface, from his Longus manuscript, Paul Louis Courier's puddle of ink,[2] he would commit an act of vandalism, for that ink stain is a literary celebrity.[3]

To speak of more ordinary examples: one often finds on a book or print, a signature or inscription which may sometimes be an autograph well worth preservation.[4] I very rarely efface signatures or the notes of early, unknown owners; I find it pleasanter to respect these souvenirs of the past. In the same way, some curious objects have certain defects which, I think, add to their interest. For example, a statuette of the Virgin, in silver or ivory, of which the features and hands are half effaced by the frequent contact of pious lips. Restore such worn parts, and the sentiment is stripped from a relic of past ages. It is far better to leave untouched such scars, which attest the antique piety of the cloister. A vellum Book of Hours of the Fifteenth Century, worn and soiled through prayer, has, to my mind, acquired a venerable patina. Here, a spot of yellow wax; there, the head of a saint blemished by the star-print from a tear of devotion: are not these stains which should be respected? On the other hand, a blot of ink or an oily smear point only to carelessness and should be removed.

About the year 1846, I was invited by M. A Farrens,[5] a skilful restorer of old books, to see in his work-shop a Dance Macabre in quarto, imprinted on paper, at Paris, toward the end of the Fifteenth Century; a rare volume which he was restoring for M. Techner.[6]

The portions already cleaned and restored, compared with those still untouched, excited my admiration. The numerous worm holes, the tom places, had disappeared through an application of paper-paste, so well joined, so well blended in the mass, that I could hardly detect the boundaries of the restorations. The letters and wood-cuts suffering from lacunae had been reformed with great skill on a new foundation. The soiled surfaces of the pages had entirely disappeared before I know not what scraping or chemical action. In a word, M. Farrens was putting into use every secret of restoration to give again to this volume its original lustre.

Ah well! today, I confess, that if I possessed this book in the dilapidated state in which I saw it, I would leave it just as it stood, and limit myself to the indispensable repair of a new and solid binding. Its worn and soiled condition came, very probably, from the frequent and pious turning of its pages, in that monachal perseverance of prayer of which our century knows nothing. Its shocking and decrepit condition had, to my eyes, a secret in harmony with all books of the kind, which, from each page, recall to us our insignificance.

Notes to Introduction

1 A. Bonnardot, *Essai sur l'art de restaurer les estampes et les livres; ou, Traité sur les meilleurs procédés pour blanchir, détacher, décolorier, réparer et conserver les estampes, livres et dessins*, 2nd ed. (Paris: Castel, 1858). The first edition appeared in 1846, and a German translation appeared in 1859. Another translation by J. Fr. Schall, *Ausführliche Anleitung zur Restauration vergelgter, fleckiger und beschädigter Kupferstiche u.s.w.* (Leipzig: Rudolph Weigel, 1863), was also available.

2 Suzanne Karr Schmidt with Kimberly Nichols, *Altered and Adorned: Using Renaissance Prints in Daily Life* (Chicago: Art Institute of Chicago, 2011), 6; see also Suzanne Karr Schmidt, "Printed Bodies and the Materiality of Early Modern Prints," *Art in Print* 1, no. 1 (May–June, 2011): 25–31.

Notes to Reading

1 [The incident, called the St. Bartholomew's Day Massacre, began in Paris on August 23, 1572, when French Huguenots were attacked by Roman Catholics in a frenzy of mob violence that spread to other parts of the country.—Ed.]

2 [This infamous mishap of Paul Louis Courier (1773–1825) occurred in the Laurentian Library in Florence and led to his official expulsion from Tuscany. The story is often repeated either as an example of historical evidence worth preserving or as a justification of the prohibition of ink in reading rooms, for example, by Sir Hilary Jenkinson (1882–1961), who cautions "that even fountain-pens and stylographs of the best makes in the most careful hands sometimes blot and that one blot may be infinitely disastrous. Students are apt to discredit this last suggestion and should have their attention directed to the tale of Paul-Louis Courier and the MS. of *Daphnis and Chloe*: it would be indiscreet to quote a more modern example." Hilary Jenkinson, *A Manual of Archive Administration* (Oxford: Clarendon Press, 1922), 53.—Ed.]

3 In 1809 Paul Louis Courier discovered at Florence a complete manuscript of Daphnis and Chloe, containing a long passage in Part I which was missing in all texts known until that time, and the existence of which, as a connecting passage, had long been a subject of speculation among scholars. Unfortunately, he had hardly more than completed a transcript of his discovery when he accidentally upset a bottle of ink over the original manuscript, partly obliterating the passage. The incident caused a bitter controversy among scholars. Courier was violently attacked and, although he had fifty copies of his text printed for special distribution, was even accused of purposely spilling the ink in order to render his transcript unique. M. S. B.

4 M. R. Yve-Plessis, elsewhere quoted, suggests that it may sometimes be desirable to strengthen the ink of some valuable and desirable signature, instead of removing it, and for this purpose recommends a mixture of: Tannin, six grammes; alcohol, thirty-five grammes; distilled water, one hundred grammes; applied with a small brush and the part afterwards brushed over several times with clear water. This operation, however, should certainly not be undertaken except in extreme cases where the signature appeared ready to entirely fade out. M. S. B.

5 ["M. A Farrens": A comparison with Bonnardot's original work indicates that this is Messr. A. Farrenc. No further information about the restorer is available.—Ed.]

6 ["M. Techner": In the French edition the spelling is *Techener*, referring to Jacques Techener (1802–1870), a well-known Parisian bookseller.—Ed.]

Reading 79

CHRIS CAPLE

Conservation Skills: Judgement, Method and Decision Making (2000)

Chris Caple (1958–), archaeologist, conservator, and senior lecturer at Durham University, gives as a positive example of "suitable" decisions this case study of the treatment of the cartoon The Virgin and Child with Saint Anne and Saint John the Baptist *by Leonardo da Vinci (1452–1519). While few conservators would have any hesitation regarding the appropriateness or success of the treatment, the reasoning behind it, at least as dictated by the drawing's "values," is debatable, since there can be no doubt that the work's fame and astounding financial worth attracted intense public scrutiny.[1] No doubt because of these factors, the bar for acceptable risk was lower.*

It is interesting to consider how arguments in favor of privileging evidence of the drawing's past use as a cartoon (functional value) or of retaining the traces of damage caused by the 1987 shotgun blast (historical value) might have swayed treatment decisions. Clearly an appreciation of the work's original function as a cartoon—a full-scale preparatory drawing for a painting or fresco—was suppressed as early as the eighteenth century when the charcoal drawing was mounted to linen, which at the time was probably a practical means of facilitating handling and ensuring its survival. Many cartoons not only sustain wear and tear through normal usage—their designs are typically transferred onto the painting support by means of pricking or incising—but also have a precarious existence because of their unmanageable size.[2] Today cartoons are valued for the technical information they provide about historical studio practices and compositional changes during the course of the creation of the final painted work.[3] It is expected that cartoons will naturally bear the traces of their former function and that such "damages" will be preserved.

From CHRIS CAPLE, *Conservation Skills: Judgement, Method and Decision Making* (London: Routledge, 2000), 196–99.

As for retaining evidence of its defacement by a shotgun blast, one wonders if a motivating political event, such as the bombing of Guernica, might have been memorialized through nonintervention as opposed to an act perpetuated by a mentally ill person. Again, the work's aesthetic and financial values preempted alternative approaches.

Regardless of its treatment, the paper conservator Salvador Muñoz Viñas (1963–) (reading 95) argues that the "true" or authentic condition of Leonardo's cartoon has not been restored at all, since the "expected, imagined or preferred state of an object is not real unless it coincides with the existing object."[4] While he does not disagree with the necessity of the restoration, or the success of the results, Muñoz Viñas cautions conservators not to justify their treatments using the flawed notion that an object's preferred state can ever be regained once it is lost.

13A Case study: Cartoon—Leonardo da Vinci—*The Virgin and Child with St Anne and St John the Baptist* (Harding and Oddy 1992)

The cartoon *The Virgin and Child with St Anne and St John the Baptist* was drawn by Leonardo da Vinci in 1507–8 using chalk and charcoal on paper. It was purchased by The National Gallery in 1962–3 from the Royal Academy. On 17 July 1987 a man entered the National Gallery and fired a shotgun at the picture. He was subsequently detained in Broadmoor high-security hospital. The picture, which had been protected behind laminated glass, sustained an approximate 100-mm circle of extensive damage. The National Gallery employs some of the world's finest conservators of oil paintings amongst their conservation staff. However, they have less expertise in dealing with works of art on paper and, after seeking advice from a number of experts, seconded Eric Harding, Chief Conservator of Western Pictorial Art at the British Museum (and previously a member of staff at the National Gallery) to conserve the damaged picture.

Detailed observation established that none of the shotgun pellets had penetrated the plastic laminate layer in the glass, although the force of the blast had pushed fragments of glass into the paper, fragmented the paper and caused pigment loss in the damaged area. Small fragments of the detached paper and pigment were analysed, as was the construction of the picture itself. The cartoon is drawn in white chalk and black charcoal, without any binder, on to a piece of paper composed of six smaller sheets stuck together. The surface of the paper had been coloured with a mixture of chalk, soot and iron oxide which had been rubbed into the surface to give a mid-red-tone background on to which the image of the picture was drawn. At a later date, possibly in the eighteenth century, the cartoon had been stuck on to canvas secured over a wooden stretcher. The paper had been poorly attached to the canvas with a number of tears and wrinkles clearly visible. Prior to the attachment to the canvas there had been a number of repairs and patches applied to support tears and cover missing areas of the cartoon; the picture had clearly received poor treatment earlier in life. Several of these repairs were of poor quality. It was desirable to remove the cartoon from its now distorted canvas backing, relax the paper so removing the creases and wrinkles, reback it on to a stable support, redo the earlier repairs and de-acidify the paper which had become brittle because of acidity, a condition which is considerably increased by the adhesive and canvas backing. However, it was felt that such remedial conservation action could not be undertaken since it would risk greater damage to the object:

- Any attempt to remove the paper from the canvas would risk weakening the adhesive holding the six sheets of paper together and thus offsetting part of the original image.
- Unbacked the paper would be difficult to handle and very weak thus there would be a considerable risk of exacerbating some of the existing tears.
- The moisture would differentially affect the pigments dislodging some, whilst binding others closer to the paper. There could consequently be considerable alteration to the visual effect of the image.
- Any dry backing would require working on the picture, face down, with consequent loss of surface pigments.
- Any non-aqueous de-acidification would again require the use of a solvent which would have some effect on the pigment particles and thus alter the visual effect of the image. The chalk highlights appeared particularly vulnerable to alteration through aqueous or solvent treatment.

In view of these factors it was decided that only very localized and limited conservation would be undertaken. After collecting all the shattered pieces of paper and removing all pieces of glass embedded in the paper, the distorted canvas was flattened through localized humidification, using an ultrasonic humidifier and modified suction tables. When flattened the canvas was removed from the wooden stretcher and backed on to another piece of paper and canvas using wheat-starch paste. All the pieces of shattered paper from the damaged area were readhered in place, eventually only leaving approximately 1 cm^2 empty. This was filled with a piece of new paper visually similar to the original which was cut to shape and inserted. There were many thin lines of pigment loss as well as the new piece of paper. Powdered chalk, charcoal and other pigments were used to tone down these areas to harmonize with the rest of the image. Applied dry to the surface this pigment could readily be removed by future conservators if required. The cartoon, mounted in a new frame, again protected with laminated glass, was returned to display in May 1989. The damage is not visible to the casual observer, only detectable upon detailed inspection (see Figure 1).

This conservation work, whilst restoring the image, demonstrated the careful consideration which is needed by a conservator. There was clear recognition of the limitations of their own skills by the conservators of the National Gallery and the benefits of involving more specialized and experienced colleagues clearly paid

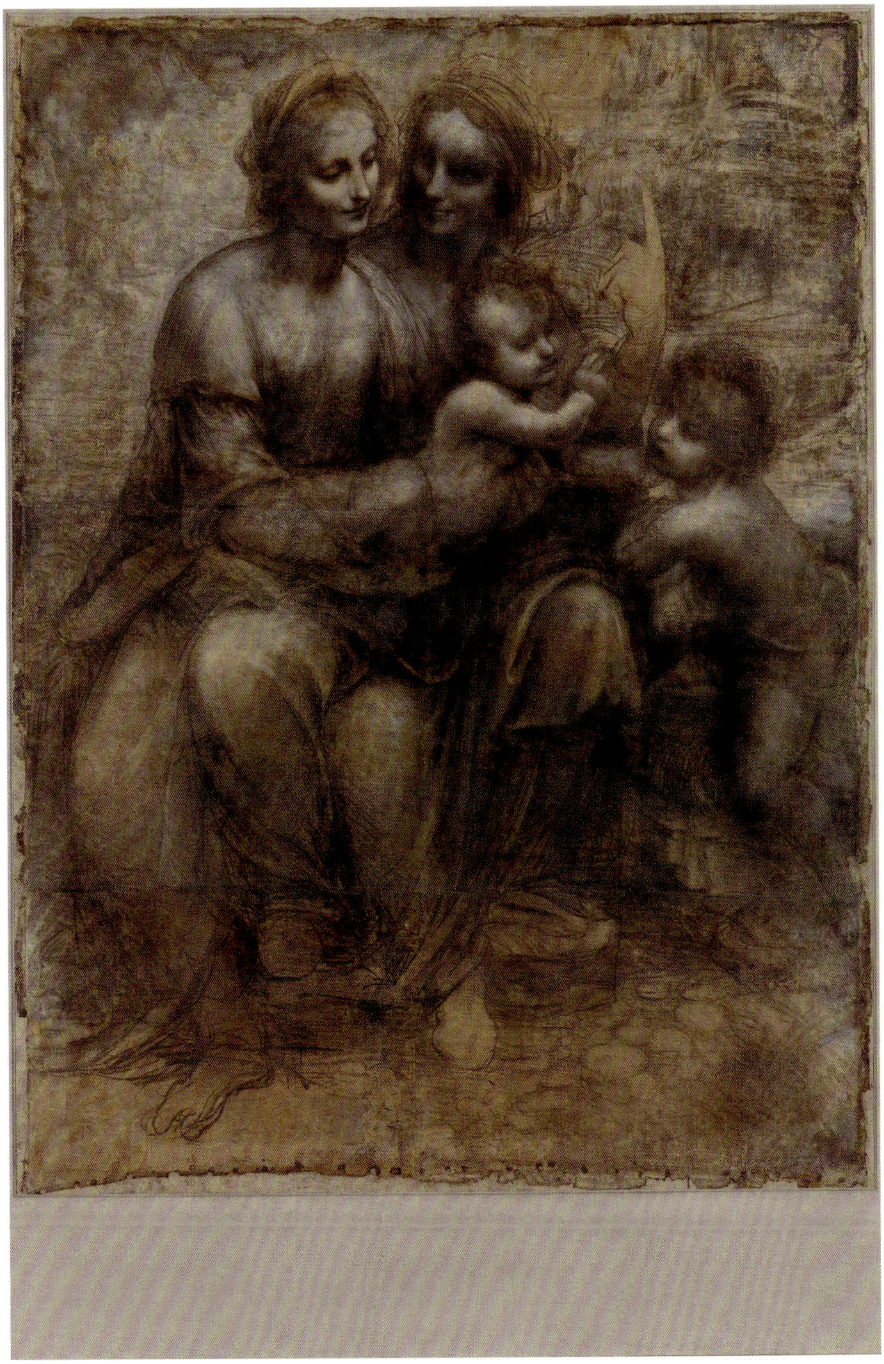

Figure 1
Leonardo da Vinci (1452–1519). *The Virgin and Child with Saint Anne and Saint John the Baptist*, about 1499–1500. Black chalk and touches of white chalk on brownish paper, mounted on canvas, 141.5 × 104.6 cm. Purchased with a special grant and contributions from The Art Fund, the Pilgrim Trust, and through a public appeal organised by The Art Fund, 1962 (NG6337). National Gallery, London, Great Britain © National Gallery, London / Art Resource, NY.

dividends. Examination gave clear information about the materials and techniques used to create this work of art, providing a clear indication of the sensitivity of the materials used. There was very careful judgement made in the extent of conservation, with the potential improvement to the stability and longevity of the cartoon balanced against the risk of further damage.

Notes to Introduction

1 The work was purchased by the National Gallery using publicly raised funds when it came up for sale in 1962. See also Eric Harding and Andrew Oddy, "Leonardo da Vinci's Cartoon *The Virgin and Child with St. Anne and St. John the Baptist*," in *The Art of the Conservator* (London: British Museum, 1992), 28–42; Eric Harding, Allan Braham, Martin Wyld, and Aviva Burnstock, "The Restoration of the Leonardo Cartoon," *National Gallery Technical Bulletin* 13 (1989): 5–28.

2 Interestingly, the Leonardo cartoon is neither pricked nor incised for transfer.

3 Credit for information on the increased appreciation of the evidential value of cartoons is given to Carman Bambach, *Drawing and Painting in the Italian Renaissance Workshop: Theory and Practice, 1300–1600* (London: Cambridge University Press, 1999).

4 Salvador Muñoz Viñas, *Contemporary Theory of Conservation* (Oxford: Butterworth Heinemann, 2005), 93.

Reading 80

ALFRED WHITMAN

The Print-Collector's Handbook (1903)

This reading from Alfred Whitman (1860–1910), keeper of prints at the British Museum, begins with an endorsement of the "kindly touch of Time" on both paper and ink and is followed by an exhortation that a restoration process should not be "carried too far," lest the print's rich and mellow appearance be compromised, even if it had a "March-wind" appearance when first created. The paper conservator Keiko Keyes (1939–1989) makes a similar plea, likening the restoration of original appearances to "forcing an old woman to wear bright clothes and behave like a school girl" (see readings 85 and 87).[1] *As stated in the introduction to Part V, from experience conservators can visualize a work's original appearance and virtually feel and hear what must have been its previous physical and chemical characteristics. They are so habituated to visualizing an artwork's original condition that a proposed conservation treatment can be rationalized by the extent to which it will restore an item's "original" physical, chemical, and visual properties: an embrittled paper becomes more flexible, an acidic paper more neutral, or a mat burn less apparent. But this hypothetical original state must not be conflated with the desired outcome of a conservation treatment, since that goal must be the restoration of a historically appropriate state. The art historian Ernst van de Wetering (1938–) elaborates on people's comfortable relationship with signs of age in objects: "A knowledge of what we now somewhat abstractly call 'natural aging' is acquired early in life. We use this knowledge daily in order to find our way around objects. Although we may not be aware of it, it is specifically the signs of natural aging and wear and tear that often provide us with significant information about the material of which an object is made. These signs also provide instant information about the meaning of an object and about the ways and means in which it was used; they even let us know the extent to which it is valued—or neglected."*[2]

From ALFRED WHITMAN, *The Print-Collector's Handbook* (London: George Bell and Sons, 1903), 100, 104–7.

Ironically, paper conservators have often observed that less latitude in terms of signs of advanced age is accorded modern twentieth-century works on paper.

Some works of art on paper confound our expectations of natural aging. The dazzling white papers of Giambattista Tiepolo (1696–1770) and the chocolate-colored papers of Winslow Homer (1836–1910) have raised the specter of overaggressive bleaching and drastic deterioration of paper, respectively. In fact, it is probable that the appearance of these papers has barely changed at all.[3]

An impression (paper as well as ink) ripens or mellows in the same way as fruit. If anyone, when turning out an old cabinet, comes across a letter written, say, half-a-century ago, he will notice that the ink has changed its tint, and the paper has softened its tone and become quite different in appearance from the note-paper in current use. This mellowing process takes place in the paper on which engravings have been printed, as well as in that upon which letters have been written; and there is little doubt but that the old stipple and mezzotint prints originally presented some of the same March-wind[1] appearance possessed by modern impressions. Time has undoubtedly laid a kindly touch upon well-preserved prints, as upon old violins, and has added much to their richness and beauty. How one shrinks from reflecting that the poor quality of the paper of to-day will cause Time to destroy rather than enrich, as we find to be the case with the hand-made papers of the seventeenth and eighteenth centuries!

[. . .]

We have in a previous chapter alluded to the restoration of prints. The collector will need to know a reliable restorer to whom he can entrust his treasures to be cleaned and repaired, and *care must be taken that the work done to the prints is not carried too far*. The word reliable is used advisedly, for a restorer who is not an expert and thoroughly conversant with his trade may work much mischief that may not disclose itself until several years after the print has been restored. Then for the safety of the collection it will be well to have the prints mounted upon cardboards of one or more regular sizes, and they should not be fastened down to the boards, but only attached to them on one side with a paper hinge, so that the backs of the prints may be examined when necessary. It is also a good plan, where a print is of great value, to hinge a second sheet of card (having an opening cut through it a trifle smaller than the size of the print) to the one upon which the engraving has been attached, so as to save the surface of the print from lateral friction.

If the amateur wishes to get the most enjoyment from his possessions, he may do so by having a number of frames made, with movable backs, of sizes to range with his mounted prints, and then by simply from time to time changing the prints they contain, he can without any trouble alter the appearance of a room, for a series of etchings may give place to a set of Dürer engravings, or a batch of mezzotints may be exchanged for a group of stipple prints.

A mistake sometimes made by amateurs is when, after visiting a public collection and seeing the strong methods employed there for the preservation of the prints, they go home and imitate the style for their own use. It should always be remembered that the needs for strength in a public collection, where the prints are being constantly handled by students, do not apply in a private cabinet where the owner usually handles (or, shall we say, should handle) the prints himself while showing his treasures to his friends. Therefore, while every precaution ought to be taken to suitably preserve his collection, the amateur will not need to imitate the mounting and placing methods which prevail in a public institution. By bearing this fact in mind he will be spared space for storage, unnecessary weight, and expense. For private use, and where the wear and tear is small, it is by no means a bad plan to back modern prints, which usually are on a rough paper, with tissue, for then when the prints are placed in a pile one on another the tissue softens the pressure, and reduces the risk of injury. The late Henry Vaughan did not devote attention to the mounting of his prints, but took the great precaution of preserving his most valuable treasures in strong iron boxes.

All amateurs are so familiar with what are known as 'sunk mounts,' that they may never have thought there was a time when sunk mounts were unknown, or ever wanted inventing. It will therefore interest many to learn how the first sunk mount came into existence. Some forty years ago, when the Print Department of the British Museum was still quite small, the drawings were kept loosely in a couple of portfolios. There they were subject to rubbing each time the portfolios were handed to students; and though it was felt that injury was constantly being done, a way to obviate the wear and tear could not for a long time be devised. At last one of the officials, Mr. W. M. Scott, thought he could overcome the evil by placing each drawing on a cardboard mount, so made that the surface of the drawing would be below that of the surrounding card. This result he obtained by pasting a sheet of cardboard, in which an opening had been cut, upon a second and uncut sheet of cardboard. The drawing was then mounted in the cut opening, and, being thus below the surface of the surrounding card, escaped all wear and tear. In this way the first sunk mount was made or 'invented,' and it was at once found to be in every way successful. The system has now come into universal use wherever prints and drawings are mounted, and the cutting of mounts has become a special trade employing a great number of hands.

Notes to Introduction

1 Keiko Keyes, "The Unique Qualities of Paper as an Artifact in Conservation Treatment," *Paper Conservator* 3 (1978): 7.

2 Ernst van de Wetering, "The Surface of Objects and Museum Style," originally presented at the Third International Seminar for Restorers, Veszprém, Hungary, 1981; published under the title "Die Oberfläche der Dinge und der museale Stil," *Maltechnik Restauro* 2 (1982): 417.

3 The high-quality fibers and presence of calcium carbonate in Tiepolo's papers contribute to their excellent state of preservation (author's observation). Homer's dark brown papers are pigmented with anthracite coal and charcoal (Roy Perkinson, "Observations on the Drawings of Winslow Homer," *Book and Paper Group Annual*, American Institute for Conservation of Historic and Artistic Works, vol. 5, 1986, 2.)

Note to Reading

1 [There is no listing for "March-wind" in *Color: Universal Language and Dictionary of Names*. However, Sherwin Williams, a leading paint manufacturer, offers a cool blue-gray paint by that name.—Ed.]

Reading 81

Carlo James

Old Master Prints and Drawings: An Art-Historical Survey of Collecting and of Preservation Methods (1992)

As a paper conservator at the Fondation Custodia in Paris, Carlo James (1940–) has benefited from working with the more than 7,000 drawings and 30,000 prints collected by Frits Lugt (1884–1970), many of which remain in their original presentation albums and on their decorative mounts.[1] *By demonstrating the evidentiary value of accumulated accretions and markings as the works changed hands, James has forced conservators to reevaluate the seemingly benign practice of removing extraneous, often nonarchival, substances from prints and drawings in the name of preservation. Practicing paper conservators naturally sympathize with the instinct to remove clumsy repairs and poor-quality mounts when confronted with the negative consequences of their use. But the scattered scraps of a rose-colored album page on the verso of a drawing by Francisco Goya (1746–1828) point to previous ownership by his son Javier.*[2] *And similarly, the military-issue folder stock onto which Paul Klee (1879–1940) routinely mounted his drawings has been frequently removed, deacidified, split, and lined, or disposed of entirely, for fear of acid migration (see reading 54).*[3] *Today conservators acknowledge that many of the prophylactic procedures undertaken to preserve prints and drawings or to make them easier to handle unintentionally remove valuable evidence of the history of collecting, which has become a stand-alone academic discipline.*[4]

The paper conservator Marjorie Shelley offers another perspective on the value of retaining historical mounting formats by pointing out that our current disapproval of flattened laid-down drawings may once have been the opposite: "Those who mounted these works were perhaps closer to the aesthetics of the old masters than are we. This is similarly the case with pastels of the late seventeenth and eighteenth

From Carlo James, "Old Master Prints and Drawings: An Art-Historical Survey of Collecting and of Preservation Methods," 86–90. This is an excerpt from an article that was originally published in *Conference Papers, Manchester 1992*.

centuries. Their original format, paper stretched over a wooden strainer, not only provided a resilient surface for the artist to work on but it gave the pastel the stature of an oil painting, the medium with which it was perennially in competition. The mounting tells something about how a drawing or pastel was used and displayed, and about its cultural context, and for this reason alone it should be preserved."[5]

Many decisions to dispose of mounts, mats, and other housing materials are not conservation driven but are dependent on shifting decorating schemes, especially the preference for "purer" presentation styles. This has led to the removal of the garish gold foil mats used by the photographer and art dealer Alfred Stieglitz (1864–1946), the pastel-colored decorative mats favored for French eighteenth- and nineteenth-century drawings, and the hand-painted frames of the American Modernist John Marin (1870–1953) and the Post-Impressionist Charles Prendergast (1863–1948). Also, for practical reasons, albums of prints and drawings are frequently disassembled for ease of exhibition or for increased profit at sale.[6] *This latter practice results in the loss of information regarding the significance of sequence, patronage, and, of increasing importance, legal provenance. Ominously, once an album is disbound, the likelihood of overexposure to light of its more colorful and hence more vulnerable pages is drastically increased, for example, the most striking species contained in* Birds of America *by John James Audubon (1785–1851) or exotic varieties in* Les roses *by Pierre-Joseph Redouté (1759–1840). Some institutions, such as the Teylers Museum in Haarlem, The Netherlands, have intentionally returned to storing their prints and drawings in custom-made albums that conform to strict conservation criteria.*

Introduction

[. . .]

At first sight, conservation and art history would seem to have but tenuous links. Nowadays, preservation is the most frequent and pressing topic discussed by the keeper, art historian and the conservator-technician. When one looks at the history of the preservation of works of art on paper, one is struck by the fact that many conservation problems stem from the preservation techniques employed. Prints and drawings suffered as a result of being mounted on board or in albums (poor quality paste, mounting paper or board, surface damage, etc.). Furthermore, many of these fragile works have been mounted and dismounted a number of times. The principal, and not always obvious reason for this, as I maintain in this paper, seems to have been that changing ideas about the interpretation and classification of works of art have led to the rearrangement of many collections. This ill-defined and complex field, which lies somewhere between art history and preservation, has begun to be explored only recently.[1]

For centuries now, works of art on paper have been regarded as works of art in their own right: today we prefer to see them with no extraneous interference, that is without any of the once-fashionable decorated mounts and richly ornamented frames. No longer does "Bonnat[2] bound lightly up the staircases of the Louvre simply to see what mounts Proust,[3] the museum's mounter of drawings, has devised for its Michelangelos or its Leonardos."[4] However, signs, marks, 'traces' left by collectors, which have hitherto seldom been interpreted, and have now lost their original function, provide precious evidence of taste and methodology. Increasingly, they are being considered as an integral part of the prints and drawings with which they are associated and old mounts are no longer regarded as mere means of protection and embellishment.

Changes in the aesthetics and methodology of preservation have led, almost inevitably, to the loss of the oldest mounts, to the break-up of albums and the dispersal of collections. However, in view of this troubled history, it is surprising that so many material remains still exist.

Preservation Methods

The simplest and most obvious way for a collector to protect his works of art was to slip them between the pages of a manuscript or book in his library. As early as the beginning of the sixteenth century, Hartmann Schedel[5] was sticking prints on to the blank pages of his books and embellishing them with coloured borders. It was then but a short step to bind blank pages together, making a volume into which prints and drawings could be stuck or slid between the leaves. The first volumes of this kind date from the early sixteenth century;[6] however, the oldest books of drawings are undoubtedly the model-books.

> That a surprisingly large number of model-books and model-book pages survive from around 1400 indicates both that they were a particularly popular working tool at that time, and that they were made to be preserved. They were intended to contain definitive studies which would still be useful to the later generations who inherited the artist's workshop, and were venerated not only as reservoirs of *motifs* but also as yardsticks of the workshop's tradition. These aspects of the use of model-books suggest that they held a fundamentally important and prestigious place in the workshops of the early *quattrocento*.[7]

It must be emphasised that these model-books were made up of bound drawings or of sketchbooks *avant-la-lettre* and not of drawings stuck on to the leaves of an album, as was the case later on. Works could not be removed without the whole album being taken apart.

Schedel's practice of embellishing his works with ornamental borders soon spread throughout Italy. The best-known, and in one sense unique, example is provided by the famous collector Giorgio Vasari.[8] In fact, T. Vigneau-Wilberg, comparing Gaddi's mounts with those of Vasari, from whom he took his inspiration, wrote:

> The collector—in Vasari's case this was also the artist who drew the decorated washed mounts—remains completely subservient to his works of art, whereas Gaddi's mounts are deployed as a sort of collector's mark—the motto [*'tan che gie viva'*] showing clearly who is the owner.[9] The mounts are all designed in the same manner and are unrelated to the drawings to be placed in them. Thus any drawing in the collection could be put in any mount. The function of the washed mount is no more than that of a *passe-partout*.[10]

Mounts quickly became simpler and simpler, disappearing altogether in Leopoldo de' Medici's collection.[11] This collection was arranged by Filippo Baldinucci[12] and he had no hesitation in doing away with the Vasarian mounts[13] to meet an aesthetic need we would recognise today. However, a little later on J. Richardson Sr.[14] complained bitterly:

> [. . .] know myself that this Ornament is of his [that is Vasari's] own hand, or of some of those Assistants which He employ'd, having seen the whole Books that He so often quotes, for the most part still unbroken up, which were in ye possession of the famous Collector, Mons. *Crozat* at *Paris*;[15] He having bought them Entire at *Florence*; tho' He had taken to peices [*sic*] some of the Books, which I well rememb' I abus'd him for, as a sort of sacriledge in the *Vertù*.[16]

I'm not sure how far we can trust Richardson's accuracy, but clearly at the beginning of the eighteenth century people were already beginning to realise the inestimable loss resulting from the break-up of the volumes of famous collectors and the destruction of their annotations and mounts. However, this did not prevent Richardson from breaking up another famous collection, the one put together and annotated by Padre Resta for Bishop Marchetti.[17] In other words there are plenty of reasons, as we shall be seeing later on—not simply a lack of respect for the taste and knowledge of others—prompting collectors to break up and reorganise older collections.

Because collectors wanted to have their finest drawings on show, to be able to pass them around freely when connoisseurs called, and to frame them, they came up with the idea of giving them a backing that was stronger than a single sheet of paper and of embellishing them in order to set them off. Here too, collectors had no hesitation in taking drawings off older mounts and putting them on their own, in order to make it clear whose they were.

The first of these mounts seem to be French and to date from the first half of the seventeenth century. The style soon spread to England, and then to Holland. In Italy, the practice of mounting on board was sporadic in the nineteenth century and no ornamental borders were used.

We know that Mariette[18] kept his drawings mounted and in portfolios. Richardson wrote initials and numbers on the backs of his mounts to show their place in the portfolios, or 'art books', and on which shelf of the specially adapted cabinets they belonged. In Holland, drawings seem to have been kept mainly stuck in albums, and it was only in the eighteenth century that the practice of mounting on 'elephant' paper[19] with a very restrained border was adopted. These mounts were kept either in 'art books' or in portfolios.

The Dispersal of Collections

Albums of prints and drawings were broken up for many reasons. Some were economic, or were to do with preservation or security, but the main reason was that it was felt necessary to rearrange the collection. There might also have been a combination of several factors which could vary considerably according to when they arose. It is all too easy to imagine the untold harm that successive and inexpert dismounting and remounting will have done to these fragile works: loss of part of the drawing itself, fading, tears, corners or borders ripped off—in a word, a significant number of the problems we deal with every day. At first sight, it would seem that in the eighteenth and nineteenth centuries—not to mention today—albums were taken apart for economic reasons. However, it was not as simple as that; works of art on paper, originally conceived as means of imparting information along with the written word, gradually came to be sought-after for their intrinsic artistic value, in part thanks to the fame of the great masters. It became well nigh impossible, even for a wealthy collector, to acquire an important album as prices had become prohibitive. Furthermore, there were serious handling problems connected with some albums "which are so big they are like church books".[20]

A large number of print collections, up until the beginning of the nineteenth century, were stored in albums and the prints were stuck in by their edges, by their corners or with paste all over their backs. This had one serious drawback: in order to save space many prints were trimmed to the plate mark.

To give an idea of the scale of the problem collectors were up against here are some figures:

> When the Abbé de Marolles' (1600–81)[21] collection was acquired by Louis XIV in 1667 it comprised an incredible 123,000 prints in 520 volumes; and he managed to collect a further 110,000 sheets in 237 volumes!
>
> Prince Karl Paar (1777–1819)[22] collected 60,000 prints which he kept in 150 portfolios.
>
> Count Moriz von Fries (1777–1826)[23] owned more than 100,000 prints and drawings.
>
> Lord Fitzwilliam (1745–1816)[24] bequeathed his collection to the city of Cambridge; it contained, among other things, 500 richly bound volumes of prints.

It was not (is not), therefore, surprising to find all too often sheets stacked in their thousands and with no protection in 'bulging portfolios' (E. de Goncourt).[25] Moreover, one of the main inconveniences of sticking prints and drawings in albums was that it became impossible to insert new acquisitions where they belonged in the collection. Often, therefore, a collection reached its definitive form only at a very late stage of its completion. As early as 1565 "Quiccheberg[26] recommends placing prints on shelves and in drawers at first [. . .]. But prints are finally to be bound into volumes once each series is filled out, leaving blank spaces for additions where appropriate."[27]

It is hard to understand why Leopold de' Medici's practice of mounting his drawings on unbound sheets of the same size—as he mentions in a letter to Bernardino della Penna[28]—was not adopted more often: "[. . .] I'm involved in sticking and adjusting my drawings in order to put each one into a book." 'Book' here would seem to mean portfolio.[29] The meaning of the word 'adjust' is only too clear . . .

The organisation of collections of drawings advanced more rapidly than that of print collections; this is because drawings were hard to come by, because they were highly sought-after even before printing was invented, and, let's face it, because of their price. Nor were collections of prints and collections of drawings organised in the same way.

P.W. Parshall, in his detailed study of the collection of Ferdinand, Archduke of Tyrol (d. 1595), today in the Kunsthistorisches Museum in Vienna, describes its deplorable state: "Prints are mounted with a lack of care comparable to the construction of Ferdinand's collection prints often trimmed or folded, many removed or damaged through use";[30] this would seem to imply that the function of the collection was more didactic and practical than speculative, with the notable exception of the three Dürer volumes, the first of which probably comes from one of his pupils; in this volume the prints have been carefully mounted and a simple ink border has been drawn around them.

During the second half of the sixteenth century, collections of graphic art were normally part of a library or of a *Kunst und Wunderkammer*, whose function was essentially encyclopaedic, and they fitted into the general organisation of the collection chiefly by their illustrative aspect. There seems to be no other logical explanation for their classification. Furthermore, the existence of large numbers of albums of ornaments suggests a purely utilitarian function. Moreover:

> The preference for old masters was already prefigured among humanist collectors even in the lifetime of Dürer. And collecting with some historical dimension is intrinsic to the major *Kunstkammer* which tended to keep antiquities separated from later materials. Furthermore, the topical or encyclopaedic aspect of graphic collecting persisted long after the Renaissance period, indeed well into the nineteenth century when museum print cabinets were still often

> arranged as topographical repositories rather than according to styles, schools, or individual masters.[31]

In this context, therefore, an undertaking like Vasari's does not seem entirely unwonted; but an historical approach of this scope could only have been the brainchild of a painter and an intellectual of his stature.

Although albums of works by Dürer, Altdoner and Lucas of Leyden were to be found, collections were still essentially made up of what was available, that is to say, the output of the great publishers like Hieronymus Cock of Antwerp, Antonio Salamanca and Antonio Lafrery in Rome, who commissioned drawings from well-known artists and had them copied by engravers with varying degrees of skill. These series of prints of allegorical or biblical subjects, and of Roman antiquities, etc., were exactly in tune with the spirit of the times.

In the seventeenth century the organisation of collections of drawings grew more refined and began to show variations. The learned and scrupulous Baldinucci followed Vasari's example, whereas Padre Resta used various highly elaborate methods to organise the different volumes of drawings that he put together during his lifetime. These, as their sometimes delightful titles suggest, were complete in themselves: one was in chronological order, illustrating the development of the art of drawing, (*Parnasso della Pittura* [. . .]); another, preserved intact at the Biblioteca Ambrosiana in Milan, is called simply *Galleria Portatile*; an other, *Felsina Vindicata contra Vasarium*, was meant to show the superiority of the Bolognese School over the Florentine.[32] These principles of organisation, which were still essentially on historical lines, were also widely followed in France (Everhard Jabach).[33]

During the second half of the seventeenth century the arrangement of print collections developed considerably, but was by no means standardised. Moreover, print collections still had a utilitarian and didactic function. A series of prints taken from the works of a famous painter was supposed to provide a source of inspiration and a model for an aspiring painter. When such prints were to go into a library, they would be arranged, for example, according to historical criteria.[34] Prints were also arranged according to schools of painting, and this method became increasingly widespread towards the end of the century. Print collecting was one of the intellectual pursuits of a 'gentleman', and the different methods of arrangement sometimes called for a collector to own several copies of the same print and required highly complex reference systems. The wide variety of the classifications was due in large measure to the differing functions attributed to collections.

Prints in seventeenth-century collections were not only stuck into albums; they were also grouped in series and bound into volumes, often according to a publisher's prescription. Indeed, publishers produced their prints with wide margins to enable collectors to have them bound and place them in their library.

During the early part of the eighteenth century, Pierre Crozat was still using an essentially historical method of arrangement; however, this was a time when other methods were beginning to appear. The strictly chronological classification adopted by Lord Somers, probably on the advice of J. Richardson Sr., was disastrous for the Marchetti collection, assembled by Padre Resta and acquired by Somers in 1711, despite Richardson's great respect for the Padre's attributions and commentaries.

> In Somers' collection (and in Richardson's), the principle of stylistic development within schools was sacrificed for the sake of precise chronology irrespective of provenance. Furthermore, this sort of universal progress was of such importance that every volume of the collection was a self-contained illustration of it from beginning to end.[35]

Thus, Italian, French, Flemish and English drawings might end up in the same volume, occasionally making for some strange bedfellows. English collectors, following the French example, were starting to use board mounts and this enabled them to rearrange their collections at will. About this time, A.J. Dézallier d'Argenville began to favour arrangement by subject, a method which was widely practised a century later.[36] (Fig. 1.)

Also at this time print collections were being hived off from libraries; their intrinsic artistic merit was at last being acknowledged and they were considered on a par with collections of paintings—a status which collections of drawings already enjoyed. Indeed, although prints were becoming increasingly sought-after, the method of arranging print collections still lagged behind that of collections of drawings and retained a certain ambivalence:

> Prints at this time were gaining in prestige, witness the fact that collectors were concentrating on acquiring engravings and etchings. The purpose was to put together as complete a collection as possible and to provide the receptive mind with a means of encyclopaedic education with regard to the universe and to other realms of knowledge as well. Thus, reproduction prints had a key position in the conception of a collection, alongside an original print e.g. a Rembrandt etching. Collecting drawings could not satisfy such a need; in this case the ideal solution was to collect works of famous artists as evidence of their creative activity.[37]

The second half of the eighteenth century saw prints and drawings beginning to be classified by school and artist, first chronologically, then alphabetically; these criteria were to take over at the beginning of the nineteenth century and lead to the wholesale break-up of most of the albums in public and private collections. The instigator of this revolution, whose system is the forerunner of the one used today,

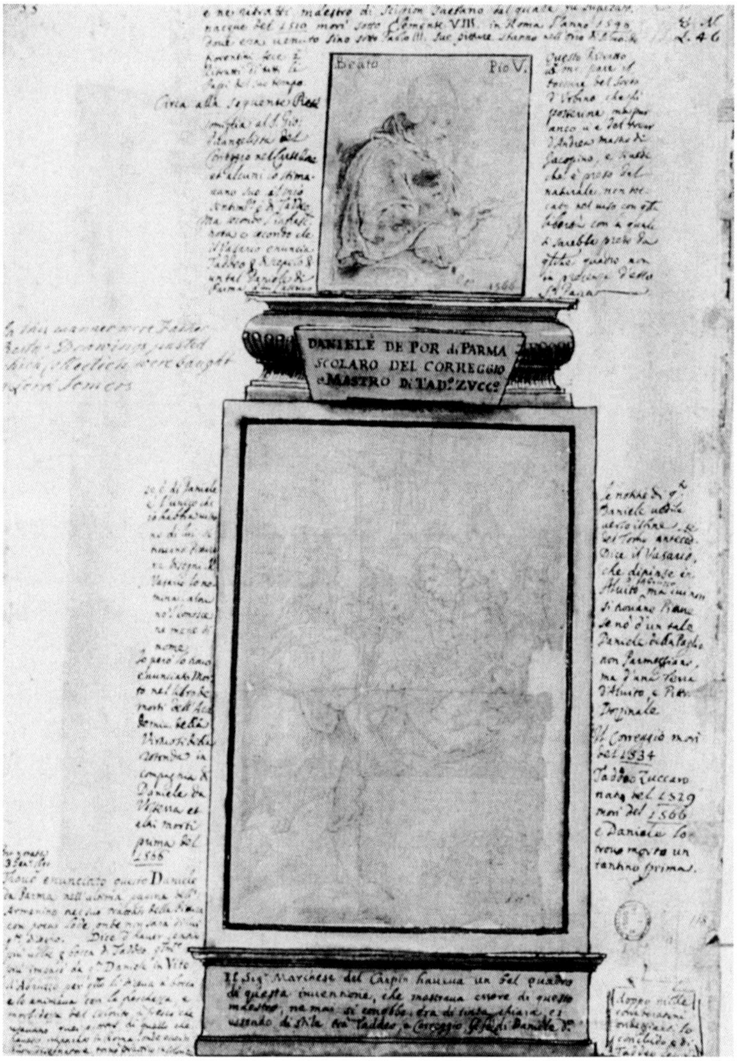

Figure 1
Album page of the Marchetti collection, with the annotations by Father Resta on the drawings. In the upper right-hand corner are Lord Somers' and J. Richardson Sr's shell numbers. On the left, in Richardson's hand, is written: 'In this manner were Father Resta's Drawings posted which collection were baught [*sic*] by Lord Somers.' (British Museum, London, inv. no 1900-2-14-1 b.)

was Pierre-Jean Mariette, whom Adam Bartsch thought the most learned man of his century on the subject.

It was the increasing importance attached to the executant (be this the artist, engraver or a copyist) which led to this new classification system. Nevertheless, the old systems continued to exist alongside the 'modern'. An artist's drawings were grouped in one or, if there were many of them, several 'art books'; or drawings of different artists were grouped by subject: landscapes, seascapes, figures, still lifes, topographical views, etc.[38] This idea, which proliferated throughout the eighteenth

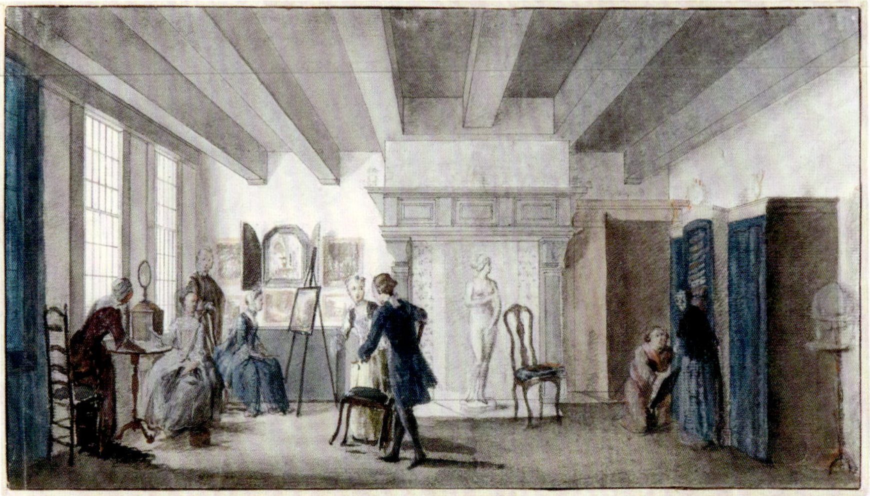

Figure 2
G. van der Mijn, Interior of the 'art room' of the collector Ploos van Amstel with on the right the cupboards containing the 'art books' between the leaves of which were stored his prints and drawings collection. Reproduced courtesy of Fondation Custodia, Collection Frits Lugt, Paris.

century, of systematically classifying drawings according to schools and artists, and prints according to schools and *inventores*—the latter abandoned at the turn of the century, in favour of the *graveurs*—was a timely issue, witness the fact that a collector like Cornelis Ploos van Amstel, thoroughly conservative by nature and upbringing, could, at the end of his life, have envisaged rearranging his collection—classified by subject—in a more rational manner (fig. 2).[39] Scientific classification, if not artistic value, gained precedence once and for all over arrangement by content of a work of art.

I have tried to give some examples of these 'traces' and of the historical facts I mentioned at the beginning. I believe it is important for restorers to have a strong grasp of the historical background and of the facts which have led to the problems they find themselves dealing with, just as they know and understand the physical and chemical conditions that bring about the deterioration of works of art on paper.

Notes to Introduction

1 Frits Lugt is best known for his encyclopedic reference work, *Les marques de collections de dessins et d'estampes*, published in 1921, followed by a supplement published in 1956. A revised and enlarged version is available at www.marques decollections.fr.
2 As in *Solo Porque le pregunta, si esta buena su madre se pone como un Tigre*, Thaw Collection, The Morgan Library & Museum; EVT 84.
3 In fact, due to its clay filler and highly polished, nonabsorbent facing paper, no damage to the Klee drawings caused specifically by his favorite cardboard has ever been observed. See Elizabeth Kaiser Schulte, Margaret Holben Ellis, and Antoinette

King, "An Approach to the Conservation Treatment of Paul Klee Drawings," *Book and Paper Group Annual*, 5 (1986): 2; see also Elizabeth Coombs Leslie, "Removing Mounts from Works of Art on Paper: Case Histories to Illustrate the Decision-Making Process," *Book and Paper Group Annual* 13 (1994): 29–30; Nina Quabeck, "Change the Frame and You Change the Game? Research and Reevaluation of the Presentation Formats of the Kunstsammlung Nordrhein-Westfalen's Paul Klee Collection," *Book and Paper Group Annual* 31 (2012): 69–80.

4 See Peter Parshall, "Art and the Theater of Knowledge: The Origins of Print Collecting in Northern Europe," *Harvard University Art Museums Bulletin* 2, no. 3 (Spring 1994): 7–36; also, Antony Griffiths, "Print Collecting in Rome, Paris, and London in the Early Eighteenth Century," *Harvard University Art Museums Bulletin* 2, no. 3 (Spring 1994): 37–59. This multidisciplinary enterprise is represented in many art history faculties; supporting scholarly resources include the *Journal of the History of Collections* published by Oxford University and the Center for the History of Collecting hosted by the Frick Collection, New York.

5 Marjorie Shelley, "Old Master Drawings: An Approach to Conservation," *Book and Paper Group Annual* 8 (1989): 62 – 66; Shelley has written extensively about the history of the pastel medium. Katherine Baetjer and Marjorie Shelley, *Pastel Portraits: Images of the 18th Century* (New York: The Metropolitan Museum of Art, 2011).

6 Happily, the Spencer Albums, compiled by the famous Mariette family, remain deliberately intact in nine volumes containing 3,588 prints. Marjorie B. Cohn. *A Noble Collection: The Spencer Albums of Old Master Prints* (Cambridge, Mass.: Harvard Art Museums, 1992). (See reading 80.)

Notes to Reading

1 Apart from the monumental work by Lugt, F., *Les marques de collections de dessins & d'estampes*, Amsterdam, Vereenigde drukkerijen 1921. *Supplément*, The Hague, Martinus Nijhoff 1956.

2 [Léon Joseph Florentin Bonnat (1833–1922), French painter.—Ed.]

3 [While the French author Marcel Proust (1871–1922) spent hours in the Louvre at this time, he does not appear to be related to a mounter named Proust.—Ed.]

4 Chennevières, P. de, *Souvenirs d'un Directeur des Beaux-Arts*, Paris, Arthena 1979, V, p. 79.

5 [The print collector Hartmann Schedel (1440–1514) was the author of the *Nuremburg Chronicle*, first published in 1493.—Ed.]

6 One of the oldest known volumes, and which is believed to have been compiled by and for a collector, is to be found at the Fondation Custodia (Coll. F. Lugt) in Paris; it came from the painter Antonio Il Badile (1424–c. 1507) and is dated 1500. It remained in the collection of the Moscardo family until the beginning of the twentieth century when the drawings were dispersed.

7 Ames Lewis, F., *Drawing in Early Renaissance Italy*, New Haven and London, Yale University Press 1981, pp. 63–64.

8 [Giorgio Vasari (1511–1574), Italian artist, architect, and writer.—Ed.]

9 [It is said that the Florentine nobleman Niccolo Gaddi (1537–1591) owned five of Vasari's albums of drawings as well as many others. The motto "tan che gie viva" is taken from the family's coat of arms. T. Vignau-Wilberg, "'Qualche deseigni d'importancia': Joris Hoefnagel als Zeichnungssammler," *Münchner Jahrbuch der bildenden Kunst* 38 (1987): 196.—Ed.]

10 Vignau Wilberg, T., "'Qualche deseigni d'importancia". Joris Hoefnagel als Zeichnungssammler', *Münchner Jahrbuch der bildenden Kunst*, XXXVIII, 1987, pp. 185–214; 196.

11 [Leopoldo de' Medici (1617–1675), Florentine cardinal, scholar, and patron of the arts. His collection of drawings is now in the Uffizi.—Ed.]

12 [Filippo Baldinucci (1624–1697), Florentine art historian, connoisseur, and curator of the Medici collections.—Ed.]

13 Chastel, A.; Lorgues Lapouge, C.; Monbeig Goguel, C., *Giorgio Vasari, les vies des meilleurs peintres, sculpteurs et architectes*, XII, Paris, Berger-Levrault 1989, p. 17.

14 [Sir Jonathan Richardson Sr. (1665–1745) was an English artist, writer, and drawings collector. A product of the Enlightenment, he believed that anyone, regardless of pedigree, could become a connoisseur through close scrutiny of original artworks.—Ed.]

15 [Pierre Crozat (1661–1740) was a wealthy French collector of drawings, including those owned by Vasari, which were sold at auction shortly after his death.—Ed.]

16 See Popham, A.E., *Catalogue of the Fenwick Collection*, Cheltenham, privately printed for T. Fitzroy Fenwick 1935, p. 63.

17 [Sebastiano Resta (1635–1714) was a collector in Rome who amassed approximately 3,500 drawings in over thirty albums, only a few of which have survived intact. Giovanni Mattei Marchetti (?–1704) was Resta's most important patron, for whom he assembled nineteen albums containing almost 2,500 drawings. See Genevieve Warwick, "The Formation and Early Provenance of Padre Sebastiano Resta's Drawing Collection," *Master Drawings* 34, no. 3 (1996): 239, 242.—Ed.]

18 [Pierre-Jean Mariette (1694–1774) was a French collector of prints and drawings who cataloged many of Crozat's drawings. Mariette and Vasari designed elaborate decorative mounts and systematically assembled and arranged drawings into albums.—Ed.]

19 ["'elephant' paper" refers to a standard size of British paper measuring 23 × 28 in., which could be easily folded in half to make a folio.—Ed.]

20 Quoted by Petrioli Tofani, A., in *Restauro e conservazione delle opere d'arte su carta*, exhibition catalogue, Florence, Casa Editrice Leo S. Olschki 1981, p. 166.

21 [Michel de Marolles, French monk and translator of classical Latin poets, is known for his collection of engravings, acquired for Louis XIV by Colbert. It is considered the foundation of the print department of the Bibliothèque nationale de France.—Ed.]

22 [Karl Paar (1772–1819) assembled over 150 albums of prints by such masters as Schongauer, Dürer, and Rembrandt. See Lugt #2009, www.marquesdecollections.fr.—Ed.]

23 [Count Moritz von Fries (1777–1826) was a Viennese banker and patron of the arts to whom Beethoven dedicated his Seventh Symphony. See Lugt #2903, www.marquesdecollections.fr.—Ed.]

24 [Richard FitzWilliam, 7th Viscount FitzWilliam of Merrion, bequeathed to the University of Cambridge his works of art and library, together with funds to house them, to further "the Increase of Learning and other great Objects of that Noble Foundation." www.fitzmuseum.ac.uk/about/museumhistory/html.—Ed.]

25 ["E. de Goncourt": Edmond de Goncourt (1822–1896), French literary and art critic, social commentator, and book publisher.—Ed.]

26 [Samuel Quiccheberg (1529–1567) published a detailed classification system for organizing collections of both natural specimens and man-made artifacts in 1565, the *Inscriptiones vel tituli Theatri Amplissimi*.—Ed.]

27 Parshall, P.W., 'The Print Collection of Ferdinand, Archduke of Tyrol', in *Jahrbuch der Kunsthistorischen Sammlungen in Wien*, 78, 1982, pp. 139–90: 183.
28 [A correspondent of Leopoldo de' Medici.—Ed.]
29 Petrioli Tofani, *op. cit.*, p. 167. Hendrik Busserus (1701–82?) used this method for his *Atlas van Amsterdam*, which was acquired after his death by C. Ploos van Amstel. The maps, prints, drawings and documents referring to Amsterdam and its environs were mounted on unbound sheets which were kept between the leaves of 'art books'; this meant they could be reclassified.
30 Parshall, *op. cit.*, p. 143.
31 Ibid., p. 144.
32 Gibson Wood, C., 'Jonathan Richardson, Lord Somers's collection of drawings and early art-historical writing in England', *Journal of the Warburg and Courtauld Institutes*, 52, 1989, pp. 167–87: 177.
33 [Everhard Jabach (1618–1695) accumulated over 4,000 drawings in twenty-six portfolios, which he left to the Louvre. Jabach was known to retain artisans to rework many of the drawings in his collection.—Ed.]
34 Waals, J. van der, *De prentschat van Michiel Hintopen*, exhibition catalogue, Amsterdam, Rijksprentenkabinet 1988, p. 14.
35 Gibson Wood, *Op. cit.*, p. 180.
36 Dézallier d'Argenville, A.J., 'Lettre sur le choix & l'arrangement d'un Cabinet curieux', *Mercure de France*, XII, pp. 1295–1330.
37 Kuhrmann, D., *Zeichnungen aus der Sammlung des Kurfürsten Carl Theodor*, exhibition catalogue, Munich, Staatliche Graphische Sammlung 1984, p. 8.
38 Laurentius, T; Nierneijer, J.W.; Ploos van Amstel, G., *Cornelis Ploos van Amstel, 1726–1798, Kunstverzamelaar en Prentuitgever*, Assen, Van Gorcum 1980, p. 11.
39 Ibid.

Reading 82

Thea Burns

Preserving Master Drawings: A Cultural Perspective (2001)

Expanding on James's discussion (reading 81), the paper conservator and art historian Thea Burns considers the influence of cultural attitudes, especially traditional drawings connoisseurship, on treatment decisions.

Current Practice

Master drawings are exploratory or preparatory drawings made in the West from the fourteenth to the nineteenth century; regarded as 'the foundation of all art, [they have been] raised to semi-divine status equal to the creative idea in the mind of God'.[1] In an article written in 1989, Marjorie Shelley clearly summarized the approach currently taken by responsible practitioners to the conservation of master drawings. This approach advocates a fairly limited range of treatments which may be required where drawings have been soiled, stained or added to (adhesives, backings, mends, hinges).[2] Minimal intervention is deemed appropriate because of the stable nature of the materials used, the straightforward way in which they are combined and the traditional methods of storage and presentation which have protected them. Yet the rationale for this approach does not reflect the complexity of the situation, for master drawings are sometimes complicated constructions and may have suffered, to varying degrees over time, from inherent vice and reworking and from damaging conditions of storage, display and treatment, all of which may be brought to the attention of paper conservators.[3] This paper will explore these discrepancies and relate approaches to the preservation and conservation treat-

---

From Thea Burns, "Preserving Master Drawings: A Cultural Perspective," *Paper Conservator* 25 (2001): 107–12.

ment of master drawings, in a preliminary way, to cultural attitudes. Identifying certain drawings as 'master drawings' introduces a hierarchy of relative values into the conservation decision-making process and a particular expectation of the final appearance of the work of art.

1. Issue of condition

The following are typical examples of condition in collections of master drawings. Drawings executed in iron-gall ink are probably the most obvious example of the use of unstable materials which have required major and irreversible intervention by conservators.[4] An example of a drawing in iron-gall ink is *Bacchanal Dance with Fates Above*. With time its white highlights have darkened unevenly and the corrosive ink has eaten through the paper increasing its fragility and the illegibility of the image. A large loss at the lower left corner was earlier repaired with a fragment cut from another drawing (Fig. 1).[5] An example of reworking is *The Israelites Wrestling with Giant Snakes* which was initially drawn by an unknown artist in pen and wash; Rubens pasted the drawing down, reworked it, cut it into pieces, rearranged them over a larger surface, altering details and filling out in brush and brown wash, white and grey body colour and black chalk.[6] Master drawings mounted in albums, a traditional form of storage and presentation, could have been attached to secondary supports by daubs of adhesive placed at each corner. Often the drawings were roughly dismounted, skinning the paper surface and tearing the corners; the torn corners might have been trimmed off diagonally and made up by paper insertions which did or did not originally match the original paper support.[7] The adhesives used for attaching drawings, sometimes applied overall, to secondary supports have

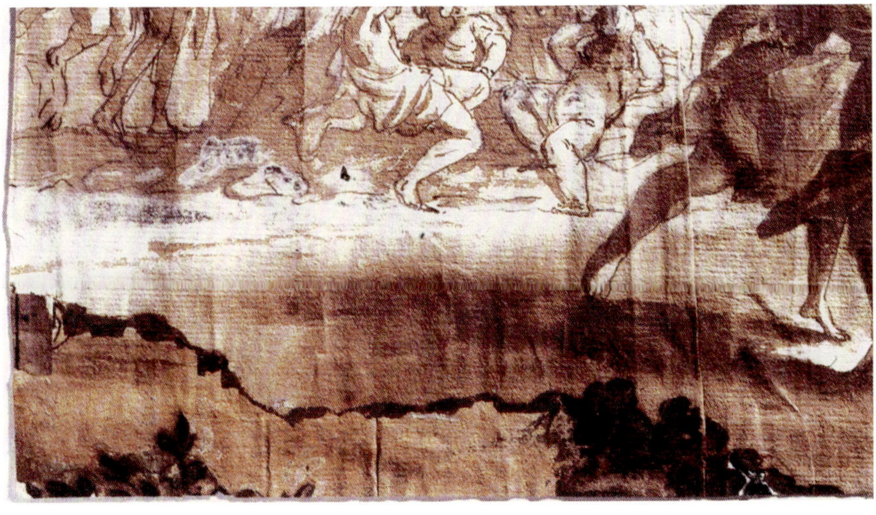

Figure 1
Anonymous, *Bacchanal Dance with Fates Above* (detail). Private Collection. Photo: M. Jürgens.

often discoloured severely with natural ageing, staining and embrittling the primary support and making the image increasingly illegible.

Thus the rationale for the current methodology does not reflect the reality of the situation for conservators, for master drawings are often not materially stable or structurally simple, nor is their condition on ageing necessarily straightforward to deal with. While paper conservators may choose to intervene as little as possible or 'partially',[8] at times, individual drawings must sometimes be subjected to a full range of interventive conservation treatments.[9] The complexity of the decision-making processes involved is suggested in many articles;[10] elucidating the role played by cultural attitudes can be helpful in understanding the choices and their validation.

2. *Cultural attitudes*

There is a substantial record of discussion of issues of condition and of what is being preserved and why in the painting conservation literature. This is not the case for master drawings. This lacuna becomes particularly clear if we separate the historical literature on the treatment of drawings from that of prints. Recent writings on the history of the restoration and presentation of prints and drawings have considered the two types of works together because they share paper as a support.[11] But, in fact, prints and drawings make use of different image-forming materials and culturally-based attitudes toward and evaluation of them, as reproduction and original, have differed historically as well; in addition, acceptable treatments, to the extent that they are recorded in the historical literature or can be divined from the evidence of the objects themselves, may have differed.

To appreciate the basis for the rationale presented by Shelley and others, we must consider cultural attitudes. Master drawings have been perceived and received not only as physical objects but, throughout their history, they have been subject to culturally-based evaluation. Although their physical condition at a particular time is a given, how it has been dealt with—what has been preserved and how it has been presented—is not. From the time of the Renaissance, the study of master drawings has been marked by the central importance of 'the draftsman's touch'. Drawings are 'the first ideas of the painter where one discovers the fire of his imagination and the spirit of his touch'.[12] Thus a connection with the artist's creative genius is made directly available to the viewer in its purest form. The study of prints, on the other hand, was recommended by the painter and writer Jonathan Richardson Senior (1665–1745) because meticulous reproductive prints conveyed 'a master's manner of thinking without the interference of an original's aesthetic distractions'.[13] For the conservator of master drawings, cultural attitudes have, in fact, shifted the basis for decision-making from what needs to be done to treat the physical drawing to its desired appearance after treatment.

Connoisseurship and the Study of Master Drawings

Given that drawings conservators work for collections and museums built on connoisseurship, it is not surprising that the central approach to the conservation of master drawings is rooted in connoisseurship traditions. Since the early sixteenth century, when interest in identifying the 'hand of the master' arose, connoisseurship has always been 'the central concern' of the study of master drawings.[14] The connoisseur views drawings as unique documents generated close to the heart of the creative process, promising a glimpse into the private world of genius—'the inner sanctuary'.[15] '[In] these first thoughts . . . the great artists make us experience impressions similar to those which they themselves have felt'.[16] The connoisseur's interest, according to Max Friedlaender and Jakob Rosenberg, both writing in the mid-twentieth century, is directed towards a judgement of quality and authorship and the unravelling of the history of a drawing's attribution and ownership.[17]

The eighteenth-century writer and connoisseur Antoine Dézaillier-d'Argenville (1680–1745) approached the study of drawings as follows: first he recognized the aesthetic quality of a drawing; next he distinguished the name and school of the master responsible; and last whether the drawing was an original or a copy.[18] Although changing attitudes and ideas about art have guided the precise weighting given to these various issues at different times, connoisseurial concerns continue to characterize much recent master drawing scholarship.[19]

Attribution is founded on the idea that artists imprint their individual mark (style, *manière*, *maniera*) on their work and that there is, therefore, something common to each artist's *oeuvre*.[20] 'The style [of the artist] reveals itself absolutely in drawings.[21] His particular touch which is like a written character unique to each one . . . identifies the author of a drawing'.[22]

In order to attribute drawings, connoisseurs have given priority to the immediate visual apprehension of their style. This is possible because, for connoisseurs, drawings are materially and structurally simple. According to Dézaillier-d'Argenville drawings give us the painter's 'first ideas', sketches made with one stroke of the pen or crayon by a free hand.[23] For Friedlaender, '[the drawing] came into being relatively quickly, as a result of spontaneous action, and did not have to take the long and toilsome road through craftsmanlike procedure'.[24] Nothing changed from the moment the artist lifted his hand from the sheet. 'A drawing has hardly ever suffered from distortion, subsequent alteration, restoration or falsification'.[25] The distinction between creation and production has worked to divorce drawings from their materiality for connoisseurs, collectors and society, refiguring technique as the hand or mark of the master.

The 1797 exhibition of master drawings at the Galerie d'Apollon, Musée Central des Arts, Paris, was the first exhibition exclusively devoted to master drawings to take place in the new institution of the public museum.[26] The installation, symmetrically arranged 'to satisfy the eye', rendered the drawings aesthetically equivalent despite their great variety.[27] 'Matted and framed uniformly under glass as in a modern exhibition, they were presented and appreciated as spontaneous, unmediated expressions of artistic genius'.[28] The unified display also drew attention to the fact that they were all owned by the French state.

Connoisseurship and the Conservation Treatment of Master Drawings

Early written instructions aimed at the restoration of drawings are very limited in their scope. Hecquet, in 1752, for example, described how to dismount poorly mounted wash drawings using poultices in order to remount them properly.[29] His subsequent instructions dealing with stain removal would have put most drawings at risk and he recommended the procedures for use only with prints.

The nature of Hecquet's instructions accords with Goguel and Viatte's observation that the finish and the dimensions of a drawing were important to the seventeenth-century collector's evaluation of it.[30] Significant drawings were classified, organized, numbered, recorded in written inventories and made presentable by appropriate mounting. For instance, drawings in the collection of Everard Jabach (1607–1695), a collection which was acquired by Colbert for Louis XIV in 1671, were mounted on cream card decorated with gold fillets. This presentation format is thought to have been suggested by the mounts of an earlier collector, Desneux de la Noue.[31] Each time important drawings changed hands they were dismounted and then remounted in a fashion which would advertise their new ownership: Jabach even had drawings from Vasari's *Libro dei Disegni* dismounted and Vasari's decorated mount replaced with his own.[32] The practice of mounting, dismounting and remounting drawings with changes in ownership continued into the nineteenth century and beyond.[33]

Recently, evidence that Jabach had many drawings retouched while they were in his possession came to light. His intention seems to have been to accentuate their degree of finish and disguise evidence of fading and other damage.[34] Pierre-Jean Mariette (1694–1774) also altered the dimensions of drawings in his collection and completed losses to make them appear more finished.[35] Thus finished drawings were not only more highly valued but it was considered acceptable practice to 'restore' damaged and cropped works to increase their value and interest.[36] Drawings of more modest subject and format ('dessins de rebut') in Jabach's collection, were considered less valuable and were not mounted.[37] The preference for finish seems to contradict the claim, first made during the Renaissance, that the less a drawing was finished, the more rewarding it might be as a work of art.[38]

Carlo James could identify no early books dedicated to the conservation of master drawings and cited Bonnardot's *Essai sur l'art de restaurer les estampes et les livres* of 1846 as the first comprehensive study.[39] However, Bonnardot's primary focus was printed materials. Perhaps little has been written about the early care and repair of drawings because their value as a direct link to the creative act which produced them required that they be in untouched condition. No contemporary document breathes a word about Jabach's extensive retouching of drawings in his collection.[40] This does not mean that treatments recommended for prints—washing, sizing, removing dirt and stains, pressing and even bleaching—were never performed on drawings. Bonnardot's text suggests, indirectly, that they were: 'All that has already been said about cleaning prints can, in many cases, be applied to drawings'.[41] However, perfect condition ('d'une conservation parfaite'; 'd'une très belle conservation') was repeatedly stressed in eighteenth-century inventories of drawings.[42] For Meder, who collected drawings 'to get closer to the creative principle and power', it was 'a serious question indeed whether any valuable original, whatever its condition, should in any case be touched'.[43]

Connoisseurs, curators and art historians (the three are not mutually exclusive) have depended on evidence of the untouched 'original' to attribute a drawing: only a drawing which has not been tampered with can be a reliable guide to an artist's style. Connoisseurs have always recognized that the material condition of a drawing can complicate their task; in particular, discolorations, stains, losses and restorations could make it difficult to study the image. Meder, a connoisseur for whom 'drawings disclose[d] the artist's soul', went so far as to lament the 'very grave damage' done to drawings by artists while using them, as intended, in the artistic process (for example, squaring, pouncing, or counterproofing).[44] Sensitivity to the aura of the 'original' drawing led Max Schweidler, a German restorer writing in the 1930s, to recommend that the cleaned drawing, from which disturbing stains had been removed, must under no circumstances give the impression that it had been treated.[45] 'Beautiful and clean' has been a coveted evaluation of condition for master drawings. This is not an attitude of the past; aesthetic reasons are today still invoked as a justification for intervention.[46]

Conservators today lack a deeper understanding of the effect of cultural attitudes on treatments accorded in times past to later accretions. Although for us they may have a primary historical value as '[the] signs, marks and traces left by collectors which . . . provide precious evidence of taste and methodology', they have been seen by connoisseurs as obscuring the primary aesthetic significance of a drawing.[47] Bonnardot wrote that drawings merited conservation treatment where connoisseurship interests might benefit: 'for example, where an old drawing lacking identifying inscriptions had been strengthened with a backing, its removal might reveal, inscribed on the verso of the drawing, its subject, a date or the name of the artist.[48] He insisted that some marks of dilapidation, for example, a missive addressed to Charles IX during the night of Saint Bartholomew, impregnated with the traces of bloody fingers, should be preserved as evidence of use, while advocating that others be removed. His decision-making was an aesthetic judgment based on final appearance: 'An ink spot, oily grime attest only to negligence and must be removed'.[49]

Connoisseurship and Technical Studies of Drawings

Connoisseurship has been historically the major approach to the study of master drawings and, remarkably, it remains the motivating force, though unacknowledged, in today's 'pragmatic' approach to their conservation which concentrates on 'minimal intervention, making treatments reversible and trying to preserve everything'.[50] Technical investigations of master drawings have long been dominated by watermark studies, the results of which are used to support attribution. Ultraviolet and transmitted light, beta-radiography and infrared reflectography all can reveal new information. Typically, this is used in the service of connoisseurship. For example, by focusing on the perceived graphic qualities of the lines in the reproduction, discussions of photographic reproductions and infrared reflectograms in the drawings literature assimilate the new technical information generated to the enterprise of connoisseurship.[51] In 1991, at the Musée du Louvre, Goguel and Dumas made use of infrared and ultraviolet photographs and X-radiographs to look beneath heavily applied later retouchings of important Italian Renaissance drawings originally in Jabach's collection, retouchings which were thought to cover 'the original lines' and 'mask the idiosyncrasies of the drawings' authors'.[52] Maryan Ainsworth calls her own substantial contributions to our knowledge of underdrawings in paintings revealed through infrared reflectographic examination, 'a new and evolving area of drawings connoisseurship', and sees them being used to re-evaluate the attributions of drawing on paper.[53] In a 1984 review of Francis Ames-Lewis' pioneering study, *Drawing in Early Renaissance Italy*, the art historian Martin Kemp claimed that 'much of value can be said [about master drawings] outside the strict confines of connoisseurship' and that 'new generations are asking new questions and attacking [their study] from different angles'.[54] However, in practice, the promise of a broader approach has had limited impact. The authentication, dating and definition of an artist's corpus of drawings continues to take precedence over broader investigations of the role of drawing and the changing relationships between function and form, and use and technique.

Conclusion

As paper conservators we must evaluate critically, make explicit and question the cultural attitudes which underpin our practice. Knowledge and clear articulation of conservation choices will empower us to function more effectively in our advocacy for cultural property. Changes in our attitudes to the conservation treatment of master drawings entrusted to our care reflect shifts that have taken place in the role of museums and in the goals of the conservation profession as a whole.[55] When I trained as a paper conservator 25 years ago, our focus was the preservation of individual objects; since that time our focus has twice shifted, first to the preservation of objects within the broader context of the collections of which they form a part and, more recently, to the care of objects within the multidisciplinary context of cultural heritage.

Notes to Reading

1 Konrad Oberhuber, 'Preface', *Drawings Defined*, eds. Walter Strauss and Tracie Felker (New York: Abaris, 1987) 8; Francis Haskell, *The Ephemeral Museum* (New Haven; London: Yale University Press, 2000).

2 Marjorie Shelley, 'Old master drawings: An approach to conservation', *Book and Paper Group Annual* 8 (Washington, DC: American Institute for Conservation, 1989): 62–66. See also: Carlo James *et al.*, *Old Master Prints and Drawings: A Guide to Preservation and Conservation*, trans. and ed. Marjorie B. Cohn (Amsterdam: Amsterdam University Press, 1997).

3 Joseph Meder, *The Mastery of Drawing*, vol. 1, trans., and ed. Winslow Ames (New York: Abaris, 1978) 495–498; 504–511. Carlo James, 'Old master prints and drawings—An art-historical survey of collecting and of preservation methods', *Conference Papers Manchester* 1992, ed. Sheila Fairbrass (Leigh: Institute of Paper Conservation, 1992) 86–90.

4 Thea Jirat-Wasiutynski, 'Sprayed poly(vinyl acetate) heat seal adhesive lining of pen and iron gall ink drawings on tracing paper', *Journal of the American Institute for Conservation* 19 (1980): 96–102. Nikki Ralston, 'Iron-gall ink meeting', *Paper Conservation News* 96 (2000): 5–6. Postprints of papers presented at this conference held at the University of Northumbria, 4–5 September 2000, will be published.

5 According to A. Bonnardot, *Essai sur l'art de restaurer les estampes et les livres*, 2nd ed., repro (New York: Franklin, 1967) 139, old papers of all types and periods coming from old books and the margins of bad prints, including an assortment of skies and shadows were used for mending larger holes in prints and drawings.

6 Egbert Havercamp-Begemann, *Creative Copies: Interpretive Drawings from Michelangelo to Picasso* (New York: Sotheby's, 1988) 87–89.

7 For example, Federico Barocci, *Head of the Virgin*, Windsor, Royal Library 5231. Various methods of attachment have been described by Catherine Hicks, 'Early approaches to conservation of works of art on paper', *Early Advances in Conservation*, British Museum Occasional Paper No. 65, ed. Vincent Daniels (London: British Museum, 1988) 7–8.

8 Shelley; Joanna Kosek, 'Restoration of art on paper in the west: A consideration of changing attitudes and values', *Restoration: Is it Acceptable?*, British Museum Occasional Paper No. 99, ed. Andrew Oddy (London: British Museum, 1994) 47.

9 Laila Hackforth-Jones, 'Conservation and restoration of *A Roman Triumph*, a drawing by Rubens', *The Conservator* 5 (1981): 23–26. Victoria Astley, 'Conservation and restoration of *Madonna and Child with Cat*, a drawing by Leonardo da Vinci', *The Conservator* 8 (1984): 22–25, Alan Donnithorne, 'The technical examination and conservation of two drawings pasted together', *Studies in Conservation* 27 (1982): 161–172, Florence Herrenschmidt and Claude Laroque, 'A propos d'un carton du Musée du Louvre: une approche de la conservation des dessins de grand format', *Traitements des supports: travaux interdisciplinaires* (Paris: Association des restaurateurs d'art et d'archéologie de formation universitaire, 1989) 229–239.

10 Elizabeth Coombs Leslie, 'Removing mounts from works of art on paper—case histories to illustrate the decision-making process', *Book and Paper Group Annual* 13 (Washington, DC: American Institute for Conservation, 1994): 25–30. Daria Keynan, 'On material(ism) or why paintings conservators have all the fun', *The Broad Spectrum: The Art and Science of Conserving Colored Media on Paper: Abstracts*, eds. Harriet Stratis and Britt Salvesen (Chicago: Art Institute, 1999) 67.

11 Hicks; Kosek; Donnithorne; James, *Old Master Prints and Drawings*, James, 'Old master prints and drawings'.

12 Antoine Dézaillier-d'Argenville, *Mercure de France* (1727): 1316–1317, 'Ce sont les premières idées d'un Peintre ou l'on découvre le feu de l'imagination et l'esprit de sa touché.'.

13 Carol Gibson-Wood, *Studies in the Theory of Connoisseurship from Vasari to Morelli* (New York: Garland, 1988) 110–111.

14 Julius Held, 'The early appreciation of drawings', *Latin American Art and the Baroque Period in Europe* (Princeton: Princeton University Press, 1963) 3:72–95.

15 Max Friedlaender, *On Art and Conoisseurship* (London: Bruno Cassirer, 1943) 218.

16 Comte de Caylus, 'Discours sur les dessins', *Le Cabinet de l'amateur et de l'antiquaire; revue* (Paris: Techener, 1846) 401: '[Dans] ces premières pensées . . . les grands artistes nous font éprouver des impressions semblables à celles qu'ils ont eux-mémes ressenties'.

17 Friedlaender; Jakob Rosenberg, *On Quality in Art* (London: Phaidon, 1967).

18 Catherine Monbeig Coguel, 'La science d'un connoisseur', *Hommage à Philip Pouncey. L'oeil du connoisseur: dessins italiens du Louvre* (Paris: Réunion des Musées Nationaux, 1992) 26–27, citing Dézaillier-d'Argenville.

19 For example: Ann Sutherland Harris, 'Guido Reni's "First Thoughts"', *Master Drawings* 37 (1999): 2–24; and E. Haverkamp-Begemann, 'Rembrandt as a draughtsman: The changing image', *Master Drawings* 27 (1989): 105–110. Connoisseurship, itself, has a history: see, for example, Gibson-Wood.

20 Antoine Dézaillier d'Argenville, 'Lettre sur le choix et l'arrangement d'un cabinet curieux', *Mercure de France* (1727): 1316–1317; André Chastel, 'Editorial. Identité du tableau: ambitions et limites de l'attribution', editorial, *Revue de l'art* 42 (1978): 4–14; David Rosand, 'Rembrandt's reach and other observations on the phenomenology of drawing', *Drawing* 14:1 (1992): 2.

21 Caylus 404: 'La manière [de l'artiste] se dévoile absolumment dans les dessins.'.

22 Dézaillier-d'Argenville 1317: 'Sa touche particulière, qui est comme un caractère d'écriture singulier à chacun, lequel fait reconnoître l'Auteur du dessein.'.

23 Dézaillier-d'Argenville 1316.

24 Friedlaender 218.

25 Friedlaender 218–219.

26 Andrew McClellan, *Inventing the Louvre: Art, Politics, and the Origins of the Modern Museum in Eighteenth-Century Paris* (Berkeley; Los Angeles: University of California Press, 1994) 127.

27 *'Variétés', La décade philosophique* 33 (l'an V, 30 Thermidor/17 Aug 1797) 365: 'pour la satisfaction de l'oeil'. 'Exposition du dessin de différentes écoles', *La décade philosophique* 36 (l'an V, 30 Fructidor/16 Sept 1797) 560.

28 McClellan 128.

29 R. Hecquet, *Catalogue de l'oeuvre de F. De Poilly . . . avec un secret pour décoller les Desseins à l'Encre de la Chine, & au Bistre, &c* (Paris: Duchesne, 1752) 135.

30 Catherine Monbeig Coguel and Françoise Viatte, 'Repertoire systématique des fonds: 1. La Collection Jabach', ts, 1978, 8–9, Cabinet des Dessins, Musée du Louvre, Paris.
31 Goguel and Viatte 2, 13.
32 Goguel and Viatte 8–9. Catherine Monbeig Goguel, 'Taste and trade: the retouched drawings in the Everard Jabach Collection at the Louvre', *Burlington Magazine* 130 (1988): 826–827.
33 Hicks 10.
34 Goguel 'Taste and trade', 821–835. Catherine Monbeig Goguel and Lola Faillant Dumas, 'Intervention du Laboratoire des Musées de France pour la lecture des oeuvres graphiques: les retouches à la gouache blanche', *Sauvegarde et conservation des photographies, dessins, imprimés et manuscrits* (Paris: Association pour la recherche scientifique sur les arts graphiques, 1991) 163, 164.
35 James *Old Master Prints and Drawings*, 195 and Fig. 105.
36 Goguel 'Taste and trade', 830, discusses the principle of retouching accepted and widely practised at the time.
37 Goguel and Viatte 9.
38 Held 87.
39 James *Old Master Prints and Drawings*, 190.
40 Goguel and Dumas 162–163.
41 Bonnardot 194: 'Tout ce qu'on a lu précédemment au sujet du nettoyage des estampes peut, en bien des cas, s'appliquer aux dessins.'.
42 Catalogue des Tableaux, Desseins . . . Ou Cabinet de feu M. Coypel, Premier Peintre du Roi . . . (Paris, n.p., 1753) 55. Caylus 406.
43 Meder 1:478, 505.
44 Meder 1:16, 497–498.
45 Max Schweidler, *Die Instandesetzung von Kupferstichen, Zeichnungen, Büchern, usw.* (Stuttgart: Max Hettler, 1950) first published in 1938, as reported in Lilly Hollander, 'Restoration of losses in old master prints and drawings using historic techniques of European conservators', *Book and Paper Group Annual* 16 (Washington, DC: American Institute for Conservation, 1997): 45.
46 Donnithorne 170.
47 James *Old Master Prints and Drawings*, 200.
48 Bonnardot 6.
49 Bonnardot 35, 37, 38.
50 Sarah Staniforth, 'Forbes Prize Lecture. Conservation: significance, relevance and sustainability', *IIC Bulletin* 2000 6 (Dec 2000): 5.
51 Rosand 1–2.
52 Goguel and Dumas 162–167: 'le tracé original', 'masquaient l'idiosyncrasie de leurs auteurs'.
53 Maryan Ainsworth, 'Northern Renaissance drawings and underdrawings: A proposed method of study', *Master Drawings* 27 (1989): 5–38.
54 Martin Kemp, 'Studying Renaissance drawings', *Art History* 6 (1984): 110.
55 David Grattan, 'The International Council of Museums—Committee for Conservation', lecture, Dept. of Art, Queen's University, Canada, Jan 2001.

Reading 83

IRENE BRÜCKLE

Aqueous Treatment in Context (2011)

Careful debate and cautionary discourse regularly occur throughout Paper and Water, *edited by Gerhard Banik and Irene Brückle, but never more than in this reading taken from chapter 14, which is probably the most thought-provoking for experienced conservators. In it Brückle (1960–) discusses the ramifications of washing and includes some obvious and less desirable ones, such as changes in media and colorants, and also some less apparent ones—to name just one example, the dislocation but not total removal of dissolved degradation products. Many paper conservators consider partial treatments "safer" or more "conservative," but, as Brückle points out, residual degradation products have the potential to cause unpredictable deterioration and visual changes in the future. Other ramifications of the interaction of paper and water include alteration of dimensions, difficulty rewetting, and, ultimately, alteration of the object's "archeometric fingerprint."*

Two common treatment procedures—deacidification and washing—can serve as examples of the dilemma presented by treatments that yield optimal preservation based on what is known about paper chemistry versus those that will achieve an appropriate historical appearance. From a purely scientific point of view, in order for a deacidification solution to be as effective as possible, it is necessary for it to penetrate completely into the pores of the paper. This is sometimes facilitated by a prewash with alcohol. Also for maximum chemical effectiveness, water can be heated or detergents added to it in order to remove even more discoloration caused by the acidic by-products of deterioration. While such procedures are scientifically logical and certainly defensible, it is possible that they will result in dislodged media and a paper that has lost

---

From IRENE BRÜCKLE, "Aqueous Treatment in Context," in *Paper and Water: A Guide for Conservators*, ed. Gerhard Banik and Irene Brückle (Amsterdam, Boston, Oxford: Elsevier, Butterworth-Heinemann, 2011), 420–26, 431–34.

its "patina," as well as its chemical fingerprint. While neither a wholly aesthetic- nor a scientific-driven approach can be adopted when designing treatments, conservators are often caught in an uncomfortable dilemma, perhaps giving rise to the motto often uttered by frustrated conservators: "Better is the enemy of good."[1]

[. . .]

14.1 Considering Risk and Benefit of Aqueous Treatment

Many kinds of aqueous treatments are designed to intervene with aged paper, and almost none with media. The relative abundance of aqueous treatments for paper in relation to the scarcity of aqueous treatment for media can be explained by the fact that, technically, the substantial physical and chemical alterations paper suffers during its natural ageing can be ameliorated by aqueous treatment, while any alterations experienced by media are, with a few exceptions, untreatable—both for technical and ethical reasons. Therefore media are usually 'exempt' from aqueous treatment, but participate in it because they are attached to paper, which explains why many aqueous interventions aim at shielding media from alterations resulting from water exposure, and are only acceptable if they do so effectively. Generally, if all media could be liberally exposed to water without any adverse effects, humidifying and washing paper objects would be a much less complex undertaking.

Because aqueous treatment often cannot be made fully selective towards the targeted damage, one is well advised to consider in advance its risks and benefits. Risks pertain to the potential occurrence of any undesired alterations of the object that can be considered, akin to medical terminology, as side-effects of treatment as opposed to the desired effects that are the benefit of treatment (Ashley-Smith 1999). Risks and benefits associated with the treatment of objects or object components can be classified according to four parameters (Fig. 1): will they take effect immediately or in the long term, or both? Will they be discernible or not? One

	benefit	risk
immediately discernible	e.g. brightness increase	e.g. colourant bleeding
latent, discernible later	e.g. paper stability	e.g. stain formation

Figure 1
Classification of aqueous treatment effects.

may review the most typical considerations surrounding washing: the treatment usually brightens discoloured paper, a desirable and immediately visible benefit; if the paper also receives an alkaline reserve, this benefit is initially not discernible, but it will make a noticeable difference in the long term when the paper will have retained strength and brightness properties due to the treatment. As for a common immediate risk of washing, water-soluble media may bleed; a latent long-term risk results from the inhomogeneous removal of degradation products that may cause the paper to age and discolour unevenly over time. Weighing the effects of treatment by these parameters may also support the decision-making process when it comes to elucidating the anticipated effects of treatment to clients. Especially the long-term and latent effects deserve particular attention because they are less tangible than the short-term, visible effects of treatment. Treatment abstinence, too, can be rated by short- and long-term risks and benefits. For example, benefit lies in the avoidance of a treatment that carries a high risk of causing undesired alterations to the object, and the associated risk is a decreased permanence and usability of the object.

Further, the expected risks and benefits of treatment can be ranked according to their significance. Risk is highest when object features or components agreed to be of primary importance are in danger of being altered significantly as well as irredeemably; risk is lower when potential alterations concern agreed-upon minor object features, when the extent of potential alterations can be seen to be limited, and when ethically acceptable technical options exist to reduce the impact of side-effects in the event of their occurrence. Media usually make up the most significant risk group in aqueous treatment of paper objects for four reasons: they tend to be critical as regards the object's value; a number are altered by contact with water; their alteration may be drastic enough to result in a substantial decrease or even complete loss of value of the object; and any alteration suffered by media tends to be not or only incompletely undoable. Most media therefore benefit from aqueous treatment only indirectly through improvement of the paper support. The treatment benefits for the latter are understood to be numerous. The risk of incurring unacceptable permanent alterations is considerably lower in comparison to media because the effect of water on paper can be controlled and made inconspicuous through appropriate application.

Giving media and paper separate consideration is, however, only an intermediate step in any risk-benefit analysis. Media and paper must be considered together because they always constitute one composite unit for which risk and benefit exist in the same balance. When, for example, the primary risk of altering media during a washing treatment is to be reduced, aqueous exposure can be curtailed by replacing a water immersion bath with another, less intense washing treatment, which might be a moist blotter washing. This may, however, diminish the intended treatment benefit for the paper because the diffusion of degradation products out of the paper will tend to be slower in a moist capillary blotter matrix than in a bulk liquid. A treatment strategy modified to protect water-sensitive media may even generate a new risk. If only the blank paper outside of media applications are to be washed, as can be accomplished by suction table treatment, this will cause the inhomogeneous removal of degradation products and will carry the associated risk of uneven ageing across the paper plane. Treatment risk and benefit, in other words, always stand in close if not reciprocal relationship. The optimization of treatment consequently involves maximizing the benefits after a minimized risk basis has been secured.

If it has been decided that, in principle, treatment would be desired, a risk-oriented evaluation of the object focuses on the sensitivity of its components to water or respective treatment solutions. The assessment may require several stages. In the first stage, fact are gathered about the material composition of the object. It involves close visual examination, may incorporate comparison with similar objects, a literature review and consultation with other specialists, and, in selected cases, a further analysis. This step is virtually non-invasive as it does not involve sample removal. It provides initial information about the likely response of the object materials to aqueous exposure. Already at this stage, the object components may be judged to be too sensitive to allow aqueous exposure, in which case treatment is ruled out, eliminating the need for any further inquiry. If, after this initial stage, it is not clear how sensitive the questioned object components are, it may be considered appropriate to gain further insight, which often involves direct testing. Because this invasive procedure carries its own risk of causing undesired alterations of the object in the short and long term, it is implemented in a judicious manner and only on highly localized, discrete object areas so as to minimize its impact on the object, and is conducted under the provision that it is documented. If doubts persist about the treatment response of the object components after an initial testing step, the procedure may have to be intensified, always keeping in mind the implicit risk of side-effects remaining when testing is ended without subsequent treatment. Testing for aqueous solution exposure not only carries risk. It is also limited in scope because it can only reveal the most immediate visible impact of water on object components. The problem is that components may be unaffected by testing conditions but may reveal an adverse response during the prolonged and more intensive exposure required by treatment. Printing inks, for example, may contain soluble dye constituents that dissolve slowly and therefore show only a gradual bleeding effect during continued water exposure. If, in the end, treatment is to be pursued despite a minor risk remaining, a more elaborate approach complete with an emergency response plan is required. Such an approach may involve increasing the treatment intensity only gradually and planning for its rapid interruption or redirection. Further, testing of object components cannot elucidate any long-term risks of treatment. These can only be predicted by scientific studies that investigate the future effect of treatments on the sample materials that relate to the object components under consideration.

14.2 Focus on Risk Factors

Exposure to plain water or alkaline pH conditions, mixtures of water and organic solvents, and drying carry the risk of causing undesired changes to the object. Risk focus areas are discussed in the following for their significance within the whole gamut of aqueous treatment options: Media are the most universal risk group because they can be significantly altered by any water exposure. Under restricted water exposure regimes, degradation products may be moved within the paper but may not be extracted. An inevitable side-effect of water exposure is the dimensional alteration of paper, which becomes a risk only if it exceeds a limit deemed to be acceptable. At the molecular level, the cellulosic substrate of paper undergoes structural changes.

An inevitable side-effect associated with washing and deacidification concerns the fact that these treatments, though beneficial, necessarily effect a change in the original object composition, which results from substances being removed or added. Previous interventions, finally, may constitute an additional risk because they alter the response of the object to further treatment.

14.2.1 Alteration of media and colourants

Any alteration of media is undesired in aqueous treatment, and unfortunately may occur even during moderate humidification. Alterations may be instant or occur gradually if solid components transit from swollen to dissolved states or if the paper matrix decreases their mobility. Side-effects of aqueous treatment include: the swelling or dissolution of inks and dyes, resulting in a migration or 'bleeding' effect; the swelling or dissolution of binding or glazing agents, for example gum Arabic in watercolour, resulting in a loss of media coherence or gloss; the dislocation or loss of loose colourant particles such as are associated with dry drawing materials or some intaglio prints; the distortion and rupture of water-repellant media films, such as on certain silk screen prints, that cannot accommodate the swelling of the moistened paper; and the disruption of varnish coatings due to water infiltration that makes them whitish opaque, i.e. blanched. Unbalancing the pH conditions into the alkaline or acidic range increases the risk of dissolving or destroying colourants or changing their hue. Further, the migration of colourless media components out of their original location, a phenomenon especially linked with transition metal-containing inks or pigments, must be given consideration because they significantly decrease the permanence of the object in the long term.

14.2.2 Dislocation of dissolved paper degradation products

The more restricted the aqueous exposure and the shorter the intervention time, the higher is the risk that the degradation products to be removed undergo dissolution, but fail to be completely flushed out of the paper. In this case, one deals in effect with an incomplete extraction process that results in the redeposition of the dissolved compounds within the paper matrix. Once dry, the redeposited compounds tend to be less water soluble, i.e. more difficult to remove. The degradation products are likely to accumulate on paper surfaces because the evaporation of the water draws them to the air-paper boundary where, being non-volatile, they are deposited. This problem also occurs when paper is locally wetted, which causes products to accumulate at the boundary between wetter and drier areas, because migration of the products stops when there is no more mobile water available for transport. Any deposit of paper degradation products may be immediately visible on the paper surface in different patterns, including tidelines. However, as paper degradation products consist not only of coloured compounds, but also of colourless ones (Uchida et al. 2007), sometimes there is no immediate indication of this side effect.

14.2.3 *Alteration of paper dimensions*

The one inevitable side-effect of any aqueous treatment is alteration of the paper structure. Paper thickness, lateral dimension and planarity tend to be affected simultaneously by water exposure. Fortunately, for many paper types the potential of dimensional alterations can be minimized through the prerequisite controlled drying systems that are part of aqueous treatment. Remaining alterations are often subtle enough not to be considered objectionable. However, the risk increases for papers that are absorbent, short-fibred, dense, highly oriented or exhibit dried-in stress, the latter two qualities being mostly associated with machine-made paper. Structural alterations can result in additional difficulties. A significant accentuation of the paper surface structure alters the appearance of paper and media. Cockling, wavy edges and curling impair the handling properties of the object, making its care with regard to mounting, storage and display more complicated. Disassembled paper pieces belonging to one object assembly may differ in their swelling and shrinkage characteristics, which may cause them to misregister when fitted back together. An increase in sheet thickness is particularly problematic if stacks of sheets belonging to one assembly, such as a book block, must be fitted again into their original configuration. Weakened paper areas, for example strongly embossed print platemarks or mould-damaged areas, may rupture.

14.2.4 *Alteration of the cellulose molecular structure*

Any drying process increases the molecular density of the cellulosic matrix, which results in a decrease of water absorbency and may make subsequent washing treatment less efficient. Therefore, scientists recommend that intermediate drying should be avoided during an aqueous treatment process, advice that may, however, not be applicable if it is pertinent to evaluate the object in dry condition at intermediate stages of treatment. An additional fact to be considered with respect to the stability of paper is the repeated water absorption and desorption. It is associated with the local swelling and shrinkage of the cellulose matrix involving heat transfer, which in the long term after many repetitions can lead to cellulose degradation. This becomes discernible by a decrease in mechanical strength and discolouration of paper.

14.2.5 *Removal and addition of components*

Washing removes water-soluble products such as degraded gelatine or low-molecular-weight cellulose fractions, whereas deacidification deposits products of neutralization and alkaline reserve. No one will maintain that degradation products and other harmful substances should be preserved in aged paper just because they are intrinsic either to its original composition or aged yet untreated state, or that deacidification should not be undertaken because it adds foreign substances to paper. However, the study of components, including trace substances that may be present, can serve historical or forensic investigations where the materials are key to classification of cultural objects (Stevens et al. 2008, Hahn 2010). Such studies usually require objects to be unaltered in their composition. While this is not necessarily a counter-indication to treatment, the permanent alteration of paper and media composition should be considered in advance of any intervention.

14.2.6 *Previous aqueous treatment*

Previous aqueous treatments, considered as part of the pretreatment condition, may influence the direction or choice of treatment. Whenever precise documentation of treatment is lacking, as is often the case with early restoration treatments, the conservator can only identify them by tell-tale signs on the object, and must face the fact that previous intervention may not have been conducted according to the principles we observe today. Some of the effects of early interventions can be partly alleviated through treatment, others not. For example, a flattened paper structure resulting from heavy pressure during drying and wavy edges due to uneven drying may be corrected by judicious application of customized treatment regimes. Crystalline deposits present on the object surface result from the precipitation of alkaline earth carbonates introduced by excessive aqueous or non-aqueous deacidification (O'Loughlin and Witty 1999). The paper surface may be abraded where attempts were made to remove stains by rubbing or sanding; these areas may differ from the rest of the paper in their water absorbency and wet strength. The water absorbency of paper may also be locally altered through previous local application of aqueous solutions or organic solvents. An unusual hue or staining is often indicative of a previous aqueous intervention (Fig. 2). Streaks of soiling may be embedded in the paper surface if the paper was brushed in moist condition in

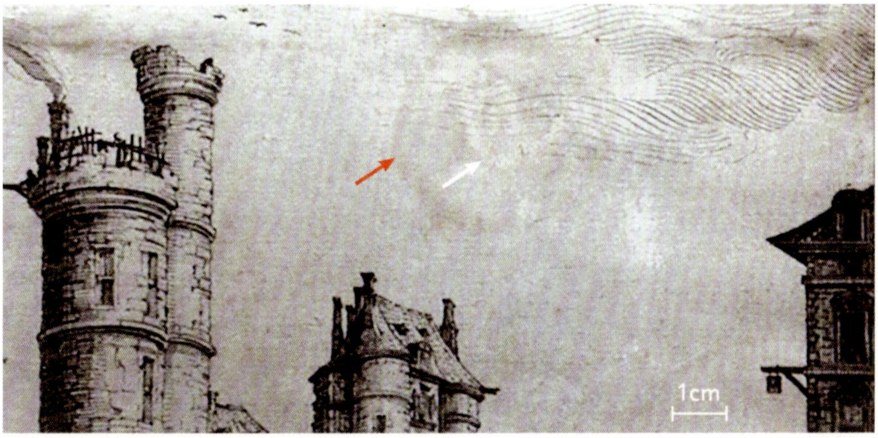

Figure 2
Effect of previous local aqueous treatment. Local staining (red arrow) was caused by locally wetting the paper in a damaged print during the course of a previous intervention (white arrow). Source: Brückle et al. (2007).

an attempt to remove soiling. Smeared or migrated media also point to an aqueous manipulation. Previous interventions may alter the response of an object towards water, and make it more difficult to predict the treatment outcome.

14.4 Strategizing Conservation Decision-Making

Objects are subject to continuous change, both as concerns their natural, even if slow, ageing; as well as their use over time. Conservation manages the object in that it intercedes with the processes of material ageing so that the object can find continued use and appreciation. This requires the scientific principles we have talked about to be embedded within an overall strategic plan. One part of it concerns the prognosis of the results of conservation intervention. Another part involves the consideration of curatorial interests. Any comprehensive conservation strategy integrates the diverse concerns surrounding the object within one framework.

14.4.1 Conservation prognostication

Because its perspective should be as long term as the life span of the object, conservation requires a far-ranging ability to foresee the results of actions to be taken as well as actions to be deliberately avoided. Prognostication is indeed a key to the strategic success of conservation, and forms a crucial part of conservation expertise. Prognosis draws from two forms of knowledge: an understanding of the scientific and technical basis of treatment such as that delineated in previous chapters, which is often considered as theoretical knowledge that can be garnered from the literature; and skill knowledge, the complementary branch of conservation exper-

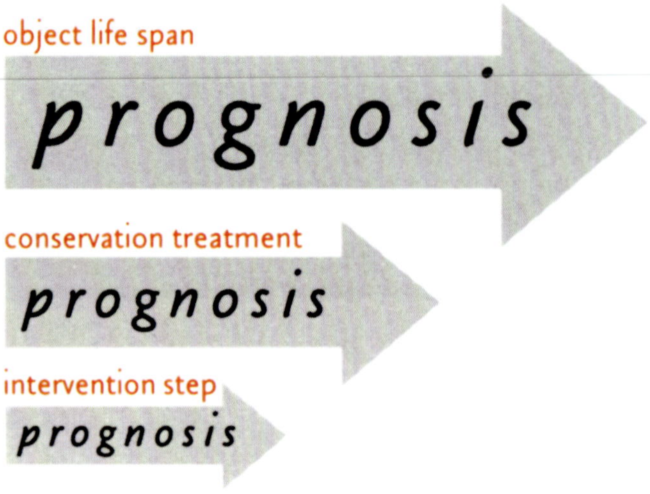

Figure 3
Levels of prognosis with respect to conservation strategy.

tise that is necessary to implement treatments successfully. Skills constitute a 'data set' equally rich as theoretical knowledge. It is built on the experiential memory of working with different objects and materials, types of treatments, variant actions as part of treatments, mistakes made and mistakes avoided, successes achieved, and modes of operation gradually altered and refined. Treatment prognostication encompasses different overlapping and connected time frames (Fig. 3). It keeps the life span of the object in perspective. It is also divided into smaller subunits that make it easier to keep control over individual intervention steps. Treatment prognostication reaches its ultimate refinement at the level where the materials, tools and the possibilities of their manipulation are continuously adjusted and re-evaluated on a minute-by-minute scale. Through this monitoring activity, treatment is customized in its intensity and direction, keeping an eye on detailed actions and their relation to the overall plan.

14.4.2 *Cultural interests*

Paper forms part of an object that may be a document, an artwork or a book; it is held in a repository such as an archive, a museum or private collection; and, besides being preserved by conservators, it is cared for and used by the custodians, owners and scholars who deal with the objects most intimately. Decisions about conservation tend to receive considerable stimulus from any of these stakeholders, who have a cultural and user-oriented interest in the object. Their concern is usually associated with what might be most universally classified as the 'intrinsic value' of the object, which implies that the original object carries certain testimonial features that augment its appreciation beyond its informational content (Menne-Haritz and Brubach 2005). Intrinsic value is, however, only an umbrella term under

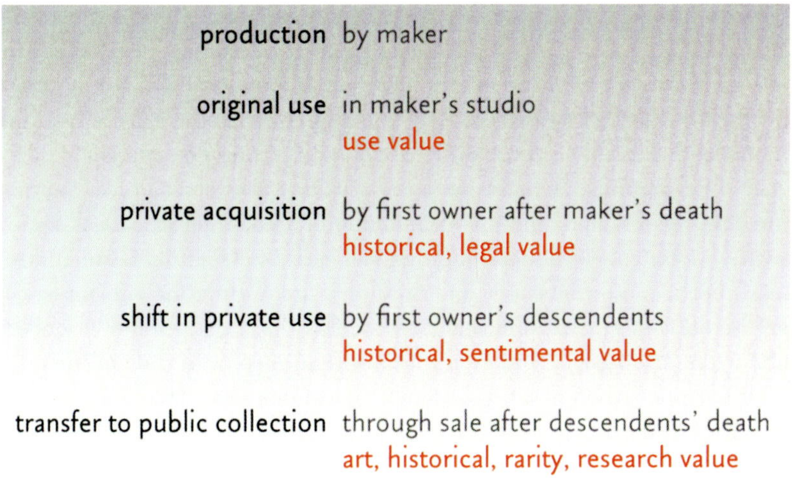

Figure 4
Evolution of values of a hypothetical object. The maker may have been an architect who created a drawing for a building construction. After his death, the building's owner valued the work in reference to his property. The drawing acquired sentimental value as it was bequeathed to his descendents. When the last descendent died, it lost its sentimental value, but acquired new values as it entered a public collection where the architect's work now was recognized as artistically innovative.

which a variety of values are subsumed, as recently pointed out by Appelbaum (2007). To mention a few of these: objects that have historical value carry evidence that makes them important as bearers of authenticity; artworks are primarily defined by their aesthetic value, but other objects originally not conceived or not yet institutionalized as artworks may also be recognized for their aesthetic appeal; use value concerns objects such as books that are still in some form of functional service; objects have research value when they continue to be investigated, which may involve uncovering information about their original, undisturbed material composition. Values of an object can be multifaceted, and they may also fluctuate over time (Fig. 4). It is always the current set of values that sets the directive at the moment of conservation, but it is prudent to consider that further re-evaluations may lie ahead in the future of the object.

14.4.3 Decision-making plans

Strategies of conservation decision-making that take into account the object context have been delineated in recent publications. They take, to name their most distinctive variations, either the form of a checklist that formulates the principal questions to be considered (Victoria & Albert Museum 2004), or a framework that elucidates the principal parameters of decision-making (Appelbaum 2007), or a guideline that explains consecutive decision-making stations (Foundation for the Conservation of Modern Art 1999). According to this latter guideline, which was written for the conservation of contemporary art, but which may also be applicable to other areas, any conservation project begins by gathering facts about the material status of the object. It compares its current, aged state with its original or intended state, which defines its cultural significance. Any discrepancies between both are noted, and on this basis a range of possible conservation strategies are generated from which then a suitable treatment approach can be formulated. If risk to the object is within an acceptable range, and the prognosis indicates that the treatment will improve the longevity and enhance the cultural appreciation of the object, then this will support a decision in favour of treatment.

Decision-making must not necessarily be a protracted process. In some cases the object assessment and the formulation of an optimal treatment strategy occur swiftly; these cases, though requiring alertness, are routine. Distinctions must be made between routine cases, and those that are not routine but can be conducted on the basis of current knowledge, as well as those that extend beyond the realm of current knowledge. And even routines that once were state of the art may lose their currency when new unanswerable questions arise. Here, one may remember that the most crucial innovation in conservation since its early inception lies in the mandate that treatment of original objects only be done on a state-of-the-art knowledge basis. According to this mandate, the conservator works within a range of options, the impact of which has already been clarified by scientific research. From this it follows that, if no research basis has yet been established for a particular intervention, the horizon of knowledge will have to be expanded through further research. In 1994, Robert Feller pointed out the need for comprehensive, funding-intensive fundamental research customized to conservation concerns, and the transfer of the results for the optimization of conservation, so that the practice could stand on the basis of verified material science (Feller 1994). This need still exists today in part because organization structures and work-loads hardly permit

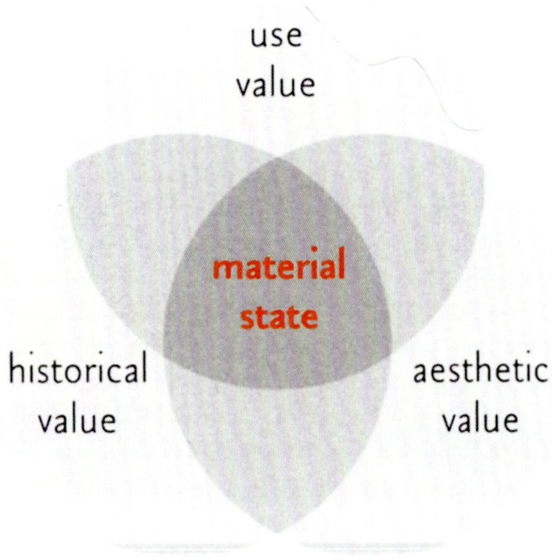

Figure 5
Interdependence of object values and the object material side.

long-term research activities. The concern for further conservation research must become a universal interest within the circle of object caretakers.

The conservator works with objects that are part of a cultural discourse that relates objects to people. Cultural values, such as aesthetic, historical or use value, are associated with the object, which stimulates its preservation. The conservator is primarily responsible for the preservation of the material integrity of an object, which may very well require a compromise to be struck in consideration of the object values (Fig. 5). Preservation and treatment planning require the capability to interrelate disparate areas of knowledge that bridge both immaterial and material qualities of the object and bring them together in decision-making.

Note to Introduction

1 "le mieux est l'ennemi du bien": from *La Bégueule*, a 1772 poem by François-Marie Arouet de Voltaire (1694–1778). Voltaire seems to be quoting a sixteenth-century Florentine proverb, "Il meglio è nemico del bene." *Irish Monthly* 31 (1903): 180.

References

Appelbaum, B. (2007). *Conservation Treatment Methodology*. London: Butterworth-Heinemann.

Ashley-Smith, J. (1999). *Risk Assessment for Object Conservation*. Oxford: Butterworth-Heinemann.

Feller, R. (2002). Aspects of chemical research in conservation: the deterioration process. In: *Contributions to Conservation Science: A Collection of Robert Feller's Published Studies on Artists' Paints, Paper, and Varnishes* (P. M. Whitmore, ed.), pp. 637–645. Pittsburgh: Carnegie Mellon University Press.

Foundation for the Conservation of Modern Art (SBMK)/Netherlands Institute for Cultural Heritage (ICN) (1999). The Decision-making Model for the Conservation and Restoration of Modern and Contemporary Art. http://www.incca.org/theory-and-ethics/303-decision-making-model-sbmk (accessed 19.7.2009).

Hahn, O. (2010). Analyses of iron gall and carbon inks by means of X-ray fluorescence analysis: a non-destructive approach in the field of archaeometry and conservation science. *Restaurator* 31, 41–64.

Menne-Haritz, A. and Brübach, N. (2005). *The Intrinsic Value of Archive and Library Material*. Marburg: Archivschule Marburg. http://www.archivschule.de/content/292.html (accessed 13.7.2009).

O'Loughlin, E. and Witty, A. (1999). Treatment of previously deacidified paper artifacts. In: *Reversibility—Does It Exist?* (A. Oddy and S. Carroll, eds), British Museum Occasional Paper, Vol. 135, pp. 81–90. London: The British Museum.

Stevens, C. H., Barrett, T., Whitmore, P. M., Wade, J. A., Mazurek, J., and Schilling, M. (2008). Composition and condition of naturally aged papers. *Journal of the American Institute for Conservation* 47, 201–215.

Uchida, Y., Masamitsu, I. and Takayashu K. (2007). Evaluation of aqueous washing methods of paper by the measurement of organic acid extraction. *Restaurator* 28, 169–184.

Victoria & Albert Museum Conservation Department Ethics Checklist (2004). 2nd edition. http://www.vam.ac.uk/files/file_up_laod/27931_file.pdf (accessed 10.6.2009).

Reading 84

ANTOINETTE KING

The Conservation Treatment of a Collage: *Man with a Hat* by Pablo Picasso (1987)

As paper conservator from 1967 to 1995 at the Museum of Modern Art, New York, Antoinette King (1932–2014) was among the first to explore the conservation problems of cut-and-pasted papers ranging in format from collages by Pablo Picasso (1881–1973), Juan Gris (1887–1927), and Kurt Schwitters (1887–1948) to cut-outs by Henri Matisse (1869–1954).[1] *Going beyond basic questions about the deterioration of paper-based ephemera, King provokes our traditional assumptions about original versus historically appropriate appearances, noting that the ephemeral paper elements making up compound works of art rarely deteriorate in unison.*[2] *With the thoughtfulness that characterizes her writings and teaching, King frequently references artistic intent, especially as interpreted from contemporary accounts. The following passage from* Picasso, *by Gertrude Stein (1874–1946), in which she describes the artist's attitude toward the aging of his unorthodox materials, is particularly germane.*

> *During all the time he did not return to Spain but during the summer he was at Cerat or at Sorgues, the beginning of life in Montparnasse was less gay, he worked enormously as he always does. It was at the rue Schoelcher that he commenced to paint with Ripolin paints, he commenced to use a kind of wall paper as a background and a small picture pasted in the middle, he commenced to use pasted paper more and more in painting his pictures. Later he used to say quite often, paper lasts quite as well as paint and after all if it all ages together, why not, and*

From ANTOINETTE KING, "The Conservation Treatment of a Collage: *Man with a Hat* by Pablo Picasso," in *Conservation of Library and Archive Materials and the Graphic Arts*, ed. Guy Petherbridge (London: Institute of Paper Conservation, Society of Archivists, Butterworth), 87–88;
 Two illustrations appearing in the original essay have been elided and one illustration added for reasons of copyright.

> *he said further, after all, later, no one will see the picture, they will see the legend of the picture, the legend that the picture has created, then it makes no difference if the picture lasts or does not last. Later they will restore it, a picture lives by its legend, not by anything else, pasted paper is just as good for creating a legend as anything else. He was indifferent as to what might happen to his pictures even though what might happen to them affected him profoundly, well that is the way one is, why not, one is like that.*[3]

In this reading, King poses the possibility that the newspapers chosen by Picasso were already discolored from age.[4] *Like Keyes (reading 85), she is hampered by the difficulty of describing the color of paper, especially paper that is aged and of poor quality. Because of the artist's stature, considerable research has since been carried out on Picasso's papers that may provide clues to their aging proclivities.*[5]

[. . .]

There is scholarly discussion about what tonalities the newsprint pieces might have had when Picasso used them for his papiers collés. Obviously, the newsprint on this picture must now be discoloured by light because the background paper is. Unfortunately, the glue discolouration on the verso of the newsprint obscures the usual clue to its original colour (Figs. 1 and 2).

Margaret Miller established and Robert Rosenblum confirmed that the newspaper used for the *Man with a Hat* was *Le Journal* of 3 December, 1912.[1] There are several other Picasso papiers collés using this same newspaper.[2] A comparison of the aged tonalities of these newsprint pieces might lead to some conclusions. The newspaper in the papier collé is from *Le Journal* and is dated 15 March, 1913. Its tonality (somewhat orange) is quite different from the newspaper from *Le Journal* dated 3 December, 1912 (a somewhat cool light brown), but it is the same as the orange tonality of the newspaper from *Le Journal* dated 4 December, 1912, used in the papier collé, *Table with Bottle, Wineglass and Newspaper*, Paris, autumn-winter 1912, Musée National d'Art Moderne, Paris.

There is no doubt that these particular newspaper pieces have discoloured considerably since Picasso used them for these papiers collés, which have often been exhibited. They would probably not have discoloured so much in the short time between the publication of each newspaper and the creation of each papier collé. The question remains whether there were differences in hues or tonalities among the various newspapers when they were used originally. Some were probably already slightly aged before use. Are there differences in manufacture (wood types, sizing, etc.), which might have caused them to have warmer or cooler, darker or lighter, tonalities when Picasso chose them for his works? Or were they exposed to different environmental factors before use? Possibly these slight differences, if they existed, might occasionally have dictated which pieces would be chosen for a given work.

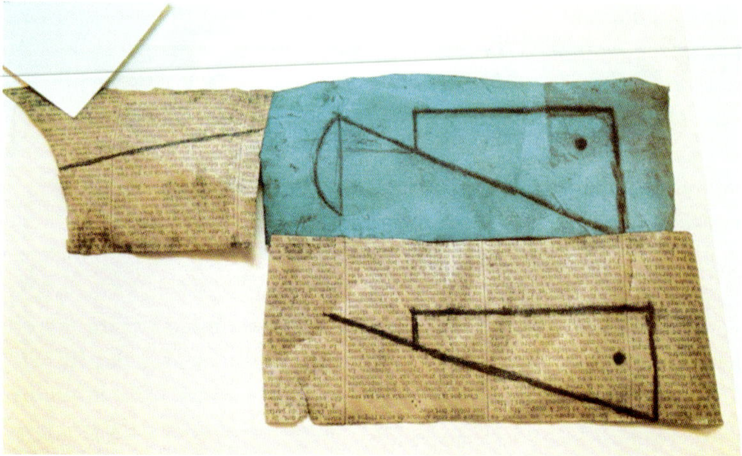

Figure 1

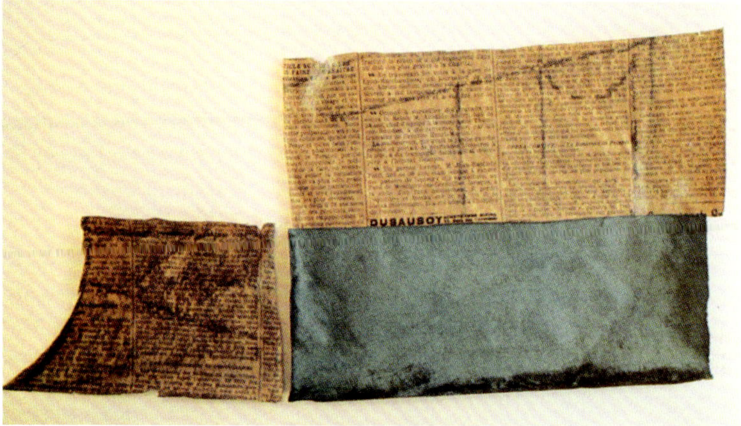

Figure 2

Picasso did keep newspaper collections in his studio for his work. John Richardson[3] has written that he was with Picasso about twenty-five years ago when the artist decided to go through a large group of portfolios, few of which had been opened since 1939, some not since 1914. One of these portfolios contained old newspapers. (Another contained beautiful papers from eighteenth-century Italy and nineteenth-century Japan. 'Far too good to use,' said Picasso.)[4]

The oldest newspaper he used for papiers collés was *Le Journal* dated 28 Mai 1883. Several papiers collés done in the spring of 1913 at Ceret[5] use this newspaper. It is a cool, buff brown, closer to the colour of the newspaper in *Man with a Hat* (Fig. 3) than that of *Bottle of Vieux Marc, Glass and Newspaper*. The same colour and tonality can be seen in *Head of a Man*, and *Guitar, Newspaper, Wineglass and Bottle*, in the collection of the Tate Gallery, London, both of which have the 1883 newspaper. The conditions of storage of the 1883 newspaper prior to its acquisition by Picasso are unknown. Also, it has not yet been possible to analyse the fibres.

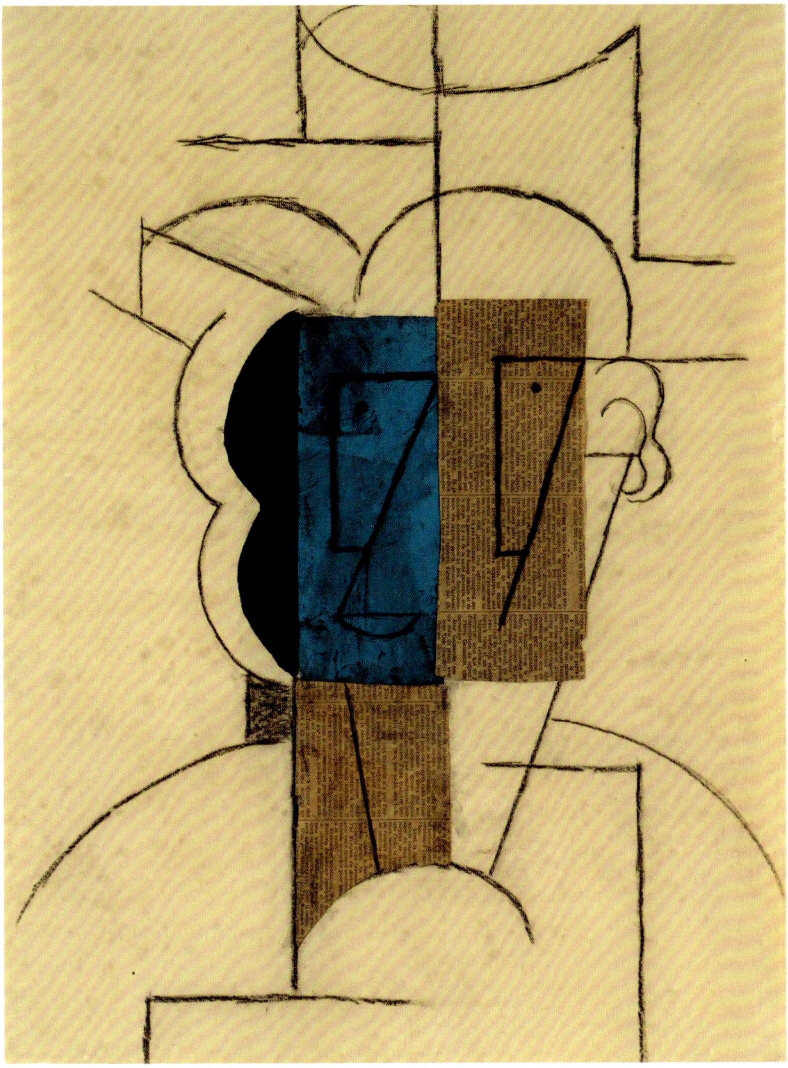

Figure 3
Picasso, Pablo (1881–1973)
© ARS, NY.
Man with a Hat.
(after December 3, 1912).
Cut-and-pasted colored paper and printed paper, charcoal, and ink on paper, 24 1/2 × 18 5/8" (62.2 × 47.3 cm). Purchase. The Museum of Modern Art. Digital image © The Museum of Modern Art/Licensed by SCALA/ Art Resource, NY.

There is evidence of careful storage and the resulting fresh appearance of the newspapers other than the portfolios viewed in Picasso's studio. One of the papiers collés found in Picasso's studio at his death was *Table with Bottle*, dated winter 1912–1913 (Fig. 4). An entire newspaper was used for this picture. The page entitled 'La Semaine Economique et Financière' comes from *Le Journal*, probably that of 8 December 1912, judging by the stock exchange report headed 'Coulisse' (the outside market) which includes two dates, 30 November and 7 December, and by the statement that the London conference would open in a few days (this conference began on 16 December 1912).[6] The date of this relatively undiscoloured

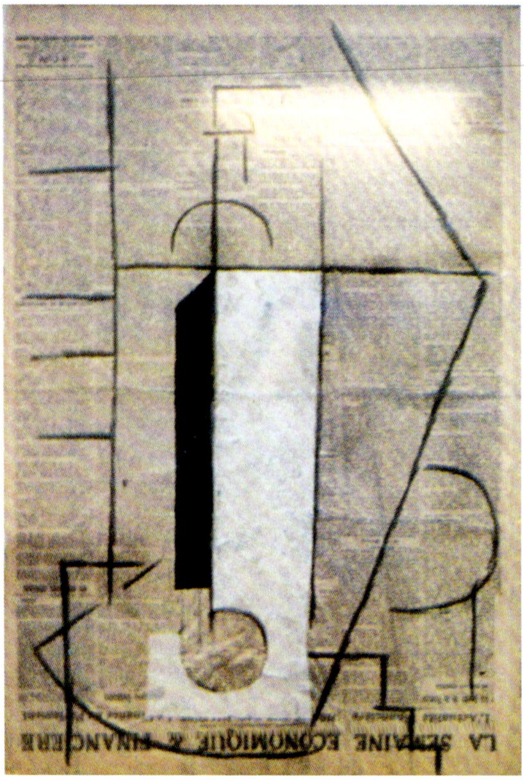

Figure 4

paper is only five days later than that used for the *Man with a Hat* (3 December 1912). However, in this picture Picasso had reversed the internal relationship of his papier collé, using newspaper as a background. Therefore, he may have wished this newspaper sheet, used as a background, to be fairly white in appearance.[7]

Notes to Introduction

1 Antoinette King, "Conservation of the Collage *Roses* by Juan Gris," in *Conservation of Paintings and the Graphic Arts, Lisbon Congress* (London: IIC, 1972), 937–41; Antoinette King, "Kurt Schwitter's *Cherry Picture*: Material Change and an Ethical Problem," *Essays on Assemblage, Studies in Modern Art* 2 (1992): 30–40; Jack Cowart et al., *Henri Matisse: Paper Cut-Outs* (St. Louis: St. Louis Art Museum, 1977), 272–77.

2 Contrast King's discomfort with uneven signs of aging with the attitude of John Ruskin: "I fear the drawing is much faded . . . and the simple colors have changed so harmoniously that I find in their faintness more to discover through mystery than to surrender as lost." *Notes by John Ruskin on His Drawings by J.M.W. Turner, R.A.* (London: Fine Art Society, 1900), 29.

3 Gertrude Stein, *Picasso* (New York: Charles Scribner's Sons, 1939), 26–27.

4 For an in-depth discussion of Picasso's use of newspapers, see Anne Baldassari, *Picasso: Papiers Journeaux* (Paris: Musée Picasso, 2003).
5 See Annabelle Ténèze, Marie-Christine Enshaïan, and Emmanuel Hincelin, "First Steps toward a Study of Papers and Watermarks in the Drawings of Picasso," *Master Drawings* 47, no. 1 (2009): 17–52; also, Scott Gerson, *Conservation and a Different Kind of Guitar God* (New York: Museum of Modern Art, April 20, 2011), www.moma.org/explore/inside_out/2011/04/20/conservation-and-a-different-ind-of-guitar-god.

Notes to Reading

1 Rubin, William (1972). *Picasso in the Collection of the Museum of Modern Art*, The Museum of Modern Art, New York, p. 209.
2 *Bottle and Wineglass*, Paris, autumn-winter, 1912. The Menil Family Collection, Houston. *Bar-Table with Bottle and Wineglass*, Paris, autumn-winter, 1912. Private Collection, Westport, Eire. *Bar-Table with Bottle and Wineglass*. Paris, autumn-winter, 1912. The Metropolitan Museum of Art, New York.
3 [John Richardson (1924–) is the leading biographer of Picasso.—ED.]
4 Richardson, John (1980). 'Your show of shows', *The New York Review*, 17 July, pp. 16–24.
5 [Céret, in southern France, was Picasso's home and studio in 1911–12.—ED.]
6 Daix, Pierre (1979). *Picasso. The Cubist Years. 1907–1916*, New York Graphic Society, Boston, p. 294.
7 Daix (1979), ibid., p. 123. 'Picasso could also reverse the internal relationship of a papier collé, using newspaper as a background where the drawing left blank paper, or putting white paper where he had previously had newsprint . . . and the system works equally well'.

Reading 85

Keiko Mizushima Keyes

The Unique Qualities of Paper as an Artifact in Conservation Treatment (1978)

Originally presented at an annual meeting of conservation students at the Cooperstown Graduate Program in the Conservation of Historic and Artistic Works on April 10–12, 1978, this reflection by Keiko Keyes (1939–1989) gently reminds budding paper conservators that the subtle physical and visual characteristics of prints and drawings can be easily altered by insensitivity and impatience.[1] *While Keyes is fully aware of the scientific underpinnings of conservation treatments, she pushes back against their wholesale application to works of art, which once again are to be differentiated from books and documents. Particularly appealing is Keyes's tendency to anthropomorphize paper; note her use of adjectives such as* tense *and* nakedness, *as well as her frequent analogies to music and the role of the paper conservator as similar to that of a solo violinist.*[2] *Not surprisingly, Keyes was a popular and beloved teacher. Her reflections on the effects of treatment on both paper and media presage Brückle's more methodical analysis of the immediate and future ramifications of treatment (see reading 83).*

[. . .]

Paper as an Artifact

So far we have only considered the material qualities and the properties of paper. As conservators we are concerned, however, with artifacts—paper, that is, which has been put to some use. Leaving aside the paper clothing and umbrellas, the

From Keiko Mizushima Keyes, "The Unique Qualities of Paper as an Artifact in Conservation Treatment," *Paper Conservator* 3 (1978): 5–8.

artifacts we encounter tend to fall into two large groups: those objects like books and documents for which paper is primarily a support for script or printing; and, objects like prints and drawings for which the paper also interacts with the media and becomes an integral part of what we will broadly call a work of art.

In most traditional painting a surface of wood, canvas, or some other material is covered with a coating called a ground. Paint is applied to the ground and adheres to it, but the paint does not normally touch or directly adhere to the supporting material. In drastic situations when a canvas has rotted, for example, the supporting material can be removed and be replaced without affecting the painted surface. This is impossible with most paper, because it is the nature of paper to absorb or engage the medium that is applied to it. Inks, chalk and colours interact with the surface of the paper and penetrate into the fibres forming a strong and sometimes inseparable bond between the medium of the image and the material of the support. This interplay of medium and paper produces certain visual effects which are usually irrelevant to the archivist, but which are essential to the appreciation of works of art. For this reason the conservator of archival materials, and the conservator of works of art on paper, approach their work from different points of view. The archival conservator is interested primarily in legibility, strength, and permanence, and therefore uses procedures that apply to paper primarily in its aspect of support, devoting skill and resourcefulness to the problems of strengthening and securing that support so the information it bears will not be lost. The conservator of works of art on paper is certainly concerned with paper in its aspects of support, but must be equally aware of and concerned with several other intimate qualities of paper. Let me give you some examples.

Certain papers are better suited and more receptive to different media. Paper with a rough or "toothed" surface will catch at a powdery medium like pastel or charcoal and allow more of the pigment particles to stay enmeshed among the paper fibres. Writing and drawing inks require a smoother surface, absorbent enough to take the ink, but sized enough to prevent it from blurring. A printing ink with an oil base functions best on a somewhat more absorbent, less sized surface, and paper with different surface qualities lend themselves more readily to the particular effects and mechanisms of etching, woodcut, or lithography.

In works of art, paper and image are integrated physically, but they are integrated visually as well. Imagine yourself as an artist, a good, sensitive one. If you know your medium and material you will choose a paper which will take your medium best, and which is complementary to the visual effect you wish to create. You will choose a receptive paper its surface may be laid or wove, rough or fine. It will have an appropriate give and absorbency. It may be buff, bright white, or coloured. You may use the white tone of the paper as a highlight, for the colour of a costume, or as a space against which your image floats. As you draw or print you are consciously and unconsciously integrating the paper into the visual effect of the image you are creating. If you are printing, the lines of etchings, drypoint and engraving project above the paper surface and the paper surface recedes. The lines of a woodcut press into the paper, making the paper surface stand in relief. These delicate variations in the tone and level of the paper surface are an inseparable part of the visual experience. The image is a part of the paper as the paper is part of the image.

These are subtle points, and although they are very obvious and important to an artist whose pleasure is in making, or to a collector or connoisseur whose pleasure is in looking, they are often overlooked and sacrificed by conservators in the process of treatment. A print can be staring white from overbleaching, or pressed so heavily that the embossed platemark is flattened, or backed with stiff paper which reduces all the delicate surface textures and variations. I have seen many examples of works of art on paper which had been violated by similar treatments. However, as a practicing conservator I know that this need not be the case. By noticing the subtleties, and becoming aware of the delicate visual properties of paper, conservators can not only preserve these qualities, but they can restore them to objects that have lost them in the course of mistreatment and the rough life they have come through.

Works of art on paper are intimate objects. They are usually small in size and were meant to be seen at close range. They are often limited in the materials they depend on for experience. Many are done in black and white or single colours. The interplay of the paper with the medium is always crucial to the visual effect. In treating works of art on paper we must be aware of the subtle qualities of texture, tone, and three-dimensionality that they have, and adjust our methods of treatment to preserve these qualities.

I have talked about the unity of the image and paper in works of art on paper. Structurally what we are faced with is a single sheet with a simple, homogeneous internal structure. Paintings have heterogeneous layers of support, ground, paint, and varnish; but paper is homogeneous even when it is layered as in the case of cardboard, and in most cases this homogeneity allows us to treat even laminated structures as single sheets.

The fact that paper is a thin sheet of a single structure presents both an advantage and a challenge in treatment. The advantage is that we can detect damage from an examination of the front and back surfaces. The examination of art on paper is basically an evaluation of surface symptoms, and we do not need complex instruments to probe a non-visible internal structure as we do in painting. Even when one surface of a sheet of paper is backed and hidden, the thin, homogeneous sheet structure allows us to fathom much of the damage from the exposed front surface. The very thinness of the material prevents us from using X-ray or other methods to examine the covered surface satisfactorily.

Paper has a front and a back, and both are visible. Every problem and everything we do to a sheet of paper is right before our eyes. Paper is a very exposed object. It does not leave us any hiding place or refuge, and our choice of possible treatment is often limited by this fact.

Consider paintings again. Because of their multi-layered structure they present problems and advantages which are quite different from paper. To evaluate a painting and determine a course of treatment, you must somehow be able to examine its internal layers to detect structural weaknesses and damage they have suffered. The examination of paintings is often a challenging and arduous task to try and fathom what is hidden and not visible from the surface.

Unlike paper, there is a variety of materials to contend with in the layered structure of paintings, each different in its properties, and the problems in painting conservation are wide ranged and complex. On the other hand, this very complexity of the structure presents a far wider range of possibilities in treatment than a simple paper structure does. A painting conservator has more choices in terms of approaches, techniques and materials to use in treatment in order to tackle his problems.

We might compare a painting to a symphony and a conservator of paintings to a conductor whose main concern is balancing the many variable factors to produce a unified, harmonious effect. Like a conductor, a painting conservator has a certain amount of leeway in treatment, interpretation, execution and detail. In an orchestra one violin can miss a note without upsetting the overall effect of the music. By analogy paper is a solo instrument and the conservator is a solo performer. Every movement, every gesture is visible, and every sour note and mistake, however small, is disturbing and obvious.

As conservators our choices in treatment are limited not only by the simple structure of the paper, but by the simplicity of the media with which the images are formed. The nature of paper allows the artist to express himself with directness and immediacy and the artist treasures this quality, choosing media which share this immediacy and directness. Media which are chosen for their freshness and spontaneity often lack the durability and permanence of painting pigments. Take pastel and charcoal for instance. They are loose, powdery media that deposit themselves on the surface of paper by friction. Though enmeshed among the hollows of the paper surface, the loosely bound pigment particles threaten to disengage themselves from the surface every time the sheet is handled. These media can be fixed with various substances, and artists often do fix charcoal drawings and pastels, but often they value the fresh, chalky texture and the vivid colours of the unfixed pigments. Faced with a damaged pastel or charcoal drawing, our choices of treatment are very limited. Fixing the pigment will alter the colour values and the texture of the media, but some damage cannot be satisfactorily treated unless the pigments are fixed. In situations like this we are faced with difficult choices; and it is important for us to remember at moments like this that our first responsibility is to the integrity of the work of art and to the artist, and we should avoid procedures that alter the visual effect of a work of art whenever possible.

What is good for the paper may be disastrous for the medium, and since the two are so intimately linked in works of art on paper, it is often necessary to restrict our treatments accordingly. Water is a benign solvent for many of the degrading factors from which paper may suffer. But frequently we are faced with media which are themselves soluble in water. In these situations, the use of water is severely limited and often it should be avoided altogether. Using a dry method in preference to moisture may mean lengthening the time we take to treat an object drastically. And at these moments we should again remind ourselves that our first responsibility is to the object, and overcome our temptation to use a dangerous shortcut in our impatience.

We encounter this sort of limitation over and over again in the treatment of paper. Since the paper and medium are one, we have to consider them as one whole and treat them accordingly. Certainly, we use our skills and ingenuity to cope with these limitations, but our attitude should be respectful to the objects and materials. If we approach problems in the spirit of conflict, and try to force and subjugate the materials to our will, we risk losing the battle. Because the margin of tolerance is so narrow in paper, losing the battle means that the work of art is irretrievably damaged, lost or ruined.

In every field of conservation we need sound judgement in treatment. Every conservator experiences a moment of truth and everyone is, in this regard, in the same arena. But considering the delicate nature of paper and the limitations it presents as an artifact and as material, I wonder if we paper conservators may not have to meet this moment of truth a bit more often. If so, we need to cultivate a frame of mind that is comfortable with risks.

With all I have said about the characteristics of paper in mind, let me now speak about a few specific points for treatment. In the treatment of any artifact, our primary concern is that it is safe for the materials involved and restores the artifact to a better state of health from the ailments it has suffered. At the same time we must remember that the very life of works of art lies in their visual effect and visual communication. If the varnish on a painting has yellowed and obscures the image, the old varnish is removed and replaced. If a delicate silver figurine tarnishes, it is polished to regain its former brilliance. By the same token, if paper is severely discoloured, and this detracts from the visual effect of the image, something has to be done to lighten it. Darker paper does not convey the original contrast in a work of art, and stains distract a viewer's eyes from seeing the work as a whole. Bleaching of the entire sheet, or of contained local areas, is a method of reducing the effects of discolouration. But bleaching is a point of much controversy and discussion in paper conservation because it is a potentially dangerous chemical procedure which, in its initial and residual reactions, may weaken paper fibres and affect some media. It is a strong measure because it deals directly with the tone of paper which is part of the visual effect of art and should not be casually undertaken. But when water and other mild methods of treatment do not reduce a disturbing discolouratlon, we may have to consider bleaching.

Granted that the paper and medium of a particular artifact allow bleaching, and we decide to use this treatment, there is a question of how white the paper should become. Certain tones of paper, which have much to do with the way paper was manufacturered in different places at different times, are associated with certain periods and places. Sixteenth and seventeenth century papers from Holland and Germany tend to be a warm off-white. Eighteenth century paper from Venice, for example, is brighter. Nineteenth century French lithographers often used a paper which is staring white. We learn to appreciate the qualities of paper tone by looking at works of art on paper. It is important for us conservators to look at works of art frequently, and with close attention, to develop a feeling for the tone and other paper qualities that are appropriate to works of different periods. Only then will we have an appropriate sense of how a picture should look before we commence bleaching or any other treatment that will alter the picture's visual properties. Comparatively few sheets of paper have survived from early centuries with the qualities they had when the sheets were formed. The colour of cellulose and sizing agents have a tendency to change somewhat with the passage of time, and we often experience these delicate changes as an agreeable "mellowing" of the colour of the paper. In many cases we simply have to admit that our experience is not broad enough to determine the original tone of the paper. It may sound unreasonable to say that we should be aware of the importance the tone of the paper plays in terms of visual effect, and at the same time to admit that we are often ignorant of what this tone was and how the paper was meant to look originally. I hope, however, that this much is clear: without being aware of the subtle qualities of art on paper and of the delicate nature of paper as a material, we may fail to produce a balanced outcome in our treatment. If we over-emphasise whiteness in paper tone, and use bleaching without sound judgement, we end up violating the integrity of the object. It is like forcing an old woman to wear bright clothes and behave like a school girl, when this will ruin her health and shorten her life.

Let us take another specific point of treatment, backing. When paper is weakened and no longer able to support itself soundly, backing may be necessary. However, we must bear in mind that backing alters the subtle but important surface qualities of a work of art on paper. No matter how thin the backing is, nor however mild the adhesive used to attach it, paper is not able to retain the surface texture it possesses in its free state when it is rigidified by another layer. The paper becomes flat and tense, and the soft, undulating surface of laid lines and felt marks; and the impression lines of woodcut and three-dimensional qualities of other media particularly suffer. Backing also affects the reflectance of light on the paper surface, and the images appear somewhat more stark and flat than they were intended. Paper is normally a soft, free, single sheet structure. The stretching and rigidifying that occurs in backing, however important these may be in painting, is foreign to the nature of art on paper. Backing is another procedure that should not be undertaken routinely or casually when you are treating works of art on paper.

The last specific point of treatment I would like to discuss is colour compensation. It is a finishing touch, comparable to "inpainting" on paintings. Its function is to minimise the distracting effect of repaired areas by toning the inserted paper to match the tone of the surrounding original surface. The term "colour compensation" may already indicate a general attitude to this process. Granted that different conservators and curators have different ideas, and that individual treatments vary, the principle of colour compensation on works of art on paper is to achieve the greatest effect with the least amount of interference and effort.

Often repaired areas may be left by simply simulating the tone of the original surface around them without attempting to recreate the missing parts of the design. This is especially true of drawings. I tend not to add missing lines unless absolutely necessary even when the continuation of the design may be clear. Undue addition of lines in the missing areas detracts from the effect of the original rather than enhancing it. Lines in drawings are so direct and personal that our additions often appear foreign, no matter how skilful we are. With prints I allow myself more freedom in recreating missing areas of design, perhaps because I am able to refer to other impressions of the same print for missing parts, and because the lines of traditional prints are fairly regular and seem easier to match than those of drawings.

It seems to me that painting conservators are much more concerned with integrating areas of loss into the original image when they are inpainting. There are various schools of thought about inpainting, and individual treatments differ, but it seems that the generally desired goal is for the repair to be "invisible". Colours are matched, lost parts of the design are often recreated. Great skill is required of the conservator, and when the inpainting is successful and the final layer of varnish coats the surface, it is sometimes quite difficult to tell that the picture had ever been damaged. Paintings, because of their layered structure, materials used, and visual complexity of their images, offer far more ground for integrating repaired areas successfully into the body of work.

Inpainting and colour compensation are a good example of the contrast between the properties of paintings and works of art on paper, and the different approaches to their treatment.

Since paper is a single-structured, exposed object, repairs are more immediately obvious. In paper there is no equivalent of varnish to give the visual and physical continuity between the repaired and original surfaces. Since paper is absorbent, the watercolours and inks used in colour compensation immediately bond with the paper. A mistake in drawing or colour is virtually impossible to reverse or improve without replacing the entire area of repair from the beginning. By the same token, if the original surface was touched-in with permeable media, it is hardly possible to remove such additions even when they are disfiguring to the visual effect of the object. Since images on paper are often simple in technique, design and colour, there is little leeway for creating optical illusions. The smaller and more intimate the object is, the more conscious the viewer becomes of the subtleties of its texture, tone and surface, and the delicate interaction between the medium and paper, and the more difficult it becomes to create the "invisible" repair. For that reason, paper conservators tend to value unobtrusive repairs using techniques that suit the nature and qualities of paper, and in their colour compensation, value taste and tact over effort and labour.

It is quite often very difficult to make a damaged paper sheet look whole again. Paper is an exposed object, unable to conceal the scars it bears. As conservators we have to accept the inherent honesty of the paper surface, and the direct and spontaneous qualities of works of art on paper, although we may hope to improve our techniques and aspire to expand the boundaries of limitations in treatment.

I have spoken about some of the aspects involved in the treatment of works of art on paper in which paper plays the role not only of a support, but participates in the overall visual effect. I have pointed out the subtle but important qualities of interplay between the image and the paper in tone, texture and three-dimensionality, of which we should be aware and try to enhance in our methods of treatment. The simple, fragile nature of paper as a material, the singularity of its sheet structure, the delicate and often impermanent media associated with paper, together with the subtle qualities of visual effect, may set narrower boundaries in the treatment of paper than are necessary in treating more robust and permanent materials with greater structural depth.

Paper conservation shares many basic principles and attitudes to treatment with other fields of conservation. At the same time the unique qualities of works of art on paper, and the unique problems that derive from these qualities, seem to suit a certain frame of mind. Would I be making an overstatement in saying that there is a certain temperament more suited to paper conservation—a temperament which can accept limitations and which functions best within these limitations, rather than in an open-ended situation with many possibilities? Perhaps the qualities required of a paper conservator are those of a solo violinist rather than a symphony conductor.

Notes to Introduction

1 The first graduate conservation program student conference was in 1974, and conferences have occurred annually since 1976. Since 1984 the meetings have been held under the auspices of the North American Graduate Programs in the Conservation of Cultural Property (ANAGPIC).

2 A sympathetic analogy is proffered by June Wayne, founder of Tamarind Lithography Workshop: "I consider printmaking to be to other media what chamber music is to the symphony. It is very disciplined—there is no place to hide and no one covers your errors." Ruth Weisberg, "The Syntax of the Print: In Search of an Aesthetic Context," *Tamarind Papers* 9, no. 2 (Fall 1986): 58.

Reading 86

ANTOINETTE KING

Technical and Esthetic Attitudes about the Cleaning of Works of Art on Paper (1986)

Antoinette King (1932–2014) continues Keyes's discussion (reading 85) of the subtle relationship between media and paper and gives several examples of the visual shifts, or, in her words, the "new set of surface relationships," that result either from the "tooth of Time" (see reading 48) or previous conservation interventions. In this respect she, like Keyes, predates Brückle's recommendation that paper conservators engage in prognostication: "Because its perspective should be as long term as the life span of the object, conservation requires a far-reaching ability to foresee the results of actions to be taken as well as actions to be deliberately avoided."[1]

What is particularly significant about King's essay is its publication in a popular but short-lived publication, Drawing, *that appealed to a broad readership of artists, curators, dealers, art historians, and paper conservators. Thus her influence extends beyond the conservation community, a role further discussed in Part VIII.*

The subtly intertwined vibrancy of media and paper which forms a drawing can present equally subtle and intertwined conservation problems as the drawing ages or is damaged. In certain ways, working for years with a collection of modern art can specifically illustrate fundamental problems entailed in preserving the artist's surface and restraining the chilling effects of time and accident. A number of acquisitions at the Museum of Modern Art have come directly from the artist, or were only a few years old when they entered the collection, and thus present a unique opportunity to assess the inevitable changes that have occurred. It is this possibility of direct assessment, keeping in mind the fallibility of memory, that clearly shows us the vicissitudes of organic materials and the difficulties of our choices for preservation.

From ANTOINETTE KING, "Technical and Esthetic Attitudes about the Cleaning of Works of Art on Paper," *Drawing* 8, no. 4 (November 1986): 79–83. Two illustrations appearing in the original essay have been elided here for reasons of copyright.

Speaking more broadly about drawings, over longer spans of time, the paper may have darkened and otherwise aged from oxidation, acid hydrolysis, and exposure to light, or may just be covered with dirt. The drawing may be partly or selectively discolored or stained from mold, pressure-sensitive tapes, water, and poor-quality framing materials, which make additional patterns on the drawing's surface, interfering with the structure of the design. The drawing media may have changed, all over or differentially, faded, rubbed off, or darkened. Therefore, a deteriorated and damaged drawing will have acquired an entirely new set of surface relationships.

The superb Matisse drawing of 1910, *Girl with Tulips* (Fig. 1), entered the collection of the Museum of Modern Art in 1970. At that time we examined it somewhat superficially through the glass and noted that there appeared to be very fine repairs along the left edge. The repairs were slightly lighter than the original paper. However, the work was so skilled that we were not sure it had even been restored. Charcoal lines extending over these lighter repaired areas were also so skilled that we doubted what we were seeing, Even a somewhat cursory look through the microscope revealed no surface disturbance in paper or charcoal medium. Curators admired the drawing's beauty and made no comment on the slightly lighter repaired areas. Several years later I was again examining the drawing with a curator who exclaimed, "What's that ugly white marking? Can you do anything about it?" There was no doubt that Matisse's paper had continued to darken, making the repair more obtrusive. Frankly, it was such a beautiful repair that we decided ultimately to leave it alone.

We have been watching the drawing over the years, but after the recent exhibition of Matisse drawings in London and New York,[1] we have again started to really examine the problems. Irregular brownish discolorations throughout are undoubtedly from water running onto or up through the paper (Fig. 2). These water stains have also darkened slightly during the past sixteen years. They were not readily apparent in the early 1970s. The [textile] conservator Jane Hutchins, in her very interesting work on water stains in cellulosic materials, found that these brown "tide lines" contain degraded, acidic material, and so one is not surprised at a continuous darkening of the previously wetted paper.[2]

The charcoal has separated in many areas into a droplet configuration which often indicates a fixative, but close microscopic examination does not reveal anything. We now think that it is entirely possible that the drawing may even have been wetted in some way in Matisse's studio and then reworked (Fig. 3). The final strokes in the eyes and the hair seem to have been drawn with wet charcoal over the other airy charcoal droplets. Such changes of quality and richness of charcoal strokes would have been consistent with Matisse's dense working methods. The water that ran into, or onto, the *Girl with Tulips* probably was not noticeable originally except as a factor changing the charcoal. This idea would certainly be supported by the repairs which must have matched the original at first because this type of exquisite repair is designed to be invisible.

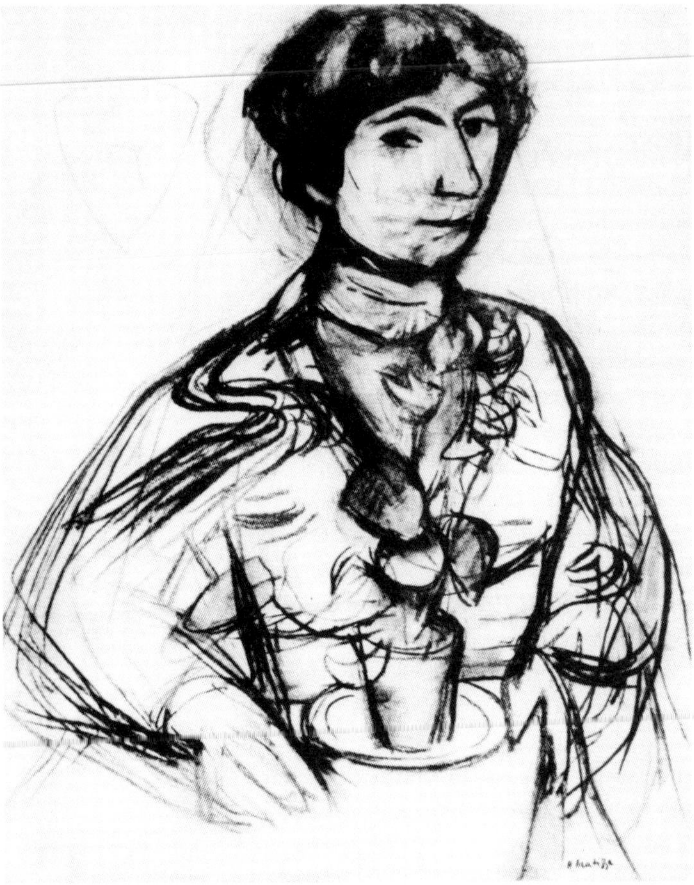

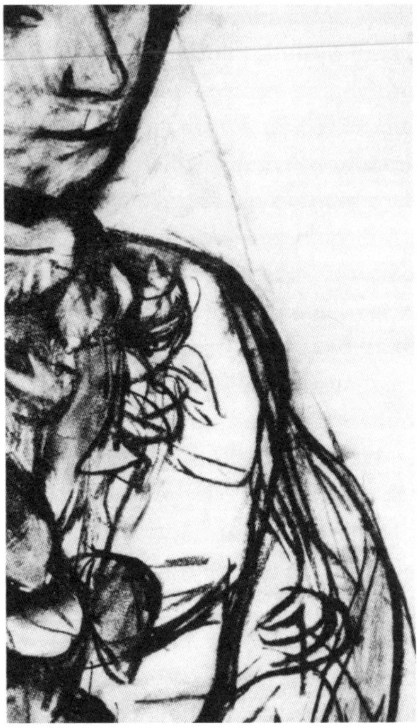

Figure 2
Matisse, *Girl with Tulips*, detail showing discolorations.

Figure 1
Matisse, Henri (1869–1954) © Copyright. *Girl with Tulips* (Jeanne Vaderin). 1910. Charcoal on paper, sheet: 28 3/4 × 23″ (73 × 58.4 cm), frame: 37 × 31″ (94 × 78.7 cm). Acquired through the Lillie P. Bliss Bequest. © Succession H. Matisse, Paris / ARS, NY. The Museum of Modern Art, New York, NY, U.S.A. Digital image © The Museum of Modern Art/ Licensed by SCALA / Art Resource, NY.

How far, though, have we come from Matisse's original. Matisse, who said, "I modulate with variations in the weight of line, and above all with the areas it delimits on the white paper. I modify the different parts of the white paper without touching them, but by their relationships."[3]

This drawing, which we examined in the laboratory, sums up one complex of drawing cleaning problems that are unsolvable at present. The charcoal cannot be fixed without changing its soft-character; any careful wetting to draw out discolorations might also change its nature as the original wetting probably did. However, we do know that the drawing will probably slowly continue to darken, though how much is unpredictable. We know that it is now different from the artist's original conception, though the charcoal retains its depth and richness, generating warmth and power, dominating the ravages of time. However, problems like this warn us that we desperately need knowledge about stabilization, and in the not too distant future. We are faced with the dilemma of possible progressive degradation of the

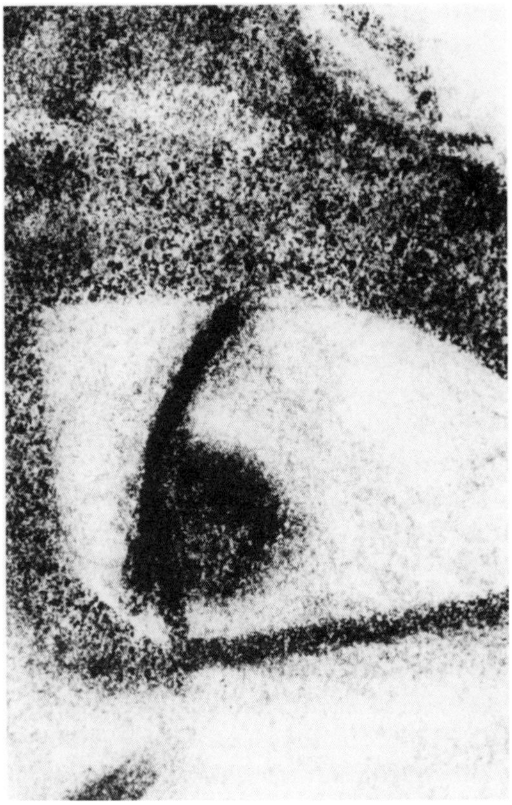

Figure 3
Matisse, *Girl with Tulips*, detail of the right eye.

paper beyond a point at which it may be returned safely to anything resembling what it was originally. We might equate this with continuous exposure of a fugitive watercolor to light; eventually it will fade.

The problems of overall cleaning of an entire drawing whose tonality may be changed by exposure to light or other degradative factors, or by the local removal of random or patterned discolorations or damages, pose very different challenges. A particular group of Picasso's studies for *Guernica*, done in May and June of 1937, were drawn on a soft, thick, good-quality wove rag paper, providing a perfect surface for the extraordinary textural range of the graphite pencil he used to express a special anguished tenderness of feeling. These particular drawings were somewhat discolored by light. Lighter edges indicate something of the paper's original luminosity. A detail of the May 8th drawing, *Composition Study*, shows how softly the pencil was rubbed and scored across and into the paper to render the tail of the horse flickering in a shadowy atmosphere.

A photomicrograph from an area of this detail shows the manner in which a paper fiber has been dislodged or intertwined with the graphite particles. Numerous fibers were raised and rubbed on the drawn side of the sheet as Picasso's hand

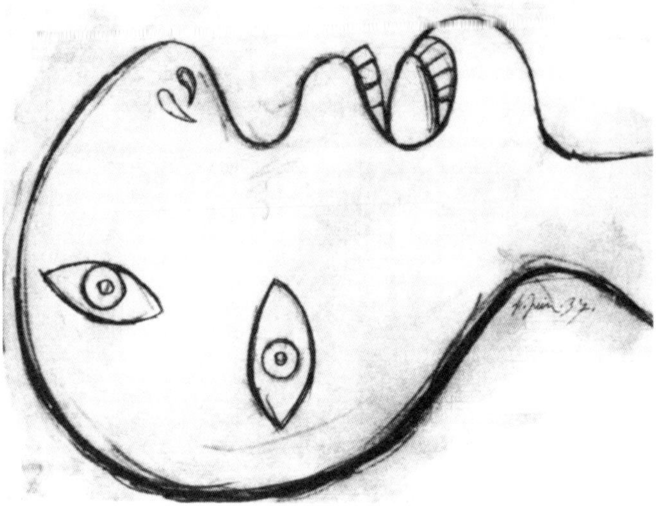

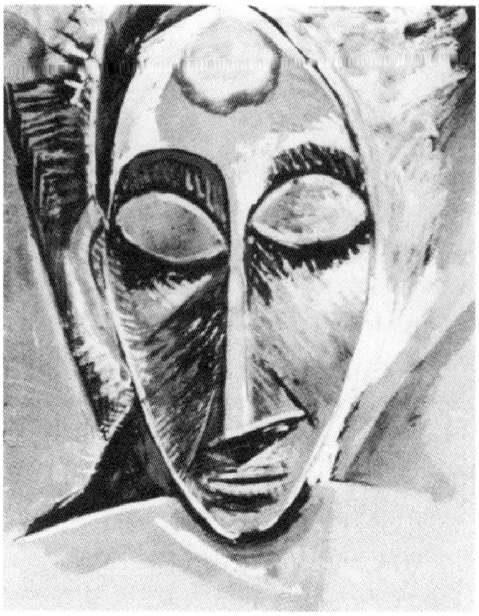

Figure 4
Pablo Picasso, *Head of a Warrior*, 1937, graphite and gouache, 11 1/2 × 9 1/4 in. (29.2 × 23.5 cm.). Museo Nacional Centro de Arte Reina Sofía © 2013 Estate of Pablo Picasso / Artists Rights Society (ARS), New York.

Figure 5
Pablo Picasso, *Head of a Boy*, 1907, gouache, 12 1/4 × 9 1/2 in. (31.1 × 24.1 cm). Before treatment. © 2013 Estate of Pablo Picasso / Artists Rights Society (ARS), New York.

passed back and forth; fibers from the soft, almost unsized surface were intertwined in new ways. It was also noticeable in all these pencil studies on this type of paper that the fibers on the undrawn side lay more smoothly on the surface.

This is clearly a case in which the delicacy of intertwined medium and paper precluded any thought of cleaning, although the loss of the paper's luminosity, in which the graphite was to sparkle, has added another small footnote of tragedy to *Guernica*.[4]

The *Head of a Warrior* (Fig. 4), a ferociously gouged graphite pencil and gouache study on a linen-weave paper, is marred by Scotch tape stains and is very subtly discolored by light. Note the tape stains vibrating in a different pattern from the violent curves of the drawing, creating a little jingly dance, quite out of keeping with the whole. The tape-adhesive stains were removed by dropping an organic solvent with a fine pipette through the stain areas, pulling the adhesive into a blotter, and using a fritted glass surface attached to a vacuum pump to prevent adhesive from spreading out through the paper. This treatment left the paper surface absolutely intact. Until the museum acquired this modest piece of equipment, designed by paper conservators working in the Oberlin Intermuseum Laboratory,[5] I was unable to remove these stains cleanly and did not attempt to work on them. I emphasize this seemingly small aspect because it so clearly shows the tremendous need for specially designed or adapted equipment for our work. Such equipment is not available commercially and we are very dependent on the few people in the profession who design and build it.

Picasso's *Head of a Boy* (Fig. 5) is an especially dramatic example of an extreme design disturbance resulting from a localized discoloration. This gouache

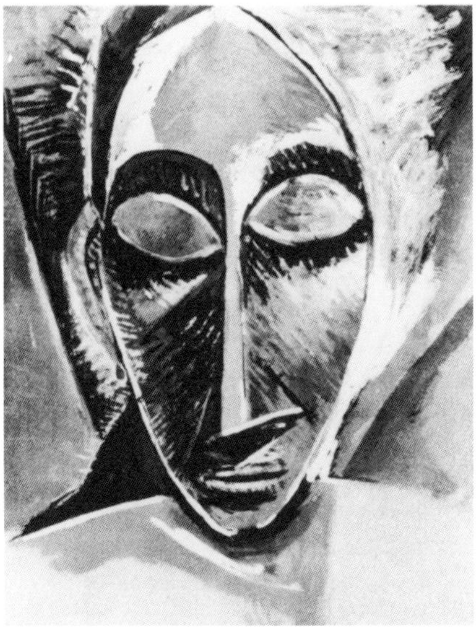

Figure 6
Pablo Picasso, *Head of a Boy*. After treatment. © 2013 Estate of Pablo Picasso / Artists Rights Society (ARS), New York.

study on a dense wove paper had been mounted on linen. The paper was considerably discolored overall from light and acidity. Water, possibly containing dirt and iron, had dropped on the forehead of the boy, also moving the paper's discoloration and thereby forming dark "tide lines." An ornate medallion seems to grace the boy's forehead as a result.

The heavy linen mounting, which was not original, had changed the surface texture of the paper and was removed so that the stain could be worked out of the paper. This was simply a matter of many, many patient hours gently dropping warm water through the paper and leaching the components of the discoloration into blotters (Fig. 6). A slight irregularity of surface tone still remained after treatment. All the discoloration could not be safely removed, but the overall composition can again be clearly read; the sweeping lines delineating the head and eyes resume their autonomy. The bold but thin impasto of extremely water-soluble gouache precluded any thought of overall cleaning.

An especially dramatic example of the return of design integrity after removal of locally patterned discolorations was observed during the cleaning of Matisse's huge cut-out *The Swimming Pool*.[6] It was covered with the brown discolorations you see in a detail from one section (Fig. 7). The discolorations seemed to literally imprison the cut-out forms by preventing the movement of blue gouache against white paper. The vibration, that subtle dance of line across a sheet, is lost, because now it vibrates with another superimposed pattern. The blue cut-out's delicate movements within the white ground were only regained after cleaning (again, the local removal of each discoloration with an organic solvent). As John Elderfield wrote of Matisse: "He had surrendered to the surface—as Cezanne had in his late

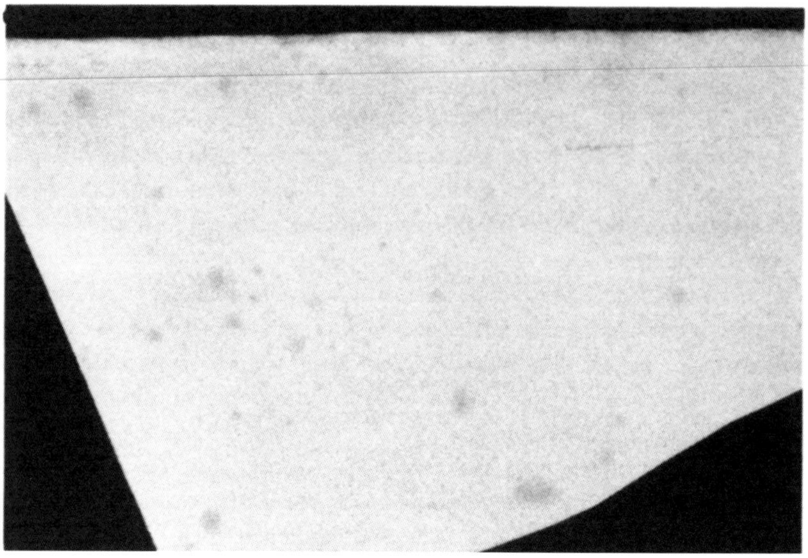

Figure 7
Matisse, Henri (1869–1954) © Copyright. *The Swimming Pool*. Nice-Cimiez, Hotel Regina, late summer 1952. Nine-panel mural in two parts: gouache on paper, cut and pasted, on white painted paper mounted on burlap, a–e, 7′ 6 5/8″ × 27′ 9 1/2″ (230.1 × 847.8 cm); f–I, 7′ 6 5/8″ × 26′ 1 1/2″ (230.1 × 796.1 cm). Mrs. Bernard F. Gimbel Fund. The Museum of Modern Art, New York, NY, U.S.A. Digital image © The Museum of Modern Art/Licensed by SCAL / Art Resource, NY. Detail of discoloration on white background paper before treatment.

work, from which this breathing openness derives—and, like Cezanne, had found that the white sheet of the surface had become virtually a medium of existence (like air, space, light, and freedom) within which objects grow and gain their life. Where does the wholeness of this image lie but in the wholeness of the surface itself?"[7]

We have been looking at two profoundly different aspects of cleaning drawings, that is, cleaning an overall change in the paper's tonality and cleaning a local discoloration or stain. There is no doubt that many curators, collectors, and conservators find an overall change more acceptable than jarring local discolorations. Also, as discussed in the beginning, paper and media are an integral unit forming a drawing. Visual deterioration of a large part of this unit—the paper—relates, as time goes on, to possible differential deterioration in other parts of the unit—the media.

Nevertheless, there are times when all concerned agree that a change in the paper's tonality, often mahogany brown, is drastic enough to warrant cleaning. In these cases, the paper's discoloration is so severe that it alone throws off the balance of the picture. There are major technical difficulties between cleaning a change in the paper's tonality and cleaning a local discoloration. The first often requires an absolute overall approach such as immersion or considerable wetting on a low pressure table, after local testing of media, sizing, and paper. It is my belief, based on my own experience and that of distinguished colleagues, that when it is determined that these types of overall treatments are safe for a given object and are done carefully, there will be no surface disturbance.

Many conservators with years of experience, working closely with concerned curators and collectors, have studied surface characteristics of drawings before and after careful cleaning and have seen no change, even when viewed with considerable magnification with photomicrographs as an aid to memory. Space does not permit the many possible illustrations of returned luminosity to sheets previously browned to a crisp by sunlight and air.

As part of our ongoing program of study and observation, samples of a variety of papers made from cotton, linen, and mixtures of chemical wood pulp, sized with gelatin, with alum-rosin, and unsized, were gently immersed together in a bath of water for half an hour, changing the water once every ten minutes. The washed papers were divided so that an example of each type of paper could be air-dried on a blotter and another example of each type dried between two blotters. No changes were observed in the papers' surfaces.[8]

The esthetic and chemical problems of cleaning drawings are inextricably interwoven as they are in any type of art object. I hope that I have clearly presented the painful dilemma of the need to preserve the delicate touches of a vast variety of media on an even vaster variety of papers and the need to prevent possible physical deterioration of these objects. There may really be a difference between the needs posed by cleaning some drawings and those posed by some paintings. If cleaning varnish from a painting is out of the question for one reason or another, the varnish may not ultimately destroy the chemical structure of the original painting. But many types of damage to drawings are integral to the entire structure and thus can be equated more closely with structural canvas damage than with darkened surface varnish. Therefore, true cleaning of drawings is a structural stabilization as well as an esthetic choice. Very little research is being done in this area to try to find better cleaning methods in general, and methods for cleaning and stabilizing the greater number of drawings that cannot be safely touched in the present state of our knowledge are desperately needed.

Max Klinger said: "Because of its particular appearance and its necessary manner of handling, every material is pervaded by a unique spirit and poetry, which in the hands of the artist greatly strengthens the character of the representation, and for which there is no substitute, in much the same way the character of a piece of music depends upon its predetermined key."[9] It is how to find the right key to unlock these problems that will always occupy us.

Note to Introduction

1 Gerhard Banik and Irene Brückle, eds., *Paper and Water: A Guide for Conservators* (Oxford: Butterworth-Heinemann, 2011), 432.

Notes to Reading

1 The Drawings of Matisse, an exhibition organized by the Arts Council of Great Britain in collaboration with The Museum of Modern Art, New York, 1985.

2 J.K. Hutchins, "Water-stained Cellulosics: A Literature Review," *Journal of the American Institute for Conservation* (vol. 22, Spring 1983).
3 Henri Matisse, "Notes of a Painter on His Drawing, 1939," *Matisse on Art*, by Jack D. Flam (New York: E.P. Dutton, 1978), p. 82.
4 [This painting by Picasso depicts civilian deaths resulting from a 1937 bombing during the Spanish Civil War.—ED.]
5 [The Oberlin Intermuseum Laboratory was founded in 1952 as a nonprofit regional conservation facility. Today it is the Intermuseum Conservation Association located in Cleveland, Ohio.—ED.]
6 Antoinette King, "Technical Appendix," *Henri Matisse: Paper Cut-Outs*, by Jack Cowart, Jack D. Flam, Dominique Fourcade, and John Hallmark Neff (The St. Louis Art Museum and The Detroit Institute of Arts, 1977).
7 John Elderfield, *The Drawings of Henri Matisse* (New York: The Museum of Modern Art, 1984), p. 40. [At the time this article was written, the art historian John Elderfield (1943–) was curator of painting and sculpture at the Museum of Modern Art, New York.—ED.]
8 I wish to thank Anne Lockwood for her careful preparation of these samples.
9 [The quotation is from Max Klinger, *Malerei und Zeichnung* (Leipzig: Verlag von Arthur Georgi, 1891), 11. See reading 14.—ED.]

Reading 87

Keiko Mizushima Keyes

Alternatives to Conventional Methods of Reducing Discoloration in Works of Art on Paper (1980)

Keiko Mizushima Keyes (1939–1989) encourages conservators to opt for "safer" conservation procedures through the use of alternative treatment approaches or "systematic gradations of treatment."[1] *Today her recommended procedures—the addition of alkaline substances and alcohols to water for more effective reduction of discoloration, the use of the then newly developed vacuum suction table, and sun bleaching—albeit somewhat modified, are routine.*[2] *However, it can be argued that these procedures are not without risks. Vacuum suction tables exert enormous mechanical stresses on paper, and sun bleaching cannot be considered "an organic alternative to chemical bleaching" since chemical bond-breaking reactions still occur. Nonetheless, this article by Keyes, like reading 85, motivates paper conservators to develop specialized tools and explore innovative ways of approaching persistent conservation problems. The adaptation of existing technologies, as exemplified by the inventive style of the ever-curious paper conservator Robert Futernick (1946–) and the discovery of new materials, for example, GoreTex, chemical compounds such as cyclododecane, and tools, such as the hot-air pencil, have been characteristic of the profession since the last decades of the twentieth century.*[3] *To use a musical analogy, which she favored in the crescendo of exciting advancements in paper conservation, Keyes plays the role of Euterpe.*[4]

[. . .]

From Keiko Mizushima Keyes, "Alternatives to Conventional Methods of Reducing Discoloration in Works of Art on Paper," in *Abstracts & Preprints: Cambridge International Conference on the Conservation of Library and Archive Materials and the Graphic Arts* (1980): 166–70.

Until recently, hardly anyone questioned the use of chemical bleaching as a standard procedure in the restorative treatment of paper objects. But today, as a result of having better information about the potential degrading effect of chemical bleaching in its immediate and latent actions on paper, the prevailing attitude among paper conservators is cautious and we tend to avoid the use of chemical bleaching in our work as much as possible. However, as conservators of works of art on paper, we are concerned with improving the visual effect of a damaged object in addition to enhancing its physical strength and longevity. If the object is disfigured by discoloration, we attempt to reduce it through our treatment to regain the visual coherence without which the art object falls short of the meaning it conveys. One challenge posed to us in treatment of works of art on paper, then, is how to achieve an effective reduction of discoloration with the safest possible means, minimizing the use of potentially dangerous chemical bleaching agents. The purpose of this paper is to discuss some methods with which we may meet this challenge.

Our standard procedure in the treatment of a discolored object is first to employ dry cleaning methods like brushing and erasures, and the application of some organic solvents which are safe to the paper and the media. When carried out with skill and ingenuity, these methods can considerably reduce certain stains and surface soil. The materials and methods for dry cleaning are discussed in various publications.[1,2,3] However, to reduce general discolorations in paper, we depend on the cleansing action of water. Water cleanses acidity, discoloration, and degradation products from paper and strengthens its structure by re-establishing bonding among fibres. Recent findings at the Library of Congress[4] caution us about the weakening effect of deionized and distilled water on paper. But the effect of water as a benign solvent for discolorations in paper is generally agreed upon.

Occasionally, chemical bleaches are resorted to prematurely when the effective limit of water's cleansing action seems to have been reached. But there are safe methods by which the cleansing action of water may be made more effective, and beyond this point there are still safe alternatives for the reduction of discoloration besides chemical bleaching. I will present three safe methods of increasing the effectiveness of water as a cleansing agent, and an organic alternative to chemical bleaching which is effective in certain instances. Many of you are already familiar with these procedures in your own practice. I will then present my observations on a traditional technique of non-chemical bleaching which I have also used experimentally in my work.

The Use of Safe Alkaline Substances in Water Cleansing: Since many of the discolorations in paper are acidic in nature, their reduction is achieved most effectively if the water used for the purpose is alkaline. Hey noted in her article on paper bleaching[5] that a deacidification procedure removes considerable discoloration from paper. She used solutions of calcium hydroxide and magnesium bicarbonate as immersion baths for this procedure.

The alkaline wash should follow a preliminary plain water wash in most cases to lessen the risk of shock to the paper. Paper conservators have observed that deacidification agents sometimes affect colours and binding of certain media, and occasionally leave a white crystalline deposit of carbonates that cloud the image surface. In cases where it is unsuitable to use a standard solution[6] of a deacidification agent, it should be used diluted in order to avoid these ill effects. Some reduction of paper discolorations can be achieved with a considerably diluted solution (e.g., magnesium bicarbonate to water 1:5, calcium hydroxide to water 1:10) used as an immersion bath. This is perhaps more effective in two baths with air-drying between each immersion. In view of the findings from the research at the Library of Congress,[7] it is also important for conservators using deionized or distilled water to add a small quantity of either a calcium or a magnesium compound to their wash water, which will prevent the over-purified water from weakening paper and may increase its effectiveness as a cleansing agent through its mild alkalinity.

Ammonium hydroxide is a weak, volatile alkali, and its salts are soluble in water. Used in weak concentrations, it is considered harmless to paper. Paper washed in water containing a small amount of ammonium hydroxide (droplets of a concentrated NH_4OH, or less than .05% of the volume of the wash water) normally looks brighter than paper washed in diluted deacidification solutions.

The same, safe alkaline substances can be used for local application to discolored areas such as foxing marks, mat burn, and other states which respond to alkalinity. In local applications, the standard solution of magnesium bicarbonate and calcium hydroxide can be used, and a slightly stronger concentration of ammonium hydroxide solution (e.g., twice the amount used for general washing) can be employed, brushed on to discolored areas a few times in succession. Additional moisture is then used to flood the area and the released discoloration is leached away with blotting paper or a vacuum suction table (see below). The released discoloration can also be washed away in an immersion bath, but this method of local application is especially useful for objects that cannot be exposed to excess moisture.

A word of caution about groundwood paper. It darkens in alkaline water, and although its acidity is more efficiently reduced by these alkaline substances, its brightness is not enhanced.

The Use of Alcohols in Water Cleansing: Among various wetting agents available for our use in the treatment of works of art on paper, alcohols appear to be the safest and most generally applicable to objects. Although alkaline water penetrates into the paper structure better than plain water, highly sized papers, especially those sized with alum rosin, and areas in paper with certain water-repellent stains will often refuse to wet thoroughly. By brushing or spraying an alcohol on the paper, or by mixing some amount in the water, the paper readily allows water to penetrate into its structure, and the cleansing effect of water is enhanced as the result.

Alcohol is particularly useful as a wetting agent when an object is cleansed by flotation over a water bath so long as no friable medium will dislodge and no colour will suffer ill effects from the more effective penetration of water.

Alcohol can be mixed in the water which is applied to discolored areas of an object that does not allow an excess use of moisture. It helps to prevent the formation of tide marks, and aids in the more effective release of discoloration as well as in faster drying of the areas.

The most common alcohol used as a wetting agent is ethyl alcohol. Methanol is more powerful and quicker in its action than other alcohols and is used in situations where such properties are called for as long as it is safe to the object.[8]

The Use of Vacuum Suction Table in Cleansing Procedures: The vacuum suction table, which was introduced by Marilyn Weidner in 1974,[9] greatly extended the range of possible treatment of paper objects. Vacuum suction tables are particularly valuable in the removal of discoloration because of their capacity to evacuate solubilized matters efficiently out of paper without immersion in a water bath, allowing water or organic solvents to flow through the paper structure by the draw caused by the vacuum suction. They are also useful in localizing moisture and solvents to areas under treatment without affecting other areas with fugitive colours or very friable medium, for example, where treatment should be avoided.

Exceptionally discolored papers and papers with thick or absorbent structures often retain dissolved discolorations within themselves even after repeated soaking in immersion baths. If these papers are simply dried, the discolorations will remain in their structure. By placing such papers over a vacuum suction table, it is possible to evacuate the discolorants out of its structure. This procedure is quite effective on objects which have been floated over a water bath. In these cases, the diffusion of discoloration into water is not as effective as in immersion baths since the mobilized discoloration on the surface of the object often cannot penetrate downward to the water through the structure of the paper. The object is placed on a vacuum suction table face upward immediately after the removal from the flotation bath, and the discoloration is evacuated by the vacuum suction, with additional water applied from the surface by spray or brush application as the object allows. This procedure avoids tide marks and the mottling effect of discoloration which often are a problem in flotation cleansing.

Localized areas of discoloration on vulnerable objects with fugitive or friable media may be treated on a vacuum suction table. Discolored areas are brushed with alkaline water, mixed with an alcohol if necessary, and evacuated through the table, and additional fresh water is applied for final cleansing.

As yet vacuum tables are not commercially available. Those in use are made with various methods and materials which reflect the needs and ingenuity of their individual makers and users. My own table was designed by Robert Futernick of the Achenbach Foundation Conservation Laboratory using a vacuum-sealed aluminium Hexal panel and a wet-and-dry vacuum cleaner as the suction pump. I have used various other methods based on the same working principles outside of my studio; and in all instances I felt it helpful to use enough cushioning material between the object and the table surface so that the pressure from the vacuum suction would not distort the textures of paper and the medium. A sheet of blotting paper on top of a felt approximately 1/2″ thick creates enough cushioning for my table.

Enzymal Reduction of Discoloration: Hardened and discolored adhesives are very tenacious and difficult to remove by ordinary means from the porous, retentive structure of paper; so is the engrimed soil which is trapped among the interstices of paper fibres. Here, the mere cleansing action from water falls short of its function. High temperature often increases the cleansing effectiveness of water in these cases, and steam heat often helps remove adhesives that react to high temperatures. But in these cases enzymes are generally a more effective cleansing agent than water.

The recent articles by Segal and Cooper[10] and by Hatton[11] describe their practical working procedures in the use of enzymes to release adhesives and to reduce adhesive stains and trapped soil in paper. More information is needed to establish the safety of enzymes in the treatment of paper, but the specificity of enzymes and the ease with which they can be denatured suggest a good safety potential for use not only as agents to facilitate the delicate and often time consuming procedure of removing the backing from an object, but also to reduce discoloration that is difficult to treat with other means.

The parameters of safety vary according to the nature of objects to be treated, and whether or not water and other agents may be used to eliminate residues and by-products of chemical substances utilized. The foregoing methods are by no means an exhaustive list of treatments possible for safe reduction of discolorations in paper, but are primary measures safe even for media that cannot withstand extensive cleansing. However, combining these means with skill and ingenuity can reduce discolorations considerably. And often in objects with fugitive or friable media, these are the only possible treatments we can apply to achieve any improvement in their visual effect.

As paper tends to retain substances or their harmful by-products to which it is exposed during the course of treatment, a large concern in using chemical bleaching agents is the elimination of harmful residues from the paper. If a bleaching method is available which does not employ a chemical bleaching agent, much of this concern may be eliminated. Bleaching with light is a method that does not employ chemical bleaching agents.

The Experimental Use of Sunlight in Bleaching of Paper Objects: The first chemical bleach, chlorine, was discovered by Karl Wilhelm Scheele in 1774.[12] Before the use of chlorine, sun bleaching was the only method of whitening the linen and cotton cloths that were used as fiber sources for papermaking. Cloth was soaked in extract of wood ashes and in sour milk, then was spread over a grassy meadow and exposed to sunlight with frequent sprinkling of water. Nascent oxygen emanating from the grass aided in the bleaching action.[13,14] Sunlight has been used to bleach discolorations in the restorative treatment of paper objects for some time. Watcher[15] used sunlight effectively to bleach black and white graphics and Egyptian papyri, spraying the objects intermittently with water to keep them moist. Plenderleith[16] describes a method to reduce tea and coffee stains by exposing the affected areas to sunlight after applying an aqueous solution of potassium perborate. Annis and Reagan,[17] in their recent study, describe the effect of sun bleaching on an old, undyed cotton fabric, exposing dry samples to the sun up to 32 hours, and report good results. The fabric samples noticeably brightened without a detrimental effect on fibre properties.

Over the past three years, I have experimented with the use of sunlight in bleaching discolored paper and would like to report my method and observations here. I generally use an immersion bath to irradiate paper with sunlight. Water mixed with magnesium bicarbonate solution (*c.* 2%) in the ratio of 5:1 is placed in a white photo-developing tray which is large enough to keep the shadow cast by the rim away from the paper. The paper should rest approximately 1″ or 2.5 cm below the surface of the water. For paper objects which cannot be placed in an immersion bath, I use a "moistening sandwich". A blotting paper slightly larger than the dimension of the paper object is moistened with the mixture of water and magnesium bicarbonate solution and placed on a Plexiglas support. The paper object is also sprayed with the mixture and placed on the moistened blotting paper and a medium weight sheet of Mylar film is placed over its surface. The entire sandwich is then placed in the sunlight. The blotting paper supplies, and the Mylar film retains, the necessary alkalinity and moisture. Although frequent spraying is not necessary since the sandwich remains moist, sunlight warms the sandwich considerably if left alone during the course of irradiation, and occasional spraying on the surface of the paper object is necessary to keep it cool. The Mylar film also avoids the harmful effect of evaporating water to cellulose at a cellulose-water-air interface.[18]

The exposure time in the sunlight varies according to the season and to the degree of discoloration of the paper but normally ranges between 2 to 4 hours. During this time the object should be turned to expose both sides.

As bleaching procedes, the solubilized discolorants are leached away into the bath or moistened blotting paper, but after the irradiation is completed the paper object is further cleansed by immersion in a clean water both or, for objects which cannot be immersed in water, by the application of moisture by spray or brush on a vacuum suction table to remove the solubilized substances remaining in paper.

Magnesium bicarbonate solution is added to the water because the presence of the magnesium ion retards the degradation of cellulose in oxidative bleaching processes by deactivating the catalytic action of metal impurities such as iron and copper present in paper.[19,20] Richter[21] found that paper treated with magnesium hydroxide solution was well-preserved after 100 hours of dry exposure to direct sunlight in Florida.

UF-3 Plexiglas, or a similar filtering material should be placed over the paper if the ultraviolet radiation in the sunlight is feared to cause an adverse effect. UF-3 filters all the ultraviolet radiation and a portion of the violet and blue range in the sun's spectrum. Even with this filtering, discoloration is reduced, although the bleaching action is slower and paper objects must be exposed to sunlight for a longer time (approximately 1 to 3 hours longer depending on the time of year and the degree of discoloration in paper) to achieve a satisfactory result.

With all bleaching methods, there is a risk of fading if colours are exposed to sunlight. However, it is possible to expose a local discolored area to sunlight and block the penetration of light to undesirable areas by masking these areas with opaque paper or aluminium foil. If the colours present are not water-soluble, such as in western colour prints, the "moistening sandwich" unit can be used after masking areas where exposure is to be avoided. If any colours are water-soluble, the moistening should be limited to the stain and the adjacent areas. Care should be taken not to form tide marks in this case by leaching away the solubilized discoloration from the surface from time to time with dry blotting paper and applying additional moisture by "feathering" toward the adjacent areas. This method may give conservators a wider latitude in treating discolorations on objects with moisture-vulnerable media since the subsequent cleansing with water is far less compared to that required when a chemical bleaching agent is applied locally. The exposure time varies tremendously, but much shorter than that of overall bleaching so as not to cause unevenness of the paper tone.

Sun bleaching is not suitable for all papers. In fact, sun bleaching should be used principally for rag papers made before the middle of the 19th century. Pure cellulose is a stable material and direct degradation of cellulose by sunlight is minimal,[22,23] and early papers were relatively free from light-sensitive impurities. But from about 1850, wood pulp was in prevalent use in papermaking. Lignin is an impurity in wood pulp and is known to be very unstable in light. It absorbs light, decomposes, and in so doing degrades paper. Papers with a high lignin content, such as groundwood papers, should not be subjected to sun bleaching as they darken rather than brighten. The presence of lignin can be tested by a simple colour reaction of phloroglucinol solution applied to a fibre sample obtained from the paper in question.[24] About the same time as the use of wood pulp, alum rosin sizing came to be used in papermaking. Richter[25] reported that the presence of rosin endangered the stability of paper in dry exposure to the sun for 100 hours. Although it is not certain that alum rosin sized papers will be appreciably affected in the short irradiation time required in this method, it is a good policy to avoid sun bleaching them. Rosin size in paper can be detected by a characteristic ring when ether or a similar solvent is dropped in a spot on the surface of paper, or by observing the resistance of ink from feathering when applied to a crease.[26]

Most papers industrially manufactured in the 20th century contain mixtures of various fibres and chemical additives used as sizing, coating and filler, and it is impossible to detect what substances may exist in a given paper without a specialist's knowledge in paper analysis. Titanium dioxide, for example, is a pigment that came into use in the 1920s, and subsequently has come into wide use as a filler to increase the brightness and opacity of industrially made papers. This substance can cause photochemical degradation of cellulose. The detection of titanium dioxide filler requires more involved chemical tests than are feasible in a normal paper conservation laboratory. Since titanium dioxide and other photosensitive substances are found in modern papers, sun bleaching should be used cautiously with ultraviolet filtering. Further studies would be helpful to establish whether potentially harmful substances are in fact harmful during the short period of filterable irradiation required in sun bleaching.

Without forgetting the cautions and limitations mentioned above, I would like to present some observations on the remarkable results of bleaching accomplished by sunlight. My observations are necessarily subjective, but some scientific tests on sun bleached paper samples are now being performed at the Cooperstown Graduate Program by Thomas Branchick under the supervision of Dr. F. Christopher Tahk, and these may result in some corroborative quantified data.

1. The use of sunlight is a very effective bleaching method. Stains such as distinct matburns, dark foxing marks, wood stains, and other pronounced discolorations that would require strong solutions of various chemical bleaching agents are satisfactorily reduced by single exposure. Since these tenacious discolorations are reduced simultaneously as the overall bleaching of the paper takes place, I have not found it necessary to expose only the local discolored areas. The only stains that do not reduce satisfactorily in sun bleaching appear to be those of metallic origin and dark pigmentation from mould growth.
2. Paper treated by this method of sun bleaching regains considerable physical strength, increasing its body and elasticity. Degraded paper that was weak and brittle before treatment is more handleable and supple, regaining its "rattle" in most instances. The recovered elasticity is observed in the ability of paper to flex more, and in the regained textural qualities of paper. The three-dimensional quality of relief and intaglio prints returns quite well. It is quite remarkable to see portions of paper which have been particularly weakened by wood stains and other highly degrading discolorations regain their normal strength and recover the appearance of the rest of paper. In chemical bleaching these areas most often persist in their weakness, and sometimes shatter in the presence of strongly alkaline bleaching agents or become spongy from the effervescence of certain bleaching agents like sodium borohydride that disrupts the already weakened fibre bonding.

3. The visual effect of sun bleached paper is most pleasing. Paper has a warm white tone rather than the staring white brightness often observed in chemically bleached paper. The brilliant appearance of ink in black and white graphics is often remarkable. This probably is due to the fact that the binding medium of the ink remains intact in the sun bleaching method, whereas in chemical bleaching the medium may be affected by the strong alkalinity of the initial bleaching agent or by the antichlor subsequently used, or by both, which cause the ink to "gray" slightly. It is also partly due, no doubt, to the congenial warm new tone of the paper interacting with the ink to create this lustrous effect. Additionally, the regained textural quality of the paper contributes to the subtle three-dimensionality of the image, enhancing its visual effect. The end result after necessary repair and gentle pressing is an object that appears natural without the forced whitened appearance of chemical bleaching. The restored object looks as it should and as it was meant to be.

No study has yet been made to my knowledge to assess the effect of sunlight on paper under the conditions I have discussed and evaluate its use as a bleaching agent for paper objects. Sunlight is a much feared light source in conservation and its degrading effects to papers and textiles have been considerably studied.[27] However, these studies are normally carried out over long irradiation period (often 100 hours or more), and usually in dry exposure conditions. We must wait for future studies to fully evaluate the soundness of this bleaching method which employs a short irradiation time and is carried out in the presence of water and magnesium bicarbonate. Until it can be proven safe, I present this method simply as an alternative possibility to be considered in paper bleaching.

The use of sunlight as an irradiation source may pose a problem for many conservators who do not have an access to it year-around as I do in California, and even where sunlight is present, air-pollution may pose a problem in urban areas. But xenon arc lamps produce a spectrum close to that of sunlight, and investigation of an artificial light source that would be practical for the use in a conservation laboratory is now underway in the project at the Cooperstown Graduate Program.

Notes to Introduction

1 T. K. McClintock, "Observations on Developments in the Treatment of Works on Paper," *Journal of the American Institute for Conservation* 31, no. 1 (Spring 1992): 114.

2 GoreTex is a vapor-permeable fabric. Keyes's recommendation for the use of enzymes to facilitate stain reduction, not included in this reading, received renewed attention in the 1970s and 1980s.

3 Robert Futernick, "New Methods, Adapted Tools and Alternate Treatments," *Abstracts, Symposium 88, Conservation of Historic and Artistic Works on Paper* (Ottawa: Canadian Conservation Institute, 1994), 21; "Methods and Makeshift,"

Book and Paper Group Annual 3 (1984): 65–74. The Book and Paper Group of the AIC began to have regular update sessions on developments in scientific research and treatments. See Leslie Kruth, "A Survey of Recent Scientific Research Which Has Caused a Re-evaluation of Commonly Used Practices in Book and Paper Conservation," *Book and Paper Group Annual* 7 (1988): 30–39; Dianne van der Reyden, "Recent Scientific Research in Paper Conservation," *Journal of the American Institute for Conservation* 32 (1992): 117–37.

4 Euterpe is the muse of music.

Notes to Reading

1 Plenderleith, H. J. & Werner, A. E. A., *The Conservation of Antiquities and Works of Art*. 2nd ed., London, 1971, pp. 78–79.

2 Clapp, F., *Curatorial Care of Works of Art on Paper*, 3rd ed., revised, Oberlin, Ohio, 1978, pp. 59–61.

3 Banks, N., Paper cleaning, *Restaurator*, Vol. 1. No. 1, 1969, pp. 52–55.

4 Tang, C. & Jones, N. M. M., The effects of wash water quality on the aging characteristics of paper. *Journal of the American Institute for Conservation*, Vol. 18, No. 2, Spring 1979, pp. 61–81.

5 Hey, M., Paper bleaching: its simple chemistry and working procedures, *The Paper Conservator*, Vol. 2, 1977, pp. 10–23.

6 It appears that there are various "standard" solutions according to the way these deacidification solutions are prepared. The standard solutions here are: Magnesium bicarbonate *c.* .2%; calcium carbonate, a saturated solution diluted with water 1:1.

7 Tang, L. C. & Jones, N. M. M., *op. cit.*

8 [The use of methanol is generally discouraged because of its toxicity. It is unclear exactly what Keyes means by "more powerful and quicker." Methanol is the more polar alcohol.—Ed.]

9 Weidner, M. K., A vacuum table of use in paper conservation, *Bulletin of the American Institute for Conservation*, Vol. 14, No. 2, 1974, pp. 115–22.

10 Segal, J. & Cooper, D., The use of enzyme to release adhesives, *The Paper Conservator*, Vol. 2, 1977, pp. 47–50.

11 Hatton, M., Enzymes in a viscous medium, *The Paper Conservator*, Vol. 2, 1977, p. 9.

12 [Karl Wilhelm Scheele (1742–1786), a German Swedish chemist, isolated several elements, including oxygen, chlorine, and manganese.—Ed.]

13 Barrow, W. J., *Permanence/Durability of the Book–III: Spray Deacidification*, Richmond, Virginia: W. J. Barrow Research Laboratory, 1964, p. 9.

14 Clapp, V. W., The story of permanent/durable book-paper, 1115–1970, *Restaurator*, Supplement No. 3, 1972, pp. 12–13.

15 Watcher, O., The cleaning and bleaching of prints, graphic arts and water-colours. English translation. *Einband- und Grafik-Pflege*. Date unknown, pp. 156–166.

16 Plenderleith, *op. cit.*, p. 86.

17 Annis, Z. K. & Reagan, B. M., Evaluation of selected bleaching treatments suitable for historic white cottons, *Studies in Conservation*, Vol. 24, No. 4, November 1979, pp. 171–178.

18 Bogarty, H., Campbell, K. S. & Appel, W. D., Some observations of the evaporation of water from cellulose, *Textile Research Journal*, 2, 1952, pp. 75–81.

19 Gilbert, A. F., Pavlovova, E. & Rapson, W. H., Mechanism of magnesium retardation of cellulose degradation during oxygen bleaching, *Tappi*, Vol. 56, No. 6, June 1973, pp. 95–99.

20 Isbell, H. S., Parks, E. W. & Naves, R. G., Degradation of reducing sugars and related compounds by alkaline hydrogen peroxide in the presence and absence of iron and magnesium salts, *Carbohydrate Research*, 45, 1975, pp. 197–204.
21 Richter, G. A., Relative permanence of papers exposed to sunlight: II, *Industrial and Engineering Chemistry*, Vol. 27, No. 4, April 1935, p. 437.
22 Padfield, T., The deterioration of cellulose: The effects of exposure to light, ultraviolet and high energy radiation, *Problems of Conservation*, London, 1969, pp. 119–120.
23 Annis and Reagan, *op. cit.*, p. 177.
24 Browning, B. L., *Analysis of Paper*, 2nd. ed., New York, 1977, p. 73.
25 Richter, *op. cit.*, p. 436.
26 Browning, *op. cit.*, p. 83.
27 Padfield, *op. cit.* List of publications as of late 1960s, pp. 159–165.

Reading 88

Carlo James, Caroline Corrigan, Marie Christine Enshaian, and Marie Rose Greca

Old Master Prints and Drawings: A Guide to Preservation and Conservation (1997)

This reading again addresses the principle of reversibility (see reading 81), a tenet still very much in force, at least philosophically, in modern conservation ethics and always an uncomfortable concept for paper conservators.[1] *Once prominently featured in the* AIC Code of Ethics, *in 1994 it was transferred into the newly created* Guidelines for Practice, *where it appears only under "Compensation for Loss."*[2] *Despite this more precise classification, the concept presents an ongoing dilemma for paper conservators, who are now guided in their treatment decisions by the principle of suitability.*

[. . .]

The idea of the reversibility of conservation treatment was born out of the recognition of the damage caused by some empirical procedures, whose long-term effects were not well understood, and by materials that age poorly. As early as 1866, Secco Suardo[1] affirmed:

> If one queries history, if one consults the art world, if one studies carefully the works that remain today still untouched, one must admit that it was neither time nor war nor fire nor the iconoclasts that destroyed most paintings, but rather the ignorant presumptions of those who pretended to clean them.[2]

The same assertion can be made about the cleaning and bleaching of drawings and prints. In the conservation of works of art on paper, especially during the last two decades, research on treatment techniques, conservation materials and reversibility

---

From Carlo James, Caroline Corrigan, Marie Christine Enshaian, and Marie Rose Greca, *Old Master Prints and Drawings: A Guide to Preservation and Conservation*, trans. and ed. Marjorie B. Cohn (Amsterdam: Amsterdam University Press, 1997), 211. Reproduced courtesy of the publisher.

has been pursued, and conservators have sought a better understanding of the irrevocableness of their profession. Because of curatorial mistrust of conservation, conservators were tempted to extend the notion of reversibility—which basically involves substances—to *all* conservatorial actions in order to reassure curators as well as themselves, which has rendered the concept meaningless. Although some conservation treatments are reversible, *a great number are not.*

Caution precludes the use of conservation techniques whose principles and long-term effects are not well understood or where the ageing properties of materials are not known. It remains true, however, that a treatment—a bleaching, a deacidification, a cleaning or even a simple bathing—while irreversible, can be necessary and beneficial to the work of art if it is done under good conditions and with an understanding of its necessity. It is ignorance of graphic media and their stability or lack of same, of chemical substances and their use, and misunderstanding of the pictorial equilibrium between support and image that are the causes of poor results. To be sure, one can never be cautious enough; but we must repeat, many treatments are possible and necessary. Systematic nonintervention is not the answer, whether intervention is reversible or not.

Notes to Introduction

1 More realistic expectations of reversibility were recommended by Barbara Appelbaum, "Criteria for Treatment: Reversibility," *Journal of the American Institute for Conservation* 26 (1987): 65–73.

2 The AIC *Guidelines for Practice*, formerly known as *Standards*, describe behaviors that promote the aspirational goals of the *Code*. Ways in which the *Guidelines* apply specifically to paper conservation are further elaborated under the *Commentaries*. See Barbara Appelbaum, "Conservation in the Twenty-First Century: Will a Twentieth-Century Code of Ethics Suffice?," *AIC News* 36, no. 4 (July 2011): 1, 7–8.

Notes to Reading

1 [Count Giovanni Secco Suardo (1798–1873) was an Italian connoisseur, restorer, and early advocate of the "scientific" approach to cleaning paintings. His popular manual, *Il restauratore di dipinti*, was originally published in 1894. www.associazion egiovanniseccosuardo.it.—Ed.]

2 G. Secco Suardo, *Il restauratore dei dipinti*, 4th ed. (Ulrico Hoepli, Milan 1927), pp. 317–318.

Reading 89

JANE MCAUSLAND

The Practicalities and Aesthetics of Retouching: Rationality versus Intuition (2002)

The paper conservator Jane McAusland (1940–) addresses here another paper conservation procedure, that of retouching, or in-painting as the procedure is more commonly known in the United States.[1] The topic of retouching has since been carefully deconstructed by the paper conservator Tina Gretta Poulsson, whose definition of retouching and differentiation of four types of "interventive" and "noninterventive" retouching are useful for putting McAusland's argument into perspective.

According to Poulsson, "Retouching is a restoration process employed to replace damaged or lost areas of media or to conceal blemishes which are not considered to be historically important. Retouching can be carried out directly on the original item or on paper infills/inserts."[2] She includes under the rubric of retouching four types:

- *Infilling missing parts of the support (noninterventive)*
- *Addition of color to these infills/inserts (noninterventive)*
- *Infilling missing media by direct application of pigments (interventive)*
- *Infilling missing media by overlays (noninterventive).[3]*

In spite of Poulsson's meticulous dissection of the practice of retouching, the ethical dilemmas that McAusland confronts, especially, she argues, for conservators in private practice, stubbornly persist. Calling for a frank discussion of the issues raised by direct, or interventive, retouching, including those of authenticity and reversibility, she does not shy away from asking difficult questions, for example, will the avoidance of retouching due to an "excess of scruple" condemn damaged artworks "to the attic where they will quietly rot away altogether"?[4] She convincingly argues

---

that paper conservators need to confront the impossibility of always conforming to inflexible standards of reversibility.[5] *As is the case with all her writing and speaking, McAusland's forthrightness is bracing.*[6]

Making decisions as to whether to retouch, carry out facsimile work, tone or stipple in missing areas of prints and drawings is extremely difficult. The subject of retouching is almost impossible to teach effectively through formal methodology, there being no substitute for an in-depth knowledge and understanding of the object, artistically, historically and materially. Once a line of intervention is chosen rationality is often discarded and subjective intuition inevitably takes its place. [. . .]

Retouching Repairs and Abrasions in Art on Paper Conservation: Making the Decisions

Most of the literature on retouching or toning in of lacunae or missing areas on a print or drawing has been written by those working in fields other than paper conservation. Although some of what has been written is relevant, a great deal does not apply to paper artefacts.[1,2,3]

The private conservator usually works for many different clients, and often on many different types of object, so decision making can be a much more complex affair than for the museum conservator, who often is required to work on collections of similar objects. The well-being of the object is, of course, paramount, but private clients naturally wish to hang their prints and drawings for display in their houses, and dealers need to present their clients with aesthetically harmonious objects that are in good condition. More often than not, non-intervention is not an option. The museum conservator is able to have a continuing dialogue with curatorial staff, and the material to be conserved is usually in a much better condition than that which the private conservator has to face; ethics and aesthetics can usually take a uniform progression as dictated by these circumstances, and also by the fact that the collection may be displayed together. Although the private practitioner has to put the needs of the object first, and often has to enter into debate with clients in order not to compromise her or his ethical integrity, there are subtle lines which can be drawn in order to achieve results both desirable to the client and conservator, as well as being best for the object. It is by drawing together all the important areas of judgement that the conservator's collective knowledge is called to muster.

Another, and not unimportant, area that the conservator in private practice has to consider is the financial value of the object. This reality can lead either to action or to inaction. However, there may be circumstances where a work of art of high artistic value (which may also often have a very high financial value), should be left untouched in spite of damage. Making broad decisions should always be a combination of the eyes and minds of with the curator or custodian of the work concerned working with those of the conservator.

Ethics

As far as ethics that are determined by museum standards are concerned, private conservators may justifiably act where museum practitioners would not, or, in fact, may have no reason to do so. Ethical decision-making is generally perceived as a 'murky' area, and is frequently the basis of the criticism of private conservators, as mentioned above.[4] It is indeed impossible, or certainly difficult, to make hard and fast rules to govern ethical decisions. The result is often excess of scruple and non-intervention. This approach must often stem, in part, from bad examples of decision-making by past conservators who may have been less than scrupulous, or even ignorant. In both these areas the knowledge, skill and critical judgement of the conservator has to play a big part. Restoration is often necessary, even vital, for the long-term conservation and general well being of the object. These are extremely difficult areas to define and I believe they really cannot be taught effectively on conservation courses—only guidelines can be given to students. The hands-off approach may be the correct decision, as long as it is checked with the over rider, that with experience and knowledge, coupled with intuitive critical judgement, decisions can be made which are essential for the long-term preservation of the object. However, if informed and intuitive steps are not taken to conserve/restore badly damaged prints, drawings or watercolours in the private sector these items may just be consigned to the attic where they will quietly rot away altogether. Conservators have to face these problems, not run away from them! As always, the final decision on whether to treat, and how far to take the conservation and restoration, must be a compromise which preserves the object and satisfies everyone concerned with a minimum of intervention, that does not jeopardize the integrity of the object.

Aesthetics

I feel very strongly that aesthetic values have been neglected in the conservation of art on paper. Again, this has roots in the 'rational versus intuitive' attitude to conservation/restoration. A drawing with large missing portions or other damages simply cannot be enjoyed if it is not restored, as it lacks visual unity. Here I will quote Philippot, father and son:[5]

> A work of art as such is not composed of individual parts but constitutes, as an image, a whole endowed with its own unity, realised in the continuity of the form—a unity therefore essentially different from that of the things represented. Any discontinuity, any interruption inevitably disturbs the reading of this rhythm. But conversely, since the unity of the form is not divisible, each

remaining fragment continues to participate in the broken unity, suggesting it to the extent that each fragment still contains the potentiality of it. The unity exists in the manner of a text where certain words are either missing or difficult to decipher. While impossible to resume the original creative process, reconstitution remains conceivable, even fully justified, if we understand it as an act of critical interpretation intended to re-establish a broken formal continuity, to the extent that this continuity remains latent in the damaged work and reconstruction brings to the aesthetic structure new clarity of reading that it has lost.

There are many routes to a harmonious visual recreation, and retouching can be carried out in many different ways ranging from full facsimile to simple toning of missing areas, without being labelled as forgery. An old master print, for example, can have the full treatment because what is missing is fully known in other examples of the multiple (Fig. 1a, b). An old master drawing, though, may

Figure 1
Albrecht Durer: woodcut. a (*left*) Damage repaired with matching paper; b (*right*) same woodcut with facsimile work carried out on lost area. Photo: Jane McAusland

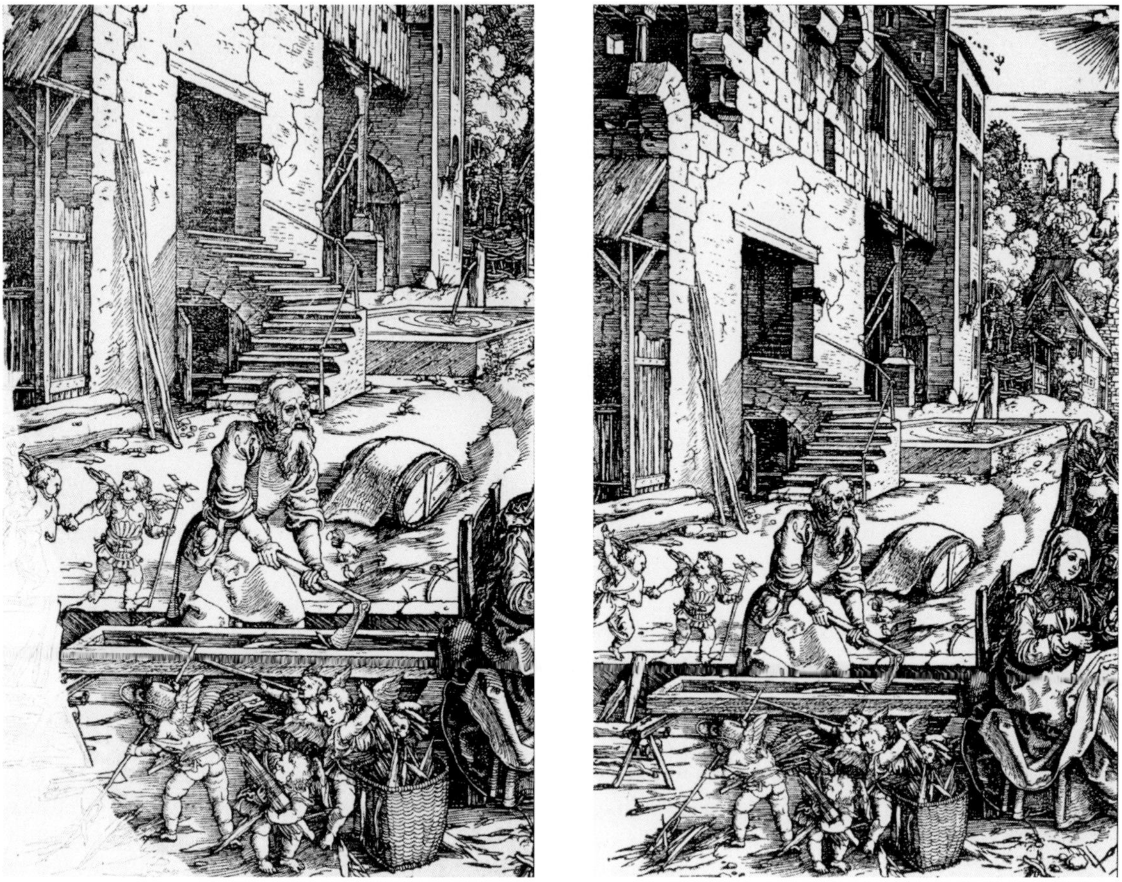

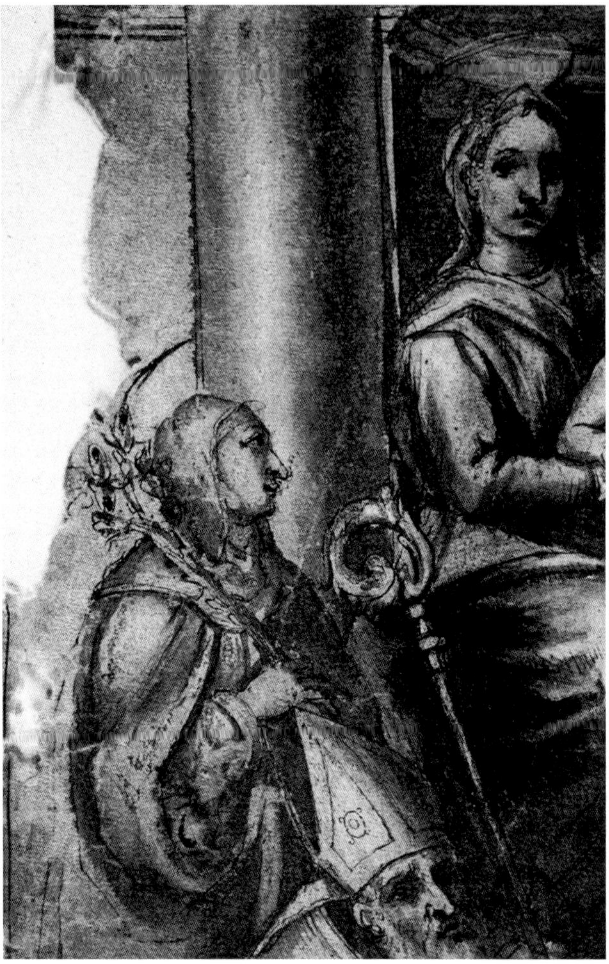

Figure 2
Old master drawing: missing area on upper left side. Photo: Jane McAusland

just have missing areas sympathetically toned in (Fig. 2). The approach to other objects can range from A to Z. I think the practising conservator must, amongst his or her talents, possess good, informed, critical aesthetic judgement. However, this will always be subjective, and therein lies the problem and the criticism from the non-interventionists. Criticism is a pivotal element: the conservator has to engage with the artist's perceived intention, but has to refrain from any personal creativity or embellishment.

The restorer-conservator Carl Schwiedler,[6] working in Berlin in the 1920s, perfected a technique of paper repairs and facsimile work on prints and drawings to such a high degree that one could consider the practice to be faking. 'Restorer-conservator' is certainly an appropriately descriptive noun to use for this type of

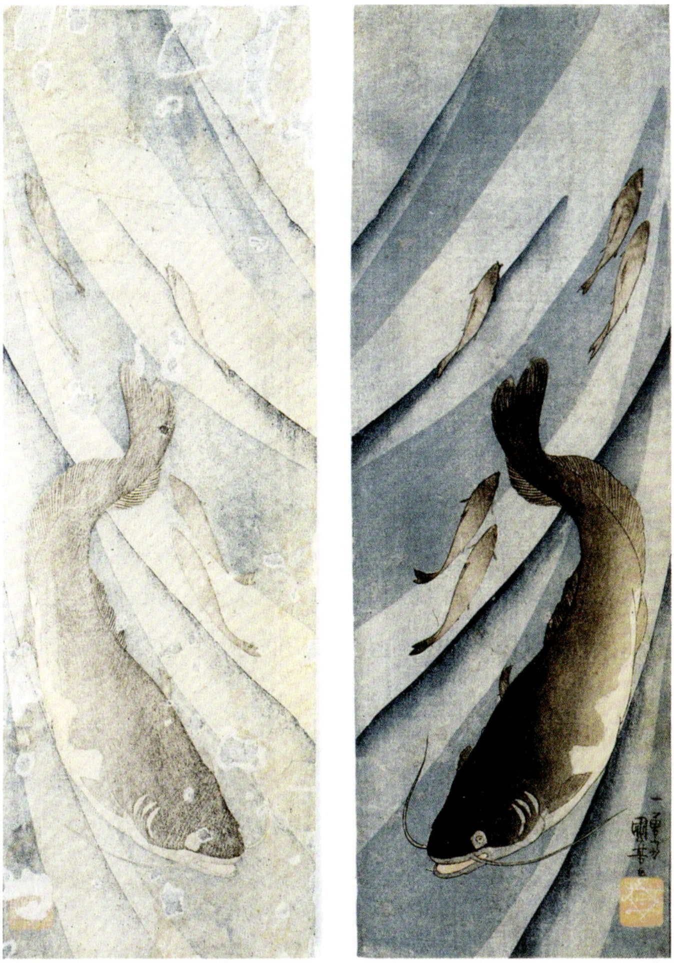

Figure 3
Kuniyoshi: Japanese wood-block print. a (*left*) Multiple repaired areas visible on verso. b (*right*) Recto of print showing retouched areas. Photo: Jane McAusland

practitioner, for, it is my opinion that there has been an overreaction against this approach, to the extent that some practitioners now seem to make it a policy to show up lacunae, thereby ignoring aesthetic appreciation, the artist's intent and the necessity for integration of the whole object. With the use of ultraviolet (UV) light, repairs can easily be made visible if the documentation is lost or not stored with the object. In any case, it may be decided it is a good policy to show repairs on the reverse side of the print or drawing, by not matching the tone of the paper, perhaps leaving it lighter if this is possible (Fig. 3).

The science of optics and the psychology that lies behind looking at a picture are of vital importance when considering the direction to be taken when toning or retouching, but these matters are beyond the scope of this article.

Historical Considerations

When considering intervention in the restoration/conservation of a work of art on paper it should be obvious that the conservator must understand and take account of the documentary evidence the work imparts. Whether the object in question is a sketch or a fully worked up presentation drawing must be considered, for example. Considerations regarding the intentions of the artist, the age of the print or drawing, and the history of the damage or deterioration must all be understood and taken account of when making decisions on restoration. Indeed, it is quite possible, for instance, that some of the damage might have been done by the artist. There are certainly objects which should have no intervention and for which inaction is the right course.

Treatments: Moving Towards a Methodology for the Toning-In and Facsimile Reproduction of Missing Areas

1. The paper support and medium

In my opinion, paper repairs carried out on a print or drawing should be executed to as high a standard as possible in order that the repaired damage recedes in dominance. The conservator must have a good knowledge of the properties and behaviour of the repair materials, in this case, paper and paste. When making a paper infill it is essential that the actual replacement paper should match, as nearly as possible, the original—in type of paper (wove or laid, oriental, etc.), thickness, and surface quality—prior to any toning or retouching. The joined area must be as flat and smooth as possible. I know it is now considered bad practice to remove any fibre of the support sheet, but as already there is a portion of this missing in some cases the edge of the loss may be pared down on the verso in order to achieve a perfect join that will be aesthetically pleasing. Even if no toning or retouching is to take place the result is so much more harmonious. The pH of repair papers must be as close to neutral as possible.

The conservator of art on paper should be able to identify the medium and technique the artist has used for the work and, if possible, to replicate this medium when retouching.

Some Practical Considerations and Examples

Knowing just how far to go with toning and retouching in the conservation of prints and drawings involves largely subjective, but informed, judgements that take into account all the considerations mentioned above. [. . .]

When confronting works with missing areas of composition, either in the support sheet or in the medium, the first step private conservators must take is to ask themselves the following list of questions:

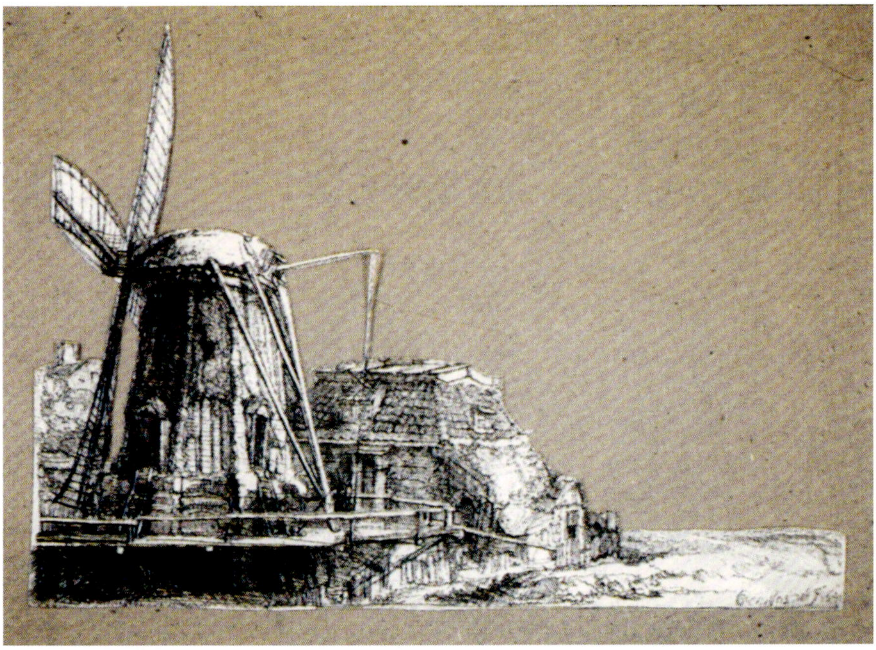

Figure 4
Rembrandt: etching. Entire etched area has been cut out. Photo: Jane McAusland

1. *What was the artist's intention?*
As broad an understanding as possible of the particular artist's working practice should be gained through research.

2. *What is the amount of loss in the sheet, and how great the damage?*
The example shown in Fig. 4 is an extreme case. The entire composition of this etching by Rembrandt has been cut round, maybe because the sheet became damaged. In order to support the work it was decided to inlay the existing piece to a sympathetic sheet. An alternative would have been to adhere it to another sheet, but by inlaying the print, its edges did not cast any shadows.

3. *What is the position on the sheet of the losses and other damages?*
Fig. 5 shows a gouache in which the missing corner disturbed the eye. A paper repair was carried out and the area retouched to tone in with the surrounding area.

4. *What method of reintegration, if any, should be used in repairing the damages?*
Sometimes decision-making is easy, but at other times hard. A large Indian miniature had very many missing areas, tears and surface losses (Fig. 6). It had a three-dimensional quality as it was undulating from being partially adhered to layers of paper that had become damp. The missing areas were harmoniously integrated after much discussion with the owner. However, on completion we both felt that

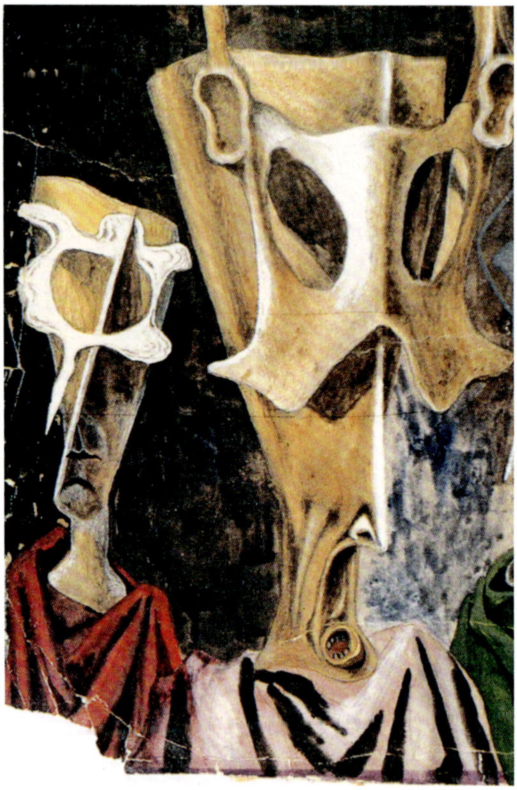

Figure 5
John Banting, gouache.
Missing corner lower left.
Photo: Jane McAusland

maybe it should have just been left as a splendid historical object in its own right, with all its flaws.

A Marco Ricci gouache had other problems (Fig. 7). The medium in the sky had begun to loosen and some areas had flaked off, leaving the animal skin support bare, making it visually disturbing. This scarring was rectified by the retouching of these areas with a similar, but not identical medium—pastel in water—that could be removed in the future, if necessary. The result was sympathetic.

There is a school of thought that a conservator should never retouch on the original work as in many cases this may not be reversible. Caution should of course prevail and, wherever possible, reversibility needs to be a first priority. However, this is not possible in many cases. Often a small non-reversible retouch is preferable to a visual distraction on the work of art. This area of the philosophy of restoration should not be a black and white one, but grey and subjective which, of course, makes it easy to opt out of altogether.

5. *What is the artistic, historic, sentimental and/or commercial value of the work to be conserved?*

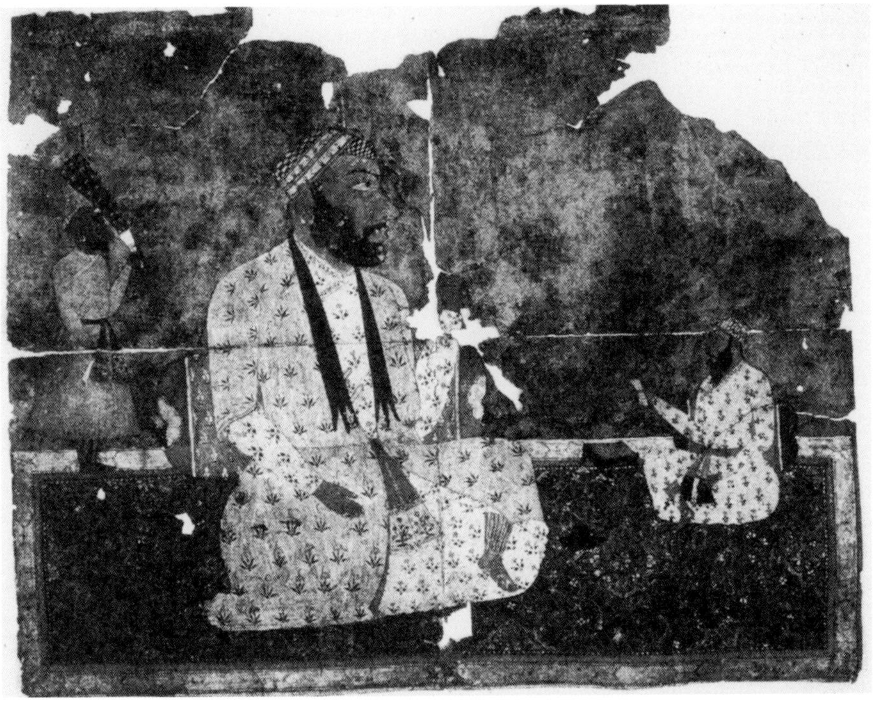

Figure 6
Indian miniature. Multiple damages revealed after removal of backings. Photo: Jane McAusland

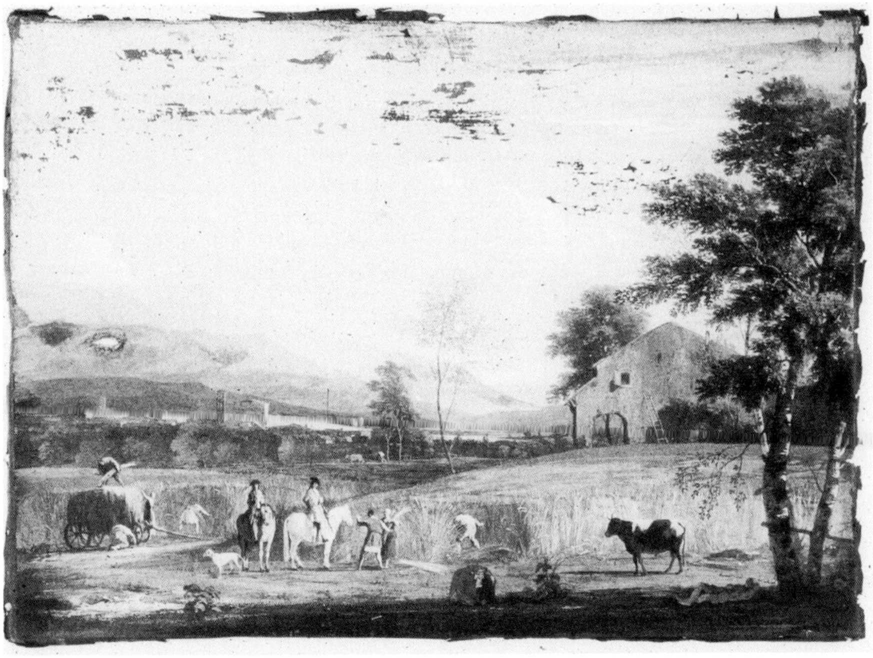

Figure 7
Marco Ricci, gouache on animal skin. Demonstration of loss of pigment. Photo: Jane McAusland

It is helpful to determine this information as a general informative tool or it can be a dispiriting experience for the conservator to work without it.

Conclusion

Paper conservators in the area of fine art need to develop a philosophy, so that a holistic approach can be taken when decisions have to be made in the conservation/restoration of works of art on paper. Everyone enjoys living with these objects, and we hang them on our walls. We must be prepared to restore beauty and harmony when it beckons, or these damaged objects may disappear forever. I hope this short article will spark off further debate in this neglected area of the work.

Notes to Introduction

1 Other terms for retouching are *loss compensation* and *reintegration*.
2 Tina Gretta Poulsson, *Retouching of Art on Paper* (London: Archetype, 2008), 2.
3 Ibid., 4.
4 One year after McAusland's article, "Inpainting and Compensation" was the topic of discussion during the AIC 31st Annual Meeting. Approximately two hundred conservators entered into lively practical and philosophical debate, concluding that inpainting decisions are highly subjective. See a summary by Judith Walsh, "Evolution in Inpainting Discussion," *Book and Paper Group Annual* 22 (2003): 105–7.
5 It should be pointed out that thoughtful commentary was added in 1997 to the *Guidelines for Practice*—Compensation for Loss, specifically addressing aspects of retouching works of art on paper. American Institute for Conservation of Historic and Artistic Works, *Guidelines for Practice*, Commentary 23. www.conservation-us.org.
6 McAusland's professional service is also noteworthy; in 1976 she was a founder of the Institute of Paper Conservation (IPC), an international professional organization under whose auspices the *Paper Conservator* was published. The IPC was merged with other British conservation-related groups to form the Institute for Conservation in 2005. Jane McAusland, "A Brief History of a Conservator of Art on Paper in the United Kingdom: 1959–2003," *Book and Paper Group Annual* 22 (2003): 41–47.

Notes to Reading

1 Jessell, B., 'Helmut Ruhemann's inpainting techniques', *Journal of the American Institute for Conservation*, vol. 17, no. 1 (1977): pp. 1–8.
2 Leavengood, P., 'Loss compensation theory and practice: A brief history', *Loss Compensation Symposium Postprints*, 1–5 (Seattle: Western Association for Art Conservation, 1993).
3 Croce, B., 'The unity and indivisibility of the work of art', *The Aesthetic as the Science of the Expression and of the Linguistic in General* (New York: Cambridge University Press, 1992).

4 ["criticism of private conservators, as mentioned above": As a conservator in private practice, McAusland begins this article by observing that the terms *commercial* and *cosmetic* have assumed negative connotations when applied to treatments carried out by the private sector.—Ed.]

5 Albert Philippot and Paul Philippot, 'Le problème de l'intégration des lacunes dans la restauration des peintures', *Bulletin de l'Institut Royal du patrimoine artistique*, 2 (1959).

6 [Carl Schweidler (1884–1962) was the brother of Max Schweidler (1885–?), author of *The Restoration of Engravings, Drawings, Books, and Other Works on Paper*, trans. Roy Perkinson (Los Angeles: Getty Conservation Institute, 2006).—Ed.]

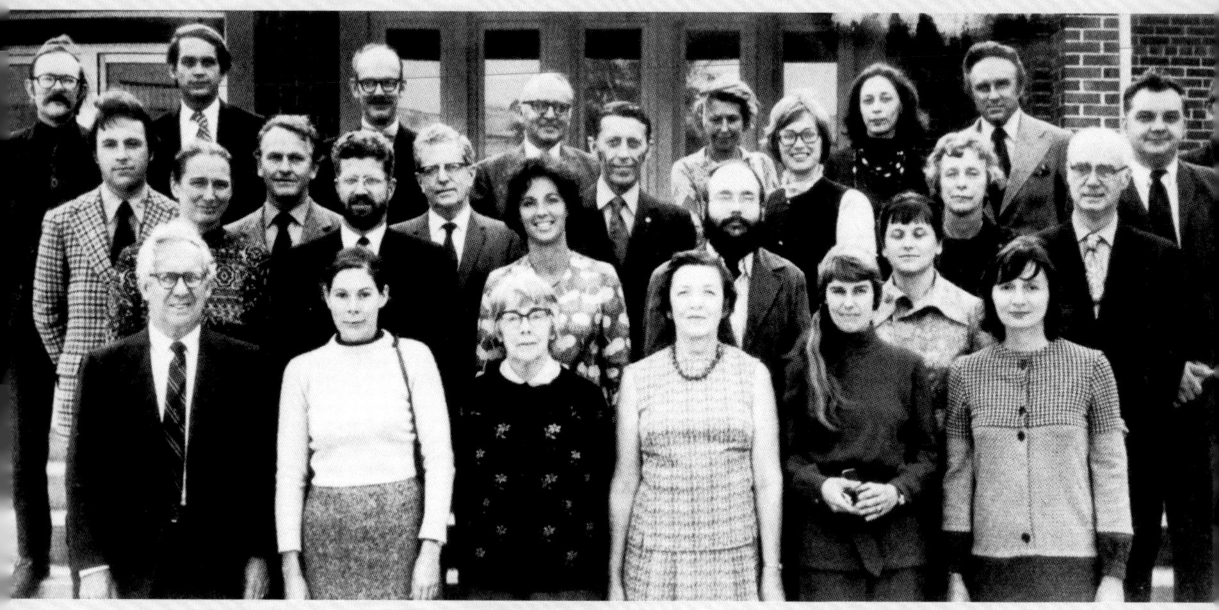

SEMINAR FOR PAPER CONSERVATORS

October 18-22, 1971

FIRST ROW: Mr. George M. Cunha, Library of the Boston Athenaeum; Miss Noel L. Kunz, Intermuseum Conservation Associatio Mrs. Stella Patri; Mrs. Carolyn Horton, Carolyn Horton & Associates, Inc.; Mrs. Helen Otis, Lincoln Center for the Performing Arts; Mrs. Antoinette King, Museum of Modern Art

SECOND ROW: Mrs. Marilyn K. Weidner, Conservation of Historic & Artistic Works; Dr. Richard D. Smith; Miss Eleanor McMillan, Smithsonian Institution; Mr. Roy Perkinson, Museum of Fine Arts; Mrs. Marjorie B. Cohn, Fogg Museum of Art; Dr. T. A. Howells, The Institute of Paper Chemistry

THIRD ROW: Mr. Douglas Kenyon, Art Institute of Chicago; Mr. Lawrence J. Majewski, Institute of Fine Arts; Dr. John C. Williams, Library of Congress; Mr. Mervyn Ruggles, National Gallery of Canada; Mrs. Mary Todd Glaser; Miss Anne Clapp, Henry Francis duPont Winterthur Museum; Dr. Donald K. Sebera, Conservation of Historic & Artistic Works

FOURTH ROW: Mr. Paul Banks, The Newberry Library; Mr. H. Wayne Eley, Jr., The New York Public Library; Mr. William Spaw American Philosophical Society Library; Mr. Harold Tribolet, R. R. Donnelley & Sons; Mrs. Christa M. Gaehde; Mrs. Martin Yamin; Mr. Merritt Safford, Metropolitan Museum of Art; Dr. George J. Reilly, The Henry Francis duPont Winterthur Museum

PART VIII

The Paper Conservator: Going beyond the Bench

Part VIII begins with a reminder of the distrust with which restorers were regarded by collectors. And this is no surprise, given the lower status accorded to works on paper relative to paintings and sculpture. For many practitioners, the craft of paper restoration ranged from a polite pastime for ladies to the practical management of a gentleman's print collection. But in the twentieth century a series of developments—some serendipitous and others carefully scripted—led to a remarkable "openness and sharing of information among conservators that came as a breath of fresh air to a field that had traditionally been rather closed and secretive, perhaps reflecting the insecurity felt by individual practitioners."[1] For many people, the first awareness of book and paper conservation emerged after the Arno breached its banks on November 4, 1966.[2] How many budding paper conservators identified with Margot Harrington, the intrepid junior conservator from the Newberry Library who flew to Florence to "help in any way I could, to save whatever could be saved, including myself."[3] The enormous damage wreaked by the Florence flood galvanized conservators around the world. Paper and book conservators in the United States had already formed a close-knit community within the American Group of the International Institute for Conservation of Historic and Artistic Works (IIC-AG). In 1971 Bertie Browning (see reading 51) and Marilyn Weidner (1928–) were instrumental in organizing the NEH-funded Seminar for Conservators of Paper Artifacts at the Continuing Education Center of the Institute of Paper Chemistry in Appleton, Wisconsin. The first day began with Marilyn Weidner, Paul Banks (1934–2000), and Harold Tribolet (1911–1993)[4] describing the conservator's problems; the remaining four days covered topics in papermaking, paper chemistry, fungi and fungicides, enzymes, and bleaching, including the popular Chloramine-T. Weidner's summary notes of this groundbreaking event capture the participants' intense engagement, preoccupations, and ready flashes of humor:

Participants in the Seminar for Conservators of Paper Artifacts at Continuing Education Center of the Institute of Paper Chemistry, Appleton, Wisconsin, October 17–22, 1971.

- "Deacidification—try Bufferin if all else fails."
- "Strength of solvents—hydrocarbon–alcohol–acetone–DMF–DMSO–pyridine–prayer."
- "Temperature—FREEZING OF P-P-P-P-PAPER."
- "Health and safety—It was noted that a one-hour treatment with dilute ethyl alcohol markedly improves the general mental (if not physical) characteristics of paper conservators."[5]

In the meantime, three degree-granting graduate programs had been established in the United States that offered specialized tracks in paper conservation.[6,7] As a result, the need for systematic curricula and agreed-upon reading lists became evident. In 1975 *Suggested Reading for a Four-Year Apprenticeship in Conservation of Works of Art on Paper* was compiled by Anne F. Clapp with suggestions by Roy L. Perkinson, followed very quickly by the more comprehensive *Reading List for Students in Conservation of Historic and Artistic Works on Paper and on Photographs*, compiled by Anne F. Clapp, Roy L. Perkinson, Antoinette King, Marjorie B. Cohn, Paul N. Banks, and Jose Orraca. Young students of paper conservation devoured over three hundred entries categorized as required or recommended for reference. Thus an official "body of knowledge" was solidified, the first step required by a concurrent certification movement that had begun as early as 1971 in the IIC-AG.[8] While certification was ultimately unsuccessful, the need for universal standards among conservators brought to the fore a thirst for information among paper conservators and an eagerness to disseminate knowledge within their growing professional ranks. In 1980 the Book and Paper Specialty Group was established within the recently formed AIC. In 1982 the *Book and Paper Group Annual* appeared, followed in 1984 by the *Book and Paper Group Paper Conservation Catalogue*.[9]

As paper conservators recognized the advantages of quantifiable credentials, whether attained through formal education or through apprenticeships, they also realized the rewards of collaboration and exchange among wider communities. To achieve these goals, however, paper conservators needed to venture beyond the bench, a journey that continues today.

Spreading the word of preservation beyond traditional repositories of paper-based collections has even greater potential rewards. We need to recognize that influencing communities beyond our own is the key to maintaining a vital and responsive profession and one that is universally acknowledged as indispensable for the preservation of our cultural heritage. Fortunately, we have easily accessible, free, convenient, and useful tools to communicate with both conservators and nonconservators. Conservation information—of every type and in every format—serves to strengthen our alliances with other professionals in heritage preservation, professionals who can, in turn, bolster our conservation platform and, in these times of cost containment and outcome-based evaluation, positively influence important policy makers. Information resources available to paper conservators have improved in three significant ways:

1. A dramatic increase in the amount and type of conservation information, ranging from the highly technical to the general;
2. A broader range of audiences targeted to receive conservation information, including fine art insurers, scrapbookers, stamp collectors, emergency responders, archive/library material and furniture manufacturers, and the general public. Building bridges with new and underrepresented communities or even those with just a slight interest or involvement with conservation increases the chances that conservators will be involved in establishing preservation policies or implementing projects that can have practical and philosophical rewards;
3. New and more effective formats to deliver conservation information. New media such as wikis and blogs make it possible to reach every sort of audience in whatever communication format it requires.

But it is naive to think that simply being a walking encyclopedia of conservation information will guarantee productive relationships with allied professionals. We cannot afford to suffer from the "Field of Dreams" syndrome: "Build it, and they will come." For this reason, the paper conservator who goes beyond the bench is the most valuable player on any conservation team.

Notes

1 Marigene Butler, Foreword to *North American Graduate Programs in the Conservation of Cultural Property* (Buffalo, N.Y.: Association of North American Graduate Programs in the Conservation of Cultural Property, 2000), vii.

2 Helen Spande, ed., *Conservation Legacies of the Florence Flood of 1966: Proceedings of the Symposium Commemorating the 40th Anniversary* (London: Archetype, 2009).

3 Robert Hellenga, *Sixteen Pleasures* (New York: Random House, 1995).

4 The book conservator Harold Tribolet received a presidential commendation for his work in Italy following the Florence flood.

5 Marilyn Weidner, Unpublished notes from the Seminar for Conservators of Paper Artifacts, October 18–22, 1971, Institute of Paper Chemistry, Appleton, Wisconsin.

6 For a history of these programs, see Irene Brückle, ed., *North American Graduate Programs in the Conservation of Cultural Property* (Buffalo, N.Y.: Association of North American Graduate Programs in Conservation, 2000).

7 Training by apprenticeship was also systematized. See Mary Todd Glaser, "Basic Minimum Standards for Training Paper Conservators by the Apprenticeship System: A Proposal," *Bulletin of the American Group, International Institute for Conservation of Historic and Artistic Works* 11, no. 2 (April 1971): 1–7.

8 One goal of the IPC seminar was to develop "criteria for competence," which "would facilitate a move, now underway in IIC-AG, for certification of paper conservators." Marilyn Weidner, Unpublished application to the National Endowment for the Humanities, April 5, 1971, 2.

9 Catherine I. Maynor, "Publications of the AIC Book and Paper Group," *JAIC* 31 (1992): 103–6.

Reading 90

George Cumberland

An Essay on the Utility of Collecting the Best Works of the Ancient Engravers of the Italian School . . . (1827)

The watercolorist George Cumberland (1754–1848) is only one of many authors who placed print restorers in the same disreputable category as "Picture cleaners." An air of distrust is stubbornly persistent in many handbooks addressed to collectors, advising them to be ever watchful for unscrupulous and deceptive practices. Unhappily, the uneasy relationship between caretakers and conservators persists today, to wit Carlo James and colleagues' reference to "curatorial distrust" (see reading 88) and their observation, "The relationship between the curator and the conservator vacillates between excessive confidence and antagonism; like all human relations, it is sensitive."[1] Annamaria Petrioli Tofani, former director and curator of drawings at the Uffizi Galleries, diplomatically rejoins, "Conservation . . . should be understood as a special moment in a much more extended discourse: a discourse that has preservation as its theme and, above all, prevention as its constant chord. And fortunately these conclusions are accepted by an ever growing number of curators and collectors."[2] The historical underpinnings of this interpersonal situation are complex, but surely two remedial trends already under way are shared decision making and increased transparency in treatment procedures, issues that are considered in the readings that follow.

Again of late years a most pernicious practice has prevailed, I mean cleaning old stained Prints, by washing them with weak acids, and the inventor of this unfortunate act has but too well succeeded in persuading many to submit their good old prints to this fatal operation, which, although it certainly removes stains, corrodes the edges of the finest lines; converting the old ink from a brown tint to a blue one, roughening the surface of the paper, and weakening the whole effect.

From George Cumberland, *An Essay on the Utility of Collecting the Best Works of the Ancient Engravers of the Italian School . . .* (London: W. Nichol, 1827), 23–24.

These Print cleaners, like the Picture cleaners, have done more injuries than time to the most valuable objects, and cut down the extent of fame to the old masters inconceivably; for prints thus treated hasten rapidly to decay. There is but one way safely to purify a dirty impression, and that is to expose it to the sun's rays,[1] under a shallow surface of water, in a leaden trough for some days.

Let those, therefore, who possess good old collections, even when some are well bronzed by time, be never persuaded to submit their treasures to this destructive operation. The opposite practice also once did nearly as much mischief, for they had recourse to tints from tobacco, and sepia, or tea, to give the appearance of age to later impressions;[2] but it has been ascertained that no art can imitate the effects of time; and, as in the false Patina of bronzes, these inventions soon discover themselves: smoke alone can give an even tender brown tint.

Notes to Introduction

1 Carlo James, Caroline Corrigan, Marie Christine Enshaian, and Marie Rose Greca, *Old Master Prints and Drawings: A Guide to Preservation and Conservation*, trans. and ed. Marjorie B. Cohn (Amsterdam: Amsterdam University Press, 1997), 213.

2 Annamaria Petrioli Tofani, Preface to James et al., *Old Master Prints and Drawings*, n.p.

Notes to Reading

1 [Exposure of prints to sunlight was a proven means of brightening paper as early as 1631. Mark Stevenson, "Print Restoration in Northern Europe: Development, Traditions, and Literature from the Late Renaissance to the 1930s," *Conservation Research: Studies in the History of Art* 2 (1995): 110–27; Mark Stevenson, "A Seventeenth-Century Manual for the Restoration of Prints," *Print Quarterly* 7, no. 4 (December 1990): 420–24.—Ed.]

2 [There is also the possibility that artists deliberately tinted their papers or selected aged papers for their mellow color. The Rembrandt expert Martin Royalton-Kisch believes that the artist toned many of his white drawing papers. See the online *Catalogue of Drawings by Rembrandt and His School in the British Museum*: www.britishmuseum.org/research/publications/online_research_catalogues/rembrandt_drawings/drawings_by_rembrandt.aspx.—Ed.]

Reading 91

MAX SCHWEIDLER

The Restoration of Engravings, Drawings, Books, and Other Works on Paper (1950)

Among paper conservators, Max Schweidler (1885–?) was renowned for his ability to make virtually invisible cast pulp fills, usually on Old Master prints, that fooled many collectors and bedeviled paper conservators, one of whom is Roy Perkinson (1940–), formerly of the Museum of Fine Arts Boston. Before its translation by Perkinson in 2006, Schweidler's treatise was only periodically consulted by American paper conservators eager to master the magic behind "Schweidler-ized" prints.[1] Alternatively, reconstituted paper pulp techniques could be learned from German- or Viennese-trained paper conservators, including Christa Gaedhe (1922–2002) and Lilly Hollander (1928–2011).[2]

This reading reveals Schweidler's desire for more transparency in treatment procedures. How much detailed information to reveal about treatments has always been a thorny issue in the field of conservation and would seem to vary from country to country. For example, as noted in the introduction to this part, American paper conservators have been eager to share information, albeit in a controlled manner in professional conferences and academic publications, since the 1970s. Other conservators remain reluctant. In a book with a similarly narrow audience, Paper and Water, *the editors, Banik and Brückle, deliberately decided against using actual works on paper as demonstration pieces in case they would stimulate "thoughtless imitation on the part of the uninformed" or "raise controversy about preferred treatment approaches among professionals."[3] In the fact sheet for the DVD, it is unequivocally stated, "No conservation treatment is featured." The handful of tattered prints, pieces of sheet music, and newspapers that do appear in the videos are described as "a few judiciously selected expendable paper objects."[4]*

From MAX SCHWEIDLER, *The Restoration of Engravings, Drawings, Books, and Other Works on Paper*, trans. and ed. Roy Perkinson (Los Angeles: Getty Conservation Institute, 2006), 204–5.

It seems that Banik and Brückle are responding to a widely held prohibition against publishing detailed descriptions of conservation treatments. As Chris Caple exhorts, "Advocacy can be a powerful tool; however, there are risks on describing conservation work, as unqualified individuals may attempt to copy some of the things mentioned and through lack of skill and understanding damage objects. Consequently, conservators should provide minimal details of materials and practices for widespread public consumption."[5]

For the Banik and Brückle publication, in lieu of stained artworks, highly soluble red food coloring was used on blank sample papers to demonstrate the mechanics of water diffusion and transfer, reminding one of virtual dissections of frogs or the practice of CPR on manikins. Paradoxically, the use of mock-ups has resulted in case studies that demonstrate the interaction of paper and water (i.e., how to remove efficiently discoloration from paper or how to flatten paper successfully) far more clearly and reliably than actual but physically and chemically unique works of art ever could do. In reality, Paper and Water *is filled with forbidden "details of materials and practices" but in a way that deftly deflects accusations of irresponsibility. Whether or not one agrees with the prohibition is an entirely different matter. Surely increased professionalism and shared decision making would go a long way toward addressing conservators' fear of the consequences of overdisclosure.*

Summary

I have arrived at the end of my book but certainly not at the end of all I could say about restorations. Caution prevented me from talking about some things. I wanted to avoid making a bad situation worse. My book should not be regarded as a cookbook that contains instructions for all circumstances. A restorer has to be a very sensitive person who avoids harmful methods of treatment and knows how to adapt useful treatments to his object. This is how my remarks should be understood and used. The reader is free to try other methods on his own. He should, however, do this as a responsible connoisseur and lover of the arts and experiment with his new methods first on worthless things and try them over and over again before he treats valuable works and perhaps exposes them to destruction. The reader will have noticed that I took a similar road. My friends have a great advantage: I have revealed to them the experiences of my fifty years of professional life. The responsibility of the restorer to objects of cultural value is the same as that of the physician to human beings. My book cannot be a cookbook, for the cook can throw out the soup if it does not taste good. But when the physician or the restorer has made an error, frequently all is lost. And we restorers often deal with a dangerously ill patient and are happy when we succeed in bringing it back to life again, when we succeed in saving a culturally valuable work of art not just for decades but for the foreseeable future.

Some time ago, in a monthly publication for bibliophiles, an article appeared under the title "Clinics for Antiquities." This is a title after my own heart. The restorer should feel in his workshop—his clinic—like a medical chief of staff who recognizes with an experienced eye whether treatment or an operation is indicated. There is a certain amount of basic knowledge that definitely has to be followed by the physician; this is also true of the restorer. In medicine much depends on the experience of the physician and everything on the constitution of the patient, and the same applies in restoration. The restorer has to collect experience and possess it. His success or lack of success often is dependent on the object he is working on. A lack of success is not always his fault; it is also not a consequence of the methods of treatment recommended here but can depend on the treated material with its mortal weaknesses.

Only he can become a restorer who enjoys his occupation, who treats the objects of his work with love. This love of the object caused me to write down my experiences. It has forced me to explain why I work a certain way and what dangers I have avoided, to help others and to spare them lost efforts and destructive methods.

I emphasized in the foreword that I would like to exchange experiences. I would be happy if in a later edition I could distribute suggestions that others have worked out and would like to make them available to the public just as I have done with mine. In the path toward the preservation of works of cultural value there should not be any professional secrets, for we owe it to our nation and mankind to preserve the few treasures that have been handed down to us from past times.

Notes to Introduction

1 From Irene Brückle, "Der notwendige Blick zurück: Max Schweidler," *Papier Restaurierung* 8, no. 3 (2007): 9–13.
2 Otto Wächter, "Methods of Restoring Old Prints, Documents and Drawings Using Liquid Paper Pulp," in *Conservation of Paintings and the Graphic Arts: Preprints of Contributions to the Lisbon Congress, 9–14 October 1972* (London: International Institute for Conservation of Historic and Artistic Works, 1972), 971–74.
3 Gerhard Banik and Irene Brückle, eds., Introduction to *Paper and Water: A Guide for Conservators* (Oxford: Elsevier, 2011), xx. See also readings 77 and 83.
4 Ibid., accompanying DVD, Fact Sheet.
5 Chris Caple, *Conservation Skills: Judgment, Method and Decision Making* (London: Routledge, 2000), 196. See also reading 79.

Reading 92

ANNE F. CLAPP

Curatorial Care of Works of Art on Paper (1973)

"Curatorial Care of Works of Art on Paper" by the paper conservator Anne F. Clapp (1910–2000) was published in 1973. It was reissued in 1974, 1978, and 1987 and became, along with her 1975 Reading List, *an indispensable reference for a new generation of paper conservators. Clapp was among the first to recognize the greater and more cost-effective rewards of preventive conservation administered by a team of caretakers. Thanks to scientific research and broad dissemination of its results to the library, archive, and museum communities, the responsibility of how to eliminate or retard deterioration has been more equitably allocated. Conservators, other collections care professionals, administrators, and curators have assumed prominent roles in the long-term preservation of works on paper through the design, implementation, and administration of holistic collections care programs. By working together, they can prevent or minimize damage to collections. The French restorer Robert Lepeltier (1913–1996) stated the danger to works on paper clearly:*

> *The principal and indomitable enemy of all works of art—and by far the most destructive—remains man himself. He is responsible in one way or another for the conditions in which paper is kept and cared for. He is the villain behind every crime committed on works of art, either directly or by allowing the damage to be done.*[1]

SECTION I

Factors Harmful to Paper

A curator's awareness of harmful factors, together with his determination to do all in his power to avoid or control them, is of prime importance.[1] He must apply his

From ANNE F. CLAPP, "Curatorial Care of Works of Art on Paper," (Oberlin, Ohio: Intermuseum Laboratory, 1973), 2. Reprinted with the permission of ICA-Art Conservation.

knowledge to the actual situations in his particular building, specifically, in the present concern, to the care of paper within it. Often this requires imagination and a lot of bullheadedness. On the one hand, he should know by his own use of monitoring instruments or by consultation with professional engineers what are the conditions of atmosphere throughout the year in rooms where paper is kept, how these conditions compare to the optimum, how to use and modify them; on the other hand, he should have some knowledge of paper as a material, its structure and principal components, its characteristic responses to environment, how modern papers differ from old, etc. His constant vigilance, rather than unquestioning reliance upon equipment and past assurances, is a definite advantage that should guard against harmful changes—a malfunctioning air conditioner, a new factory in the neighborhood, a product whose manufacture has been altered.

Note to Introduction

1 Robert Lepeltier, *The Restorer's Handbook of Drawings and Prints*, trans. Anne Ward (New York: Van Nostrand Reinhold, 1977), 34.

Note to Reading

1 [Not exempt from Clapp's directive were curators, who on its first publication were charged with monitoring the environment. By the 1987 revision, that task was assigned to the conservation technician, a new position in the emerging discipline of collections care.—Ed.]

Reading 93

MARJORIE B. COHN

Editorial (1979)

In this editorial in the JAIC, *Marjorie B. Cohn (1939–) seeks to rally the membership around the recent passage of a revised Code of Ethics and Standards of Practice. While these core documents of the AIC continue to be refined in response to changes in the profession, Cohn's enthusiasm for the growing professionalism of the organization's membership is palpable. Her appeal to conservators to contribute their considerable knowledge to establishing collections care policies seems to have catalyzed a surge in productivity especially among paper conservators: the AIC Book and Paper Specialty Group was established one year later. The first* Book and Paper Group Annual *appeared in 1982, and the first chapters (Humidification, Mending, Drying and Flattening) of the* Book and Paper Group Conservation Catalogue *were published in 1984. Today the Book and Paper Specialty Group counts over nine hundred members, the largest specialty group of the AIC. While it is safe to say that Cohn's challenge to the next generation to match the accomplishments of the first has been met, the extent to which conservators have assumed a "larger voice in establishing curatorial policy" remains debatable.*

When I say that our revised Code of Ethics is an achievement of the first generation of conservation, I do not mean to imply "a certain age" in us who have proposed and accepted it, but rather the age—the epoch—of the profession itself. The first generation was the half-century that has elapsed since restoration was brought out of the closet and into the light of objectively tested, documented and shared materials, techniques and procedures. Our ambition, the ideal embodied in our Code

---

MARJORIE B. COHN, Editorial, *Journal of the American Institute of Conservation* 19, no. 1 (Fall 1979): 1–2. Reprinted with the permission of the American Institute for Conservation of Historic and Artistic Works, 1156 15th Street NW, Suite 320, Washington, DC 20005, info@conservation-us.org, www.conservation-us.org.

of Ethics, became one of service: to the perfectly understood artistic or cultural object, perfectly perserved to the limits of its past history and its ideal future natural life, perfectly accessible to its present audience and forever accompanied by a perfect record of our ethical, skillful, scholarly, tactful, imaginative and reversible utmost efforts on its behalf.

An absurd formulation? I do not mean it to be. I think that in our heart-of-hearts we hope for this. We feel as strongly as we do because we see so often in objects brought to us a legacy of neglect, abuse, ignorance, mysterious and unrecorded substances, shoddy practice and false economies. Not only do we want to do better, not only do we want to do our best. We want to do the best for each and every thing.

To that end our professional association has not only adopted a Code of Ethics but also established means to disseminate information about materials, techniques and procedures of objects and their treatment. While these include various publications, conferences and the FAIC, I refer here to the *Journal*. Usually the subjects of articles submitted to this magazine attempt to advance our ideal as I have defined it, and properly so; for the first generation's mandate can never be fulfilled: each new professional advance is a step to the next.

Inevitably some day another generation will write its Code of Ethics. Its direction is suggested by practical discussions already begun in the literature on, for example, museum climatology, salvage operations after disaster, field archeology procedures, etc. But the philosophic basis of mass treatment has not yet received the consideration in print that it requires before it can be considered a mature and accepted ideal. What guidance does the conservator have today, for example, in planning a mass treatment when there is no doubt that its effect on a measurable percentage of objects within the collection under consideration will be harmful rather than beneficial? There are no loopholes in our Code of Ethics: we must identify and exempt from treatment this fraction of the whole.

Yet as intelligent professionals, we also foresee that ethical considerations could themselves preclude this; for the value of a collection as a statistical, conceptual or historical entity can dictate that each element in it, however precious in itself, must preserve its traditional relationship to the whole. Even economic considerations can have an ethical aspect, if the expense of professionally acceptable intensive treatment of selected objects in a collection prohibits extensive treatment of the whole.

There is no doubt that during the whole history of our profession conservators have seriously considered these problems. There is also no doubt, however, that the conservator has traditionally deferred to the curator or other custodian of the collection to provide the larger framework for treatment decisions; this has often been inevitable and appropriate. But it is beginning to be realized that the professional conservator may be in possession of a greater understanding of the physical facts, causes and effects which affect not only the preservation but the interpretation of a collection. This conservator can and, I believe, should have a larger voice in establishing curatorial policy. We have proved our essential value as conservators of individual objects; we should have a comparable role in shaping decisions which inevitably precede the selection of these objects for our attentions.

But rather than rely upon each of us, independently, in isolation, to establish this dialogue with curators and the administrators of collections, should we not discuss the ethical issues as well as practical methods among ourselves? There is much to think about and much to say before the next generation can match the achievement of the first.

Reading 94

MARJORIE B. COHN

Change, We Hope for the Better (2001)

In a backward glance, Marjorie B. Cohn (1939–) makes note of historical developments in paper conservation, including several mentioned in previous readings. She goes on to observe two positive trends: the education of and concomitant elevation of professionalism among paper conservators and the recognition that a distinction must be made between marks of history and marks of deterioration.

As I see it, the great change in conservation over the past two generations is caricatured—favourably—by the present-day cachet of no conservation. Yet I am still a pawn of the traditional hierarchy that puts painting on top, and I think first of the high value, both aesthetic and commercial, now attached to the unvarnished or unlined canvas. But 'unwashed' is now also a sales point in catalogue descriptions of old master prints, and I know one dealer who mounts his wares (washed or not) on time-patinated, antique-laid album sheets, to provide the subliminal suggestion that the prints too have miraculously survived untouched from the eighteenth century. This trend had begun 25 years ago, the nominal limit of consideration for this essay, but the vanishing point of my perspective extends further back, to the early 1960s. At that time the bleaching discussion focused on which hypochlorite to use. The debate between oxidizing and reducing bleaches had yet to begin, much less end, as it has now, given today's highly restricted use of any bleach at all, at least in institutional settings. W. J. Barrow was the prophet of the coming age,[1] and the pH of Whatman drawing papers[2] (which were in what I realize now was perfect condition) was anxiously assessed. Folding strength was the measurement

From MARJORIE B. COHN, "Change, We Hope for the Better," *Paper Conservator* 25 (2001): 101–5.

of choice, and I asked only myself, under my breath, whether it was likely that the drawings would ever be folded.

The 1950s were, at least in the United States, years of technological faith and hope. Just as nuclear weapons and missiles, and not communicative diplomacy, would ensure survival on a global scale, it was believed, scientific treatments would offer a new level of cure and rehabilitation for works of art on paper. By the 1960s, proprietary bleaching mysteries of the past had been displaced by chlor and antichlor, and sodium borohydride promised an entirely different regimen. The next new deacidification formula leapfrogged the last. When we examined a print that had been restored in the bad old days, marvelling at its invisible repairs, rather than reflecting on our incapacity to replicate them, we congratulated ourselves on the modern marriage of science and morality. Absolute chemistry had replaced absolute artistry.

But that was then and now is now. What changed, and why? Inevitably, paper conservation shared the developing ethos of the last third of the twentieth century in developed nations, and everywhere around us we can see similar transformations. Speaking strictly from my personal experience as an American: city centres were blasted by the urban renewal schemes of the 1960s. In the 1970s what was fortunate enough to survive was pounced upon by preservationists. Then the preservationists themselves evolved from their first perfectionism, which had taken every structure back to its origins. By the late 1980s accretions of time and use could be acknowledged as valid expressions of history. Remembering Ruskin's despair at the restorations of medieval structures and then the blossoming of the Arts and Crafts movement, it seems that, *mutatis mutandis*, the end of 'our' century mirrored the end of the last.[3]

Likewise in medicine, the miracle drugs and miraculous salvation from the scourges of my childhood—I am thinking of scarlet fever and, above all, polio—produced by discrete scientific inventions and not the silent betterments of public health measures gave us hope in the 1960s that cancer could be cured. President Nixon proposed a national research initiative, with federal funding. Decades later, biological research has continued to multiply its marvels (Dolly!), but the rates of certain cancers escalate, cures are uncertain, and alternative medicine has stolen the scientist-doctors' mystique. Well-being has become more a personal responsibility, to the extent that beyond our individual fitness we now take account of the good health of the *soma* of the entire earth.

A parochial facet of this concern for the health of the environment, whether our immediate surroundings or the planet, surely can be seen in the change in conservators' attitude toward chemical treatments. When I began work at the conservation department of the Fogg Museum, a gas bleaching apparatus survived in a storeroom, an ominous souvenir of my predecessors, which I never attempted to resuscitate. One by one over the past three decades options for fumigation have been deemed too hazardous, until now there are no facilities remaining at my institution, Harvard University, and good housekeeping rather than extermination is the mode. Likewise, carcinogenic solvents have been minimized in the conservators' repertory, and ventilation and disposal are now the first engineering priority in the design of facilities.

Although it may be conceded that developments in paper conservation over the past 35 years have simply participated in the larger trend, seminal moments still can be sought in the history of the discipline itself. I can propose two, and welcome their augmentation by others. First, there is the pair of conjoined changes in the population of paper conservators. Through this period, the original cadre of specialists employed by institutions as *conservators*, rather than preparators or binders whose skills and duties extended to restoration, retired. Understandably, these founders of our discipline had been positivists. Though trained by the age-old apprenticeship system, they were full participants in the new world of technology and, as pioneers, were out to demonstrate the validity and necessity of their own careers. Scientific method, examination verified by instrumentation, technical publications, the institution of academic training and societies, conferences, associations and journals with an international reach—all of these were the hallmarks of their transformation of a craft into a profession.

The pioneers were replaced by the pupils that they had educated in their new schools. At the beginning, this new generation was, perhaps, even more convinced than their elders that a formula or a machine could resolve the problems that presented themselves: acidity, mould stains, paper losses all would find cures in technology. My most vivid memory of these early generations of graduates from the new academic conservation programmes who came to work at the Fogg Museum as interns was their exclusive focus on problems. If they saw a spot on a work of art, functionally speaking that was all they saw, and they did not always have the cautious, non-rational discipline fostered by craft practice to restrain them from a formulaic cure.

Yet their academic background usually also gave the younger generation a more sophisticated appreciation for art history. With diplomas to prove their status, they were more readily accepted within the professional culture of curators, connoisseurs and scholars, and gradually they made common cause with the latter. This resulted in what I identify as the second factor for change in our discipline, one in which I was fortunate to play a role.

In 1967, as the paper conservator at the Fogg Museum I was the immediate overseer of a unique gathering of a large number of Ingres portrait drawings, under the aegis of our drawing curator Agnes Mongan, who was a champion of the archival care of works of art on paper and who had in a graduate seminar in 1961 introduced me to the study of the materials of old master drawings. Some of these Ingres[4] drawings had not left the hands of the family for whom they had been drawn, and many of them were on their original mounts. These and some of the others were mounted in acidic boards. Miss Mongan's desire for exact technical descriptions in the catalogue that she was preparing, together with her horror at the physical circumstances of many of the works, prompted her to solicit from the owners of all of the drawings scheduled for exhibition permission for me to unframe and examine them, and treat them if necessary.

It was a unique moment: the monetary value of the drawings had not reached the astronomical heights of subsequent decades; their private and even some of their institutional owners had not yet heard the gospel of acid-free and were ripe for conversion; museums, and especially conservation departments, had not suckled at the teat of federal grants. We seemed to have the motivation, the knowledge, the licence and even the leisure for discovery. Without any sanction beyond the owners' permission and the curator's encouragement, I could allow my imagination to assemble the clues presented by the sheets into a new concept of the original format of Ingres' portrait drawings. When I explained to Miss Mongan that it was quite different from the modern notion of a pencil sketch on a sheet of paper, she and her co-author Hans Naef readily invited me to write a completely unforeseen and unprecedented essay for their exhibition catalogue.[5]

Since then, conservators' essays in exhibition catalogues have become the norm. Art historians more hidebound than Agnes Mongan have also come to appreciate the potential of technical examination and scientific analysis. Most important, however, and critical for changes in conservation practice, conservators themselves have realized the essential importance of historical evidence on and in the works of art themselves. An adhesive stain that might formerly have been analysed solely in terms of its potential for removal is now examined as a sign of artistic practice or provenance. Marks of burnishing or tracing, creases and other disfigurements are preserved whenever possible and in any case are documented as characteristics rather than flaws. The appurtenances of the artworks—mounts, frames, inscriptions, stamps, labels—are taken into account.

The two decades between 1967 and 1987 were crucial. Speaking personally, I wrote another technical appendix and then organized an entire exhibition on the historical development of a technique.[6] Others were engaged in similar studies; I think in particular of Antoinette King's[7] research into the cut and pasted paper artworks of early modernism.[8] I select 1987 as a moment of climax because in that year the Fogg Museum acquired the Spencer Albums, nine volumes containing well over 3,000 old master prints. They had been assembled in the early eighteenth century by the Mariette family, pre-eminent Parisian print dealers.[9] I am convinced that without the preceding decades' re-evaluation of historical format and evidence, I would have dismantled the albums and mounted the prints in acid-free boards. Perhaps for ease and economy, I would have soaked the prints off their mounting sheets, whether bathing was required as a conservation treatment or not.

Certainly prints mounted individually could have been integrated more easily into the museum's rationalized storage scheme, which organizes works by size, nationality, date, and artist. And they could have been more easily installed on exhibition and made individually available to study room visitors. But we would have lost so much. Collectively, the tangible evidence of an earlier scheme of cataloguing would have been dissipated. The 'Dutch' album does not differentiate, for example, between Dutch and Flemish prints, and it gives Italianate artists the same status as the realists, who since the intensification of nationalism in the mid-nineteenth century have been perceived as somehow more authentic to the Netherlands, north and south.

But above all, we would have lost the earlier mode of viewing prints. We would not have the privilege and pleasure of laying an album on a table in good north light and turning its pages. The plane surfaces of the prints flex slightly and constantly change their angle relative to the light, so that the imperceptible relief of their intaglio lines, and the lustrous texture of the plate area, planished by the printing process, seen in concert with the more nubbly album papers all contribute to our subliminal appreciation for the life of the page. No print rigid in its mount ever looks so real.

More particular to the interest of the paper conservator, we would have lost an invaluable document of healthy ageing. None of the prints had been treated by any modern process: they were installed in the album several generations prior to the application of chlorine to bleaching. Pierre Mariette II, father and grandfather of the albums' compilers, inscribed his name and dates on some of the prints. Particularly telling is the deep brown-black of his inked inscriptions, untouched by bathing or light exposure since they were written well over 300 years ago. A Mariette signature has long been accepted by collectors as an indicator of quality, but the examples heretofore known to me, on prints that had passed through the trade many times, were always a pallid, rusty brown.

In sum, the prints have been subjected in no significant way to the environmental hazards of modern art life: they have not been exposed to light at least for the last 270-odd years of their existence; the Mariettes used the finest papers available for the albums; the prints were mounted well within the volumes so that they have not been touched by the inevitable browning at the pages' edges; that the bindings remain tight and their pages unsmudged suggests that the Spencers did not spend many evenings poring over their print collection. In particular the etchings by Castiglione,[10] though very early impressions which must antedate their mounting in the album by 70 or more years, seem as fresh as the day they were printed, with their lines and platemarks sharply three-dimensional and their papers a bright, mellow white.

In fact, it is the colours of the prints in the albums, which range through all shades of blacks and whites, that offer food for conservatorial thought. Not every print is in good condition. Several by Della Bella[11] are browned, embrittled and even broken up in a manner that suggests that they must have deteriorated after mounting from some inherent vice. The papers of the Waterloo[12] etchings are discoloured with a peculiarly disagreeable reddish mottling, which is identical to that seen on the same prints by Waterloo in albums also assembled by the Mariettes now in the collection of the Graphische Sammlung Albertina, Vienna. Was this a particular edition of the prints, or are these damages evidence of early eighteenth-century restoration?

Though the colour of the damaged papers is thought provoking, it is the colour of the prints that are in excellent condition—the vast majority—which is most instructive. We should saturate our visual memory with this warm, gleaming white, so that we cherish it when we find it in old master prints. And we should acknowledge the variations: the Italian papers do not look like the Dutch papers, though both are a clear white. It seems to me that today more and more paper conservators, even those in private practice who are sometimes forced by the demands of the trade to work toward different ends than museum conservators, recognize the imperative to preserve this overall appearance, even when local accident has disrupted the surface. It is no accident that, with the hope of encouraging standard nomenclature, two paper conservators working for the Print Council of America produced a paper sample book utilizing old sheets of paper and not modern sheets in gradated tones and hues.[13]

I date the profession's maturation in this regard to the seminal article on alternatives to chemical bleaching published by Keiko M. Keyes in 1980.[14] One could fit her renunciation of chemical bleaching in favour of water washing or sun bleaching into the rising ecological consciousness of the time. One could equate it with the contemporary search by artists and conservators for non-toxic materials and methods. One could even say that the curative imagination of an Asian conservator worked in harmony with an Asian approach to medicine and good health. I am not suggesting any of these 'deep' causes for her recommendation of sun bleaching, which was for me the significant beginning of a now profession-wide appreciation for the aesthetic integrity of the historical work of art on paper. I am simply saying that Mrs Keyes was bold enough to renounce a certain kind of faith in the substances and procedures of science, endemic in the 1950s and 1960s, which had overtaken artistic and craft values. She did not return to the methods of the bad old days (when the print to be bleached was laid 'sur une pièce de gazon tout récemment fauchée'—a newly mown lawn[15]); she adapted their principles to conscientious modern practice. Thanks to her and to others working with the same perspective, conservation has today arrived at a better balance between disease and cure, damage and authenticity. If, as a second-order effect, this has caused the aura of fashion to surround the unconserved work of art, at least that is a sin of omission.

Notes to Reading

1 [William J. Barrow (1904–1967) was a paper chemist; see reading 54.—ED.]
2 [Although many varieties exist, Whatman drawing paper typically refers to a sturdy wove paper made in England starting in the late eighteenth century.—ED.]
3 [John Ruskin (1819–1900) advocated conservation over restoration of medieval monuments; he endorsed the Arts and Crafts movement, which stressed fine motor skills.—ED.]
4 [John-Auguste-Dominique Ingres (1780–1897), French painter.—ED.]
5 Marjorie B. Cohn, 'Technical appendix', *Ingres Centennial Exhibition*, by Agnes Mongan and Hans Naef (Cambridge, MA: Fogg Art Museum, 1967) 240–249.
6 Marjorie B. Cohn, 'A note on media and methods', *Tiepolo, a Bicentenary Exhibition*, by George Knox (Cambridge, MA: Fogg Art Museum, 1970) 210–221; *Wash and Gouache: A Study of the Development of the Materials of Watercolor* (Cambridge, MA: Fogg Art Museum, 1977).
7 [For Antoinette King, see readings 84 and 86.—ED.]
8 Antoinette King, 'Technical appendix', *Henri Matisse: Paper Cut-Outs*, by Jack Cowart *et al.* (St Louis; New York: St Louis Art Museum, 1977) 272–277. See also her 'Conservation of the collage *Roses* by Juan Gris', *Conservation of Paintings and the Graphic Arts*, IIC: Preprints of contributions to the Lisbon Congress, 1972, 937–941; 'The conservation treatment of a collage: *Man with a Hat*, by Pablo Picasso', *Conservation of Library and Archive Materials and the Graphic Arts*, ed. Guy Petherbridge (London: Institute of Paper Conservation; Society of Archivists; Butterworths, 1987) 85–96; 'Kurt Schwitters's *Cherry Picture*: Material change and an ethical problem', *Essays on Assemblage, Studies in Modern Art* 2 (New York: Museum of Modern Art, 1992) 30–41.
9 See Marjorie B. Cohn, *A Noble Collection: The Spencer Albums of Old Master Prints* (Cambridge, MA: Fogg Art Museum, 1992).
10 [Baldassare Castiglione (1478–1529), Italian painter and printmaker.—ED.]
11 [Stefano Della Bella (1610–1664), Italian printmaker.—ED.]
12 [Antonie Waterloo (1609–1690), Dutch painter.—ED.]
13 Elizabeth Lunning and Roy Perkinson, *The Print Council of America Paper Sample Book: A Practical Guide to the Description of Paper* (n.p.: Print Council of America, 1996).
14 Keiko Mizushima Keyes, 'Alternatives to conventional methods of reducing discolouration in works of art on paper', *Conservation of Library and Archive Materials and the Graphic Arts*, ed. Guy Petherbridge (London: Institute of Paper Conservation; Society of Archivists; Butterworths, 1987) 166–170, republished in AIC, Book and Paper Group: Postprints of 10th Annual Meeting, Milwaukee, 1982, 2nd ed. (Washington, DC: American Institute for Conservation, c.1984) 100–104. See also her 'The unique qualities of paper as an artifact in conservation treatment', Postprints, Conference of Students in Art Conservation, Cooperstown, NY, 1978, 7–17, also published in *The Paper Conservator* (1978): 5–8. [For Keyes, see readings 85 and 87.—ED.]
15 Georges Meusnier, *De l'entretien et de la restauration des tableaux, gravures, aquarelles et pastels* (Paris: Librairie Artistique, 1909) 61.

Reading 95

SALVADOR MUÑOZ VIÑAS

Contemporary Theory of Conservation (2005)

Salvador Muñoz Viñas (1963–) is a paper conservator and professor at the Polytechnic University of Valencia. Here he explores several of the so-called classical principles of conservation ethics, such as reversibility and minimal intervention. After a consideration of the various uses, values, and meanings of paper—its "powers"—he describes the advantages of collaborative decision making.

Conclusion: A Revolution of Common Sense

Contemporary theory of conservation calls for 'common sense', for gentle decisions, for sensible actions. What determines this? Not truth or science, but rather the uses, values and meanings that an object has for people. This is determined by the people. Predicting the chemical alterations in an artwork, for instance, is perfectly valid, but in conservation, it is far more important to predict how people will be affected by a conservation process. Allocating resources to determine how a cellulose monomer ages is socially acceptable because it affects people; for instance, the substitution of the oxhydril groups in a cellobiose unit is supposed to have an effect upon the tensile strength and the folding endurance of a sheet of paper, which may compromise its integrity, and which, in turn, might affect the user of that sheet for intangible, symbolic reasons. The ultimate goal of conservation as a whole is not to conserve the paper, but to retain or improve the meaning it has for people. Sure, this might mean keeping the paper in its current state, but then

From SALVADOR MUÑOZ VIÑAS, *Contemporary Theory of Conservation* (Oxford: Elsevier, Butterworth-Heinemann, 2005), 212–14.

again, it may mean modifying it: cleaning stains, repairing tears, removing adhesive tapes, reinforcing and flattening it, etc. Researching the oxidation of the OH groups is just a means to an end, not an end in itself. From a conservator's point of view, the results of these efforts should be primarily judged not by their contribution to increasing the human body of knowledge, but by their contribution to increasing the satisfaction of the people for whom an object has meaning—or for whom it fulfils a symbolic function or has symbolic value.

In fact, conservation is a means, and not an end in itself. It is a way of maintaining and reinforcing the meanings in an object; it is even a means through which the appreciation for what an object symbolizes is expressed.

However, contemporary conservation ethics is not conservation *à la carte*: neither the conservator nor the public are given *carte blanche* to act as they choose. Conservators (or empowered decision-makers) cannot just do what they think honest or what they would do if the object were theirs.

Contemporary ethics offers better answers than classical theories, but this does not mean that these are easy answers—quite the contrary. Contemporary theory says to anyone involved in conservation decision-making that things are not as easy as classical theories suggest: decision-making does not consist of deciding how to implement a number of well-known directives. Decision-makers are not just limited to studying how to enforce truth within the object. Contemporary ethics asks them to consider the different meanings that an object has for different groups of people, and to decide not just which meanings should prevail, but also how to combine them to satisfy as many views as possible.

Even though this is a difficult task, the intent of contemporary conservation ethics is not to eliminate discussion, but to encourage it. It also intends for this discussion to take place *before* the actual conservation process begins. Contemporary conservation ethics provides more efficient, down-to-earth, reflection tools for the understanding of ethical problems in conservation, and thus for their actual solution. Since it is based upon people's needs and feelings, it does not imply a radical change in current conservation practice. Science, material fetishism, *rigattini*[1] or neutral reintegrations and other features common to current conservation practice will still be valid in many cases; as long as the affected people view these features as valuable and pertinent, they will be valuable and pertinent.

This is another advantage of contemporary conservation ethics: it is adaptive. It adapts to the needs of the users, revealing that these needs are ethically much more relevant than any scientific principle, revealing that objective or scientific principles are relevant only as long as they are accepted as such by the users of the object.

Therefore, there is no reason why contemporary conservation ethics should substantially modify the practice of conservation, except in those cases in which the affected people might feel offended by its practice. This is hardly a revolution, but if it were, it would be the revolution of common sense: the revolution of understanding why, and for whom, things are conserved.

Note to Reading

1 [*Rigattini* is a highly disciplined technique of inpainting usually used in the color compensation of early Italian panel paintings and frescoes.—Ed.]

Reading 96

Richard Hemmings

Restoration (2004)

This poem by Richard Hemmings (1959–) describes the treatment of philatelic materials and evokes the intense awareness and heightened sensibility experienced by many paper conservators during the course of their work. Its subject matter, postage stamps, serves as a reminder that conservators must pay attention to underrepresented paper-based collections and their collecting communities.[1]

Restoration

each file cracks
a scent of old
detonates atmosphere
the church tower
strikes two
dehumidifier humming
air conditioner conditioning
HEPA filters trapping
the unwanted new
sanctioned away from these old sheets
letterheads dancing the rag
expiating for their shameless affair
with juicy ink

watch as patches of benzene
color the air like a hallucination
each leaf has a gauge

Richard Hemmings, "Restoration," unpublished, 2004. Reproduced courtesy of the author.

in microns
a date
a watermark
identity in this cabinet
absolute as a photograph
a carte de visite
a daguerreotype
the timecard you punch in on
the restorer thumbs through catalogs
dog-eared lists of fonts typesets and types
of binding of security paper of newsprint
china clay content acid free
onion skin
coated uncoated
gummed bundled
frisked wound

this is alchemy
this is wind and the windmill and water quality
this is as valuable as a phone number scratched
on a cocktail napkin
in Afghanistan cancel
means to bite to tear
to rip out from a piece of paper
watch as shades of lake
chrome yellow sugar white
turn to the past pressure of prejudice

with age comes a march of insects
the Ides of March
then papyrus
bibles and illuminations
embossed as a notary embosses
licenses taxes
transmittals on fax paper
airplanes to the recycler
origami and kites the last motes
on our minds

paper wrinkles
every language dies before it is dead
the science of history
spills and spreads
this immortal stain.

Note to Introduction

1 Frequently mentioned conservation problems of philatelic materials include foxing, unstable nineteenth-century dye-based inks, and damage resulting from poor-quality album pages and adhesives. John B. Sharkey, "Chemistry of Postage Stamps," *International Journal of Forensic Document Examiners* 5 (1999): 345–53.

Further Reading

Part I: The Powers of Paper

Baker, Cathleen. *From the Hand to the Machine. Nineteenth-Century American Paper and Mediums: Technologies, Materials, and Conservation*. Ann Arbor: Legacy Press, 2010.

Balston, John N. *The Elder James Whatman: England's Greatest Papermaker (1702–1759)*. Vols. I and II. West Farleigh: J. N. Balston, 1992.

———. *The Whatmans and Wove (Velin) Paper, Its Invention and Development in the West*. West Farleigh: J. N. Balston, 1998.

Barrett, Timothy. "European Papermaking Techniques, 1300–1800." In *Paper through Time: Nondestructive Analysis of 14th- through 19th-Century Papers*. Iowa City: University of Iowa, 2012. http://paper.lib.uiowa.edu/index.php.

———. *Japanese Papermaking: Traditions, Tools, Techniques*. New York: John Weatherhill, 1983.

Basbanes, Nicholas A. *On Paper: The Everything of Its Two-Thousand-Year History*. New York: Alfred A. Knopf, 2013.

Bidwell, John. *American Paper Mills, 1690–1832: A Directory of the Paper Trade, with Notes on Products, Watermarks, Distribution Methods, and Manufacturing Techniques*. Hanover, N.H.: Dartmouth College Press, 2013.

———. *Introduction to Early American Papermaking: Two Treatises on Manufacturing Techniques Reprinted from James Cutbush's American Artist's Manual (1814)*. New Castle, Del.: Oak Knoll Books, 1990.

Bloom, Jonathan. *Paper before Print: The History and Impact of Paper in the Islamic World*. New Haven, Conn.: Yale University Press, 2001.

Blum, André. *On the Origin of Paper*. New York: R. R. Bowker, 1934.

Butler, Frank O. *The Story of Paper-Making: An Account of Paper-Making from Its Earliest Known Record down to the Present Time*. Chicago: J. W. Butler Paper Company, 1901.

Centre de Recherche sur la Conservation des Collections. *Khartasia*. http://khartasia-crcc.mnhn.fr/en.

Clapperton, Robert H. *Paper: An Historical Account of Its Making by Hand from the Earliest Times down to the Present Day*. Oxford: Shakespeare Head Press, 1934.

———. *The Paper-Making Machine, Its Invention, Evolution and Development*. Oxford: Pergamon Press, 1967.

Collins, Thomas, and D. Milner. "A New Chronology of Papermaking Technology." *Paper Conservator* 14 (1990): 58–62.

Febvre, Lucien Paul Victor. *The Coming of the Book: The Impact of Printing, 1450–1800*. London: New Left Books, 1976.

Hughes, Sukey. *Washi—The World of Japanese Paper*. Tokyo: Kodansha International, 1978.

Hunter, Dard. *Papermaking: The History and Technique of an Ancient Craft*. New York: Dover Publications, 1978.

Hunter, Dard, and Gerhard Gerlach. *Papermaking by Hand in India*. New York: Pynson Printers, 1939.

International Association of Paper Historians. www.paperhistory.org.

Lalande, Joseph-Jérôme Le Français de. *Art de faire le papier: Description des arts et métiers*. Paris: Chez Desaint & Saillant, 1761.

Loeber, E. G. *Paper Mould and Mouldmaker*. Amsterdam: Paper Publications Society, 1982.

Loveday, Helen. *Islamic Paper: A Study of the Ancient Craft*. London: Don Baker Memorial Fund, 2001.

Munsell, Joel. *A Chronology of Paper and Papermaking*. Albany: J. Munsell, 1864.

Piette, Louis. *Traité de la fabrication du papier*. Paris: F. G. Levrault, 1831.

Polastron, Lucien X. *Le Papier: 2000 Ans d'histoire et de savoir-faire*. Paris: Impr. Internationale Editions, 1999.

Powers, William. *Hamlet's Blackberry: Why Paper Is Eternal*. Cambridge, MA: Joan Shorenstein Center on the Press, Politics and Public Policy, Harvard University, 2007. www.williampowers.com/wp-content/uploads/2010/05/Hamlets_Blackberry_William_Powers.pdf.

Sansom, Ian. *Paper: An Elegy*. New York: William Morrow, 2012.

Sellen, Abigail, and Richard Harper. *The Myth of the Paperless Office*. Cambridge, Mass.: MIT Press, 2001.

Stedman, Ebenezer Hiram. *Bluegrass Craftsmen: Being the Reminiscences of EHS*. Lexington: University of Kentucky Press, 1959.

Stromer, Ulman, and Der Verband deutscher Papierfabriken (Bonn). *Püchel von mein Geslecht und von Abentewr*. Stuttgart: Müller und Schindler, 1990.

Sutermeister, Edwin. *The Story of Papermaking*. Boston: S. D. Warren Co., 1954.

Tschudin, Peter. *Grundzüge der Papiergeschichte*. Stuttgart: Anton Hiersemann, 2002.

Turner, Silvie, and Brigit Skiöld. *Handmade Paper Today*. New York: Frederic C. Beil, 1983.

Part II: The Mastery of Drawing

Ambers, Janet, Catherine Higgett, and David Saunders. *Italian Renaissance Drawings: Technical Examination and Analysis*. London: British Museum, 2010.

Ames-Lewis, Francis. *Drawing in Early Renaissance Italy*. New Haven, Conn.: Yale University Press, 1982.

AMIEN: The Art Materials Information and Education Network. www.amien.org.

Baetjer, Katharine, and Marjorie Shelley. *Pastel Portraits: Images of 18th-Century Europe*. New York: Metropolitan Museum of Art, 2011.

Bambach, Carmen. *Drawing and Painting in the Italian Renaissance Workshop: Theory and Practice, 1300–1600*. London: Cambridge University Press, 1999.

Berry, Ian, and Jack Shear. *Twice Drawn: Modern and Contemporary Drawings in Context*. Saratoga Springs, N.Y.: Frances Young Tang Teaching Museum and Art Gallery, Skidmore College, 2011.

Buchberg, Karl. "Seurat: Materials and Techniques." In *Georges Seurat: The Drawings*, 31–41. New York: Museum of Modern Art, 2007.

Burns, Thea. *The Invention of Pastel Painting*. London: Archetype, 2007.

———. *The Luminous Trace: Drawings and Writing in Metalpoint*. London: Archetype, 2012.

Chapman, Hugo. *Italian Renaissance Drawings*. London: British Museum Press, 2010.

Cohn, Marjorie. *Wash and Gouache: A Study of the Development of the Materials of Watercolor*. Cambridge, Mass.: Fogg Art Museum, 1977.

De Tolnay, Charles. *History and Technique of Old Master Drawings: A Handbook*. New York: Hacker Art Books, 1972.

Ellis, Margaret Holben. "Materials, Tools, and 'Technics': Works on Paper by Jackson Pollock." In *No Limits, Just Edges: Jackson Pollock Paintings on Paper*. Berlin: Deutsche Guggenheim, 2005.

Ellis, Margaret Holben, and Lindsey Tyne. "Roy Lichtenstein's 'Mechanical Drawings' in the Age of Mechanical Reproduction." In Isabelle Dervaux, Graham Badder, Thomas Crow, Margaret Holben Ellis, Lindsey Tyne, and Clare Bell, *Roy Lichtenstein: The Black-and-White Drawings, 1961–1968*. New York: Morgan Library & Museum, 2010.

Fuga, Antonella. *Artists' Techniques and Materials*. Los Angeles: J. Paul Getty Museum, 2006.

Goldman, Paul. *Looking at Prints, Drawings, and Watercolours: A Guide to Technical Terms*. London: British Museum Publications, 1988.

Gottsegen, Mark D. *The Painter's Handbook: A Complete Reference*. New York: Watson-Guptill, 2006.

Holcomb, Melanie. *Pen and Parchment: Drawing in the Middle Ages*. New York: Metropolitan Museum of Art, 2009.

James, Carlo. *Visual Identification and Analysis of Old Master Drawing Techniques*. Florence: Olschki, 2010.

James, Carlo, and Marjorie B. Cohn. *Old Master Prints and Drawings: A Guide to Preservation and Conservation*. Amsterdam: Amsterdam University Press, 1997.

Koschatsky, Walter. *Die Kunst der Zeitnung: Technic, Geschichte, Meisterwerk*. Munich: Deutscher Taschenbuch Verlag, 1999.

———. *Watercolor History and Technique*. Trans. Mary Whittall. New York: McGraw-Hill, 1970.

Krug, Margaret M. *An Artist's Handbook: Materials and Techniques*. New York: Abrams Studio, 2007.

Lambert, Susan. *Reading Drawings: An Introduction to Looking at Drawings*. New York: Pantheon Books, 1984.

Mannick, Annette, and Antoinette Owen. "Bringing Back Something Fine." In *John Singer Sargent: Watercolors*, 205–24. Boston: Museum of Fine Arts Publications; New York: Brooklyn Museum of Art, 2013.

Mayer, Deborah, and Pamela Vandiver. "Red Chalk: Historical and Technical Perspectives. Part 2: A Technical Study." In *Drawings Defined*, ed. Walter Strauss and Tracie Felker, 171–80. New York: Abaris Books, 1987.

Mayer, Ralph. *The Artists' Handbook of Materials and Techniques*. 5th ed. New York: Viking Press, 1991.

Mayhew, Timothy, Margo Ellis, and Supapan Seraphin. "Natural Black Chalk in Traditional Old Master Drawings." *Journal of the American Institute for Conservation* 49, no. 2 (2010): 83–95.

———. "Steatite and Calcite Natural White Chalk in Traditional Old Master Drawings." *Journal of the American Institute for Conservation* 51, no. 2 (2012): 175–98.

Mayhew, Timothy, Steven Hernandez, Philip Anderson, and Supapan Seraphin. "Natural Red Chalk in Traditional Old Master Drawings." *Journal of the American Institute for Conservation* 53, no. 2 (2014): 89–115.

Melli, Lorenza. *I disegni italiani del Quattrocento nel Kupferstich-Kabinett di Dresda*. Florence: Centro Di, 2006.

Museum of Fine Arts, Boston. *CAMEO—Conservation and Art Materials Encyclopedia Online*. http://cameo.mfa.org.

Petherbridge, Deanna. *The Primacy of Drawing: Histories and Theories of Practice*. New Haven, Conn.: Yale University Press, 2010.

Petrioli Tofani, Annamaria. *Il disegno: Forme, tecniche, significati*. Cinisello Balsamo: Silvana Editoriale, 1991.

Philadelphia Museum of Art. *Drawing Materials*. www.philamuseum.org/collections/22-394-217.html.

Read, Leonard. *I, Pencil*. Irvington-on-Hudson, N.Y.: The Foundation for Economic Education, 1958. http://www.econlib.org/library/Essays/rdPncl1.html.

Reynolds, Graham. *Watercolors: A Concise History*. New York: Thames and Hudson, 1985.

Shelley, Marjorie. "The Drawing Materials Used by Vincent van Gogh in Provence." In *Vincent van Gogh: The Drawings*. New York: Metropolitan Museum of Art, 2005.

Stratis, Harriet K., and Britt Salveson, eds. *The Broad Spectrum: Studies in the Materials, Techniques, and Conservation of Color on Paper*. London: Archetype, 2002.

Strauss, Walter, and Tracie Felker, eds. *Drawings Defined*. New York: Abaris Books, 1987.

Tedeschi, Martha, Kristi Dahm, Ruth Fine, Charles Pietrazewski, and Christine Conniff-O'Shea. *John Marin's Watercolors: A Medium for Modernism*. Chicago: Art Institute of Chicago, 2011.

Tedeschi, Martha, Kristi Dahm, Judith Walsh, and Karen Huang. *Watercolors by Winslow Homer: The Color of Light*. Chicago: Art Institute of Chicago, 2008.

University of Glasgow. *ArtMatters: International Journal for Technical Art History*. www.artmattersjournal.org.

Watrous, James. *The Craft of Old-Master Drawings*. Madison: University of Wisconsin Press, 1957.

Weekes, Ursula. *Techniques of Drawing from the 15th to 19th Centuries*. Oxford: Ashmolean Museum, 1999.

Part III. Printmaking: Multiple Originals

Antreasian, G. Z., and C. Adams. *The Tamarind Book of Lithography: Art and Techniques*. Los Angeles: Tamarind Lithography Workshop, 1971.

Ayres, Julia. *Printmaking Techniques*. New York: Watson-Guptill, 1993.

Ballinger, Raymond. *Art and Reproduction: Graphic Reproduction Techniques*. New York: Van Nostrand Reinhold, 1977.

Bann, Stephen. *Parallel Lines: Printmakers, Painters and Photographers in Nineteenth-Century France*. New Haven, Conn.: Yale University Press, 2001.

Benson, Richard. *The Printed Picture*. New York: Museum of Modern Art, 2008.

Biegeleisen, J. I. *Screen Printing: A Contemporary Guide to the Technique of Screen Printing for Artists, Designers, and Craftsmen*. New York: Watson-Guptill, 1971.

Brown, Kathan. *Ink, Paper, Metal, Wood: How to Recognize Contemporary Artists' Prints*. San Francisco: Point Publications, 1992.

Brunner, Felix. *A Handbook of Graphic Reproduction Processes*. 6th ed. New York: Visual Communications Books, 1984.

Cate, Phillip, and Sinclair Hamilton Hitchings. *The Color Revolution: Color Lithography in France, 1890–1900*. New Brunswick, N.J.: Rutgers University Art Gallery, 1978.

Chamberlain, Walter. *The Thames and Hudson Manual of Etching and Engraving*. London: Thames and Hudson, 1972.

Colbourne, Jane, and Reba Fishman Snyder. *Printed on Paper: The Techniques, History, and Conservation of Printed Media*. Newcastle upon Tyne: Arts and Social Sciences Academic Press, Northumbria University, 2009.

Curwen, Harold. *Processes of Graphic Reproduction*. 3rd ed. London: Faber and Faber, 1963.

D'Arcy Hughes, Ann, and Hebe Vernon-Morris. *The Printmaking Bible: The Complete Guide to Materials and Techniques*. San Francisco: Chronicle Books, 2008.

Digital Art Practices & Terminology Task Force. *Glossary of Digital Art and Printmaking*. www.dpandi.com/DAPTTF.

Dyson, Anthony. *Pictures to Print: The Nineteenth-Century Engraving Trade*. London: Farrand Press, 1984.

Flader, Louis. *Achievement in Photo-Engraving and Letter-Press Printing*. Chicago: American Photo-Engravers Association, 1927.

Gascoigne, Bamber. *How to Identify Prints: A Complete Guide to Manual and Mechanical Processes from Woodcut to Ink Jet*. London: Thames and Hudson, 2004.

Further Reading

Grasselli, Margaret M. *Colorful Impressions: The Printmaking Revolution in Eighteenth-Century France*. Washington, D.C.: National Gallery of Art, 2003.

Griffiths, Antony. *Prints and Printmaking: An Introduction to the History and Techniques*. Berkeley: University of California Press, 1996.

Hackleman, Charles W. *Commercial engraving and printing; a manual of practical instruction and reference covering commercial illustrating and printing by all processes for advertising managers, printers, engravers, lithographers, paper men, photographers, commercial artists, salesmen, instructors, students and all others interested in these and allied trades*. Indianapolis: Commercial Engraving Pub. Co., 1924.

Heller, Jules. *Printmaking Today: An Introduction to the Graphic Arts*. New York: Holt, Rinehart and Winston, 1958.

Hind, Arthur M. *A History of Engraving and Etching*. 3rd ed. New York: Dover, 1963.

Image Permanence Institute. *Digital Print Preservation Portal*. http://dp3.graphicsatlas.org.

———. *Graphics Atlas*. www.graphicsatlas.org.

Ivins, William, M., Jr. *Notes on Prints*. New York: Da Capo Press, 1967.

———. *Prints and Visual Communications*. New York: Da Capo Press, 1969.

Ivins, William M., and Marjorie B. Cohn. *How Prints Look: Photographs with Commentary*. Boston: Beacon Press, 1987.

Jürgens, Martin. *The Digital Print: Identification and Preservation*. Los Angeles: Getty Conservation Institute, 2009.

———. *Identification of Digital Prints*. http://aic.stanford.edu/sg/emg/juergens.

Jussim, Estelle. *Visual Communication and the Graphic Arts: Photographic Technologies in the Nineteenth Century*. New York: R. R. Bowker, 1983.

Lalanne, Maxime. *Traité de la gravure à l'eau-forte*. Paris: Cadart et Luquet, 1866.

Lalanne, Maxime, and Sylvester R. Koehler. *A Treatise on Etching: Text and Plates*. Boston: Page Co., 1880.

Lambert, Susan. *The Image Multiplied: Five Centuries of Printed Reproductions of Paintings and Drawings*. London: Trefoil Publications, 1987.

———. *Printmaking*. London: Her Majesty's Stationery Office, 1983.

Landau, David, and Peter Parshall. *The Renaissance Print, 1470–1550*. New Haven, Conn.: Yale University Press, 1994.

Lengwiler, Guido. *A History of Screen Printing: How an Art Evolved into an Industry*. Cincinnati, Ohio: ST Media Group International, 2013.

Lumsden, E. S. *The Technique of Etching*. New York: Dover Publications, 1962.

Mayor, A. Hyatt. *Prints and People: A Social History of Printed Pictures*. New York: Metropolitan Museum of Art, 1971.

Moser, Joann. *Singular Impressions: The Monotype in America*. Washington, D.C.: Smithsonian Institution Press, 1997.

Mukerji, Chandra. "Pictorial Prints and the Growth of Consumerism: Class and Cosmopolitanism in Early Modern Culture." In *Graven Images: Patterns of Modern Materialism*, 30–78. New York: Columbia University Press, 1983.

Museum of Modern Art. *Printmaking Techniques*. http://moma.org/explore/collection/ge/techniques/index.

Nadeau, Luis. *Encyclopedia of Printing, Photographic, and Photomechanical Processes: A Comprehensive Reference to Reproduction Technologies*. Vols. 1 and 2. New Brunswick, N.J.: Atelier Luis Nadeau, 1994.

Nichols, Kimberly. "Physical Qualities of Early Prints." In *Altered and Adorned: Using Renaissance Prints in Daily Life*, 99–105. Chicago: Art Institute of Chicago, 2011.

Primeau, Thomas. "The Materials and Technology of Renaissance and Baroque Hand-Colored Prints." In *Painted Prints: The Revelation of Color in Northern Renaissance &*

Baroque Engravings, Etchings and Woodcuts. University Park: Pennsylvania State University Press, 2002.

Rhein, Erich. *The Art of Printmaking: A Comprehensive Guide to the Graphic Techniques*. Trans. Neil Morris and Roswitha Morris. New York: Van Nostrand Reinhold, 1976.

Riegger, Wilhelm. *Die Technik der graphischen Künste*. Weissenhorn: Konrad, 1979.

Ross, John, and Clare Romano. *The Complete Printmaker*. New York: Free Press, 1972.

Saff, Donald, and Deli Sacilotto. *Printmaking: History and Process*. New York: Holt, Reinhart and Winston, 1978.

Salter, Rebecca. *Japanese Woodblock Printing*. Honolulu: University of Hawai'i Press, 2002.

Senefelder, Alois, and Friedrich Schlichtegroll. *A Complete Course of Lithography: Containing Clear and Explicit Instructions in All the Different Branches and Manners of That Art: Accompanied by Illustrative Specimens of Drawings. To Which Is Prefixed a History of Lithography, from Its Origin to the Present Time*. London: R. Acker-mann, 1819.

Simmons, Rosemary, and Katie Clemson. *The Complete Manual of Relief Printmaking*. New York: Knopf, 1988.

Sotriffer, Kristian. *Printmaking: History and Technique*. New York: McGraw-Hill, 1968.

Spencer Museum of Art. *Print Connoisseurship*. Lawrence: University of Kansas. www.spencerart.ku.edu/collection/print/connoisseurship/.

Stiber, Linda S., Elmer Eusman, and Sylvia Rodgers Albro. "The *Triumphal Arch* and the *Large Triumphal Carriage of Maximilian I*: Two Oversized, Multiblock, 16th Century Woodcuts from the Studio of Albrecht Dürer." *Book and Paper Group Annual* 14 (1995): 63–85.

Stijnman, Ad. *Engraving and Etching, 1400–2000: A History of the Development of Manual Intaglio Printmaking Processes*. London: Archetype, 2012.

Suzuki, Sarah. *What Is a Print? Selections from the Museum of Modern Art*. New York: The Museum of Modern Art, 2011.

Thompson, Wendy. "The Printed Image in the West: History and Techniques." In *Heilbrunn Timeline of Art History*. New York: Metropolitan Museum of Art, 2000–2013. www.metmuseum.org/toah/hd/prnt/hd_prnt.htm.

Tyler, Kenneth. Glossary. In *Printmaking Collection*. Canberra: National Gallery of Australia, 2013. www.nga.gov.au/InternationalPrints/Tyler/Default.cfm?MnuID=9.

University of Iowa Libraries. *The Atlas of Early Printing*. http://atlas.lib.uiowa.edu.

Zorach, Rebecca, and Elizabeth Rodini. *Paper Museums: The Reproductive Print in Europe, 1500–1800*. Chicago: David and Alfred Smart Museum of Art, 2005.

Part IV. "Paper Is Part of the Picture"

Agoston, G. A. *Color Theory and Its Application in Art and Design*. 2nd ed. Berlin: Springer Verlag, 1987.

Albro, Sylvia Rodgers. "Printed on Fabriano Paper." In *Printed on Paper*, ed. Jane Colbourne and Reba Fishman Snyder, 2–13. Newcastle upon Tyne: Arts and Social Sciences Academic Press, Northumbria University, 2009.

American Paper and Pulp Association. *The Dictionary of Paper*. 4th ed. New York: American Paper Institute, 1980.

Ash, Nancy, and Shelley Fletcher. *Watermarks in Rembrandt's Prints*. Washington, D.C.: National Gallery of Art, 1998.

Barrett, Timothy. *Paper through Time: Nondestructive Analysis of 14th- through 19th-Century Papers*. Ames: University of Iowa. http://paper.lib.uiowa.edu.

Bell, Lillian A. *Papyrus, Tapa, Amate and Rice Paper: Papermaking in Africa, the Pacific, Latin America and Southeast Asia*. McMinnville, Ore.: Liliaceae Press, 1988.

Berger-Shunn, Anni. *Practical Color Measurement*. New York: John Wiley and Sons, 1994.

Berns, Roy S. *Billmeyer and Saltzman's Principles of Color Technology*. 3rd ed. New York: Wiley-Interscience, 2000.

Bidwell, John. *Fine Papers at the Oxford University Press.* Cheltenham: Whittington Press, 1999.

———. "The Study of Paper as Evidence, Artefact, and Commodity." In *The Book Encompassed*, ed. Peter Davison, 69–82. Newcastle, Del.: Oak Knoll Press, 1998.

Bower, Peter. "All the Colors of White: The Changing Nature of White Papers." In *The Broad Spectrum*, ed. Harriet Stratis and Britt Salvesen, 128–35. London: Archetype, 2002.

———. "Blues and Browns and Drabs: The Evolution of Colored Papers." In *The Broad Spectrum*, ed. Harriet Stratis and Britt Salvesen, 42–48. London: Archetype, 2002.

———. *Turner's Later Papers: A Study of the Manufacture, Selection and Use of His Drawing Papers, 1820–1851*. London: Tate Gallery; New Castle, Del.: Oak Knoll Books, 1999.

———. *Turner's Papers: A Study of the Manufacture, Selection and Use of His Drawing Papers, 1787–1820*. London: Tate Gallery, 1990.

———. "Who Made This Rubbish? The Historical Investigation of Particular Twentieth-Century Papers." In *Modern Works, Modern Problems?*, ed. Alison Richmond, 50–59. London: Institute of Paper Conservation, 1994.

Briquet, C.-M. *The Briquet Album: A Miscellany on Watermarks, Supplementing Dr. Briquet's Les filigranes*. Hilversum, Holland: Paper Publications Society, 1952.

Briquet, C.-M., and Allan Stevenson. *Les filigranes: Dictionnaire historique des marques du papier dès leur apparition vers 1282 jusqu'en 1600 [par] C.m. Briquet*. Amsterdam: Paper Publications Society, 1968.

Brückle, Irene. "Aspects of the Use of Alum in Historical Papermaking." In *Conference Papers Manchester, 1992*, ed. Sheila Fairbrass, 201–6. Leigh: Institute of Paper Conservation, 1992.

———. "The Historical Manufacture and Use of Blue Paper." *Book and Paper Group Annual* 17 (1993): 20–31.

Coural, Natalie, ed. *Le papier à l'œuvre*. Paris: Musée du Louvre, 2011.

Garlick, Karen. "A Brief Review of the History of Sizing and Resizing Practices." *Book and Paper Group Annual* 5 (1986): 94–107.

Gaskell, Philip. *A New Introduction to Descriptive Bibliography*. New Castle, Del.: Oak Knoll Press, 2000.

Harris, Theresa Fairbanks, Paul Sandby, and Scott Wilcox. *Papermaking and the Art of Watercolour in 18th-Century Britain: Paul Sandby and the Whatman Paper Mill*. New Haven, Conn.: Yale Center for British Art in association with Yale University Press, 2006.

Holik, Herbert. *Handbook of Paper and Board*. Weinheim: Wiley-VCH, 2006.

Irving, Joan. "Construction Paper: A Brief History of Impermanence." *Book and Paper Group Annual* 16 (1997): 47–54.

Jahans, Gordon A. *Paper Testing and Chemistry for Printers: A Valuable Book on the Raw Material of the Printers' Craft*. London: Sir I. Pitman and Sons, 1931.

Johnston-Feller, Ruth. *Color Science in the Examination of Museum Objects: Nondestructive Procedures*. Los Angeles: Getty Conservation Institute, 2001.

Kelly, Kenneth L., and Deane B. Judd. *Color: Universal Language and Dictionary of Names*. Washington, D.C.: National Bureau of Standards, 1976.

Kolseth, Petter, and J. Bristow. *Paper Structure and Properties*. New York: M. Dekker, 1986.

Koops, Matthias. *Historical Account of the Substances Which Have Been Used to Describe Events: And to Convey Ideas, from the Earliest Date, to the Invention of Paper. Printed on the First Useful Paper Manufactured Soley [sic] from Straw*. London: Printed by T. Burton, 1800.

Krichels, Engelbert. "Analysis with the Aid of the Fluorescence of Ultra-Violet Light." *Paper Maker* 2, no. 6 (1933): 109–10.

Krill, John. *English Artists' Paper: Renaissance to Regency*. New Castle, Del.: Oak Knoll Press, 2002.

Laroque, Claude. "History and Analysis of Transparent Papers." *Journal of Paper Conservation* 28 (2004): 17–32.

Loeber, E. G. *Supplement to E. J. Labarre: Dictionary and Encyclopaedia of Paper and Papermaking*. Amsterdam: Swets and Zeitlinger, 1967.

Lofton, Roscoe E. *A Measure of the Color Characteristics of White Papers*. Washington, D.C.: Government Printing Office, 1923.

Mosser, Daniel, et al. *Puzzles in Paper: Concepts in Historical Watermarks*. New Castle, Del.: Oak Knoll Press, 2000.

Mustalish, Rachel. "Optical Brighteners: History and Technology." In *Tradition and Innovation: Advances in Conservation; Contributions to the Melbourne Congress, 10–14 October* 2000, ed. Ashok Roy and Perry Smith, 133–36. London: International Institute for Conservation of Historic and Artistic Works, 2000.

———. "Plastics in Paper: History of Synthetic Polymers in Papermaking." In *Edinburgh Conference Papers 2006: Proceedings from the Fifth International Conference of the Institute of Paper Conservation and the First International Conference of the Institute of Conservation, Book and Paper Group, 26–29 July* 2006, ed. Shulla Jaques, 141–46. London: Institute of Paper Conservation, 2007.

Perkinson, Roy. "Statistics without Anesthesia: Interpreting Color Data." In *The Broad Spectrum*, ed. Harriet Stratis and Britt Salvesen, 230–35. London: Archetype, 2002.

———. "Summary of the History of Blue Paper." *Book and Paper Group Annual* 16 (1997): 75.

Proteaux, A. *Practical Guide for the Manufacture of Paper and Boards*. Philadelphia: Henry Carey Baird, 1866.

Rance, H. F. *The Structure and Physical Properties of Paper*. Amsterdam; New York: Elsevier Scientific, 1982.

Robison, Andrew. "An Aesthetic History of Paper in Prints." In *Paper: Art and Technology*, ed. Paulette Long and Robert Levering, 60–69. San Francisco: World Print Council, 1979.

———. *Paper in Prints*. Washington, D.C.: National Gallery of Art, 1977.

Rogers, Bruce. *Specimens: A Stevens-Nelson Catalogue*. New York: Stevens-Nelson, 1953.

Rohlfing, Christian. "A Universal Color Language." In *Museum News Technical Supplement*, no. 8 (1965): n.p.

Scott, William E., James C. Abbott, and Stanley Trosset, eds. *Properties of Paper: An Introduction*. 2nd ed. Atlanta: TAPPI Press, 1995.

Slavin, John, et al., eds. *Looking at Paper: Evidence and Interpretation. Symposium Proceedings, Toronto, 1999*. Ottawa: Canadian Conservation Institute, 2001.

Spector, Stephen. *Essays in Paper Analysis*. Washington, D.C.: Folger Shakespeare Library, 1987.

Stevenson, Allan. *Observations on Paper as Evidence*. Lawrence: University Press of Kansas, 1961.

Tschudin, Peter. *International Standard for the Registration of Papers with or without Watermarks*. Basel: Swiss Paper Museum, 1997.

Turner, Silvie. *The Book of Fine Paper*. New York: Thames and Hudson, 1998.

———. *Which Paper? A Guide to Choosing and Using Fine Papers for Artists, Craftspeople and Designers*. New York: Design Books, 1994.

Voorn, Henk. "A Brief History of the Sizing of Paper." *Papermaker* 30, no. 1 (1961): 47–53.

Würster, C. "Paper Testing." *Journal of the Society of Chemical Industries* 7 (1898): 338.

Part V. Deterioration and Change: Paper "Yellow'd with Their Age"

Area, Maria Cristina, and Hervé Cheradame. "Paper Aging and Degradation: Recent Findings and Research Methods." *BioResources* 6, no. 4 (2011): 5307–37.

Aribert, M., and F.M. Bouvier. "Durability, Preservation, and Deterioration of Paper and Analysis of Methods of

Manufacture of Ancient Papers to Determine the Factors That Produce Durability." *Paper Makers Monthly Journal* 58, no. 6 (1920): 187, 189.

Baer, N. S., and Norman Indictor. "Use of the Arrhenius Equation in Multicomponent Systems." In *Preservation of Paper and Textiles of Historic Value*, ed. J. C. Williams, 336–51. Washington, D.C.: American Chemical Society, 1977.

Banik, Gerhard, and Irene Brückle, eds. *Paper and Water: A Guide for Conservators*. Oxford: Elsevier, 2011.

Barrow, William. *Manuscripts and Documents: Their Deterioration and Restoration*. Charlottesville: University Press of Virginia, 1955.

Barrow, W. J., and Randolph W. Church. *Deterioration of Book Stock: Causes and Remedies; Two Studies on the Permanence of Book Paper*. Richmond: Virginia State Library, 1959.

Baty, J. W., C. L. Maitland, W. Minter, M. A. Hubbe, and S. K. Jordan-Mowery. "Deacidification for the Conservation and Preservation of Paper-Based Works: A Review." *BioResources* 5, no. 3 (2010): 1955–2023.

Beckwith, Theodore D., W. H. Swanson, and Thoman M. Iiams. "Deterioration of Paper: The Causes and Effect of Foxing." *Publications of the University of California at Los Angeles in Biological Sciences* 1, no. 13 (1940): 299–356.

Bradley, S., ed. *The Interface between Science and Conservation*. London: British Museum, 1997.

Browning, B. L. *The Analysis of Paper*. New York: Marcel Dekker, 1969.

Dupont, Anne-Laurence. *Gelatine Sizing of Paper and Its Impact on the Degradation of Cellulose during Aging: A Study Using Size-Exclusion Chromatography*. Amsterdam: University of Amsterdam, 2003.

The Effects of Air Pollutants on the Accelerated Ageing of Cellulose Containing Materials: Paper: Step Project Ct 90-0100. Delft: TNO, 1994.

Fellers, Christer, Tommy Iversen, Tom Lindström, Thomas Nilsson, and Mikael Rigdahl. *Ageing/Degradation of Paper: A Literature Survey*. Stockholm: FoU—projektet för papperskonservering, 1989.

Florian, Mary Lou. *Heritage Eaters: Insects and Fungi in Heritage Collections*. London: James & James, 1997.

Gallo, Fausta. *Biological Factors in Deterioration of Paper = Facteurs biologiques de détérioration du papier*. Rome: ICCROM, 1985.

Hall, Gösta. "Permanence of Paper." *Paper Trade Journal* 82, no. 14 (1926): 52–58.

Hanson, Fred S. "Resistance of Paper to Natural Aging." *Paper Industry and Paper World* 20 (1939): 1157–63.

Havermans, Johannes B. G. A. *Environmental Influences on the Deterioration of Paper*. Rotterdam: Barjesteh, Meeuwes & Co., 1995.

Hendriks, K. B. "Permanence of Paper in Light of Six Centuries of Papermaking in Europe." In *Environment and Conservation of Writing, Images, and Sound: Proceedings of ARSAG's Second International Study Days, 16–20 May 1994*, 131–37. Paris: ARSAG, 1994.

Hon, David N. S. "Discoloration and Deterioration of Modern Papers." In *Science and Technology in the Service of Conservation: Preprints of Contributions to the Washington Congress*, 89–92. London: IIC, 1982.

Inaba, M., and R. Sugisita. "Permanence of Washi (Japanese Paper)." In *The Conservation of Far Eastern Art: Preprints of the Contributions to the Kyoto Congress, 19–23 September 1988*, ed. J. Mills, 1–4. London: International Institute for Conservation of Historic and Artistic Works, 1988.

J.G. "The Deterioration of Paper on Aging." *Nature* 130, no. 3278 (1932): 320.

Kimberly, Arthur Evarts, and Adelaide Lancaster Emley Minogue. "A Study of the Deterioration of Book Papers in Libraries." *Occasional Publication of the National Bureau of Standards*, no. 140. NBS, 1933.

Kosek, Joanna, and David Jacobs. "What Happens to Enclosed Paper?" In *Art on Paper: Mounting and Housing*, 29–37. London: The British Museum/Archetype, 2005.

Launer, H. F., and W. K. Wilson. "Photochemical Stability of Papers." *National Bureau of Standards* 30 (1943): 55–74.

Luner, Philip, ed. *Paper Preservation: Current Issues and Recent Developments*. Atlanta: TAPPI Press, 1990.

Mendenhall, G. D., George B. Kelly, and John C. Williams. "The Application of Several Empirical Equations to Describe the Change of Properties of Paper on Accelerated Aging." *Preservation of Paper and Textiles of Historic and Artistic Value II*, 177–88. Washington, D.C.: American Chemical Society, 1981.

Meynell, G. C., and R. J. Newsam. "Foxing, a Fungal Infection of Paper." *Nature* 274, no. 3 (August 1978): 466–68.

Okayama, Takayuki, and Kei-ichi Sato. "A New Method for the Assessment of Deteriorated Paper Documents." In *Edinburgh Conference Papers 2006: Proceedings from the Fifth International Conference of the Institute of Paper Conservation and the First International Conference of the Institute of Conservation, Book and Paper Group, 26–29 July 2006*, ed. Shulla Jaques, 219–26. London: Institute of Paper Conservation, 2007.

Pedersoli, Luiz José, Jr., Frank J. Ligterink, and Giovanna Di Pietro. "Browning of Paper: Fluorescence Changes Accompanying the Discolouration of Artificially Aged Paper." *PapierRestaurierung: Mitteilungen der IADA* 1, Suppl. (2000): 47–54.

Porck, Henk J. *Mass Deacidification: An Update of Possibilities and Limitations*. Amsterdam and Washington, D.C.: European Commission on Preservation and Access and Commission on Preservation and Access, 1996.

Sée, Pierre. *Les maladies du papier piqué*. Paris: O. Doin et Fils, 1919.

Smith, Richard Daniel. "Paper Impermanence: pH and Storage Conditions." *Library Quarterly* 39, no. 2 (1969): 153–95

Society of Arts, Committee on the Deterioration of Paper. "Report." *Journal of the Society of Arts* 46 (1898): 597–601.

Stephens, Catherine H., Timothy Barrett, Paul M. Whitmore, Jennifer A. Wade, Joy Mazurek, and Michael Schilling. "Composition and Condition of Naturally Aged Papers." *Journal of the American Institute for Conservation* 47 (2008): 201–21.

Strofer-Hua, E. "Experimental Measurement: Interpreting Extrapolation and Prediction by Accelerated Ageing." *Restaurator* 11 (1990): 254–66.

Walton, Robert P. *Causes and Prevention of Deterioration in Book Materials*. New York: New York Public Library, 1929.

Warda, Jeffrey. "Made of Paper: Motherwell's Materials in the 1940s." In *Robert Motherwell: Early Collages*, by Susan Davidson, with contributions by Megan M. Fontanella, Brandon Taylor, and Jeffrey Warda, 54–67. New York: Solomon R. Guggenheim Foundation, 2013.

Waters, Peter. "An Assessment of Lamination and Encapsulation." *Cambridge 80: International Conference on the Conservation of Library and Archive Materials and the Graphic Arts, Cambridge, England*, ed. Guy Petherbridge, 74–77. London: Society of Archivists and Institute of Paper Conservation, 1980.

Weidner, Marilyn. "Damage and Deterioration of Art on Paper due to Ignorance and the Use of Faulty Materials." *Studies in Conservation* 12, no. 1 (1967): 5–24.

Whitmore, Paul. "Determination of the Cellulose Scission Route in the Hydrolytic and Oxidative Degradation of Paper." *Restaurator* 15, no. 1 (1994): 26–45.

———. "Paper Ageing and the Influence of Water." In Gerhard Banik and Irene Brückle, eds., *Paper and Water: A Guide for Conservators*, 219–54. Oxford: Elsevier 2011.

Whitmore, Paul, and John Bogaard. "The Effect of Oxidation on the Subsequent Oven Aging of Filter Paper." *Restaurator* 16, no. 1 (1995): 10–30.

Williams, John C. "Metallic Catalysts in the Oxidative Degradation of Paper." *Preservation of Paper and Textiles of Historic and Artistic Value*. 1977.

———. "A Review of Paper Quality and Paper Chemistry." *Library Trends* 30 (Fall 1981): 203–24.

Wilson, William. "An Analysis of the Aging of Paper: Possible Reactions and Their Effects on Measurable Properties." *Restaurator* 3 (1979): 37–61.

———. "Reflections on the Stability of Paper." *Restaurator* 1, no. 2 (1969): 79–86.

Wilson, William K., and B. W. Forshee. *Preservation of Documents by Lamination.* National Bureau of Standards, U.S. Department of Commerce, Monograph 5. Washington, D.C.: Government Printing Office, 1959.

Workshop on the Effects of Aging on Printing and Writing Papers: Proceedings, July 6–8, 1994, Philadelphia, Pa. Philadelphia: ASTM Institute for Standards Research, 1994.

Part VI. Deterioration and Change: Media "Even in Their Partial Ruin Marvelous"

Banik, G. "Decay Caused by Iron-Gall Ink." In *Iron-Gall Ink Corrosion: Proceedings of the European Workshop on Iron-Gall Ink Corrosion, June 16–17, 1997*, 21–26. Amsterdam: Netherlands Institute for Cultural Heritage, 1997.

———. "Discoloration of Green Copper Pigments in Manuscripts and Works of Graphic Art." *Restaurator* 10 (1989): 61–73.

Banik, Gerhard, and Johann Ponahlo. "Some Aspects of Degradative Phenomena of Paper Caused by Green Copper-Containing Pigments." *Paper Conservator* 7 (1982–83): 3–7.

Biggs, Julie L., Yasmeen R. Khan, Sylvia R. Albro, Claire Dekle, Mary E. Haude, and Cyntia Karnes. "Treatment Trees for Iron-Gall Ink on Paper: Using Flow Charts to Develop Treatment Protocols." In *Edinburgh Conference Papers 2006: Proceedings from the Fifth International Conference of the Institute of Paper Conservation and the First International Conference of the Institute of Conservation, Book and Paper Group, 26–29 July 2006*, ed. Shulla Jaques, 211–18. London: Institute of Conservation, 2007.

Brommelle, Norman. "The Russell and Abney Report on the Action of Light on Watercolours." *Studies in Conservation* 9, no. 4 (1964): 140–45.

Brown, A. J. E., ed. *The Iron Gall Ink Meeting: 4th & 5th September 2000, the University of Northumbria, Newcastle upon Tyne: Postprints.* Newcastle upon Tyne: Conservation of Fine Art, University of Northumbria, 2000.

Brunelle, Richard L., and Robert W. Reed. *Forensic Examination of Ink and Paper.* Springfield, Ill.: C. C. Thomas, 1984.

Carvalho, David N. *Forty Centuries of Ink; or, a Chronological Narrative Concerning Ink and Its Backgrounds, Introducing Incidental Observations and Deductions, Parallels of Time and Color Phenomena, Bibliography, Chemistry: Poetical Effusions, Citations, Anecdotes and Curiosa together with Some Evidence Respecting the Evanescent Character of Most Inks of To-Day and an Epitome of Chemico-Legal Ink.* New York: B. Franklin, 1971.

Daniels, Vincent. "Aging of Paper and Pigments Containing Iron and Copper: A Review." In *The Broad Spectrum: Studies in the Materials, Techniques, and Conservation of Color on Paper*, ed. Harriet K. Stratis and Britt Salvesen, 116–21. London: Archetype, 2002.

Duffy, J. I. *Printing Inks: Developments since 1975.* Park Ridge, N.J.: Noyes Data Corp., 1979.

Ellis, Margaret Holben. "The Porous Pointed Pen as Artistic Medium." In *Conference Papers, Manchester 1992*, 11–18. n.p.: Institute of Paper Conservation, 1992.

Ellis, Margaret Holben, and M. Brigitte Yeh. "The History, Use, and Characteristics of Wax-Based Drawing Media." *Paper Conservator* 22 (1998): 48–55.

Garfield, Simon. *Mauve: How One Man Invented a Color That Changed the World.* New York: Norton, 2001.

Iron-Gall Ink Corrosion: Proceedings, European Workshop on Iron-Gall Corrosion, June 16–17, 1997. Rotterdam: Museums Boijmans van Beuningen, 1997.

Iron-Gall Ink Meeting Postprints, September 4th & 5th, 2000. Ed. A. Jean Brown. Newcastle: University of Northumbria, 2001. Transitional Metals in Paper (MIP). http://miponline.org.

Kolar, Jana, and Matija Strlic. *Iron Gall Inks: On Manufacture, Characterisation, Degradation, and Stabilisation.* Ljubljana: National University Library, 2006.

Krekel, C. "The Chemistry of Historical Iron Gall Inks." *International Journal of Forensic Document Examiners* 5 (1999): 54–58.

Light and Water-Colours: A Series of Letters Addressed to the Editor of "the Times" London. n.p., 1887.

Lussier, Stephanie M., and Gregory D. Smith. "A Review of the Phenomenon of Lead White Darkening and Its Conversion Treatment." *Reviews in Conservation* 8 (2007): 41–53.

Neeval, Johan G., and Birgit Reissland. "The Ink Corrosion Project at NICH: A Review." *Abbey Newsletter* 21, nos. 5–6 (1997): 88.

Ormsby, Bronwyn A., Joyce H. Townsend, Brian W. Singer, and John R. Dean. "British Watercolour Cakes from the Eighteenth to the Early Twentieth Century." *Studies in Conservation* 50, no. 1 (2005): 45–66.

Reissland, Birgit. "Ink Corrosion, Aqueous and Non-aqueous Treatment of Paper Objects: State of the Art." *Restaurator* 20, nos. 3–4 (1999): 167–80.

________. "Iron-Gall Ink Corrosion: Progress in Visible Degradation." In *Contributions to Conservation: Research in Conservation at the Netherlands Institute for Cultural Heritage (ICN)*, ed. Jaap A. Mosk and Norman M. Tennent, 113–18. London: Earthscan Ltd., 2002.

Ryan, Cindy Connelly, Elmer Eusman, Holly Krueger, Linda Stiber Morenus, and Heather Wanser. "Optimizing Ink Corrosion Treatment Protocols at the Library of Congress." In *Edinburgh Conference Papers 2006: Proceedings from the Fifth International Conference of the Institute of Paper Conservation and the First International Conference of the Institute of Conservation, Book and Paper Group, 26–29 July 2006*, ed. Shulla Jaques, 195–202. London: Institute of Conservation, 2007.

Stratis, Harriet K., and Britt Salvesen, eds. *The Broad Spectrum: Studies in the Materials, Techniques, and Conservation of Color on Paper*. London: Archetype, 2002.

Taylor, J. Scott. *A Descriptive Handbook of Modern Water-Colour Pigments Illustrated with Seventy-Two Colour Washes Skilfully Gradated by Hand on Whatman's Drawing Paper. With an Introductory Essay on the Recent Water-Colour Controversy*. London: Winsor & Newton, ca. 1887.

Van Gulik, R., and N. E. Kersten-Pampiglione. "A Closer Look at Iron Gall Ink Burn." *Restaurator* 15 (1994): 173–87.

Wiborg, Frank B. *Printing Inks: A History*. New York: Harper and Brothers, 1926.

Part VII. Treatment: Limits and Limitations

Appelbaum, Barbara. *Conservation Treatment Methodology*. Oxford: Butterworth-Heinemann, 2007.

Blunn, Denis, and Guy Petherbridge. "Leaf Casting: The Mechanical Repair of Paper Artifacts." *Paper Conservator* 1 (1976): 26–32.

Brown, A. J. E., and A. Bacon. "Perspectives on Image Reintegration." *Paper Conservator* 26 (2001): 5–12.

Brückle, Irene. "Bleaching Paper in Conservation: Decision-Making Parameters." *Restaurator* 30, no. 4 (2009): 321–32.

Brückle, Irene, Andrea Pataki, and Gerhard Banik. "On the Significance of Science for Conservation Treatment." In *Edinburgh Conference Papers 2006: Proceedings from the Fifth International Conference of the Institute of Paper Conservation and the First International Conference of the Institute of Conservation, Book and Paper Group, 26–29 July 2006*, ed. Shulla Jaques, 105–12. London: Institute of Conservation, 2007.

Daniels, Vincent. "Factors Influencing the Wash-Fastness of Watercolors." *Paper Conservator* 19 (1995): 31–40.

Evans, Debra. "Loss Compensation in Paper." In *Loss Compensation Symposium Postprints, WAAC Annual Meeting, 1993*, ed. Patricia Leavengood, 25–28. Seattle: Western Association for Art Conservation, 1994.

Feller, R. L., and D. B. Encke. "Stages of Deterioration: The Example of Rubber Cement and Transparent Mending Tape." In *Science and Technology in the Service of Conservation*, 19–23. London: International Institute for Conservation of Historic and Artistic Works, 1982.

Futernick, Robert. "Leaf Casting on the Suction Table." *Journal of the American Institute for Conservation* 22, no. 2 (1983): 82–91.

Hey, Margaret. "The Washing and Aqueous Deacidification of Paper." *Paper Conservator* 4 (1979): 66–79.

Keyes, Keiko M. "A Manual Method of Paper Pulp Application in the Conservation of Works of Art on Paper." *Paper Conservator* 1 (1976): 33–34.

Klinger, Jane E. "The Importance of the Provenance of Damage." In *Edinburgh Conference Papers 2006: Proceedings from the Fifth International Conference of the Institute of Paper Conservation and the First International Conference of the Institute of Conservation, Book and Paper Group, 26–29 July 2006*, ed. Shulla Jaques, 51–58. London: Institute of Conservation, 2007.

Lienardy, Anne, and Philippe van Damme. "Paper Washing." *Paper Conservator* 14 (1990): 23–30.

McAusland, Jane. "Facsimile Paper Repairs for Works of Art on Paper." *Paper Conservator* 3 (1978): 28–32.

———. "Re-integration of Missing Areas in Old Master Prints and Old Master Drawings: Some Different Approaches." In *The Postprints of the Image Re-integration Conference: 15–17 September 2003, Northumbria University, Newcastle upon Tyne, UK*, ed. A. J. E. Brown, 59–64. Newcastle upon Tyne: Northumbria University Press, 2007.

———. "Techniques of Lining for the Support of Fragile Works of Art on Paper." *Paper Conservator* 4 (1979): 33–44.

McClintock, T. K. "Observations on Developments in the Treatment of Works of Art on Paper." *Journal of the American Institute for Conservation* 31, no. 1 (1992): 107–15.

Oddy, A., and S. Carroll. *Reversibility—Does It Exist?* London: British Museum, 1999.

Oddy, W. A. *Restoration—Is It Acceptable?* London: British Museum, 1994.

Oddy, W. A., and Sandra Smith. *Past Practice—Future Prospects*. London: British Museum, 2001.

Pearlstein, Ellen, D. Cabelli, A. King, and N. Indictor. "The Effect of Eraser Treatment on Paper." *Journal of the American Institute for Conservation* 22 (1982): 1–12.

Perkinson, Roy. "What You See Is Not What You Get: Deceptive Repairs and Alterations in Old Master Prints." In *Printed on Paper: The Techniques, History, and Conservation of Printed Media*, ed. Jane Colbourne and Reba Fishman Snyder. Newcastle upon Tyne: Arts and Social Sciences Academic Press, Northumbria University, 2009.

Poulsson, Tina Gretta. *Retouching of Art on Paper*. London: Archetype, 2008.

Rayner, Judith, Joanna M. Kosek, and Birthe Christensen. *Art on Paper: Mounting and Housing*. London: Archetype Publications in association with the British Museum, 2005.

Santucci, Ludovico. "The Application of Chemical and Physical Methods to Conservation of Archival Materials." In *Recent Advances in Conservation, Contributions to the IIC Rome Conference 1961*, ed. G. Thomson, 39–47. London: IIC, 1963.

Schaeffer, Terry Trosper, Mary Baker, Victoria Blyth-Hill, and Dianne van der Reyden. "Effects of Aqueous Light Bleaching on the Subsequent Aging of Paper." *Journal of the American Institute for Conservation* 31, no. 3 (Fall–Winter 1992): 289–311.

Smith, Christine. "Inpainting/Overpainting Paper Art: Using Mixed Dry Pigments and Complementary Colors." *Book and Paper Group Annual* 17 (1998): 103–9.

Souguir, Zied, Anne Dupont-Laurence, and E. René de la Rie. "Formation of Brown Lines in Paper: Characterization of Cellulose Degradation at the Wet-Dry Interface." *Biomacromolecules* 9, no. 9 (2008): 2546–52.

Sugarman, Jane E., and Timothy J. Vitale. "Observations on the Drying of Paper: Five Drying Methods and the Drying Process." *Journal of the American Institute for Conservation* 31, no. 2 (1992): 175–97.

Tang, Lucia C. "Stabilization of Paper through Sodium Borohydride Treatment." *Historic Textile and Paper Materials: Conservation and Characterization*, ed. Howard L. Needles and Haig S. Zeronian, 427–41. Advances in Chemistry 212. Washington, D.C.: American Chemical Society, 1986.

Tang, Lucia C., and Norvell M. M. Jones. "The Effects of Wash Water Quality on the Aging Characteristics of Paper." *Journal of the American Institute for Conservation* 18, no. 2 (1979): 61–81.

Windon, Katrina. "The Right to Decay with Dignity: Documentation and the Negotiation between an Artist's Sanction and the Cultural Interest." *Art Documentation: Journal of the Art Libraries Society of North America* 31, no. 2 (Fall 2012): 142–57.

Part VIII. The Paper Conservator: Going beyond the Bench

AIC Ethics and Standards Committee. *Code of Ethics and Guidelines for Practice*. Washington, D.C.: American Institute for Conservation, 1994.

Appelbaum, Barbara. *Guide to Environmental Protection of Collections*. Madison, CT: Sound View Press, 1991.

Bachmann, Konstanze, ed. *Conservation Concerns: A Guide for Collectors and Curators*. Washington, D.C.: Smithsonian Institution Press, 1992.

Bearman, Frederick. "Conservation Principles and Ethics: Their Origins and Development." In *IPC Conference Papers London 1997: Proceedings of the Fourth International Conference of the Institute of Paper Conservation, 6–9 April 1997*, ed. Jane Eagan Leigh, 83–89. London: Institute of Paper Conservation, 1998.

Bewer, Francesca. *A Laboratory for Art: Harvard's Fogg Museum and the Emergence of Conservation in America*. Cambridge, Mass.: Harvard Art Museum; New Haven, Conn.: Yale University Press, 2010.

Bonnardot, A. *Essai sur l'art de restaurer les estampes et les livres*. New York: Burt Franklin, 1967.

Bower, Peter. "Powerful Connections: Interdisciplinary Collaboration in the Investigation and Discovery of Works of Art on Paper." In *Conference Papers 2006: Proceedings from the Fifth International Conference of the Institute of Paper Conservation and the First International Conference of the Institute of Conservation, Book and Paper Group, 26–29 July 2006*, ed. Shulla Jaques, 41–50. London: Institute of Conservation, 2007.

Bradshaw, Susan, and Helen Lindsay. "Professionalizing Preventive Conservation." In *Edinburgh Conference Papers 2006: Proceedings from the Fifth International Conference of the Institute of Paper Conservation and the First International Conference of the Institute of Conservation, Book and Paper Group, 26–29 July 2006*, ed. Shulla Jaques, 33–40. London: Institute of Conservation, 2007.

Brückle, Irene, and Jana Dambrogio. "Paper Splitting: History and Modern Technology." *Journal of the American Institute for Conservation* 39, no. 3 (2000): 295–325.

CAC. *Code of Ethics and Guidance for Practice*. Ottawa: Canadian Association for Conservation of Cultural Property, 2000.

Caple, Chris. *Preventive Conservation in Museums*. Leister Readers in Museum Studies. London: Routledge, 2011.

Clapp, Anne F. *Curatorial Care of Works of Art on Paper*. New York: Nick Lyons, 1987.

Donnithorne, Alan. "Early Approaches to the Conservation of Works of Art on Paper: Cleaning, Repair and Restoration." Early Advances in Conservation, ed. Vincent Daniels. *British Museum Occasional Paper No. 65* (1988): 15–25.

ECCO. *Professional Guidelines and Code of Ethics*. Brussels: European Confederation of Conservator-Restorers' Organisations, 2002–4.

Ellis, Margaret Holben. *The Care of Prints and Drawings*. Nashville, TN: AASLH, 1987; Walnut Creek, CA: Altimira Press, 1995.

Hicks, Catherine. "Early Approaches to the Conservation of Works of Art on Paper." In *Early Advances in Conservation*, ed. Vincent Daniels. *British Museum Occasional Paper No. 65* (1988): 7–12.

Higginbotham, B. B. *Our Past Preserved: A History of American Library Preservation, 1876–1910*. Boston: Hall, 1990.

ICOM-CC. *The Conservator-Restorer: A Definition of the Profession*. Copenhagen: ICOM-CC, 1984. www.icom-cc.org/47/about-icom-cc/definition-of-profession/.

James, Carlo, Caroline Corrigan, Marie Christine Enshaian, and Marie Rose Greca. *Old Master Prints and Drawings: A Guide to Preservation and Conservation*. Trans. and ed.

Marjorie B. Cohn. Amsterdam: Amsterdam University Press, 1997.

Kosek, Joanna M. "Restoration of Art on Paper in the West: A Consideration of Changing Attitudes and Values." In *Restoration: Is It Acceptable?*, ed. Andrew Oddy. *British Museum Occasional Paper No.* 99 (1994): 41–50.

Long, Jane. *Caring for Your Family Treasures*. New York: Harry N. Abrams, 2000.

Lucanus, Fr. G. H. *Die Praxis des Restaurators*. Halberstadt: Verlag von Rudolf Schönherr, 1929.

Marwick, Claire S. *An Historical Study of Paper Document Restoration Methods*. Washington, D.C.: American University, 1964.

National Committee to Save America's Cultural Collections. *Caring for Your Collections*. New York: Harry N. Abrams, 1992.

Ogden, Sherelyn. *The Storage of Art on Paper: A Basic Guide for Institutions*. Champaign: Graduate School of Library and Information Science, University of Illinois at Urbana-Champaign, 2001.

Print Council of America. *Guidelines for Lending Works of Art on Paper*. New York: Print Council of America, 1995.

Shelley, Marjorie. *The Care and Handling of Art Objects: Practices in the Metropolitan Museum of Art*. New York: Metropolitan Museum of Art, 1987.

Smith, Theresa. "An Evaluation of Historical Bleaching with Chlorine Dioxide Gas, Sodium Hypochlorite, and Chloramine-T at the Fogg Art Museum." *Restaurator* 33 (2012): 3–4.

———. Historical Paper Conservation: Washing with Boiling Water Followed by Sun Bleaching. *Book and Paper Group Annual* 26 (2007): 167–70.

Staniforth, Sarah, ed. *Historical Perspectives on Preventive Conservation*. Readings in Conservation. Los Angeles: Getty Conservation Institute, 2013.

Stevenson, Mark. "Print Restoration in Northern Europe: Development, Traditions, and Literature from the Late Renaissance to the 1930s." *Conservation Research, Studies in the History of Art* 2 (1995): 110–27.

———. "A Seventeenth-Century Manual for the Restoration of Prints." *Print Quarterly* 7, no. 4 (December 1990): 420–24.

United Kingdom Institute for Conservation. *Code of Ethics and Rules of Practice*. London: UKIC, 1996.

Walsh, Judith. "The Use of Paper Splitting in Old Master Print Restorations." *Print Quarterly* 27, no. 4 (2000): 383–90.

About the Editor

Margaret Holben Ellis holds a BA in art history from Barnard College, and an MA in art history and an advanced certificate in conservation from the Institute of Fine Arts, New York University. In 1977 she joined the paper conservation staff at the Metropolitan Museum of Art. In 1987 she was appointed Sherman Fairchild Chairman of the Conservation Center, Institute of Fine Arts, New York University, where she now serves as the Eugene Thaw Professor of Paper Conservation, teaching the conservation treatment of prints and drawings, as well as technical connoisseurship seminars for art historians. She is also director of the Thaw Conservation Center at the Morgan Library & Museum in New York. Her awards include the AIC Rutherford John Gettens Merit Award, the AIC Sheldon and Caroline Keck Award, and the Rome Prize, from the American Academy in Rome.

Index

Note: *Italicized* page numbers refer to illustrations.

J

K

L

M

T

U

V

W

Y

Z